'A wide ranging, comprehensive and innovative mapping of the geographies of the energy-society interface, *The Routledge Research Companion to Energy Geographies* is an essential compendium for the scholar of 21st century energy systems. Breaking apart the silos of so much of what passes as the analysis of our hydro-carbon civilization, this collection casts it gaze not only to the prospects and challenges of a transition to a low carbon future but compellingly confirms the need for a complex and sophisticated geographical analysis of what one might call the contemporary energy complex. An indispensable text.'

Michael Watts, Class of 63 Professor of Geography and Development Studies,
University of California, Berkeley, USA

'This timely collection puts questions of energy firmly back on the geographical agenda. Rather than being a taken for granted element of everyday life or a distant geopolitical concern, contributors critically examine how energy matters – access, security, politics, economics, sustainability and justice to name just a few – configure socio-spatial relations. As the richly varied chapters explore, energy is constitutive of some of the fundamental entities of geographical enquiry – from the state to landscape, infrastructure to development. *The Routledge Research Companion to Energy Geographies* offers a key reference point for the discipline and showcases the contribution geography can make to debates across the social sciences in this rapidly evolving domain.'

Harriet Bulkeley, Department of Geography, Durham University, UK

'Energy starts with physics, but the rest is geography. Where is the supply, where is the demand, and how does it get from one to the other? How does human geography drive production, use, substitution, waste, and myriad transformations from one form to another? How does economic geography – why people do what they do where they do it – impact primary production, transportation, goods, and services? Reading *The Routledge Research Companion to Energy Geographies*, one might exclaim, "Who knew energy and geography could be so complicated?" Or, one might read the entire book and never think twice about how much complexity is presented painlessly.'

Jerome Dobson, Department of Geography & Atmospheric Science,
University of Kansas, USA

The Routledge Research Companion to Energy Geographies

Energy has become a central concern of many strands of geographical inquiry, from global climate change to the effects of energy decisions on our lives. However, many aspects of the 'black box' of relationships at the energy-society interface remain unopened, especially in terms of the spatial underpinnings of energy production and consumption within nations, cities and regions. Debates focusing on the location and nature of energy flows frequently fail to consider the multiple geographical networks that illustrate and explain the distribution of fuels and services around the world.

Providing an integrated perspective on the complex interdependencies between energy and geography, *The Routledge Research Companion to Energy Geographies* offers a timely conceptual framework to study the multiple facets of energy geography, including security, space and place, planning, environmental science, economics and political science. Illustrating how a geographic approach towards energy can aid decision-making pathways in the domains of social justice and environment, this book provides insights that will help move the international community toward greater co-operation, stability, and sustainability.

This timely Companion will be of interest to scholars, students and professionals working in geography, environmental studies, environmental anthropology, sustainability, development studies and politics.

Stefan Bouzarovski is Professor of Geography at the University of Manchester, UK. He is also an External Professor at the Institute of Geography, University of Gdańsk, Poland, and a Visiting Professor at the University of Bergen, Norway.

Martin J. Pasqualetti is Professor at the School of Geographical Sciences and Urban Planning, Co-Director of the Energy Policy Innovation Council and Senior Sustainability Scientist in the Julie Ann Wrigley Global Institute of Sustainability at Arizona State University, USA.

Vanesa Castán Broto is Senior Lecturer and Director of Research at the Development Planning Unit at University College London, UK.

The Routledge Research Companion to Energy Geographies

Edited by
Stefan Bouzarovski,
Martin J. Pasqualetti and
Vanesa Castán Broto

Routledge
Taylor & Francis Group

LONDON AND NEW YORK

First published 2017 by Routledge
4 Park Square, Milton Park, Abingdon, Oxon OX14 4RN

and by Routledge
605 Third Avenue, New York, NY 10158

Routledge is an imprint of the Taylor & Francis Group, an informa business

First issued in paperback 2020

British Library Cataloguing-in-Publication Data
A catalogue record for this book is available from the British Library

Library of Congress Cataloging-in-Publication Data
Names: Bouzarovski, Stefan, editor. | Pasqualetti, Martin J., 1945- editor. |
Broto, Vanesa Castán, editor.
Title: The Routledge research companion to energy geographies / edited by
Stefan Bouzarovski, Martin J. Pasqualetti and Vanesa Castan Broto.
Description: Abingdon, Oxon ; New York, NY : Routledge, 2017. | Includes index.
Identifiers: LCCN 2016055058| ISBN 9781472464194 (hardback) |
ISBN 9781315612928 (ebook)
Subjects: LCSH: Power resources. | Energy development. | Energy security. |
Environmental geography.
Classification: LCC HD9502.A2 R685 2017 | DDC 333.79—dc23
LC record available at https://lccn.loc.gov/2016055058

ISBN: 978-1-4724-6419-4 (hbk)
ISBN: 978-0-367-66005-5 (pbk)

Typeset in Bembo and Stone Sans
by Florence Production Ltd, Stoodleigh, Devon, UK

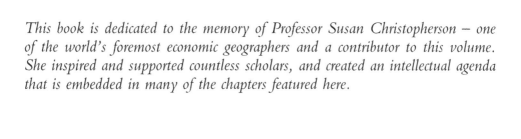

This book is dedicated to the memory of Professor Susan Christopherson – one of the world's foremost economic geographers and a contributor to this volume. She inspired and supported countless scholars, and created an intellectual agenda that is embedded in many of the chapters featured here.

Contents

Contents

Figures

Figures

Tables

Contributors

Sharmistha Bagchi-Sen is Professor in the Department of Geography at the State University of New York at Buffalo. Her research career has focused on the study of foreign direct investment; innovation in biotechnology, biopharmaceuticals and bioenergy; and urban geography. She served as the Editor-in-Chief of *The Professional Geographer* (2005–10), one of the flagship journals of the American Association of Geographers. Her research has been supported by several grants from the National Science Foundation since 2001. She has published over 70 refereed journal articles and book chapters. She has also co-edited three books including a recent publication on *Innovation Spaces in Asia*. Her first book (co-authored), *Shrinking Cities: Understanding Decline and Shrinkage in the United States*, has just been published by Routledge. She served as the co-Director of SUNY-Buffalo's Gender Institute in 2009–10 and was a Fellow of the American Council on Education in 2006–7.

J. Brad Barnett is a PhD candidate in Michigan Technological University's Environmental and Energy Policy programme. His dissertation work focuses on developing decision-making tools to assist policy makers incorporate social values for ecosystem services in natural resource management planning. He holds a Masters of Public Administration from Eastern Kentucky University and a Bachelor's degree in Political Science from the University of Kentucky. Brad is a regional planner for the Western Upper Peninsula Planning and Development Region in Houghton, Michigan where he works on energy planning and community development projects.

Karen Bickerstaff is Associate Professor in Human Geography at Exeter University, UK. She has research interests in geographies of risk, energy system transformations and socio-ecological justice. She has pursued many of these concerns through a focus on locating techno-environmental hazards – notably in relation to questions of democracy and the siting of nuclear waste facilities. She has also worked extensively on energy justice – leading a recent research network on energy systems, equity and vulnerability (www.lancaster.ac.uk/lec/sites/incluesev/) and co-editing a collection on *Energy Justice in a Changing Climate* (Zed, 2013).

Kean Birch is a Senior Associate at the Innovation Policy Lab, University of Toronto and Associate Professor at York University, Canada where he is a member of the Science and Technology Studies, Sociology, and Geography Graduate Programmes. His recent books include: *We Have Never Been Neoliberal* (2015, Zero Books); *The Handbook of Neoliberalism* (2016, Routledge – co-edited with Simon Springer and Julie MacLeavy); *Innovation, Regional Development and the Life Sciences: Beyond Clusters* (2016, Routledge); and *Business and Society: A Critical Introduction* (2017, Zed Books – with Mark Peacock, Richard Wellen, Caroline Hossein, Sonya Scott and Alberto Salazar); and *A Research Agenda for Neoliberalism* (forthcoming 2017, Edward Elgar).

Stefan Bouzarovski is Professor of Geography at the University of Manchester, where he convenes the Collaboratory for Urban Resilience and Energy within the Manchester Urban Institute. He is also an External Professor at the Institute of Geography, University of Gdańsk, Poland, and a Visiting Professor at the Department of Geography, University of Bergen, Norway. Stefan Bouzarovski also chairs the European Energy Poverty Observatory. He is an internationally leading expert in energy and urban policy, with a particular focus on energy poverty in Europe. He has undertaken research, consultancy, advocacy and policy-orientated work on this subject for more than 15 years. He has previously held full-time appointments at Universities of Oxford, London and Birmingham, and visiting professorships at universities in Prague, Berlin, Stockholm, Bruges and Brisbane. His work been funded by a wide range of governmental bodies, charities and private sector organisations (in more than 60 different projects or consultancy engagements), and has been published in more than 80 scientific and policy papers, including the books *Energy Poverty in Eastern Europe* (Ashgate, 2007) and *Retrofitting the City* (IB Tauris, 2015). Its outcomes have informed the work of the European Union, World Bank and International Energy Agency. He is a member of the Editorial Boards of the journals *European Urban and Regional Studies*, *Energy and Buildings* and *Energy Research and Social Science*, and was one of the founding members of the UK Royal Geographical Society's Energy Geographies Research Group.

Christian Brannstrom is Professor of Geography and Associate Dean for Academic Affairs in the College of Geosciences at Texas A&M University. He has published on various aspects of wind power in Texas and Brazil and on municipal policies for hydraulic fracturing in journals such as *Environment and Planning A*, *Annals of the Association of American Geographers*, *Geoforum* and *Renewable and Sustainable Energy Reviews*. He teaches a course on Geography of Energy. His work also has focused on land cover change and environmental history.

Catherine Butler is an Advanced Research Fellow in environment and sustainability in the Geography Department at University of Exeter. Her research sits at the interface between human geography, political theory, and science and technology studies and addresses responses to socio-environmental change. She is convenor of the British Sociological Association Climate Change Study Group, Secretary of the Royal Geographical Society Energy Geographies Research Group, and a member of the UK's Economic and Social Research Council Peer Review College. She is currently Principal Investigator on two major grants examining different dimensions of policy relating to climate change and low carbon transitions.

Kirby Calvert is an Assistant Professor of Geography at the University of Guelph (Ontario, Canada). Dr Calvert's research program has three core themes: renewable energy mapping and spatial planning using GIS; the relationship between energy and land-use/landscapes policies, planning, and value systems, and community energy planning. Kirby is co-Director the Community Energy Knowledge-Action Partnership (www.cekap.ca); a national partnership of universities and non-academic partners with a shared interest in building more resilient and sustainable communities through community energy planning

Matthew Campo focuses on understanding tools and methods to assess the sustainability and resilience of built environments and transportation systems. He has worked with the Department of Homeland Security, National Oceanic and Atmospheric Administration, and US Environmental Protection Agency to enhance planning capabilities and risk management strategies for coastal communities and working waterfronts. In addition, supports the New Jersey Climate Adaptation Alliance and Mid-Atlantic Regional Council on the Ocean in developing tools and guidance for enhancing climate resilience in the Mid-Atlantic United States. Prior to his work at Rutgers, Matt worked as a consultant for public agencies and private companies to develop growth and consolidation strategies for commercial, residential and industrial real estate assets. Matt is a member of the American Planning Association, Association of Collegiate Schools of Planning, and is chair of the Transportation Research Board Freight and Marine Young Members Committee.

Vanesa Castán Broto is a Senior Lecturer at the Bartlett Development Planning Unit (University College London) where she directs the MSc Environment and Sustainable Development. In her research, Vanesa Castán Broto addresses the conundrum of how to achieve simultaneously universal energy access and a transition to a low carbon society. She advocates participatory, bottom up approaches to environmental governance. Her work has been funded by the UK's Engineering and Physical Sciences Research Council, the UK's Economic and Social Research Council, the UK Department for International Development, and the Institution of Civil Engineers. She has worked with international organisations such as UN-Habitat, UNEP and the WHO. Most recently she was one of the lead chapter authors for the 2016 World Cities Report (wcr.unhabitat.org). Her books include *An Urban Politics of Climate Change* (with Harriet Bulkeley and Gareth Edwards; London: Routledge) and *Participatory Planning for Climate Compatible Development in Mozambique* (London: UCL Press). In 2013, her work about planning for climate change in Mozambique was recognised as a Lighthouse Activity for the Urban Poor by the UNFCCC. In 2016 Vanesa was awarded the Philip Leverhulme Prize for contributions to Geography.

Susan Christopherson was Chair of the Department of City and Regional Planning at Cornell University. She was a geographer whose career was based on a commitment to the integration of scholarly work and public engagement. She was the 2016 recipient of a Lifetime Achievement Award from the American Association of Geographers. From 2010, she received a series of grants to conduct research on the economic and social consequences of shale development in the US. She was a member of the 2013–14 National Academy of Sciences panel on Risks and Risk Management in shale gas development in the US. She was Chair of the Editorial Board of the Regional Studies Association/Taylor Francis book series on Cities and Regions and a founding member of the editorial board for *The Cambridge Journal on Regions, Economy, and Society*.

Molly Coon is Programme Coordinator for Partner Energy, an energy efficiency and sustainability company in Southern California. She earned a master's degree in City and Regional Planning from Rutgers University's Bloustein School, where her focus included urban design, land use and environmental planning. While a student at the Bloustein School, Molly worked with an after-school programme in the New Brunswick Public Schools, interned at the New Jersey Department of Environmental Protection, and worked as a research aid providing data analysis and support. Molly also has BA in Architecture from the University of Kentucky and is a Teach for America Alumni.

Richard Cowell is Professor of Environmental Planning at the School of Geography and Planning, Cardiff University. Richard's research interests cover theoretical and political policy aspects of the relationship between public policy and sustainable development, with particular reference to land use planning, public engagement and energy. His current work focuses on the relationship between scale, place and transitions to more sustainable energy, including renewable energy expansion, infrastructure decision-making, and the relationship between infrastructure and host communities. His work has been published widely in international energy and planning journals, and he regularly provides advice on energy and environment issues to the National Assembly for Wales.

Elvin Delgado is the founding Director of the Institute for Integrated Energy Studies and Assistant Professor in the Department of Geography at Central Washington University. Dr Delgado is a Fulbright scholar, and is former Chair of the Energy and Environment Specialty Group of the American Association of Geographers. His teaching and research interests lie in the areas of energy, political ecology, political economy and nature, and critical resources geography. Dr Delgado's research critically explores changing patterns of human-environment interactions in the context of fossil fuel production in Latin America. Empirically, his research analyzes processes of socio-political struggles over resource development, the multi-scalar interconnections of the material flows of natural resources, and the socio-ecological transformations associated with resource extraction industries. He has published in the areas of energy, health and GIS in different scholarly outlets. His new research studies the political economy and political ecology of hydraulic fracturing in northern Patagonia in Argentina. Recently, he has been appointed by Governor Inslee of Washington State to be a member of the Board of Directors of the Joint Centre for Deployment and Research in Earth Abundant Materials.

Patrick Devine-Wright is a Professor at the Geography Department, University of Exeter. He is a leading scholar on place attachment and place identity, investigating their implications for the social acceptance of climate change mitigation technologies, including renewable energy projects and power lines. His 2013 book on Place Attachment was granted an Achievement Award by the Environmental Design Research Association (co-authored with Dr. Lynne Manzo, University of Washington) and he was a Distinguished Visiting Scientist at CSIRO, Australia in 2012–2013. He has acted as leader of the Environment and Sustainability Research Group at the University of Exeter since 2013, as well as leading and contributing to multi-disciplinary large-scale research projects. Author of more than 60 articles, he sits on the editorial board of journals including *Global Environmental Change, Energy Research and Social Science*, and *Environment and Behavior*. He has given advice to the UK government on energy policy, as a member of the Social Science Expert Panel, and as Lead Expert to the Foresight project on sustainable energy. He is also an invited member of the National Advisory Group for EirGrid, the grid operator for the Republic of Ireland.

Bohumil Frantál is a senior research fellow at the Department of Environmental Geography, Institute of Geonics, Czech Academy of Sciences. His academic background includes degrees in Sociology, Andragogy and Regional Geography and Regional Development. In his research he focuses on social-spatial contexts of ongoing energy transition, particularly the renewable energy development and related land use conflicts, risk perceptions, urban renewal and brownfields regeneration, local identity processes, quality of life and spatial models of behaviour. His principal recent publications address the issues of public perceptions and social acceptance of wind energy, nuclear power and coal mining, and sustainable regeneration of brownfields. He also works as a lecturer at Palacký University in Olomouc and Masaryk University in Brno. His current teaching profile includes 'An Introduction to Energy Geography', 'Environmental Sociology' and 'Advanced Statistical Methods in Geographical Research'.

Marina Frolova is Senior Lecturer at the Department of Regional and Physical Geography, and scientific secretary of the Research Institute of Regional Development, University of Granada.

She is an experienced contributor to and leader of multi-disciplinary research projects. Her research interests include landscape analysis, landscape policies and, more recently, renewable energy policies, relationships between landscape and renewable energy and their relevance for the social acceptance of energy technologies. She has published and edited several books, book chapters and articles on landscape – especially energy-related – issues. She was co-editor of *Renewable Energies and European Landscapes. Lessons from Southern European Cases* (2015).

Matthew Fry received his doctorate in Geography from the University of Texas at Austin and is an associate professor in the Department of Geography and the Environment at the University of North Texas. His research focuses on energy and resource governance, socioeconomic and environmental impacts of resource production and consumption, and land-use change. His work on urban shale gas production, municipal distance ordinances, and concrete and cement production/consumption has appeared in *Environmental Science & Technology*, *Energy Policy*, *Ecological Economics*, *Geoforum*, *Annals of the Association of American Geographers* and *Environment and Planning A*.

Sara Fuller is a Senior Lecturer in the Department of Geography and Planning at Macquarie University, Australia. Her research explores concepts and practices of justice and democracy in the field of the environment, with an empirical focus on grassroots, community and activist responses to climate change. Prior to joining Macquarie University, she held postdoctoral positions at Durham University, UK and City University of Hong Kong where she conducted research on low carbon transitions and climate governance; NGO discourses of energy justice; low carbon communities and social justice; and energy vulnerability in communities. Her current research investigates the politics and governance of urban climate justice across the Asia-Pacific region.

Jessica K. Graybill received her PhD in Geography and Urban Ecology from the University of Washington in 2006 and is currently Associate Professor of Geography and Director of the Russian & Eurasian Studies Programme at Colgate University. Additionally, she has served as the Director of First-Year Seminars and the Global Engagements Programme for the Core Programme. Her primary research interests are in urban and socio-environmental transformations, especially as they relate to the impacts on communities of resource extraction and climate change. Her focus in Eurasia is on the Far East and North and on the cultural and socioeconomic changes in shrinking cities in other contexts, specifically in multicultural and refugee-repopulated Utica, New York. She is past president of the Russian, Central Eurasian and East European Specialty Group and current Vice President of the Polar Geography Specialty Group of the American Association of Geographers. She is a recipient of American Council for Learned Societies (2009–10) and Fulbright (2014–15) fellowships, both awarded for research related to resource use, extraction, and climate change in the Russian Far East and North. Currently, she is a co-Principal Investigator on two National Science Foundation grants (Arctic Frontiers of Resources and Sustainability and Arctic Coast) and lead Investigator on a Colgate-sponsored grant awarded to understand how all-terrain vehicle tracks on tundra impact social and ecological communities in the Russian Arctic. Recent publications include "Urban climate vulnerability and governance in the Russian North" (2015), "The suburban bias of American society?" (2016) and *Cities of the World*, sixth edition (2016).

Michael Greenberg is distinguished professor and associate dean of the faculty of the Edward J. Bloustein School of Planning and Public Policy, Rutgers University. He studies environmental health and risk analysis and has written more than 30 books and more than 300 articles. His most recent books are *Protecting Seniors Against Environmental Disasters: From Hazards and Vulnerability to Prevention and Resilience* (Earthscan, 2014) and *Explaining Risk Analysis: Protecting Health and Safety* (Earthscan, 2016*)*. Currently, he is completing an assignment as director of a Committee for the US Senate and House Appropriations Committees examining the US DOE's prioritisation of human health and safety in its environmental management programmes. Dr Greenberg served as area editor for social sciences and then editor-in-chief of *Risk Analysis: An International Journal* during the period 2002–13, and continues as associate editor for environmental health for the *American Journal of Public Health*.

Ralitsa Hiteva is a Research Fellow at the Science and Technology Policy Research Unit (SPRU) at the University of Sussex and a member of the Sussex Energy Group. Ralitsa works mainly in the areas of energy, infrastructure and governance. She has trained in economic geography and environmental governance. Her doctoral research focused on the role of intermediation in energy governance and change, in the context of transmission, distribution and supply of natural gas and renewable energy across the EU. Her most recent research is on low carbon infrastructure policy and innovation, and examines business model innovations for the delivery of inclusive and sustainable infrastructure, and infrastructure interdependencies at multiple scales. Ralitsa's research interests include the EU, UK, Eastern Europe (especially Bulgaria and Romania) and East Asia (China, Japan, South Korea and Taiwan).

Sylvy Jaglin is a Professor of geography and urban studies at the École d'Urbanisme de Paris, Université Paris-Est (France). Her research focuses on urban service delivery policies in Sub-Saharan African, looking in particular at issues related to utility politics, infrastructure transition and service access. It examines the patterns of sociotechnical change in relation with urban transformations and shifts in local governance. Her publications include (with A. Dubresson) *Eskom: Electricity and Technopolitics in South Africa*, Cape Town, UCT Press, 2016.

Scott Jiusto is Associate Professor of Geography at Worcester Polytechnic Institute (WPI) and former Director of the WPI Cape Town Project Centre (http://wp.wpi.edu/capetown). He works on sustainable community development in both the United States and South Africa, with ongoing collaborative 'Shared Action Learning' projects engaging students in efforts with communities and local governmental agencies, civic organisations, businesses, academics and others. Scott publishes regularly on the diverse outcomes and insights arising from this project-based, experiential learning approach to applied research. His work on informal settlement upgrading strategies in Cape Town seeks to advance collaborative, integrated strategies for addressing topics such as public participation, water and sanitation, urban drainage, energy, housing, early childhood development and micro-entrepreneurship.

Corey Johnson is Associate Professor and Department Head of Geography at the University of North Carolina at Greensboro. Corey's research and teaching areas include the political geography of Europe and Eurasia, borders and border security, natural resources and energy geopolitics, and Germany. In 2011–12 he was the Joachim Herz Fellow at the Transatlantic Academy in the German Marshall Fund of the United States in Washington, DC. Originally from Emporia, Kansas, Corey holds a PhD in geography from the University of Oregon and a BA in geography (honours) and German from the University of Kansas.

Phil Johnstone is a research fellow at the Science Policy Research Unit (SPRU) University of Sussex, and a Human Geographer gaining his PhD at the University of Exeter. His recent work looks at disruptive innovation in energy systems drawing on institutional theory and political economy. Previously he was working at SPRU on a project looking at the Governance of Discontinuation in technological systems and the democratic implications of phasing out certain technologies. Planning reforms, and understanding the nature and role of the nation state in sustainability transitions is also a central research interest. He has a long-standing interest in nuclear energy issues including a focus on the democratic and spatial implications of the UK governments commitments to nuclear energy, the economics of nuclear, the comparative politics of nuclear policy, and the ensuing linkages between military and civilian nuclear activities.

Peter Kedron is an Assistant Professor of Geography at Oklahoma State University, whose research programme focuses on understanding how geographic differences in economic activity are related to social and ecological aspects of regions. His ongoing research into the emergence of renewable energy industries examines how innovation, policy, and incumbent response shape geographic patterns of adoption and commercialisation. Dr Kedron also contributes to the development of new spatial analytical methods, with a current focus on technological recombination, cluster detection, and downscaling prediction.

Joshua Kirshner is a Lecturer in Human Geography in the Environment Department, University of York. His current research focuses on the challenges of energy access and urban infrastructure in low income countries. His interests are in the political dimensions of energy infrastructures, international co-operation and aid in energy, extractive industry and spin-off urbanisation, transitions to low carbon, and the interconnected aspects of global, regional and local infrastructure systems. Joshua also has a long-standing interest in urban migration, urban settlement and the ensuing politics of difference. His research draws on Southern Africa and Latin America, while he is keenly interested in cross-regional comparison. His previous posts include Researcher in the Department of Geography at Durham University (UK) and Lecturer in Human Geography at Rhodes University (South Africa). He holds a BA in Anthropology from Harvard University, an MA in Urban Planning from UCLA, and a PhD in City and Regional Planning from Cornell University (USA).

Olivier Labussière holds a research position at the French National Scientific Research Centre (CNRS) dedicated to 'Energy and Society' issues. He had been senior lecturer in geography and land planning at the Joseph Fourier University in Grenoble for five years (2010–15), following post-doctoral work at EHESS and CIRED in Paris (2008–10) and a PhD at Pau University (2007). His work deals with the sociotechnical, spatial and temporal changes in access to energy resources (wind, solar, coalbed methane). He is co-leading with Alain Nadaï the Social Sciences Energy Group of the Alliance Athena, which contributes in France to the emergence of a community of researchers on energy issues by fostering interdisciplinary work, and the international research network ENGAGE which gathers researchers and NGOs from France, United Kingdom and Canada on these issues. He is a member of the Executive Board of the laboratory PACTE (Grenoble) to which he is affiliated.

Warren Mabee is the Canada Research Chair of Renewable Energy Development and Implementation, and Professor and Head of the Department of Geography & Planning at Queen's University in Ontario, Canada. He is also Director of the Queen's Institute for Energy and Environmental Policy. He has published extensively in the fields of energy policy, resource management and biofuel systems. All of his degrees are from the Faculty of Forestry at the University of Toronto.

Stephen McCauley is an Assistant Teaching Professor in the Interdisciplinary and Global Studies Division at Worcester Polytechnic Institute. A geographer by training, Dr McCauley's work focuses on urban sustainability, community and regional planning, and energy-society interactions. He links sustainability policy agendas with commitments to environmental justice, community-based practice, and a grassroots approach to innovation and sustainability planning. He also embraces the practice and pedagogy of project-based learning. Through WPI's Global Projects Programme, Dr McCauley advises immersive, action-oriented student projects related to appropriate technology, human-centred design and community engagement. He is the Co-Director of WPI's Melbourne Project Centre and has advised student projects in Massachusetts, Morocco, South Africa and India.

Alain Nadaï is a sociologist and Research Director at CIRED, the International Research Centre on Environment and Development, which is part of the French CNRS. His research activity has been centred on environmental controversies as well as on environmental, energy and landscape policies. Applied fields for his research work have been: climate change policy, EU pesticide regulation, EU product eco-labelling policies, landscape policies and, more recently, renewable energy policies and CCS (Carbon Capture and Sequestration). He contributed as a leading author to the recent IPCC Special Report on Renewable Energy Sources and Climate Change Mitigation (SRREN).

Karen Parkhill joined the Environment Department at the University of York in 2015 as a lecturer in human geography. Prior to this she was a lecturer at Bangor University and a research fellow within the Understanding Risk Research Group at Cardiff University. She explores how the public and stakeholders engage with, or resist, notions of low carbon lifestyles and low carbon transitions, including examining how they themselves perceive energy issues. She is also interested in risk perception and how the public socially constructs and engages with environmental and technocratic risks. She is Chair of the Royal Geographical Society Energy Geographies Research Group.

Martin J. Pasqualetti is Professor of Geography in the School of Geographical Sciences and Urban Planning; Senior Sustainability Scientist in the Julie Ann Wrigley Global Institute of Sustainability; and Director of the Energy Policy Information Council (EPIC), all at Arizona State University. He co-founded the Energy and Environment Specialty Group within the American Association of Geographers in 1979. Since then, he has examined several themes, including energy as a social issue, the food/energy/water nexus, energy landscape development, public acceptance of renewable energy technology. He served two Arizona governors as Chairman of the Arizona Solar Energy Advisory Council, and was a founding member of the Arizona Solar Centre. He has advised several government agencies, including the US Department of Energy, the National Renewable Energy Laboratory, the US Office of Technology Assessment, and the US Academy of Science. In 2015, he received the Alexander and Ilse Melamid Gold Medal from the American Geographical Society for his international contributions to the study of energy. His latest book is *The Renewable Energy Landscape: Preserving Scenic Values in our Sustainable Future* (Routledge).

Saska Petrova is a Lecturer in Human Geography at the School of Environment, Education and Development at the University of Manchester, and Research Co-ordinator of the Collaboratory for Urban Resilience and Energy (CURE) within the Manchester Urban Institute. Her main research interests lie in intra-community relations and vulnerabilities as they relate to natural resource management, energy flows, social justice and local governance. Saska has published extensively on these issues, including a monograph on *Communities in Transition* (Ashgate, 2014) as well number of articles in leading scientific journals such as *Environment and Planning A, Urban Studies, Geoforum, Area, Geojournal* and *Energy Policy*. She has been involved in several interdisciplinary research projects funded by the Royal Geographical Society, Engineering and Physical Sciences Research Council, Cheshire Lehmann Fund and Higher Education Academy. She also has an extensive professional background as a public advocate and consultant for a range of global government institutions and think tanks. She is currently Treasurer of the UK Royal Geographical Society's Energy Geographies Research Group.

Jonathan Silver is an urban geographer based at the Urban Institute, University of Sheffield with experience in a range of cities including Accra, Cape Town, Kampala, Manchester and Philadelphia. He is an undertaking comparative work considering the urban political ecologies of infrastructure, particularly in relation to spatial forms of racialised capitalism and environmental injustice. He currently holds a Leverhulme Early Career Researcher Fellowship.

Barry D. Solomon is Professor Emeritus of Geography at Michigan Technological University, where he directed the graduate programme in environmental and energy policy. He has also worked for the US Environmental Protection Agency and Department of Energy. He has published widely across a range of journals in geography, ecological economics, and energy and environmental policy studies. His books have included *Sustainable Development of Biofuels in Latin America and the Caribbean* (Springer, 2014 edited with Rob Bailis), *Renewable Energy From Forest Resources in the United States* (Routledge, 2009 edited with Valerie Luzadis), *The International Politics of Nuclear Waste* (Palgrave Macmillan, 1991 with Andy Blowers and David Lowry) and he was an associate editor for the six volume *Encyclopedia of Geography* (edited by Barney Warf, Sage, 2010). Also, he sits on the editorial board of *Sustainability*.

Eric Verdeil is Professor of Geography and Urban Studies at Sciences Po Paris. He studies the current transformations of urban policies, notably the links between urban infrastructure (energy, water, solid waste), urban inequalities and local politics. His fieldworks are mainly located in the Middle East, particularly in Lebanon, and more recently in Jordan and Tunisia. He works on a manuscript whose title is The Political Ecology of Urban Energies in South and Eastern Mediterranean Cities.

Jennifer Whytlaw's research interests focus on hazards research including climate impacts on vulnerable populations and preparedness activities in communities to help mitigate impacts on critical infrastructure including senior facilities. She assisted the US Environmental Protection Agency Region II in developing a database of facilities to aid in prioritizing pollution prevention technical assistance for facilities vulnerable to flood hazards. She is the Bloustein lead member of the development team on a New Jersey Climate Adaptation Alliance initiative to develop a NJ-based all-climate hazards web-based application and associated website (www.NJADAPT. org). Prior to joining Bloustein in 2009, she worked in the fields of maritime safety, environmental policy and planning, and environmental engineering. Jennifer is an active member of MAC URISA, the Chair of the Executive Committee of the NJ Geospatial Forum, and a member of the Natural Hazard Mitigation Association (NHMA).

Acknowledgements

Collectively, the three editors of this volume are indebted to the Energy Geographies Working Group of the Institute of British Geographers with the Royal Geographical Society, and the Energy and Environment Specialty Group of the American Association of Geographers, for providing the vibrant intellectual fora that helped spawn the ideas and initiatives leading to this book. We also wish to thank all authors and editorial team at Routledge for their outstanding co-operation and patience throughout the publication process. We are grateful to all copyright holders for granting us permission to use the numerous images included in the book. Special thanks are due to Sylvie Douzou for kindly providing royalty-free access to the photograph that is displayed on the front page of this volume.

Vanesa Castán Broto would like to thank the ESRC for supporting the project Mapping Urban Energy Landscapes that has allowed her to dedicate time to projects related to urban energy (grant number ES/K001361/1). Thanks to Yvonne Rydin and Muki Haklay for constant mentorship and support. Thanks to Harriet Bulkeley and Niki Frantzeskaki for helping with crisis management. Thanks to the urbanisation and energy access community for engaging with her ideas, particularly: Idalina Baptista, Josh Kirshner, Yacob Muguletta, Long Seng To, Julia Tomei, Lucy Stevens, Emmanuel Ackom, Dana Abi Ghanem, Lucy Baker. Thanks to Ping Huang for his continuous support.

Martin J. Pasqualetti thanks Tom Wilbanks and Gary Dirks for their encouragement and example. Thanks to Andy Blowers for providing an example of how to apply academic interests in energy toward the service of the greater public good. Thanks to Benjamin Sovacool for his indefatigable interest in matters of energy and society. Thanks to Jerry Dobson, Barry Solomon and Marilyn Brown for helping develop energy geography into a robust and important sub-discipline. Thanks to Will Graf for providing an unwavering backstop.

Stefan Bouzarovski wishes to thank the European Research Council and the European Union's Seventh Framework Programme for supporting him with a Starting Grant (FP7/2007–2013/ERC grant agreement number 313478) that has proven priceless for his professional development, and creating a network of researchers on the geographies of energy vulnerability. Special gratitude is also due to colleagues at the geography discipline at Manchester University's School of Environment, Environment Education and Development; the Collaboratory for Urban Resilience and Energy within the Manchester Urban Institute; the Department for Economic Geography at the University of Gdańsk; and the Department of Geography at the University of Bergen.

Introduction

Stefan Bouzarovski, Martin J. Pasqualetti
and Vanesa Castán Broto

Growing concerns about the impacts of climate change and the security of hydrocarbon supplies have foregrounded the political and societal importance of 'global energy dilemmas' (Bradshaw, 2013). Concurrently, the social and cultural aspects of energy flows have come to wide attention as it has been increasingly clear that conventional energy technologies are only part of the answer to the challenging demands of decarbonization. Geographic considerations have played an important role in this shift, both in terms of the complex spatial distribution of energy activities, and the ability of energy flows to shape the more abstract formations of place and territory. Together, energy and geography have become the staging ground for a diverse range of theoretical perspectives and empirical contexts, from the wider spatial aspects of transitions to low carbon futures to the everyday lived experiences of energy use (Pasqualetti and Brown, 2014; Castáh Broto, 2016; Bouzarovski and Tirado Herrero, 2017).

This book aims to contribute towards the development of an explicitly geographical perspective on energy. We explore the scale- and place-dependent nature of the relationship between the material and technical properties of energy circulations, on the one hand, and different types of social formations – individuals, households, communities – on the other. One of our starting points is that the geography of energy is relevant at every scale and in every realm. This leads us to ask questions about the types of places and spaces that are associated with increased greenhouse gas emissions, as well as the patterns of uneven development and resource consumption that drive this disparity. The location of energy reserves has affected national, regional and local economies, as well as the balance of political power. Energy circulations are implicated in shaping our lives, values and cultures, while being enmeshed with the grain of everyday life: Which forms of social interaction can be linked to energy consumption? How might the changing nature of energy demand promote the development of an alternative socio-technical future, predicated upon more sustainable forms of energy consumption and production?

These and other questions are increasingly becoming the central concern of many strands of geographical inquiry. 'Energy' is explicitly recognized as a structural force in discussions of issues such as climate change, environmental impacts, urban restructuring and identity formations. At the same time, there is a rising engagement between the interdisciplinary field of energy research, on the one hand, and spatial issues and themes, on the other – as evidenced by the prominence of issues of place and territory in debates on low carbon transition and security. Geographical considerations have also become more apparent in international debates on carbon regulation, regional policy and market reform. The black box of demand and culture in geographical energy research is increasingly becoming unpacked, especially in terms of the every-day routines that underpin energy consumption within different material sites. Debates around locations and processes of energy production take into account the complex energy chains that allow various fuels and services to be distributed across wide geographical areas.

With such developments in mind, this volume aims to provide an integrated perspective on the relationship between energy and geography. The chapters presented here form the contours of a geographical framework for studying the multiple facets of security, sustainability, space and place as they relate to energy, while foregrounding such themes in current debates within the disciplines of human geography, planning, environmental science, economics and political science. A range of policy relevant strands emerge from the application of a geographical perspective on energy in the domains of security, social justice, and environment. As such, the Companion responds to an important lacuna in the academic, educational and publicist litera-ture, where book contributions on energy topics are mostly focused on issues of supply, and emphasize either technical, engineering or regional aspects of energy use.

The remainder of this chapter offers a critical exploration of existing scholarship in the emergent domain of energy geographies. We highlight the need for a *sui generis* disciplinary perspective on the wider political and spatial levers that underpin energy recovery, transmission and consumption. At the same time, the chapter seeks to reframe existing debates on the subject, in terms of the intrinsic material and social properties of energy as they relate to the production of place and space.

Early energy geography (1950–2000)

It is not much of a stretch to assert that our survival depends on understanding geographies of energy at several scales. For example, Local: How are we going to meet our energy needs? National: What is the role of energy acquisition and use as a part of nation building? Global: To what degree does humanity's survival depend on mitigation of the impacts of energy use on climate change?

Whatever the form of energy – from digestion of food to the conversion of matter in nuclear reactions – knowing the location of energy resources is always a fundamental part of part of our social lives. Early studies that fell under the rubric 'geography of energy' were ones that proffered the rather straight-forward goal of simply knowing the location of the resources we wished to exploit. As early as 1950, for example, Pratt and Good published a book on the world geography of oil, with the sponsorship of the American Geographical Society.

By the latter half of the 20th century, geographers were writing energy books that fell largely into two general types. One focused on the energy resources in specific countries or groups of countries. These included books on New Zealand (Farrell, 1962), Ghana (Hart, 1980), the Caspian Basin (Croissant and Aras, 1999), China (Kuby, 1995; Smil, 1976, 1988), and the former Soviet Union (Hooson, 1965; Dienes and Shabad, 1979; Hoffman and Dienes, 1985; Dienes *et al.*, 1994). Notably, no one tackled the large and complex energy patterns of the United States.

Taking a hint from the early AGS work mentioned above, a second group of geographers added to the literature by concentrating on different types of energy. These included books on oil (Odell and Rosing, 1980; Gever *et al.*, 1991), renewables (Pryde, 1983), recreation (Knapper *et al.*, 1983), modelling (Lakshmanan and Nijkamp, 1980, 1983; Lakshmanan and Johansson, 1985), world history (Smil, 1994), and ecological economics (Hall *et al.*, 1992).[1]

These two approaches eventually merged into a series of books that concentrated on the role – rather than just the location – of different types of energy resources in different countries. Such books included those on fossil fuels in the UK (Fernie, 1980; Chapman, 1975; Hogg and Hutcheson, 1975; Manners, 1981). In addition, Morgan and Moss (1981) contributed to our understanding of fuel wood in the humid tropics, while on the continent Odell (1986) regularly updated his global survey of the geopolitics of oil. Keith Chapman (1991) returned with another book on oil just after John Chapman (1989) in Vancouver addressed the complexities of commercial energy systems.

The contribution of geographers to energy studies remained modest until the accidents at Three Mile Island in 1979 and Chernobyl in 1986. Immediately, energy geographers started mining the rich vein of spatial question it offered. These included:

- Risk perceptions and behavioral responses (Pasqualetti and Pijawka, 1984; Blowers and Peppers, 1987).
- Power plant siting (Openshaw, 1986; O'Riordan et al., 1988).
- The social costs of decommissioning (Pasqualetti, 1990).
- Lessons about democratic principles (Gould, 1990).
- The distribution of nuclear power plants (Mounfield, 1991).
- The transportation and disposal of nuclear waste (Openshaw et al., 1989; Jacob, 1990; Beaumont and Berkhout, 1991; Blowers, Clark, and Smith, 1991; Blowers, Lowry, and Solomon, 1991; Flynn et al., 1995; Pasqualetti, 1997).

In a reflection of the rising interests in the geography of energy, professional organizations began demonstrating a targeted interest in the subject. The American Association of Geographers (AAG), for example, published at least four studies on energy (Cook, 1977; Calzonetti and Eckert, 1981; Sawyer, 1986; Kuby, 1996). Also under the auspices of the AAG, Martin Pasqualetti and Jerome Dobson organized the Energy Specialty Group – now the Energy and Environment Specialty Group (www.eesg.org/) – at the close of the 1970s.[2] Subsequently, at least two comparable groups surfaced in Europe, including the Energy Geographies Research Group of the Royal Geographical Society (https://energygeographiesworkinggroup. wordpress.com/), and Ak Geographische Energieforschung in der Deutschen Gesellschaft für Geographie (Geographical Energy Research in the German Society for Geography) (www. geographische-energieforschung.de/). The existence of such formalized groups reflects the growing recognition that an understanding of the geography of energy is vital to the interests of a vast swath of governments, companies and individuals. So much so, in fact, that several organizations have posted maps of energy conditions on the internet, such as Energypedia (https:// energypedia.info/wiki/Portal:Countries). The themes that have interested geographers most included understanding the workings of the energy economies of countries, the distribution of energy resources, facility siting, regional governance, and the environmental costs of energy production, transportation and use.

Over the years, several books have grabbed the title *The Geography of Energy*, and have configured that publications to reflect a somewhat different interpretation of the phrase. As time went on, they became become progressively more inclusive and varied in tone, and they have grown in maturity and complexity (George, 1950; Manners, 1964; Guyol, 1971; Wagstaff, 1974; Sevette, 1976; Cook, 1976; Chapman, 1989).

As contributions by geographers mounted, we saw the inevitable appearance of summaries in edited volumes and encyclopedia (Calzonetti and Solomon, 1985; Solomon and Pasqualetti, 2009; 2004; Solomon et al., 2004). The 21st century has brought us a pair of over-arching books entitled *The Geography of Energy* (Brücher, 2009; Mérenne-Schoumaker, 2008). That both authors are European suggests the growing recognition of the important interactions of geography and energy in that part of the world.

Energy geography enters the 21st century

The turn of the 21st century saw an acceleration in the interest in the geography of energy, and a greater sophistication in the understanding of the role of energy in world events and globalization. People everywhere had come to realize the interdependencies of energy supply and the global consequences of energy consumption. Within this context, climate change has been one of the most active areas for energy geographers, both from those who believe quick remediation is needed (Parry *et al.* 2007; Knight, 2010) and those whose analysis suggests it is not as much of a problem as people think (Balling, 1992).

With the growing attention to the rise in carbon dioxide in the atmosphere, in 2009 Chevalier produced a thoughtful treatment on the costs of carbon-based energy (Chevalier, 2009). His book employed a regional approach to the many energy challenges of the rising level of atmospheric carbon, including examples from Asia, the Caspian Basin, Russia, the Middle East and North Africa, and the United States. That it emphasizes the geopolitics of energy is not surprising; almost all of the authors are affiliated with the Centre de Géopolitique de l'Energie et des Matières Premières. Two chapters in Chevalier's book address energy poverty and energy security and their links to climate change.

Both these 21st century topics fall strongly into the domain of social sciences, as they went about tacitly demonstrating how geographers have been digging into the rising concern not just about such things as supply, distribution and consumption, but direct ties between energy and such topics as happiness, well-being, health, migration and ethics. The work of four geographers stands out in this realm: Watts's (2010) *The Curse of Black Gold,* Buzar's (2007) *Energy Poverty in Eastern Europe,* Labban's (2008) *Space, Oil and Capital,* and Le Billon's (2005) *The Geopolitics of Resource Wars: Resource Dependence, Governance and Violence.*

Central themes of energy justice echo back and forth among these books. These include that (1) long-term political stability depends more on establishing the basic needs and security for local populations than simply on increasing energy supplies; (2) dialectical tensions between transnational oil corporations and resource-owning states help determine the profitability of oil production and the availability of oil in the world market; (3) matters of energy justice can drown out all other discussions in developing countries such as Nigeria, Ecuador, and Angola; (4) there is a direct link between the energy crises experienced by poor countries and the social aspects of domestic energy use (also see Bouzarovski *et al.*, 2017).

For its part, energy security provides a scaffolding for the geographic discussion of supply, demand, cost, environmental impacts and energy transitions. It logically and fruitfully embraces spatial scales from house to globe, and temporal scales from the immediate to the distant future. In addition, energy security can provide "cover" for those individuals, companies and government that wish to justify drilling in fragile environments military interventions where there is promise of great supplies and profits.

Geographers are contributing strongly in this space. Recently, participants in a three-day workshop in Singapore discussed possible indicators of energy security. Several geographers were involved in this conference and interactions were always suffused with geography, including climate change, sustainable development, public policy options, environmental costs, energy poverty and social development, and energy efficiency (Sovacool, 2010).

If the 1980s witnessed the turn of attention to nuclear issues, the 21st century has provided a rich literature on our imminent turn more towards renewables as part of the next energy transition (Smil 2010). The reason is clear: Several resources – including geothermal, wind, hydro and tidal – are site-specific, thereby leading to more land-use conflicts than those energy resources that can be moved around more readily like oil or coal. Of all such renewables, wind energy has been attracting the most public fire.

Several factors contribute to this attention, including their large size (up to almost 300 feet), their inconsistent motion and noise, and their threats to birds, bats and other wildlife. However, the dominant public issue is their visual presence in the landscape, as identified by several geographers (Pasqualetti *et al.* 2002). It has continued to grow and evolve in many countries, most recently taking on the mantle of what has come to be called "energy landscapes" (Nadaï and van der Horst 2010). We treat such energy landscape in much more detail later, but suffice to say here that it has been emerging as a major theme of energy geographers. One reflection of how this theme has struck a responsive chord is found in the several dozen papers on the subject delivered in Dresden in September of 2015. Many of these papers were produced by geographers and those in aligned disciplines such as environmental history and landscape architecture (e.g., http://lrg2015.ioer.info/index.html). It is in many ways a reflection of the sub-discipline of cultural geography that predominated in some institutions such as the University of California at Berkeley for over half a century.

As energy gains in importance and interest, the contribution of geographers is growing quickly. For interested students just starting out, a good place to start is to examine the range of energy books published by Vaclav Smil, especially his books on the societal dimensions of energy (Smil 2008, 2010) and energy transitions.

Since 2010, the final year of the survey that is represented in Figure I.1, several additional polished and significant books have been published, including those by Bridge and Le Billon (2012), Bradshaw (2013), Ekins and Bradshaw (2015) and Huber (2013), and they are progressively taking an edgier approach.

Future directions for research in energy geographies

With the negotiations at the COP21 in Paris, and the United Nations commitment to energy for all, global environmental politics are likely to have a marked influence on the directions that energy geographies research may take in the coming years.

Fossil fuel geographies will continue to have great relevance, both because it is unlikely that there will be an agreement on banning fossil fuel emissions but also, because fossil fuels are thought of play a key role in the future energy mix. Few countries contemplate a future without fossil fuels. Ireland, for example, whose electricity network has one of the highest proportion of renewables, relies on natural gas to stabilize the supply of energy. This also means that, taking a pragmatic view, many countries are likely to continue financing fossil fuels, often with climate funds, under the assumption that this will be the most efficient way to foster a low carbon transition. Japan, for example, caused a stir in early 2015 after the first round of negotiations of the Green Climate Fund, when they declared that they would continue to fund efficient technologies for coal-energy production as a means to reduce carbon emissions, because more efficiency coal-generation may lead to carbon reductions.

In this context, geographers must play a key role both in situating the energy realities within different contexts while also pointing out the contradictions that shape current global environmental politics (Castán Broto, 2015). Here, there is a dual role in being a forceful critic while opening up pathways for possible sustainable environmental futures. For example, while a blanket ban on all fossil fuels is both unlikely and undesirable, a global agreement to limit its production may have the highest impact yet in emissions reductions. Part of the problem in making this a reality is that there is little understanding of how it could be possible not just in terms of how to reach a geopolitical agreement, but also, in terms of understanding what the economic and environmental consequences of such a ban could be. This calls simultaneously for geographies of energy that are global in scope but that also engage with the specificities of the political and environmental context in which such limits could be possible.

Another theme that emerges from global energy policy is energy access, and the extent to which current imperatives for decarburization support or hinder programmes which attempt to reduce energy poverty and facilitate access to different energy services. With their Sustainable Energy for All program, the United Nations emphasize the linkages between providing universal energy access while promoting both renewables and energy efficiency. There is plenty of evidence that renewables may provide energy solutions for the energy poor, for example, by providing access to electricity with microgrids to remote areas not connected to centralized networks or by providing an independent energy source, such as the micro-solar panels that are bringing lighting and phone charging facilities to the residents of informal settlements (Kumar, 2015). However, the evidence that renewables can actually facilitate energy access is patchy and, often, anecdotal. There is a lack of understanding with regards to the effectiveness of different technologies, the practical difficulties that emerge during their implementation, which factors facilitate them, but also, as well as their appropriateness in different contexts. Research on the feasibility and effectiveness of microgrids is particularly urgent.

The same goes for the discourse on improving efficiency. In developing countries, upgrading programs for informal settlements, for example, encounter many difficulties. The extent to which they benefit the actual urban poor is contested. Partly, there is great difficulty in characterizing what is poverty, because it depends on multiple variables. This is also true for the characterization of energy poverty. Would a connection to the grid facilitate energy access to poor populations or would this access pose unbearable costs that may impact them in the long term? In many cases, the question is not what makes people energy poor but rather, which are the institutional systems that provide energy and how certain business models facilitate the exclusion and inclusion of different groups. These kinds of observations are not limited to developing countries: retrofitting studies have revealed the multiple ways in which interventions in the built environment may affect people's perceptions of their energy needs and, sometimes, determine their possibilities to access energy (Bouzarovski, 2015; Hodson and Marvin, 2015).

Such questions also relate to the mechanisms that embed energy in society and that establish a certain 'lock-in', that is, certain stability that is hard to challenge with traditional environmental policy means of regulation, incentives, technological change or activism (Bouzarovski *et al.*, 2016). In some studies of infrastructure and sustainability been take for granted that energy infrastructures – especially electricity – are 'invisible' (Leigh Star, 1999). Yet, these infrastructures are very visible and present in people's lives and experiences. Instead, they have become part of the lived experience of the house, the city, the landscape (Vannini and Taggart, 2013). Such relationships are periodically disturbed – such as in the case of wind energy landscapes – but in essence, our dependence from energy remains embedded in our culture.

Studies of practice have explored how the use of different forms of energy is normalized within everyday life. This line of work has challenged that an individual consumer-citizen who can change energy-consumer lifestyles could be identified. Instead, they see energy-dependent practices emerging within complex assemblages of objects and meanings (Bouzarovski *et al.*, 2015; Debizet *et al.*, 2016). A strand emerging in relation to the theory of practice is the undertanding of the multiple energy publics that emerge around energy geographies, the increase disaggregation of 'publics' and the growing interest in how publics are constructed within particular energy debates (Walker and Cass, 2007). A particularly important area of work relates both the critical analysis of energy in everyday life fostered by the theory of practice and the growing interest on energy publics with the geopolitics of energy, and the need to highlight how the everyday life of energy should be considered in broader debates of energy, dominated by discourses of energy efficiency and the energy mix.

In practice, everyday uses of energy have been politicized through the application of psycho-analysis and other psychological therapies. For example, with the conceptualization of our society as being 'addicted to oil' and the use of terminology commonly used to treat personal addictions to work out how to escape the oil addiction. This approach has been particularly effective in mobilizing networks of communities to enable collective action, for example, in Transition Towns, while finding the means to create carbon subjects which attempt to find new means of self-regulation within the existing low carbon society (Seyfang and Haxeltine, 2012; Bulkeley *et al.* 2014). This practical example suggests that there could be great potential for critical explorations of energy cultures and the symbolism of energy in society, particularly with the application of psychoanalysis and Lacanian theory. While this has been widely influential in recent years, to explain everything from the economic crisis to social mobilization, its application to understand energy geographies is practically absent. Here, interdisciplinary studies which relate energy geographies with critical studies in architecture and planning may foster innovation within the field.

In this book, we aim to chart the current state of knowledge in energy geographies to provide a reference guide for the development of future research. The aim is to develop a collection of examples of research which engages with the topic critically. In doing so, this body of research challenges established assumptions about how to respond to energy challenges and in doing so, they contribute to cutting edge policy thinking. The companion is divided in three parts: the first develops a conceptual toolbox highlighting by focusing on studies of how energy flows in relation to specific social and political systems. This part emphasizes on learning from past and present transitions to understand spatial patterns in relation to specific trajectories of change. The second follows on energy as a contested subject. In particular, this part emphasizes how energy uses relate to the use of different resources, and what that means in different locations and for different people. Impacts related to fossil fuel extraction, electricity generation or trans-missions cannot be understood without referring to processes of conflict and resolution. Finally, Part 3 explores geographical perspectives that both seek to re-signify and open up discussions beyond established fields of research, building upon political ecology approaches, consumption geographies and assemblage approaches.

Note

1   A related, although less familiar, literature flowed in the form of technical reports from US research laboratories. This was particularly true of Oak Ridge National Laboratory in the late 1970s and 1980s, when about a dozen energy geographers were active there at the same time. These reports, although not published commercially, illustrated the work of geographers in energy analysis, and they were often applied in the development of energy policy and legislation. Much of this section is an edited version of that found in Pasqualetti (2011), to which the reader is referred for more detail.

References

Beaumont, J. R., and F. Berkhout. 1991. *Radioactive waste: Politics and technology*. London and New York: Routledge.

Beaumont, J. R., and P. Keys. 1982. *Future cities: Spatial analysis of energy issues*. New York: Wiley.

Blowers, A. T., and D. Peppers. 1987. *Nuclear power in crisis*. London and New York: Routledge.

Blowers, A. T., M. Clark, and D. Smith, eds. 1991. *Waste location: Spatial aspects of waste management, hazards and disposal*. London: Routledge.

Blowers, A. T., D. Lowry, and B. D. Solomon. 1991. *The international politics of nuclear waste*. New York: Palgrave Macmillan.

Bouzarovski, S. 2015. Retrofitting the City: Residential Flexibility, Resilience and the Built Environment. London: IB Tauris.

Bouzarovski, S., L. Sỳkora, and R. Matoušek. 2016. Locked-in post-socialism: rolling path dependencies in Liberec's District Heating system. Eurasian Geography and Economics, http://dx.doi.org/10.1080/15387216.2016.1250224.

Bouzarovski, S., M. J. Bradshaw, and A. Wochnik. 2015. Making territory through Infrastructure: The governance of natural gas transit in Europe. Geoforum 64: 217–228.

Bradshaw, M. 2013. *Global energy dilemmas*. London: Polity.

Bridge, G., and P. Le Billon. 2012. *Oil*. London: Polity.

Brücher, W. 2009. *Energiegeographie*. Gebr. Borntraeger Verlagsbuchhandlung. Berlin and Stuttgart.

Bulkeley, H. A., V. Castán Broto, and G. A. S. Edwards. 2014. An Urban Politics of Climate Change: Experimentation and the Governing of Socio-Technical Transitions. London: Routledge.

Buzar, S. 2007. Energy Poverty in Eastern Europe: Hidden Geographies of Deprivation. Aldershot, UK: Ashgate.

Calzonetti, F. J. and M. S. Eckert. 1981. *Finding a place for energy: Siting coal conversion facilities*. Resource Publications in Geography. Washington, DC: Association of American Geographers.

Calzonetti, F. and B. D. Solomon, eds. 1986. D. Reidel Publishing Company, Dordrecht, The Netherlands.

Castán Broto, V. 2015. Contradiction, intervention, and urban low carbon transitions. Environment and Planning D: Society and Space 33 (3): 460–476.

Castán Broto, V. 2016. Innovation territories and energy transitions: energy, water and modernity in Spain, 1939–1975. Journal of Environmental Policy & Planning 18 (5): 712–29.

Chapman, J. D. 1989. *Geography and energy: Commercial energy systems and national policies*. New York: Longman.

Chapman, K. 1975. *North Sea oil and gas*. London: David & Charles.

Chapman, K. 1991. *The international petrochemical industry*. Malden, MA: Blackwell.

Chevalier, J.-M. (ed.). 2009. *The new energy crisis: Climate, economics and geopolitics*. Basingstoke: Palgrave Macmillan.

Cook, E. F. 1976. *Man, energy, society*. San Francisco, CA: Freeman.

Cook, E. F.1977. *Energy, the ultimate resource?* Resource Papers for College Geography, No. 77–4, Washington, DC: Association of American Geographers.

Croissant, M. P., and B. Aras. 1999. *Oil and geopolitics in the Caspian Sea Region*. Westport, CT: Praeger.

Debizet, G., A. Tabourdeau, C. Gauthier, and P. Menanteau. 2016. Spatial processes in urban energy transitions: Considering an assemblage of socio-energetic nodes. Journal of Cleaner Production 134: 330–341.

Dienes, L., and T. Shabad. 1979. *The Soviet energy system: Resource use and policies*. New York: Wiley.

Dienes, L., I. Dobozi, and M. Radetzki. 1994. *Energy and economic reform in the former Soviet Union: Implications for production, consumption and exports*. New York: Palgrave Macmillan.

Ekins, P., and M. Bradshaw. 2015. *Global energy: Issues, potentials, and policy implications*. Oxford: Oxford University Press.

Farrell, B. H. 1962. *Power in New Zealand: A geography of energy resources*. Wellington, New Zealand: A.H. & A.W. Reed.

Fernie, J. 1980. *The geography of energy in the United Kingdom*. Englewood Cliffs, NJ: Prentice-Hall.

Flynn, J., J. Chalmers, D. Easterling, R. Kasperson, H. Kunreuther, D. K. Mertz, A. Mushkatel, K. D. Pijawka, and P. Slovic. 1995. *One hundred centuries of solitude: Redirecting America's high-level nuclear waste policy*. Boulder, CO: Westview.

George, P. 1950. *Géographie de l'energie* [Geography of energy]. Paris: Genin.

Gever, J., R. Kaufmann, and D. Skole. 1991. *Beyond oil: The threat to food and fuel in the coming decades*. Boulder, CO: University of Colorado Press.

Gould, P. 1990. *Fire in the rain: The democratic consequences of Chernobyl*. Baltimore, MD: Johns Hopkins University Press.

Guyol, N. B. 1971. *Energy in the perspective of geography*. Englewood Cliffs, NJ: Prentice-Hall.

Hall, C., C. Cleveland, and R. Kaufmann. 1992. *Energy and resource quality: The ecology of the economic process*. Boulder, CO: University of Colorado Press.

Hart, D. 1980. *The Volta River project: A case study in politics and technology.* Edinburgh, UK: Edinburgh University Press.

Hodson, M., and Simon M. 2015. Introduction. In Retrofitting Cities: Priorities, Governance and Experimentation, edited by M. Hodson and S. Marvin. London: Routledge. pp. 1-10.

Hoffman, G., and L. Dienes. 1985. *European energy challenges: East and West.* Durham, NC: Duke University Press.

Hooson, D. J. M. 1965. *Introduction to studies in Soviet energy.* Vancouver, BC, Canada: Tantalus Research.

Huber, M. 2013. *Lifeblood: Oil, freedom, and the forces of capital.* Minneapolis, MN: University of Minnesota Press.

Jacob, G. 1990. *Site unseen: The politics of siting a nuclear waste repository.* Pittsburgh, PA: University of Pittsburgh Press.

Knapper, C., L. Gertler, and G. Wall. 1983. *Energy, recreation and the urban field.* Waterloo, ON, Canada: University of Waterloo, Department of Geography.

Knight, G. 2010. Climate change: The health of a planet in peril. *Annals of the Association of American Geographers* 100:1036–1045.

Kuby, M. 1995. *Investment strategies for China's coal and electricity delivery system.* Report No. 12687-CHA. Washington, DC: World Bank.

Kuby, M.1996. *Population growth, energy use, and pollution: Understanding the driving forces of global change.* Washington, DC: Association of American Geographers.

Kumar, A. 2015. Cultures of lights. Geoforum 65: 59–68.

Labban, M. 2008. *Space, oil, capital.* London: Routledge.

Lakshmanan, T. R., and B. Johansson, eds. 1985. *Large-scale energy projects: Assessment of regional consequences.* New York: North Holland.

Lakshmanan, T. R., and P. Nijkamp, eds. 1980. *Economic, environmental and energy interactions: Modeling and policy analysis.* Hingham, MA: Martinus Nijhoff.

Lakshmanan, T. R., and P. Nijkamp, eds.1983. *Systems and models for energy and environmental analysis.* London: Gower.

Le Billon, P., ed. 2005. *The geopolitics of resource wars: Resource dependence, governance and violence.* London and New York: Routledge.

Manners, G. 1964. *The geography of energy.* London: Hutchinson University Library.

Manners, G. 1981. *Coal in Britain: An uncertain future.* London: Allen & Unwin.

Mérenne-Schoumaker, B. 2008. *Géographie de l'énergie – Acteurs, lieux et enjeux.* Paris: Belin.

Morgan, W. B., and R. P. Moss. 1981 *Fuelwood and rural energy, production and supply in the humid tropics.* Dublin, Ireland: Tycooling International.

Mounfield, P. R. 1991. *World nuclear power.* London: Routledge.

Nadaï, A., and D. van der Horst. 2010. Introduction: Landscapes of energies. *Landscape Research* 35(2): 143–155.

Odell, P. 1977. *Energy: Needs and resources* (Aspects of geography). 2nd revised ed. London: Macmillan.

Odell, P. 1986. *Oil and world power.* 8th revised ed. Harmondsworth, UK: Penguin.

Odell, P., and K. E. Rosing. 1980. *The future of oil.* New York: Nichols.

Openshaw, S. 1986. *Nuclear power: Siting and safety.* London and New York: Routledge & Kegan Paul.

Openshaw, S., S. Carver, and J. Fernie. 1989. *Britain's nuclear waste: Safety and siting.* London: Belhaven Press.

O'Riordan, T., R. Kemp, and M. Purdue. 1988. *Sizewell B: An anatomy of the inquiry.* London: Macmillan.

Parry, M. L., O. F. Canziani, J. P. Palutikof, P. J. van der Linden, and C. E. Hanson, eds. 2007. *Contribution of Working Group II to the Fourth Assessment Report of the Intergovernmental Panel on Climate Change.* Cambridge, UK: Cambridge University Press.

Pasqualetti, M. J. 1986. The dissemination of geographic findings on nuclear energy. *Transactions of the Institute of British Geographers* 11: 326–36.

Pasqualetti, M. J. 1990. *Nuclear decommissioning and society: Public links to a new technology.* London and New York: Routledge.

Pasqualetti, M. J. 1997 Landscape permanence and nuclear warnings. *The Geographical Review* 7(1): 73–91. (Published January 1998.)

Pasqualetti, M. J. 2011. The geography of energy and the wealth of the world. *Annals of the Association of American Geographers* 101(4): 971–980.

Pasqualetti, M. J., and M. Brown. 2014. Ancient discipline, modern concern: Geographers in the field of energy and society. *Energy Research & Social Science* 1:122–133.

Pasqualetti, M. J., and K. D. Pijawka, eds. 1984. *Nuclear power: Assessing and managing hazardous technology.* Boulder, CO: Westview.

Pasqualetti, M. J., and B. Sovacool. The importance of scale to energy security. *Journal of Integrative Environmental Sciences* 9(3): 167–180.

Pasqualetti, M.J., P. Gipe, and R. Righter (eds). 2002. *Wind energy in view: Landscapes of power in a crowded world.* San Diego, CA: Academic Press.

Pratt, W. E., and D. Good, eds. 1950. *World geography of petroleum.* American Geographical Society Special Publication No. 31. Princeton, NJ: Princeton University Press.

Pryde, P. R. 1983. *Nonconventional energy resources.* New York: Wiley.

Sawyer, S. 1986. *Renewable energy: Progress and prospects.* Washington, DC: AAG Resource Publications.

Sevette, P. 1976. *Geographie et économie comparee de l'énergie* [Geography and comparative economics of energy]. Grenoble, France: Institut Economique et Juridique de l'Énergie, Université de Grenoble.

Seyfang, G., and A. Haxeltine. 2012. Growing grassroots innovations: Exploring the role of community-based initiatives in governing sustainable energy transitions. Environment and Planning C: Government and Policy 30 (3): 381–400.

Smil, V. 1976. *China's energy.* Westport, CT: Praeger.

Smil, V. 1988. *Energy in China's modernization.* Armonk, NY: Sharpe.

Smil, V. 1994. *Energy in world history.* Boulder, CO: Westview.

Smil, V. 2008. *Energy in nature and society: General energetics of complex systems.* Cambridge, MA: MIT Press.

Smil, V. 2010. *Energy transitions: History, requirements, prospects.* Santa Barbara, CA: Praeger.

Solomon, B. D. and M. J. Pasqualetti. 2004. History of energy in geographic thought. In *Encyclopedia of Energy.* Edited by Cutler Cleveland. Academic Press: San Diego, pp. 831–842.

Solomon B. D., and Pasqualetti, M. J. 2009. History of energy in geographic thought. In *Concise Encyclopedia of History of Energy.* C. Cleveland (ed). San Diego, CA: Elsevier. pp. 117–126.

Solomon, B. D., and M. J. Pasqualetti. 2013. History of energy in geographic thought. In *Reference Module in Earth Systems and Environmental Sciences,* Elsevier.http://dx.doi.org/10.1016/B978-0-12-409548-9.01282-3.

Solomon, B. D., M. J. Pasqualetti, and D. A. Luchsinger. 2004. Energy geography. In *Geography in America at the Dawn of the 21st Century.* Edited by G. Gaile and C. Willmott. Oxford: Oxford University Press, pp. 302–313.

Sovacool, B., ed. 2010. *Routledge handbook for energy security.* London: Routledge.

Star, S. L.. 1999. The ethnography of infrastructure. American Behavioral Scientist 43 (3): 377–91.

Vannini, P., and J. Taggart. 2013. Domestic lighting and the off-grid quest for visual comfort. Environment and Planning D: Society and Space 31 (6): 1076–1090.

Walker, G., and N. Cass. 2007. Carbon reduction, 'the public' and renewable energy: engaging with socio-technical configurations. Area 39 (4): 458–69.

Wagstaff, H. R. 1974. *A geography of energy.* Dubuque, IA: Brown.

Watts, M. 2008. *The curse of black gold.* New York: Powerhouse Press.

Part 1

Energy territories and transitions

Introduction

Stefan Bouzarovski

The completion of this volume coincides with the Annual Meeting of the Royal Geographical Society with the Institute of British Geographers – the largest academic gathering of geographers in the United Kingdom. At the conference, the recently established Energy Geographies Research Group is due to sponsor a record number of 24 sessions – on topics as wide ranging as the politics of energy transitions, fuel poverty and nexus thinking. The high number of energy sessions has been a feature of this event for several years, and means that it is possible to attend an energy session in every slot of the conference. It mirrors a situation that can be found at other large geography gatherings, such as the Annual Meeting of the American Association of Geographers as well as the German Geography Congress. The papers presented at these sessions display ever-increasing amounts of conceptual diversity, spatial scope and empirical richness. There is an unprecedented level of international and trans-disciplinary engagement.

The outward facing nature of energy geographies, as well as its desire to develop a coherent conceptual language, was recently reflected in two pivotal articles focusing explicitly on the development of energy geographies as a sub-discipline: Kirby Calvert's (2015) 'From "energy geography" to "energy geographies": Perspectives on a fertile academic borderland' and Matt Huber's (2015) 'Theorizing energy geographies'. Despite offering different approaches to the study of energy geographies, both papers are motivated by the recent expansion of research in this domain, while sharing a common intention to provide a conceptual overview of the field and its future directions. In this, they echo the calls made by a growing tide of contributions by geographers and cognate social scientists (Bridge et al., 2013; Pasqualetti & Brown, 2014; Rohracher & Späth, 2014), who have been arguing that space and place are not only constitutive of energy circulations and formations, but also define the wider political, economic and technical contingencies that matter in this context. With the transition away from conventional fossil fuels redefining the very material nature of, and connections among, locales of energy and production, the need for developing and mainstreaming a geographically sensitive theoretical approach to the study of energy has become all the more pertinent.

Calvert (2015) argues that energy geographies' position at a number of academic 'borderlands' affords the discipline with a degree of flexibility and openness that may prove advantageous in intellectual and practical terms. Nevertheless, he calls for the development of a conceptual roadmap that would illuminate the specific vocabulary that geographers can bring to the debate. At the same time, Huber's (2015) contribution assumes a more critical tone: he contends that recent interventions in the social sciences and humanities on energy-related topics have paid little attention to the broader theoretical implications that arise in this context. Focusing on electricity in particular, he shows how energy is 'constitutive of the metabolism of cities' while producing distinct inequalities and materialities. This leads him to consider – rarely among recent geographical work on energy – the spatialities of energy demand and consumption, via explorations of mobility, the social construction of home, as well as non-residential forms of energy use. His contribution, as a whole, makes important forays into the development of a theoretical framework that is explicitly grounded in political ecology thought while encompassing the entirety of the 'energy chain' (Chapman, 1989).

All of these papers reflect a set of concerns that are assuming pivotal importance in the emergent discipline of energy geographies. On the one hand, there is the need for a comprehensive socio-spatial theorization of energy flows and formations, resting on core geographic concepts like place, territory, scale and landscape while speaking to paradigms that have been developed at the boundaries with cognate disciplines – including resources, consumption, assemblage, geopolitics, home and inequality. On the other hand, there is a desire to build upon the conceptual apparatus of human geography while reaching out to wider scientific debates. This reflects the polyvalent nature of the discipline, whose roots combine a variety of insights from different intellectual traditions into comprehensive theoretical frameworks. Energy geographers are thus ideally placed to understand and analyse the entirety of the energy chain – from sites of resource recovery to practices of final consumption – while encompassing the territories and landscapes that energy services necessitate and create.

Chapters in this section

In light of the above, the eight chapters that follow illustrate some of the ways in which energy geographers are engaging with and rethinking key ideas in geography while providing the building blocks for an indigenous conceptual approach to understand the spatialities of energy in society. Even if constraints on space have not allowed us to explore a broader range of energy carriers and geographical spaces, the chapters that ensue have been selected in a manner that both reflects the entirety of the energy chain – from generation to consumption – while exploring a multiplicity of renewable and non-renewable resources. We have also sought to include case study examples from across the world. Collectively, the eight chapters challenge multiple established assumptions about how energy is spatially conditioned, as well as the social, cultural and economic geographies that it creates.

The section commences with Christian Brannstrom and Matthew Fry's interrogation of Texas's two parallel energy 'revolutions': (i) the potential departure from fossil fuels as the dominant source of the supply mix thanks, principally, to the expansion of electricity production from wind; and (ii) the growth of 'unconventional' hydrocarbon extraction via hydraulic fracturing and horizontal drilling. That two contradictory policy developments may take place within a single spatial realm at the same time testifies to the complexity of contemporary energy reconfigurations. The two authors unpack this paradox by asking a number of pertinent questions around the distributional benefits of each 'revolution' as well as their attendant infrastructures and politics. Ultimately, however, it is the interaction between an age old geographical concept – scale – and the implementation of energy-relevant legislation that has played a decisive role: by concentrating regulatory authority at the state level, the Texas state government has been able to mould a variety of energy developments and choices. Yet the future reach of any energy transformations is limited by suburban Texans' energy-intensive lifestyles, the absence of environmental and public health data, as well as the specificities of the state's political setting.

The enrolment of different types of energy resources in the remaking of energy geographies is a central theme in the chapter by Peter Kedron and Sharmistha Bagchi-Sen, who focus on the changing patterns of biofuel production in the United States. Here, the emphasis is on the evolution of the biofuels industry, particularly in relation to technological innovations that have the potential to alter the geographic distribution of economic activity. The two authors demonstrate that the strategies employed by firms in order to enable the shift towards cellulose-based technologies vary widely at the national scale, partly because a single model of industrial development does not exist – thus demonstrating that energy networks shape places and spaces not only through infrastructural or policy links, but also via the exchange of socio-technical expertise and innovation.

Kirby Calvert, Kean Birch and Warren Mabee's chapter on 'biomass, bioenergy and emerging bio-economies' also explores this set of energy resources, by asking a number of questions about the viability of a transition from fossil fuels to biomass, hinging upon institutions, infrastructures and technologies. The authors are particularly interested in the manner in which bioeconomies reconfigure existing socio-spatial relations, landscapes and identities, underlining the central role of the 'biorefinery' in driving this type of energy transition, while pointing to the need for novel regional-level analyses as well as systematic conceptual frameworks.

Completing the bloc of chapters focusing on bio-energy resources is Barry Solomon and John Bradley Barnett's overview of the global geographic distribution of biofuels, which, they argue, is increasingly driven by the movement away from petroleum-based fuel use alongside rural development policies and energy security concerns. The two authors highlight the characteristics of established markets for this resource – particularly the US and Brazil – as well as the entrance of new global players such as China, Argentina, Germany, France and Indonesia. Also of importance in this context is the EU's role in driving demand for biodiesel and bioethanol.

The ability of energy circulations to forge territorial interdependencies and connections across large-scale geographic realms comes through powerfully in Jessica Graybill's exploration of Eurasian energy flows. Taking Mackinder's notion of geographical pivot and heartland as a starting point, she highlights how energy production, transport and consumption patterns both foreground and strengthen the relations among physical, cultural and geopolitical contingencies. Her analysis focuses on three oil and gas regions: the Russian Arctic, Siberia and eastern Kazakhstan. By examining changes in governance, industrial development and transportation networks – as well as issues of geographic imagination and ethnic diversity – she illuminates the ability of distinct Eurasian regions to create extractive energy territories that affect how states are run and people move across space. Despite involving a historical overview, the chapter is also forward looking in its investigation of the possible future of the pivot and heartland identified in the study.

Corey Johnson also examines the flow of hydrocarbon resources – more specifically, natural gas – between Russia and Europe. Displacing the traditional preoccupation of geopolitical research on the supply of energy, his analysis concentrates principally on Europe – in its role as a significant consumer of imported energy resources. He includes a variety of material sites, from households to the planet as a whole, to consider how natural gas moves across national boundaries via pipelines whose construction and functioning provides the connective tissue between different spatial domains. This multi-scalar analysis draws upon insights from planetary urbanization frameworks in emphasizing how recent policy initiatives – such as the European Energy Union – are embedded in wider landscapes of political and economic power.

An explicit theorization of the urban scale as a locus of energy transformations is provided by Sylvy Jaglin and Éric Verdeil, using evidence from Buenos Aires, Delhi, Istanbul and Cape Town, as well as – and to a lesser extent – Sfax in Tunisia and several Turkish cities. Inspired by the shift of governance and power away from the national towards the urban scale, they are primarily interested in the defining features of urban 'transition pathways', particularly when it comes to the disputed, undetermined and politicized spatialities of energy-related reconfigurations. Their analysis contests normative and linear understandings of energy transitions, emphasizing instead that economic actors, political elites, civil society groups and other social entities all have divergent motivations and aims in these contexts.

Last but not least, Ralitsa Hiteva's contribution explicitly develops the institutional contexts through which energy systems are managed and steered, by highlighting the strategic role of intermediary organizations in low carbon transitions. These bodies enable the material flows of energy among sites of production, transmission and distribution, often substituting for activities formerly undertaken by the state. Hiteva illustrates the resulting 'geographies of energy intermediation' via case studies of the regulation of natural gas supply in Bulgaria, and the transmission of electricity generated from offshore wind plants in the UK. By exploring a relatively wide set of spatial and resource contexts, she is able to argue that intermediary activities represent a 'new reality' in the governance of energy, one that serves to blur the boundaries between private and public interests via multiple trajectories of socio-technical change.

In their entirety, the eight contributions focus on different aspects of energy connectivities – from the ways in which various types of resource use reorganize existing regional inequalities, to the impacts of energy on institutional, geopolitical and urban configurations. Nearly all chapters embed a temporal aspect alongside the explicit focus on spatial transformations: most often this involves low carbon or other types of energy transitions, but there are also references to structural changes in international relations.

References

Bridge, G., Bouzarovski, S., Bradshaw, M., & Eyre, N. (2013). Geographies of energy transition: Space, place and the low-carbon economy. *Energy Policy*, *53*, 331–340.

Calvert, K. (2015). From 'energy geography' to 'energy geographies'. Perspectives on a fertile academic borderland. *Progress in Human Geography*, *40*, 105–125.

Chapman, J. D. (1989). *Geography and Energy: Commercial Energy Systems and National Policy*. Harlow: Longman.

Huber, M. (2015). Theorizing energy geographies. *Geography Compass*, *9*, 327–338.

Pasqualetti, M. J., & Brown, M. A. (2014). Ancient discipline, modern concern: Geographers in the field of energy and society. *Energy Research & Social Science*, *1*, 122–133.

Rohracher, H., & Späth, P. (2014). The interplay of urban energy policy and socio-technical transitions: The eco-cities of Graz and Freiburg in retrospect. *Urban Studies*, *51*, 1415–1431.

1

New geographies of the Texas energy revolution

Christian Brannstrom and Matthew Fry

Introduction

Texas is pushing the leading edge of two energy revolutions. One represents an important move toward a possible transition away from fossil fuels, as Texas is (as of 2015) the leading U.S. state for wind energy production with 16,406 megawatts (MW) of installed capacity located mainly in the arid and semi-arid western part of the state, accounting for 20 per cent of the U.S. total wind capacity. The second comprises renewed efforts to extract more hydrocarbons through aggressive use of hydraulic fracturing and horizontal drilling, collectively part of "unconventional" fossil fuel development (Chew 2014), deepening a path established more than a century ago. Texas is the leading U.S. producer of oil, 3.6 million barrels per day in 2014, or nearly 40 per cent of U.S. oil production, and produces about 26 per cent of U.S. natural gas.

This chapter describes and analyzes geographical aspects of benefits, infrastructures, and politics related to these parallel developments in Texas. We first identify the origins of the recent developments in renewables and hydrocarbons, then offer comparisons in the form of three questions that help further energy geographies: what is the distribution of benefits? what new infrastructures are required? what new types of politics have emerged? We make specific reference to wind power near Sweetwater, shale gas development in the Dallas-Fort Worth region, and shale oil and condensate (a "light" form of petroleum requiring minimal refining) production from the Eagle Ford shale region. The case of Texas shows how the transition to renewable energy may be overwhelmed by continued and enormous investments in oil and gas, and how a single geopolitical entity, a sub-national territory in this case, may host contradictory energy developments.

Contradictory energy revolutions? Renewables and fossil fuels in Texas

Texas has become a major player in renewable energy. Approximately 10 per cent of the state's electricity comes from wind farms located mainly in the state's western region. The large and rapid investments in Texas wind power relied upon state legislation promoting renewable energy standards and electricity market deregulation, authorized in 1999, just a few days after then-governor George W. Bush declared his presidential candidacy. This statute resulted from Bush's desire to encourage wind—he told his top utility appointee to "get smart on wind" in 1996—

and Enron chief Kenneth Lay's desire for a deregulated electricity market in Texas, which helped create conditions for Enron's construction of the Trent Mesa wind farm near Sweetwater in 2001, signaling the start of the wind boom (Galbraith and Price 2013).

Complementing Bush's favorable stance on renewables, in 2005 governor Rick Perry authorized Competitive Renewable Energy Zones (CREZ), linking rural regions with the greatest wind energy potential to Texas's growing urban areas through $7 billion of new transmission lines (Figure 1.1). These policies, combined with favorable wind climatology, buoyant demand for electricity, little political opposition to wind-farm construction, the federal Production Tax Credit (PTC) for renewable power, and the fact that Texas offers simple permitting of wind farms (no public consultation), created powerful economic and political incentives for wind farms. County officials offered their own incentives by authorizing property tax abatements on wind turbines, arguing that positive economic benefits— increased employment and economic activity—are greater than property tax forgone through abatements.

Texas has been a major oil and natural gas producer in the US beginning with Spindletop (in Beaumont, 130 km east of Houston) and Ranger (130 km west of Fort Worth) in the early

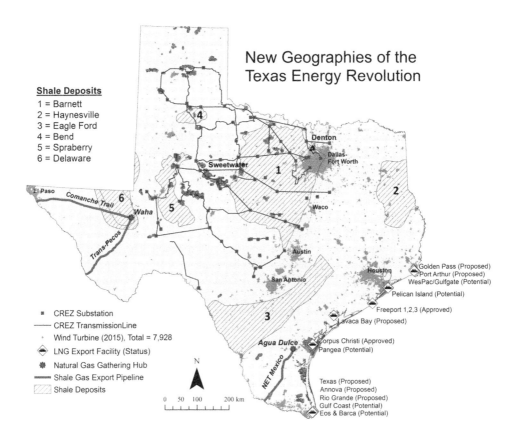

Figure 1.1 Location of major unconventional (shale) deposits, Competitive Renewable Energy Zones (CREZ) substations and transmission line, wind turbines, liquid natural gas (LNG) export facilities, and shale gas export pipelines

Source: Matthew Fry.

1900s and followed by major discoveries and investments in east Texas and Permian Basin oil fields in west Texas. The state achieved new oil and gas relevance since 2000 with the commercial production of unconventional oil and gas from hydrocarbon-bearing shale with hydraulic fracturing and horizontal drilling. Hydraulic fracturing uses a mixture of water, proppants, and chemicals injected at high pressure to free gas and oil from tight shale deposits (Figure 1.2). Horizontal drilling enables operators to tap larger areas within stratified shale layers. These new technologies opened large shale and unconventional sources of natural gas and oil in many US states, including Pennsylvania, North Dakota, Colorado, and Oklahoma (Chew 2014).

Although seven deposits currently produce shale gas in Texas, the Barnett Shale accounted for nearly 66 per cent of U.S. shale gas production in the 2000s (Figure 1.1). In terms of area, proven reserves, number of drilled wells, and total production to date, the Barnett is the largest of the Texas shale gas plays. It is also where hydraulic fracturing was first used in 1981 and where horizontal drilling began in the early 2000s. The combination of fracking, horizontal drilling, high market prices for natural gas, and relaxed federal environmental regulations encouraged firms to drill many wells, from 2616 in 2003 to approximately 17,000 in 2013. Operators generally target the northeastern region of the deposit at 2.0–2.6 km depths. This lucrative area or "sweet spot" of the Barnett shale underlies much of the Dallas-Fort Worth (DFW) metropolitan area. As a result, urban drilling has become pronounced in the Barnett.

Texas also leads the U.S. in proven oil reserves, largely from the Eagle Ford shale and other tight (unconventional) formations in west Texas. With approximately 5600 wells permitted between 2008 and 2014, the Eagle Ford produced around 1.6 million barrels per day in 2014. Reportedly, hydrocarbon firms invested approximately $28 billion in capital expenditures in 2013, making the Eagle Ford the world's largest single oil and gas development (EIA 2015; Tunstall 2015).

Benefits: new geographies of royalties

One new geographical relationship is the spatial distribution of royalties resulting from wind energy and hydrocarbon production. In Texas, nearly all wind farms are sited on private properties, requiring wind-farm developers to establish agreements with landowners regarding lease payments and royalties from electricity generated. Landowners normally seek legal counsel as they seek to maximize their return in exchange for land used for erecting turbines and building access roads, electricity substations, and maintenance facilities.

Information on precise amounts of royalties that wind farms pay to landowners is private, so a study estimated royalties on the basis of estimated output of electricity per turbine, wholesale electricity price, and royalty rate. Total royalty payments estimated for 1700 wind turbines near Sweetwater (Nolan and Taylor counties) were approximately $11.5 million per year. This works out to a mean annual royalty received per landowner per year of $47,879, and a median annual royalty of $25,756 (for four turbines on 129 hectares), which is less than median household income in either county. But only 241 landowners out of more than 8,000 total rural landowners in the two-county region have wind turbines on their land. The low percentage of landowners receiving royalties is a "property advantage," defined as the pre-existing land-tenure patterns that benefit the fraction of rural landowners who receive wind turbines (Brannstrom et al. 2015).

The distribution of royalties among landowners is uneven because of the "property advantage." The top quintile concentrates 60 per cent of royalties, with a mean of $146,189 in estimated royalties per landholding per year. This group has a mean landholding area of 842 hectares. By contrast, the fifth quintile captures less than 4 per cent of total royalties (mean = $7,656), with 1.3 turbines per landholding on average. Approximately 60 per cent of total royalties accrued to landowners residing within Nolan and Taylor counties, 30 per cent of royalties went to landowners residing elsewhere in Texas, and 7 per cent of royalties accrued to landowners residing outside Texas.

Thinking about royalties more broadly, the term "royalty paradox" describes the fact that royalties flow to a small fraction of landowners even though support for wind power is high (Brannstrom et al. 2015). The opposition to wind reported elsewhere in the U.S. is not present in west Texas. One group of social scientists reported high social support for wind power in west Texas and Iowa, with more than 70 per cent of respondents to a mail-out survey indicating that wind energy improved employment and the overall economy (Slattery et al. 2012). We are not certain how large payments to a small number of landowners translate into high acceptance of wind power. But it is likely that the landowners who benefit the most from royalties are prominent spokespersons for the wind economy, occupying key positions in organizations such as local school boards and other local civic groups that help influence opinion.

Figure 1.2 Drilling rig for hydraulic fracturing in the Eagle Ford Shale region, Texas
Source: Christian Brannstrom.

What do landowners do with their wind royalties? "Pay my notes [farm debt]" was the response of a cotton farmer near Sweetwater. Cotton farming and wind power are compatible because landowners with turbines on cotton fields lose only 3–5 per cent of their farming area to the pad site and access road. For ranchers, wind turbines were compatible with livestock grazing and the use of land for hunting leases. For example, a rancher near Sweetwater joked that royalties allowed him to "maintain my hobby," by which he meant ranching. Farming or ranching is a "pretty tough gig" and royalties help "pay bills and put kids through college." For other landowners, wind was "strictly money," but these financial resources had important effects, helping families maintain their land in difficult times.

Determining royalties from oil and gas production in the U.S. is complicated not only because royalty payments to individuals or groups are private, as in the case of wind power, but also because subsurface mineral ownership may be severed from ownership of the surface property. For this reason, a study in the City of Denton used appraised mineral property values as a proxy for royalty payments to examine where and to whom monetary benefits from shale gas development in the city accrue (Fry, Briggle, and Kincaid 2015). Denton County mineral appraisals are based on initial production rate, decline rate, gas price, lease operating expenses, productive life of well, and severance tax expense in a discounted cash flow calculation used to estimate mineral property values. The Denton study used data for 194 of the approximately 280 active gas wells in the city from 2003 through 2013. A total of 1455 property owners received royalty payments from 4923 mineral properties, valued at $66.2 million, associated with the 194 wells.

The study found that owners based outside of the City of Denton control at least 68 per cent of the city's appraised mineral values. Among this group, the single largest beneficiary (with nearly 30 per cent of the total appraised mineral value) was a retirement community developer based in Arizona. Other major absentee owners include families with legacy ranch properties. Among city-based owners, the largest local share (31 per cent) went to the City of Denton itself (which owns mineral property in 22 gas wells), followed by locally based family trusts. The third largest city share went to individual homeowners who comprise 90 per cent of the city-based owners, but received only 19 per cent of the value that stayed in the city, or 6.3 per cent of the total appraised value of Barnett shale mineral properties located under city territory.

Homeowners who owned minerals underneath Denton received a small percentage of the appraised value because most have relatively small-sized mineral holdings. Many gas wells are located close to the neighborhoods where these residents live. Proximity increases the likelihood that they will experience potential negative health effects associated with hydraulic fracturing and other gas well production activities (Shonkoff et al. 2014). This means that the 61 per cent of owners who do not live in Denton and receive the majority of the royalties from their subsurface minerals experience none of the risks associated with drilling activities. This uneven distribution of costs and benefits raises concerns about drilling in urban and suburban areas. For example, who should make decisions about the placement of shale gas wells in populated areas?

Getting energy to consumers: new distribution infrastructures

A second aspect of the new energy geographies emerging in Texas is the new infrastructure required to move electricity produced by wind farms to houses and businesses and to move unconventional oil and gas to refineries and consumers.

Wind farms and oil and gas are generally located in places distant from the final site of consumption (Figure 1.1). For west Texas wind farms, hundreds of miles separate wind turbines from large urban centers such as San Antonio, Fort Worth, Dallas, and Houston. Similarly, oil and gas must flow from wells into a distribution network supplying refineries, industries, and export terminals.

In west Texas, wind-farm investments initially relied on a transmission line that carried power from electricity generators in Odessa and Midland, which burned gas obtained by drilling for oil. This modest transmission line crossed Nolan and Taylor counties, supplying power to the Dallas-Fort Worth area long before wind turbines were built. However, production from wind farms quickly surpassed capacity to move power on this transmission line, causing many turbines to stand still on windy days. This is known by the technical term "curtailment," which reached a peak in 2009, until the new transmission lines were erected through the 2005 CREZ project. The Texas CREZ is probably the most ambitious U.S. transmission line project at 3,600 miles— of which about 600 miles were added to accommodate landowner requests for relocation. Completion in 2013 allowed for reduction of congestion and related curtailment, and the new transmission capacity should encourage more installed capacity beyond the 18,500 MW because new wind investment may be reliant on non-CREZ transmission lines.

The output and future production potential of unconventional oil and gas stimulated new oil exports from the Port of Corpus Christi and the development of liquid natural gas (LNG) export facilities and pipelines to deliver shale gas to Mexico (Figure 1.1). Before the U.S. allowed crude oil exports in December 2015, companies working in the Eagle Ford skirted the federal ban by lightly processing shale condensate then exporting from the Port of Corpus Christi. The first cargo of this type left Corpus Christi in June 2014 from what was once an oil import facility. In 2014, nearly 500,000 barrels per day were exported out of the port, which obtained a $20 million federal grant to expand rail capacity to bring condensate and unconventional oil from the Eagle Ford. The port also is working to expand in width and depth to accommodate the larger tankers that will sail through the newly expanded Panama Canal. In December 2015, the first crude oil shipment under the newly liberalized oil export regime left Corpus Christi, bound for a European refinery.

More broadly, 15 LNG facilities (two sites approved by the US Federal Energy Regulatory Commission, eight proposed sites, and six potential sites) are planned for the Texas Gulf coast (Figure 1.1). When cooled to -162 C° natural gas compresses to 0.0016 of its gaseous volume. After trans-oceanic transport by dedicated tankers, LNG is regasified at import terminals (Figure 1.3) and distributed inland through standard pipelines. Although LNG is a relatively new natural gas transport alternative, in 2013 it accounted for 31 per cent of the international gas trade. Using a high oil and gas price scenario, the U.S. Energy information Agency predicts that LNG exports will reach 7.5 trillion cubic feet, or 18 per cent of total U.S. gas production, in 2040 (EIA 2016).

With nearly 25 per cent of U.S. natural gas production in 2010, Texas is the largest gas-producing state in the nation. The state's planned LNG export facilities in Corpus Christi (2), Port Arthur (3), Freeport (3), Brownsville (5) and other coastal areas (Figure 1.2) will have an estimated capacity of 22.32 billion cubic feet per day combined, which would more than double the amount of gas Texas supplied to the U.S. in 2004.

Exporting natural gas to Mexico also represents a major new development in Texas. Approximately 430,000 miles of pipelines carry oil and gas throughout the state. The expansion of two natural gas gathering hubs, Waha in west Texas and Agua Dulce in south Texas, will serve as the starting points for three planned pipelines to export gas to Mexico (Figure 1.1). In 2014, Mexico opened its energy sector to private investment. The transport of shale gas from Texas to Mexico will be one of the first major new investments in redevelopment of Mexico's energy infrastructure.

Figure 1.3 Golden Pass LNG import facility near Port Arthur, which will expand to an export facility

Source: Matthew Fry.

Agua Dulce in Nueces County consolidates gas deliveries from ConocoPhillips' 1100-mile South Texas intrastate and gas gathering pipeline system, and ExxonMobil's 925 million cubic feet per day King Ranch processing facility. The Net Mexico pipeline is a 42-inch, 124-mile pipeline that will run from Agua Dulce to Rio Grande City in Starr County on the Mexico border. From there, new pipelines carrying Texas natural gas will serve as the foundation to expand Mexico's hydrocarbon transport infrastructure so it can support the projected growth in its unconventional production in the future.

The Waha hub consolidates natural and shale gas deliveries from the west Texas area and directs gas exports to Mexico via two proposed pipelines. The Comanche Trail Pipeline is a 42-inch, 190-mile pipeline that will transport natural gas to San Elizario near El Paso and then into the Mexican state of Chihuahua. With a capacity of up to 1.1 billion cubic feet per day, commercial operations are scheduled to begin in 2017. The 42-inch Trans-Pecos Pipeline, which will transport gas 143 miles from Waha to Presidio, is the most controversial of the three Mexico export pipelines. Scheduled to begin commercial operations in 2017, the pipeline raised concern among opposition groups in the towns of Alpine and Marfa regarding safety, health, and environmental risks associated with ruptures. Property rights advocates decry the use of eminent domain by the state government to confiscate property for encouraging international gas exports.

New geographies of opposition and support

A third aspect of the new energy geographies emerging in Texas is the new set of oppositional and supportive politics that have accompanied wind energy and unconventional oil and gas production.

In west Texas, wind power attracted strong local support because of presumed economic benefits, mainly through increased employment (Slattery et al. 2011) and royalty payments to landowners (Brannstrom et al. 2011; Jepson et al. 2012). Sweetwater, the Nolan County seat, has used the fact that it hosts approximately 2 gigawatts of wind power to brand itself as the center of wind-energy development by promoting trade shows, encouraging manufacturing related to wind power, and hosting a wind technician training program at a local community college (Figure 1.4).

Figure 1.4 Wind turbine blade greeting motorists on Interstate 20 in Sweetwater
Source: Christian Brannstrom.

Supporters of wind power argued that it had an enormous positive impact on the economy and community. According to one local official, before wind power, "Sweetwater was struggling . . . we were starting to lose population, we were seeing a large decline in our enrollment in our schools." Wind power meant that "ranchers and farmers can keep ranching and farming because there is enough confidence that the underlying amount of income [from royalties and leases] will be coming."

One of the few points of disagreement centers on the fact that county commissioners extended property tax abatements to wind-farm developers as a subsidy. Supporters argued that "we can lower our tax rate and still have more tax revenue than we ever thought possible before the wind industry came here." Other "wind welcomers" in Nolan County argued that Sweetwater was "on a downward spiral" before wind power arrived, and "cutting taxes to allow them [wind energy companies] to operate" was "all we have to offer." For these supporters, "our population stopped disappearing" because of tax abatements and wind power. Moreover, "if we had not offered them a tax abatement, then all the wind turbines would be in a different county." But opponents of abatements indicated there would be a "day of reckoning" when funds will be necessary to repair infrastructure that wind energy firms use: "You need to be collecting that tax . . . as the infrastructure degrading is going on, you got to take care of it."

A second source of opposition was temporary, as it focused on the construction phase, typically the period that produces most opposition because of increased truck traffic and other nuisances. Some observers noted that Sweetwater "turned to pure greed" during the construction phase, as house and apartment owners increased rents and evicted tenants to make way for construction crews and their larger budgets. But others viewed this a natural process and that it was a good thing that housing prices had increased. Critics were "just a typical head in the sand," according to one stakeholder, because if property values are not increasing, then "the town is dying . . . you are just filling up the cemetery."

Local elites are strongly supportive of wind power, but with equal enthusiasm they deny that wind power defines their region as an environmentalist stronghold. For example, one elected official said, "people from this area don't take the pledge of being green and being only for renewable energy," even though they are in the center of one of the largest clusters of wind turbines in North America. For this local official, everyday experience with wind power did not change views about greenhouse gas emissions or other touchstone environmental issues. The truth of the matter, for one official, was that "the ones for it [wind power] . . . are seeing money" in terms of royalties, employment, and property value: "I don't see it as an ideological thing for why we have [wind] turbines . . . if they were spitting smoke into the air, they would probably still be there." As another participant in the wind economy reminded us, "in Texas, you do not gripe about the money that is coming in from oil wells, and you shouldn't be concerned about money coming in from the wind farms."

West Texas does not have an environmental or aesthetic complaint against wind turbines, either. One person summed up this view nicely when she argued that "the fact that there are so many turbines, it's almost like going to a garden and seeing something growing . . . it actually improves the landscape." Another person placed possible aesthetic concerns into economic context, telling us that the "worst scenery would be the 'for sale' signs and all the ranches gone."

Perhaps the most coherent opponents of wind power were the engineers who observed the effects of rapid wind power expansion on a power grid that was not prepared to handle an intermittent source of electricity produced far away from large clusters of urban residents and their ever-increasing electricity needs. The core of their argument focused on the wild price swings, which sometimes produced negative electricity prices, with the entrance of wind power into a grid that had been supported by coal, natural gas, and nuclear power—all "base load" sources of electricity. One engineer argued that the cost of wind-generated power was three times more than producing electricity from natural gas, adding that wind power created "adverse impacts on system reliability" for the grid operator, the Electricity Reliability Council of Texas (ERCOT) (Zarnikau 2011). However, recent reports indicate that the negative prices reported in 2011 and 2012 disappeared in 2013 and 2014 because of reforms internal to ERCOT and increased transmission capacity through the CREZ program (EIA 2015).

In the Eagle Ford shale, oppositional politics focus on air quality, mainly because of poorly regulated flaring and accidental releases of pollutants. Releases of volatile organic compounds (VOC) and particulate matter (PM) occur from oil and gas extraction, hydraulic fracturing, flaring, compressor stations, and increased truck traffic, which state regulatory agencies do not fully address. The Texas Railroad Commission (RRC), the state-level oil and gas governing body, seldom refuses flaring permits. The Texas Commission for Environmental Quality (TCEQ) had only one air quality monitoring station in the Eagle Ford shale, but began operating another in Karnes City in late 2014.

Although early phases of the Eagle Ford shale boom were mainly unnoticed, scrutiny is increasing. One of the first reports was the 2013 *Reckless Endangerment* report produced by the non-governmental organization Earthworks, which portrayed a family suffering from acute respiratory problems in the heart of the shale boom. A collaborative report by The Weather Channel (2014) and Center for Public Integrity (2014) painted a similarly negative view of air quality impacts from oil and gas in the Eagle Ford shale. The report detailed symptoms of residents and their complaints made to the TCEQ. The TCEQ refused their requests for interviews, but an industry representative claimed that the TCEQ had a robust presence. This claim seems ill informed when contrasted with data on the paltry TCEQ air quality monitoring system for the

Eagle Ford. The overall message of the report is that the industry shows high regard for TCEQ, which is unable to monitor air quality and cannot even respond to requests for interviews.

Air quality in the San Antonio metropolitan region has been declining in parallel to unconventional oil drilling in the nearby Eagle Ford (Schade and Roest 2015). Rapid ozone formation can be caused by reactive VOCs such as ethylene and formaldehyde emitted from flaring. The *San Antonio Express* put the issue of flaring in sharp focus for its readers: flaring shows that the industry has "wasted billions of cubic feet of natural gas — enough to meet the needs for an entire year of every San Antonio-area household that relies on" natural gas (Tedesco and Hiller 2014).

Residents near drilling sites in the Eagle Ford are filing complaints with the TCEQ and are using infrared cameras to create their own findings regarding emissions from wells, fracking operations, tanks, and compression stations, as noted in *Texas Tribune* reporting for its "Shale Life" project in October 2014, which showed how a rancher used an infrared camera to show alleged emissions from oil and gas infrastructure. The *Texas Tribune* report also discussed complaints of acute respiratory distress among people who live near sites of oil and gas drilling (*Texas Tribune* 2014).

Although studies in the Barnett region indicate regional ozone pollution (Ahmadi and John 2015) and degradation of well water quality (Hildenbrand et al. 2015) from unconventional extraction activities, most opposition to Barnett Shale gas extraction focuses on noise, traffic, infrastructure damage, water use and contamination, and environmental and public health risks associated with hydraulic fracturing (Figure 1.5). The placement of gas wells near homes is a major concern for homeowners in DFW. These residents have filed numerous complaints about noise and light pollution, and report nosebleeds, nausea, headaches and other symptoms. Some homeowners have even filed and won lawsuits against drilling operators for damages.

Residents opposed to urban drilling also direct their frustrations towards city governments, which have some authority to regulate oil and gas activities within their territory. For example, since 2002, DFW municipalities have used their zoning and statutory powers to regulate how close gas wells can be placed near homes, apartments, schools, and other land uses. This distance is known as the "setback," which ranges from 300 to 1500 feet (91.4 to 457.2 m). Generally, setback distances greater than 600 feet (182.8 m) suggest greater citizen opposition to drilling. Since 2000, when drilling activity expanded throughout DFW, municipalities passed drilling ordinances with longer setback distances, often 600 feet or greater, and tighter restrictions on various aspects of drilling (Fry 2013). The city of Dallas even created a 1500-foot setback between protected uses and gas drilling (Fry, Brannstrom, and Murphy 2015).

Perhaps the most well-publicized example of how contentious setback distance became is the City of Denton, where voters banned hydraulic fracturing in November 2014. The events leading up to the Denton fracking ban highlight why 59 per cent of resident voters opposed this activity in their city. In late 2001, the city established a gas well drilling and production ordinance, later mandating a 500-foot (152.4 m) setback between gas wells and homes. Between 2001 and 2014, the number of gas wells in the city increased from seven to approximately 280. In 2009, a decision by the city government to allow three controversial gas wells close to homes, a public park, and a hospital generated considerable citizen backlash, which led to a city-imposed moratorium on drilling and the organization of a task force to help revise the gas drilling ordinance. A new ordinance was adopted in January 2013 that increased the setback distance to 1200 feet (365.8 m). However, five months after the new ordinance went into effect, an operator drilled three gas wells within 250 feet (76.2 m) of homes in a Denton neighborhood. Because the pad sites on which the wells were drilled existed before the neighborhood was built, the operator's right to drill was grandfathered or "vested" despite the new municipal distance restrictions. This

Figure 1.5 Hydraulic fracturing in the City of Denton, June 2015
Source: Matthew Fry.

meant that rights of the gas drilling firm were superior to those of the homeowners, who moved into their houses after the city had authorized the pad site. The close proximity of wells to neighborhood homes alarmed citizens, reignited debates about the adequacy of the city's ordinance, and contributed to the citizen-led campaign to ban fracking in the city. Opponents of fracking in the city used signs, newspaper advertisements, billboards, puppet shows, YouTube videos, and public events to gather support for their cause (Figure 1.6).

The day after Denton residents voted to ban fracking in the city, the Texas Oil and Gas Association (TxOGA) filed a lawsuit arguing that Denton's ban "is preempted by Texas state law and therefore unconstitutional." According to TxOGA, the outcome of the popular vote was "an impermissible intrusion" on the powers of the RRC, which regulates oil and gas drilling, and the TCEQ. The speed with which the lawsuit was filed suggests that reliance on the doctrine of preemption was well developed before the vote. In the meantime, the Texas legislature went into session just a few weeks after the Denton ban. Its members considered House Bill 40 (HB 40), written by industry attorneys in response to the Denton ban. HB 40 was approved after overcoming resistance by municipalities and activists. The governor signed HB 40 into law in May 2015, effectively banning Denton's fracking ban, but also substantially limiting the power municipalities once used to regulate urban drilling. The lawsuits, which raised moot points after HB 40 became law, were dropped.

By simultaneously expanding the power of a higher level of government and reducing the power of a lower level of government, the Texas state government used preemption legislation to rescale, redistribute, and remove city and local government regulatory authority over planning, environmental hazards, and business activities. HB 40 not only disallows Texas municipal governments from passing ordinances addressing subsurface hydrocarbon activities, but preempts any city hydrocarbon regulations deemed not "commercially reasonable," which seems

Figure 1.6 Billboard advocating Denton residents to vote for the hydraulic fracturing ban in
        November 2014

Source: Matthew Fry.

to allow the oil and gas industry (and the market prices for oil and gas) to determine the extent of municipal regulations rather than elected local officials who generally weigh health, safety, and welfare into their regulations. Although the short- and long-term effects of HB 40 on urban drilling are still unclear, fracking resumed in Denton in June 2015 – two weeks after HB 40 was signed into law and less than seven months after residents voted to end fracking in their city. The initial effects of HB 40 are most pronounced there, where opposition to HB 40 and fracking continue (Figure 1.7).

Although no other local groups opposed to urban drilling went to the extent of Denton's citizens, residents in the City of Mansfield, the Town of Flower Mound, the City of Dallas, and the City of College Station have mobilized in opposition to some aspects of unconventional drilling for oil and gas. Despite assurances from industry about their protective measures, the limited environmental impacts, and local economic benefits, not all residents support the Texas unconventional energy revolution.

Differing claims for the geographical locus of control—municipality or the sub-national state— are at the core of these debates, which center on the question of geographical proximity of drilling to suburban residences and parks. However, three geographically specific phenomena constrain oppositional politics against oil and gas. First, the energy-intensive lifestyles of suburban Texas (and the wealth these lifestyles sustain) undermine radical attempts to resist the unconventional revolution in spite of modest electricity supply from wind power. Second, the absence of environmental and public health data untainted by industry or activist bias—and immune from the highly influential lobbyists casting doubt on studies critical of hydrocarbon production— makes it difficult to engage in rational discussion about reasonable constraints to oil and gas production. Finally, the political setting in Texas is highly resistant to use of the precautionary principle. For example, instead of enacting regulations that would impose higher environmental standards on fracking, state regulators prefer "proactionary" policies: moving forward with risky industrial and extractive activities until harm can be demonstrated (Briggle 2015).

Figure 1.7 Anti-fracking protestors with "Frackulla" in Denton, Texas, July 3, 2015
Source: Matthew Fry.

Conclusion

Popular notions of the energy revolutions we discuss in this chapter vary dramatically from the actual geographies that these new developments create. Most popular views of renewable power focus on how renewables offer reductions in greenhouse gas emissions compared with other ways of generating electricity. Concern for global warming is often synonymous with support for renewable power, such as wind and solar. The oil and gas industry promotes unconventional hydrocarbons as representing U.S. technological prowess and ingenuity, energy independence (despite plans for large-scale export), and as sources of cheap energy for industries, transport, and households.

As we have shown, a geographical approach offers rather different views on renewables and unconventional oil and gas in Texas. First, energy geographies illuminate the question of who benefits from energy development—not in terms of industries and consumers, but in terms of the actual owners of the land or subsurface used to generate energy. How are benefits distributed geographically? Renewables, tied necessarily to surface ownership, offer royalties to a small fraction of landowners in rural areas suited to wind power, but these royalties may translate into overwhelming political support for wind farms through social networks. Unconventional hydrocarbons, held deep underground, may be owned by people who do not own surface land, therefore complicating immensely the issues of decision making, benefits, and impacts in populated areas.

We also highlighted a second, and rather unheralded, aspect of the new energy geographies in Texas: the transport infrastructure needed to move energy from points of production to sites of consumption or transformation. As the examples of CREZ, LNG plants, and gas collection facilities demonstrate, massive investment in transport infrastructure and conversion facilities are inevitable components of energy development. Determining the infrastructure required to make production of future renewable energy sources and new unconventional hydrocarbons economically viable will remain important to any future study of energy geographies because of their presence in land-use disputes and their importance in regional identity. For example, highly visible drilling and fracking in middle-class suburban settings (Figure 1.5) helped inspire an oppositional movement, but the movement did not ally politically with groups opposed to emissions from the enormous—and more visible—refining complexes near Corpus Christi and Houston. Although visible elements of renewable power are a source of local pride (Figure 1.4), drilling rigs for hydrocarbons are ubiquitous icons in many Texas cities where oil and gas means employment and wealth.

Finally, energy development has differing scalar-dependent effects on politics, economies, and societies. As our examples show, the politics that support (and contest) wind and urban shale gas development occur at the state, municipal, household, and individual level. For wind power, people directly impacted by wind infrastructure do not associate the turbines with broader notions of sustainable energy, but see turbines as a way to maintain traditional ways of life. Quite differently, there is pronounced opposition to unconventional oil and gas production, especially among residents who suffer the burdens but do not receive benefits.

The two energy revolutions will continue to develop in parallel, with the modest gains of renewables outpaced by the fracking boom, pushing an energy transition into the distant future. The apparent contradiction between these revolutions is resolved partially by abundant land for siting renewables in west Texas, where declining rural economies need capital infusion that wind power provides for some landholders who support renewables but are skeptical of climate change. Hydrocarbons nurture affluent suburban lifestyles, sustain personal and corporate wealth, and do not directly challenge renewables geographically or politically. The moment when suburban Texans disavow hydrocarbons will be a long time coming—even when it is, literally, in their backyards. But if that time comes, it is likely that the modest supply of electricity from wind power will be a minor factor in the face of the catastrophic environmental or economic changes that finally move political sentiment.

Acknowledgements

We thank Wendy Jepson, Victoria Knaupp, Sarah Gossett, Trey Murphy, Mary Tilton, Andrew Klein, Nicole Persons, Martin Aucoin, Adam Briggle, and Jordan Kincaid. Funding was provided by a National Science Foundation grant to Fry and Brannstrom ("Collaborative Research: A Spatial Analysis of the Determinants of Setback Distance Variation between Shale Gas Wells and Residences," #1262521 and #126526) and by the TCU-NextEra Wind Initiative grant to Jepson and Brannstrom.

References

Ahmadi, Mahdi, and Kuruvilla John. 'Statistical Evaluation of the Impact of Shale Gas Activities on Ozone Pollution in North Texas'. *Science of the Total Environment* 536 (2015): 457–467.

Brannstrom, Christian, Mary Tilton, Andrew Klein, and Wendy Jepson. 'Spatial Distribution of Estimated Wind-Power Royalties in West Texas'. *Land* 4 (2015): 1182–1199.

Brannstrom, Christian, Wendy Jepson, and Nicole Persons. 'Social Perspectives on Wind-Power Development in West Texas'. *Annals of the Association of American Geographers* 101 (2011): 839–851.

Briggle, Adam. *A Field Philosopher's Guide to Fracking: How One Texas Town Stood Up to Big Oil and Gas.* New York: Liveright Publishing, 2015.

Center for Public Integrity. 'Big Oil, Bad Air'. 2014, at http://eagleford.publicintegrity.org/ (accessed 15 August 2015).

Chew, Kenneth J. 'The Future of Oil: Unconventional Fossil Fuels'. *Philosophical Transactions of the Royal Society A* 372 (2014): 20120324.

Energy Information Administration (EIA). 'Wind Generates More than 10 per cent of Texas Electricity in 2014'. 2015, at www.eia.gov/todayinenergy/detail.cfm?id=20051 (accessed 7 June 2015).

EIA. 'Drilling Productivity Report'. 2015, at www.eia.gov/petroleum/drilling/ (accessed 15 August 2015).

EIA. 'Annual Energy Outlook'. 2016, at www.eia.gov/forecasts/aeo/er/index.cfm (accessed 8 June 2016).

Fry, Matthew. 'Urban Gas Drilling and Distance Ordinances in the Texas Barnett Shale'. *Energy Policy* 62 (2013): 79–89.

Fry, Matthew, Adam Briggle, and Jordan Kincaid. 'Fracking and Environmental (In)justice in a Texas City'. *Ecological Economics* 117 (2015): 97–107.

Fry, Matthew, Christian Brannstrom, and Trey Murphy. 'How Dallas Became Frack Free: Hydrocarbon Governance under Neoliberalism'. *Environment and Planning A* 47, 12 (2015): 2591–2608.

Galbraith, Kate and Asher Price. *The Great Texas Wind Rush: How George Bush, Ann Richards, and a Bunch of Tinkerers Helped the Oil State Win the Race to Wind Power*. University of Texas Press: Austin, 2013.

Hildenbrand, Zacariah L., Doug D. Carlton, Brian E. Fontenot, Jesse M. Meik, et al. 'A Comprehensive Analysis of Groundwater Quality in the Barnett Shale Region'. *Environmental Science & Technology* 49 (2015): 8254–8262.

Jepson, Wendy, Christian Brannstrom, and Nicole Persons. '"We Don't Take the Pledge": Environmentality and Environmental Skepticism at the Epicenter of U.S. Wind Energy Development'. *Geoforum* 43 (2012): 851–862.

Schade, Gunnar W., and Geoffrey Roest. 'Is the Shale Boom Reversing Ozone Air Quality Progress?' *EOS, Transactions American Geophysical Union* 96 21 April 2015, at https://eos.org/opinions/is-the-shale-boom-reversing-progress-in-curbing-ozone-pollution.

Shonkoff, Seth B., Jake Hays, and Madelon L. Finkel. 'Environmental Public Health Dimensions of Shale and Tight Gas Development'. *Environmental Health Perspectives* 122 (2014): 787–795.

Slattery, Michael, Eric Lantz, and Becky L. Johnson. 'State and Local Impacts from Wind Energy Projects: Texas Case Study'. *Energy Policy* 39 (2011): 7930–7940.

Slattery, Michael C., Becky L. Johnson, Jeffrey A. Swofford, and Martin J. Pasqualetti. 'The Predominance of Economic Development in the Support for Large-Scale Wind Farms in the U.S. Great Plains'. *Renewable and Sustainable Energy Reviews* 16 (2012): 3690–3701.

Tedesco, John and Jennifer Hiller. 'Up in Flames', *San Antonio Express-News* (August 2014), at www.expressnews.com/business/eagleford/item/Up-in-Flames-Day-1-Flares-in-Eagle-Ford-Shale-32626.php (accessed 15 August 2015).

Texas Tribune. 'Eagle Ford Air', *Texas Tribune* 29 October 2014 at http://apps.texastribune.org/shale-life/eagle-ford-air/ (accessed 15 August 2015).

Tunstall, Thomas. 'Eagle Ford and the State of Texas'. In *Economics of Unconventional Shale Gas Development: Case Studies and Impacts*, edited by W. E. Hefley and Y. Wang, 121–148. New York: Springer, 2015.

Weather Channel. 'Fracking the Eagle Ford Shale: Big Oil and Bad Air on the Texas Prairie' 18 February 2014 at http://stories.weather.com/fracking (accessed 15 August 2015).

Zarnikau, Jay. 'Successful Renewable Energy Development in a Competitive Electricity Market: A Texas Case Study'. *Energy Policy* 39 (2011): 3906–3913.

2

A study of technology and policy in liquid biofuel production in the United States

Peter Kedron and Sharmistha Bagchi-Sen

Introduction

Renewable energy is expected to reduce dependence on conventional energy sources and in the process alter energy landscapes to the benefit of society. The prevailing belief is that shifting energy production away from fossil fuels toward emerging renewable alternatives in the form of solar, wind, geothermal, and biomass will generate environmental benefits and economic opportunities that will carry over into future generations. However, we have yet to fully develop an understanding of how renewable energy systems evolve, how that evolution is connected to location, and what impacts these changes have across space. The location of emerging renewable energy industries is not necessarily coincident with existing centers of energy production, and the developers and producers of renewable energy are not necessarily current energy producers. To create policies capable of facilitating transition to a renewable energy system, and addressing the conflicts and impacts that transition will have, requires a geographic perspective that takes into account the complex interconnections that structure energy production, transfer, and use throughout the globe.

Geography is well positioned to provide such a spatial perspective. Bridge, Cooke, and Hayter argue that future socio-economic arrangements will integrate environmental concerns directly into industrial value-chains that reach around the world, which will make the question of renewable energy important in every transaction (Bridge 2002; Cooke 2002; Hayter and LeHeron 2002; Cooke 2004; Hayter 2004). Bridge (2008, p. 80) also notes that understanding these changes means incorporating environmental priorities into industrial development paradigms to analyze, "how forms of 'environmental governance' relate to the restructuring of economic relations, and to the production of conditions that favor some interests over others." Building on these and other geographic perspectives, energy systems research has examined how the spatial arrangement of established energy industries and the physical properties of energy itself facilitate certain forms of social relationships and modes of production (Bakker and Bridge 2006; Bridge 2010; Huber 2009, 2011). However, studies in this area have paid less attention to geographic processes of knowledge development, diffusion, and commercialization.

This chapter contributes to that emerging geographic perspective by examining spatial patterns of innovation and production in a leading U.S. renewable energy industry – biofuel. Specifically, we identify regional concentrations of biofuel innovation and connect the development of those innovations to policy initiatives and geographic variations in the structure of the biofuel industry. Understanding the emergence of regional variation in the U.S. biofuel industry sheds light on the development of renewable energy industries in general by identifying processes and policies critical to industry evolution and its connection to space. Examining the role of innovation is particularly important in renewable energy as these industries lack dominant production technologies, rely on policy support, and face substantial competition from incumbent producers of conventional forms of energy. Within the biofuel industry, the emergence of new technologies has the potential to challenge the existing organization of production by bringing new energy reserves into the industry. However, difficulty scaling-up and commercializing those technologies has limited the impact those innovations on industrial production.

Linking production and innovation in renewable energy industries

As Bouzarovski et al. suggest in the Introduction to this volume, geographic inquiry centers on the study of place, and energy connects and structures places throughout the globe. While the physical properties of energy extraction, distribution, and waste management come immediately to mind when we think about energy systems, the development of energy innovations with the potential to alter characteristics of those systems is equally important. Tied to spaces and networks of knowledge production, understanding energy innovation requires a perspective that takes seriously the ways in which geographic processes of knowledge development, diffusion, and commercialization evolve. Linking energy innovation with energy production provides insight into how and why landscapes of renewable energy change.

Connections between innovation and production are particularly important in renewable energy industries. Rather than having a single established production technology, renewable energy industries commonly have several technologies competing to become the industry's dominant design. Competition among technologies creates variation in the landscape of renewable energy production. While centers of production remain tied to resource-rich locations, the form production takes can vary with the technology adopted in a particular region. For example, examining the Brazilian biofuel industry, Compean and Polenske (2011) show how alternative labor- and technology-intensive forms of production exist in the country's north and south. Closely connected with regional centers of biofuel innovation, southern biorefiners adopted leading technologies and less labor-intensive forms of production. Cooke (2009) similarly identifies the importance connections between location-specific conditions and differing technologies have in the emergence of renewable energy production worldwide.

Alternative arrangements of production observable within and among renewable energy industries are not tied to technological or resource variation alone. To overcome competition from established energy industries and technical barriers to entry, renewable energy industries receive policy support, which protects these industries in market niches that encourage production (Smil 2005 2010; Coenen et al. 2012; Bridge et al. 2013). Geographic differences in the amount, type, and duration of policy support creates further variation in the arrangement of renewable energy industries. As Kedron and Bagchi-Sen (2011; Bagchi-Sen and Kedron 2015) demonstrate, state-to-state variation in support for corn-based biofuel has created alternative industry structures within the United States' largest biofuel producing region. States that adopted policies earlier, when production technologies remained in development and uncertain, have local biofuel industries with small capacity firms and ownership models tied to related smaller scale agricultural industry interests (e.g., co-op ownership). In contrast, corporate ownership models and larger production capacities exist in states that entered the industry later with policies oriented toward expansion of production. Similar studies of the wind, solar, and geothermal industries highlight the role of policy in regionally differentiating innovation (Stenzel and Frenzen 2008; Lehman et al. 2012), establishing local conditions for commercialization (Essletzbichler 2012; Simmie 2012), and encouraging the diffusion of innovation by key actors (Jacobsson and Lauber 2006; Hellsmark and Jacobsson 2009).

Although regional policies can influence the structure of renewable energy industries, it is often the case that innovation and production activities take place in different locations. Corsatea's (2014) patent-based examination of innovation activities in the European solar, wind, and bioenergy sectors identifies concentrations of innovation that only partially align with production centers identified by the European Commission. Moreover, different forms of renewable energy innovation take place in different regional contexts (Suurs and Hekkert 2009; Kedron and Bagchi-Sen 2011; Kedron 2015). Radical innovations requiring specific forms of high-tech knowledge related to biotechnology, electrical engineering, and materials manufacturing appear to be clustered in high-technology regions geographically distant from sites of production. In contrast, incremental innovations that improve production efficiency through process change occur at production sites, often in collaboration with regional partners. For example, Kedron and Bagchi-Sen suggest corn-based biorefiners select their production sites based not only on resource abundance and market access, but on the potential to access regional knowledge related to agricultural production through formal and informal collaboration. Those collaborations lead to the development of efficiency generating incremental innovations crucial in a renewable energy industry that produces an undifferentiated commodity product.

Given industry variation and the tendency for production and innovation to occur in different regions, connecting innovation with production in renewable energy industries means linking different locations. Understanding the pattern and concentrations of renewable energy innovation and production, the configuration of the organizations especially their networks, and the factors that led to the success of a particular configuration requires a closer examination of the entities involved in the whole value chain of a renewable energy industry. While previous research has examined a range of factors (e.g., policy, key scientists), fewer studies have thoroughly examined the role renewable energy firms play in making connections between regions. In firm-focused research, markets forces select firms, favoring those with an ability to create competitive advantages through the development and accumulation of knowledge (Penrose 1959; Martin and Sunley 2007). Competitive firms are able to dictate not only the development and diffusion of innovation (e.g., dominant design), but are instrumental in creating institutions that shape industry pathways through legitimization of organizational practices (Taylor and Oinas 2006; Bergek et al. 2008). To achieve their goals, firms often collaborate with other firms and institutions. However, how firms generate innovation and manage production depends on a range of firm-specific (e.g., size, age), industry-specific (e.g., lifecycle stage), and context-specific (e.g., location, regulation) factors.

Examining firms as connectors of centers of renewable energy innovation and production contributes to the development of hybrid geography frameworks applicable to the field of energy studies (Bridge 2008,; Hayter 2004, 2008; Patchell and Hayter 2013). An explicit focus on innovation in energy studies also adds another dimension to the ways in which energy connects and structures places. Improved understanding of those energy connections, better positions us as designers of policy and developers of firm strategy because we have more insight into how policies and firms succeed and fail in varied regional environments.

Renewable energy and biofuels

In the United States, liquid biofuels account for 22 percent of renewable energy production and consumption when measured in BTU (United States Energy Information Administration 2015a). U.S. biorefiners produce a specific form of biofuel called ethanol through the industrial scale distillation of plant sugars. In 2014, the U.S. produced 14.3 billion gallons of ethanol, representing approximately five percent of the motor vehicle fuel market (United States Energy Information Administration 2015b). First-generation technologies that are based on corn-based production currently dominate U.S. biofuel production, accounting for 98 percent of total industry output (RFA 2015). However, significant federal funding supports the development of alternative, second-generation, production technologies that convert sugars contained in other forms of non-food crops (e.g., agricultural and wood waste) into biofuels. Second-generation innovation has followed two tracks. Biochemical technologies create biofuel through the pretreatment, hydrolysis, and fermentation of cellulose, while thermochemical conversion gasifies renewable feedstocks and extracts biofuels from the resulting syngas (Sims et al. 2010; Gruenspecht 2013). However, transition to a second-generation led biofuel industry has yet to occur.

Both first- and second-generation biofuel production exist within a policy-generated market niche put in place during the early 2000s within the broader United States energy market. Initially not distinguishing between first- and second-generation biofuels, policy programs like the 2005 Energy Policy Act created renewable fuels standards that mandated production of transportation fuels blended with oxygenates to reduce aggregate emissions. Coupled with earlier state-level bans of substitute products like MTBE[1] and federal tariffs on foreign biofuel imports, these policies generated a five-fold increase in biofuel production between 2000 and 2008. However, first-generation corn-based production already operating at the industrial scale accounted for that rise in production. Corn-based production increases were further bolstered by state policy support, which lead to the development of energy industries in rural regions without significant prior involvement in those sectors.

Responding to environmental interests and public concern that rising ethanol demand would lead to an increase in corn prices, that would in-turn result in increased food prices the federal government shifted its policy focus to second-generation technologies after 2007. The renew–able fuel standard established in the 2005 Energy Policy Act was amended under the 2007 Energy Independence and Security Act to encourage a progressive shift designed to replace first-generation biofuels with second-generation alternatives. To achieve that goal, the federal government introduced research grant and loan programs to encourage the development and commercialization of second-generation production technologies. The full effect of this policy shift is still uncertain, as emerging second-generation production technologies remain unproven beyond the pilot stage.

A shift from first- to second-generation technologies could change the geographic concentration of energy reserves, in the form of preferred feedstocks, and alter the industrial geography of a leading renewable energy industry in the United States. However, second-generation technologies have yet to impact the industrial production of biofuels in the United States. The remainder of this chapter analyzes the distribution of second-generation biofuel innovation and production, and examines how this industry's structure is emerging as producers attempt to commercialize innovations developed in different locations under regionally varied policy and market conditions. Connecting innovation to production through an examination of firm activity in the U.S. biofuel industry contributes to our understanding of how and why renewable energy landscapes do or do not change. Such an understanding is essential for understanding the path toward regional investment in renewable energy with a related goal of understanding the facilitators and barriers in moving toward non fossil fuel based energy landscapes.

We use patent data gathered from the United States Patent and Trademark Office (USPTO) to measure second-generation biofuel innovation. To identify second-generation biofuel patents, we (i) searched all patent files from 1980–2012 for keywords (e.g., thermochemical, Fisher-Tropsch) associated with biofuel processing, and (ii) manually reviewed all selected patents. Since patents often have multiple inventors, we adopted the procedure established by Moreno et al. (2005) and proportionally assign a fraction of each patent to each inventor's CBSA.[2] Summing the fractional counts of all patent within a selected geographic area measures that location's patent output during the study period. We gathered second-generation biofuel production information from the United States Energy Information Administration (2014) and the Renewable Fuels Association (RFA 2015), which maintain listings of first- and second-generation biofuel producers. In total, 30 firms pursued pilot scale development of 44 biorefineries in the United States as of 2014. We compiled archival evidence on the innovation and production activities of those firms from corporate websites, and industry and government publications. Combined with policy information gathered from the congressional and state legislative records, we used this data to examine how second-generation producers linked innovation and production, and if those linkages altered the biofuel industry landscape.

Geographies of second-generation U.S. biofuel innovation and production

Innovation and production of biofuels in the United States does not occur in a single geographic location. Instead, different generations of biofuel production occur in different regions, while innovations are developed at research centers located throughout the United States. How second-generation producers connect with innovation and policy networks to commercialize differing technologies also varies throughout the country.

Geographies of second-generation biofuel innovation

Measured through the accumulation of U.S. patenting, second-generation biofuel innovations are distributed across the United States, but are also concentrated in selected states, and a small number of CBSA. Figure 2.1 presents geographic distribution of second-generation biofuel patenting between 1980 and 2012, and identifies federally funded renewable energy innovation centers. Seven states with over 100 total patents account for 58 percent of total second-generation biofuel innovation in the United States. Illinois and California represent the largest proportion of second-generation biofuel patenting, accounting for 30 percent of total patenting. However, each region of the country has at least one state with over 100 total patents. States with moderate levels of patenting, between 50 and 99 total patents, are similarly spread throughout the United States.

Matching the dispersed pattern of second-generation patenting, federally funded bioenergy research centers are likewise distributed throughout the country. To achieve substitution of second-generation production for first-generation production, following amendment of the renewable fuels standard in 2007, the federal government introduced research grant and loan programs to encourage the development and commercialization of still emerging production technologies. Included in that funding was the establishment of a system of three bioenergy research hubs supported by a network of partner institutions located throughout the United States. The USDOE Bioenergy Centers are located in California, Tennessee, and Wisconsin along with the National Renewable Energy Laboratory located in Colorado. In support of this initiative, the USDA and USDOE funded 72 projects researching the development and use of bioenergy crops for biofuel

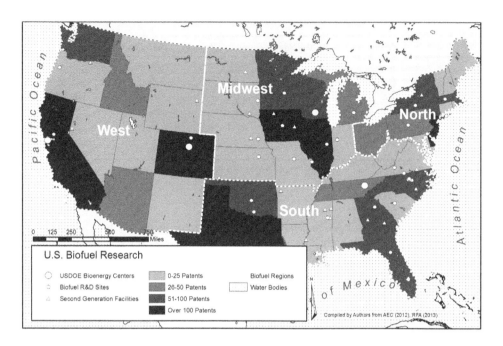

Figure 2.1 Second-generation biofuel centers and patenting

production from 2006 to 2012. Projects focused on the genetic identification and modification of wood and grass producing plants for bioenergy production. To translate that research into production, the USDOE distributed $603 million, with another $326 million committed, to 29 biorefinery projects through 2014 (United States Department of Energy 2013a, 2013b).

Although a large number of states have some involvement in patenting, second-generation biofuel innovation concentrates in a small number of urban centers. In total, 33 percent of second-generation patenting occurring in the United States takes place in five CBSA. Table 2.1 shows that the two CBSA with the greatest number of biofuel innovations are both located in the Midwestern United States. Chicago and Des Moines together account for over 200 patents. However, the two cities present differing temporal patterns. While Chicago was highly involved in patenting early in biofuel development during the 1980s, nearly all of Des Moines second-generation patenting occurred after 2000. Des Moines pattern across time appears more typical of other leading CBSA. Second-generation patenting in San Francisco, Philadelphia, and Denver is similarly concentrated after 2000.

The geographic concentration of patenting in selected states and CBSA may be linked to funded bioenergy research centers. Reexamining the geographic patterns of second-generation biofuel patenting, we see that California, Colorado, Tennessee, and Wisconsin all have above average patent totals. In each instance, patents from those states are disproportionately concentrated in the CBSAs where USDOE Bioenergy Research Centers are located. The effect of the bioenergy centers is in fact likely understated in these results given the lag time from innovation to date of granting. Established in 2007, many of the patents developed by and in collaboration with these research centers are likely still being assessed by the USPTO or are still to be submitted. Nonetheless 47 of Wisconsin's 78 second-generation biofuel patents have application dates after 2007. Similarly, 213 of California's 420 patent applications occurred after

Table 2.1 Five largest U.S. CBSA by second-generation biofuel patents

CBSA	State	1980–1989	1990–1999	2000–2012	Total
Chicago	IL	136.9	73.8	184.2	397.9
Des Moines	IA	4.3	10.1	226.9	241.3
San Francisco	CA	8.1	15.5	118.7	143.4
Philadelphia	PA	17.5	15.1	86.8	120.1
Denver	CO	13.8	29.1	76.2	119.2
United States		571.3	536.3	2,206.1	3,313.8

Source: Data collected by authors from United States Patent and Trademark Office (2015)

2007. Narrowing to the CBSA containing the USDOE Joint Bioenergy Institute led by the Lawrence Berkeley National Lab, 69 percent (99 of 143 patents) of that location's patent application occurred after 2007. That proportion of patents after 2007 is well above the overall average of 52 percent.

Geographies of second-generation biofuel production

Geographies of second-generation biofuel innovation are different from those of first- and second-generation biofuel production. Figure 2.2 demonstrates two distinct geographic distributions of first- and second- generation biofuel production in the United States. First-generation producers are located in corn-rich Midwestern states with agricultural industry expertise. The technological

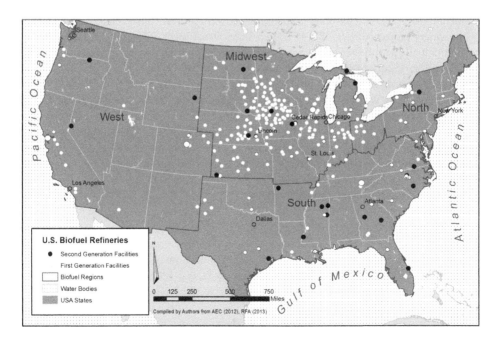

Figure 2.2 Distribution of first- and second-generation biofuel refineries

dominance of first-generation production means six states (IA, IL, IN, MN, NE, SD) account for over 70 percent of total national biofuel production (RFA 2015). In contrast, second-generation producers are found throughout the United States four geographic regions. Geographically dispersed second-generation production has potential implications for the industrial organization of biofuel production should new technologies replace first-generation corn-based alternatives. A change in production technologies would also mark a shift in the location of energy reserves, likely creating industry expansion in those locations. Successfully scaling-up second-generation innovations would make possible numerous industrial arrangements given the diversity in second-generation feedstock and national differences in the concentration of those feedstock resources.

Unlike first-generation producers, second-generation biorefiners use a range of production technologies tailored for a variety of feedstocks (e.g., switchgrass, wood waste, algae). Second-generation production facilities are located in regions with an abundance of those feedstocks. However, these technologies are operating at only the pilot or commercial scale. Only five second-generation facilities currently have annual production capacities over 22mgy. Matched with Figure 2.1, differences between locations of production and innovation become apparent. For example, California and Illinois represent 30 percent of second-generation biofuel patents, but house only six second-generation production facilities. In contrast, Mississippi and Oregon, which collectively account for less than 40 total patents, are the site of seven emerging production facilities.

How second-generation producers connect with innovation and policy networks to commercialize differing technologies also varies throughout the country. Two types of firms currently pursue second-generation biofuel production. First, large incumbent first-generation producers invest in production technologies tied to the use of agricultural residues. Integration of these technologies into existing facilities offers a possible route to extending fixed investments in first-generation production. Second, new entrants, both domestic and foreign, are generally smaller firms with histories in industries related the feedstock of a particular second-generation production technology, or newly formed ventures. The northern, southern, eastern, and western portions of the United States all have different mixtures of second-generation producers pursuing different production technologies. Table 2.2 presents the characteristics of selected firms in each region. Only two second-generation biorefieneries are in development in the northern United States. We therefore focus our discussion on the Southern, Midwestern, and Western United States.

Second-generation biofuel development in the Southern United States

New entrants manage all 17 cellulosic biorefinery projects in the Southern United States. The Southern states have a mixture of foreign and domestic entrants. Domestic entrants are small firms with venture financing, while foreign entrants are owned by large related industry interests with biofuel operations in their home markets.

Both foreign and domestic entrants in the South primarily pursue wood-based biorefineries, while producers partner with firms with experience in the forestry or pulp-paper industries. For example, wood-based biorefiner American Process leveraged knowledge as an engineering consultant for the pulp-paper industry to develop its cellulosic technology AVAPCO in collaboration with Danish enzyme producer Novozymes, biotechnology based forestry company ArborGen, and Finnish pulp-paper innovator Metso. Other forms of related industry experience observed in the region include biotechnology and chemical engineering. For example, Texas-based Kior, founded by catalyst scientists and funded by Khosla Ventures, has received $125 million from the latter to develop three biorefineries in Texas and Mississippi.

Table 2.2 Organization and collaborative characteristics of selected incumbent and new entrant 2G biofuel producers

| Name | Location | | Feedstock | Industry Experience | Collaboration |
	HQ	Facilities			
South					
American Process	GA	GA, MI	Wood Residues	Pulp-paper, Engineering	Developed technology in partnership with global biotech firms. Production collocated with paper mill.
Beta Renewables*	IT	NC	Multiple	Biotechnology	$350m Joint-venture among Italian biotech firms that are commercializing their technology in NC.
Enerkem*	QB	MS	Municipal Waste	Biotechnology	Canadian firm with multiple facilities in Quebec. Developing DOE funded ($120m) facility in MS.
Fiberright	MD	IA, VA	Municipal Waste	Chemical, Recycling	Developed technology in partnership with Novozymes, and received $25m USDA loan for facility building.
Ineos*	CH	AR, FL	Municipal Waste	Chemical	Funded by $130m from Swiss parent company.
Kior	TX	TX, MS	Wood Residues	Biotechnology	Partnerships between Kholsa Ventures and catalyst scientists based in TX. Expanding across the south.

Midwest

Abengoa*	SP	KS	Crop Residues	1G Biofuels	Funded by equity investment from global biofuel firm. Supported by $76m loan guarantees from DOE.
Green Plains	NE	IA	Algae	1G Biofuels, Refinery	2G refinery operates a joint-venture with several U.S. based biotechnology companies.
Mascoma	MA	MI, NY	Wood Residues	Chemical Engineering	University collaboration to develop process enzymes. Partnership with Valero to develop MI biorefinery
Poet	SD	IA, SD	Crop Residues	1G Biofuels	Process innovation developed with global enzyme and biotech firms. 2G refinery a partnership Royal DSM.

West

Amyris	CA	CA, IL	Crop Residues	Biotechnology	Operates 7 global biorefineries. Maintains partnership with oil refiner Total to commercialize core technology
Bluefire Renewables	CA	CA, MS	Wood Residues	Engineering	Technology development agreement with Solazyme, and DOE funding of MS biorefinery project
Zeachem	CO	CA, OR	Wood Residues	Biotechnology, 1G Biofuel, Engineering	Received $232m in government loan guarntees. Partnered with USDA funded hybrid poplar plantation.

* Indicates foreign-owned company operating within the United States

Source: Compiled by authors from corporate reports and websites

Foreign entrants that expanded their operation into the Southern United States combined parent company and government funding to establish biorefineries in resource-rich regions. These companies imported second-generation technologies proven at pilot scale facilities in their home countries. For example, the Canadian firm Enerkem operates five facilities in Canada that convert municipal wood waste to biofuels. Enerkem, which has received C$217 million in funding from the Canadian government, received $130 million from the USDA and DOE to develop a commercial scale facility in Pototoc, MS (Kedron 2015). Similarly, Italian joint-venture Beta Renewables invested over $200 million in the development of its biofuel technology prior to its investment in a North Carolina biorefinery in 2012 (AEC 2012). To build that biorefinery, Beta Renewables received $103 million in state and federal support. Foreign entrants use southern states as a point of entry into the otherwise protected United States' biofuel market and as an opportunity to commercialize 2G technologies.

Despite the success of domestic and foreign entrants, funding and related industry experiences are not a sufficient condition for development of successful operations. Range Fuels, a conglomerate venture that brought together industry experience from the oil refinery, chemical, and pulp-paper industries exited the industry in 2011 despite receiving over $145m in government support and $160m in private investment (Parker 2011). The venture failed after the company was unable to move its thermochemical process technology to commercial scale production. New Zealand based Lanzatech used the Range Fuels failure to enter the U.S. market. The company purchased Range Fuel's unfinished GA biorefinery in 2012 for $5m with the intention of converting the site for the company's second-generation technology (Bevill 2012). Kior undertook similar restructuring after filing for bankruptcy in 2014.

Second-generation biofuel development in the Midwestern United States

In the Midwestern United States, incumbent first-generation producers operate crop residue based second-generation biorefineries in traditional centers of corn based production, while new entrants primarily pursue wood based facilities in states rich in forest residues. Both types of producers use collaboration to develop production technologies, but, incumbent producers work with regional agri-processing firms and universities to add second-generation innovations into vertically integrated first-generation value chains that span the United States.

When first-generation production is an incumbent's primary business, firms focus on developing cellulosic technologies that use a single feedstock. When first-generation production is an incumbent's secondary business (e.g., oil refiners), firms invest in multiple technologies that use a range of feedstocks. For example, Poet's development of corn-stover based second-generation technologies began with development of first-generation process innovations developed in collaboration with international partners Novozymes and Satake. The company then used a five-year collaboration with the USDOE, National Renewable Energy Laboratory, and South Dakota State University to scale up its second-generation technology into a 25mgy 2G biorefinery in Emmetsburg, IA. That facility is collocated with an existing Poet refinery, and is designed to use its waste material in second-generation production.

In contrast to the experience of Poet, incumbents tied to related industries and foreign entrants pursue development of cellulosic technologies that use a variety of feedstocks. For example, Spanish incumbent producer Abengoa pursues technologies using corn stover, switchgrass, and wheat straw. The company has research partnerships with U.S. universities and research laboratories located in nine states, and biorefineries in the United States, European Union, and Brazil. Abengoa accepted $132m in government support (e.g., USDA) to develop a commercial scale switchgrass and corn stover based biorefinery in Houghton, KS. That investment depends on radical enzyme innovations developed in collaboration with biotechnology firm Dyadic, as well incremental process innovations developed with SunOpta and the USDA. In contrast, the oil refiner Valero has financed projects throughout the United States using wood residues, algae, and municipal solid waste as feedstocks for second-generation production. Valero commonly acts as an investor in projects rather than a developer. For example, in 2011 Valero invested $50 million in new entrant Mascoma's production technology, but would later withdraw that financing in 2013 (McGlashen 2013).

Finally, entrants in the Midwestern region concentrate in wood rich subregions and face substantial financial and technological barriers despite related industry expertise. The DOE-funded Bioenergy Science Center (Oak Ridge, TN) collaborates with second-generation producer Mascoma to improve efficiency of their wood degrading enzymes. Difficulty scaling up that technology has led to Mascoma's failure to begin construction of its $232 million commercial scale Kinross, MI biorefinery despite receiving over $120 million in guaranteed government financing as well as venture capital funding from Khosla Venture. To offset commercial scale costs, other new entrant producers are converting existing facilities to second-generation production. For example, Virginia based Fiberight received $28 million in government funding to convert a closed corn-based biorefinery in Blairstown, IA to second-generation production.

Second-generation biofuel development in the Western United States

New entrants with industry experience in biotechnology and chemical engineering manage the majority of the 13 cellulosic biorefinery projects in the Western United States. Western entrants use industry experience in biotechnology and chemical engineering to develop production technologies, but pursue different pathways to commercialization. Some entrants target domestic expansion within resource rich western regions using government funding. For example, after using $25 million in USDOE funding to develop process innovations, Colorado based Zeachem used $232 million in USDA loan guarantees to build a commercial scale cellulosic biorefinery in Boardman, OR. That Oregon facility is located near a government funded project to develop polar varieties suitable for conversion to biofuels.

Other western entrants pursue a national expansion strategy. For example, California based Bluefire Renewables is working toward commercialization of its licensed Arkenol technology, which converts sorted municipal and wood waste to biofuels at a Mississippi biorefinery. Selected western entrants have targeted global expansion of their cellulosic technologies. California based biotechnology company Amyris currently operates seven biorefineries in California, Illinois, Brazil, and Spain. Unlike Zeachem and Bluefire, Amyris entered into a joint-venture in 2013 with oil refiner Total to further commercialize. Industrial R&D with university partners is also common. Ceres (located in California) focuses on collaborative energy crop development—the company has received multiple research grants from the DOE and has expanded its collaborative network to include both Texas A&M and the Noble Foundation.

Learning from the experience of U.S. second-generation biofuel

While cellulosic technologies have the potential to alter innovation and production geographies within the biofuel sector, technological uncertainty makes it unclear what patterns will emerge. Successful development of innovation using crop residues could maintain production in corn-rich regions, and support expansion of first-generation interests and existing producers. In contrast, standardization of cellulosic alternatives could shift production to new regions. For example, emergence of an industrial scale wood-based production technology could shift production to forested regions, particularly southern states east of the Mississippi River. However, technological uncertainty and difficulty in the scale-up of production processes from the laboratory to the refinery makes these alternative industrial development pathways unlikely.

With technological uncertainty, no single ownership structure dominates the second-generation biofuel industry, but innovation, production, and ownership present regionally discernable patterns. In the Midwest, incumbent producers pursue cellulosic innovation that will augment their existing operations in corn rich regions, while a select number of small entrants develop wood-based technologies. In the South, domestic and foreign entrants pursue wood-based technologies. In the West, new entrants with biotechnology experience seek to commercialize their technologies outside their region using government funding and industry partnerships.

Among the number of organizational forms operating in the industry, no single form of industry expertise or feedstock focus has created industrial scale success. While government loan guarantees encourage commercialization of technologies, financing remains a problem for many firms. For additional project financing, firms seek out partnerships and joint-ventures with a range of related industry interests. Biorefineries have strong ties with foreign energy firms that act as investors to sustain domestic operations and as platforms for international expansion. Within the United States, second-generation biorefiners use joint ventures with venture capital, biotechnology companies, oil refiners, pulp-paper producers, and filtration companies to develop and commercialize cellulosic technologies. However, the exit of well-financed firms demonstrates the volatility and technological uncertainty of the cellulosic biofuel industry.

The experience of first- and second-generation biofuel innovators and producers examined in this chapter demonstrates some of the contributions a geographical perspective has to offer the wider field of energy studies. First, by analyzing the connection between innovation and production in a leading U.S. renewable energy industry, this chapter demonstrates another way in which energy connects and structures places. In addition to the physical and infrastructure changes energy industries bring to a region, we show how the pursuit and commercialization of biofuel innovations reorganizes impacts knowledge development. Although the development path of biofuel production remains unclear, development of second-generation technologies has changed the mix of players involved in the industry and the number of locations involved.

Second, this chapter highlights the uncertainty inherent in renewable energy industry development when production technologies remain in development. Incorporating a geographic perspective into the analysis of that uncertainty brings into focus some of the alternative way in which the industry could be organized. In biofuel, technological uncertainty surrounding second-generation production that uses a range of feedstocks opens a number of different possible regions for development. Emergence of a second-generation technology capable of supporting large-scale production would introduce a new energy reserve to the industry. Regions with abundant supplies of that resource, or environmental conditions capable of supporting it, could increase their biofuel operations and expand production outside the Midwest. Alternatively, development of second-generation technologies that use feedstock abundant in the Midwest (e.g., crop residues) could further regional concentration of production.

Resolution of technological uncertainty could similarly impact the politics and policies of renewable energy production, as each of the scenarios above would lead to differing constituent groups in support of biofuel. The failure of second-generation biofuel to deliver industrial scale production of renewable energy could reinforce the geographic arrangement of first-generation production as the only remaining alternative. Failure of second-generation technologies could also lead to the phasing out of policy support for biofuel in general. If policies designed to support failing second-generation technologies are ended, inter-related support for first-generation production may also end. For example, annual adjustment of the renewable fuel standard required each year second-generation technologies fail to meet production targets, could lead to elimination of the standard altogether.

Acknowledgement

We would like to acknowledge and thank Dieter Kogler for his comments and contributions to the development of our patent data set. We would also like to acknowledge the National Science Foundation which supported this work (Award No. 1338970).

Notes

1 Methyl tertiary butyl ether (MTBE) is a gasoline additive used to oxygenate automobile fuel. Beginning in 2004, state governments began progressively banning the use of MTBE to address concerns related to drinking water contamination. MTBE bans created a dedicated market for ethanol.
2 Assigning innovation generated to the inventors of the patent is advantageous because it attributes the ideas contained in the patent to the team that developed them. Many patents are developed by teams of inventors located in a number of different places, and the procedure adopted here attributes some invention to each. In contrast, assigning the invention embodied in a patent to the institution that owns it, the assignee, could obscure the source of ideas by compiling invention into one location, which may not in fact be the location of a single inventor.

References

Bagchi-Sen, S., and P. Kedron, 2015. 'Governance of biofuel production in the United States.' In Hilpert, U. (ed), *Handbook of Politics and Technology*. London: Routledge.
Bakker, K., and G. Bridge, 2006. 'Material worlds? Resource geographies and the matter of nature.' *Progress in Human Geography* 30(1): 5–27.
Bergek, A., Jacobsson, S., Carlsson, B., Lindmark, S., and A. Rickne, 2008. 'Analyzing the functional dynamics of technological innovation systems: A scheme of analysis.' *Research Policy* 37: 407–429.
Bevill, K., 2012. LanzaTech buys Range Fuels site. *Ethanol Producer Magazine*. http://ethanolproducer.com/articles/8458/lanzatech-buys-range-fuels-site.
Bridge, G., 2002. 'Grounding globalization: The prospects and perils of linking economic processes of globalization to environmental outcomes.' *Economic Geography* 78: 361–386.
Bridge, G., 2008. 'Environmental economic geography: A sympathetic critique.' *Geography Compass* 39: 76–81.
Bridge, G., 2010. 'Geographies of peak oil: The other carbon problem.' *Geoforum* 41(4): 523–530.
Bridge, G., Bouzarovski, S., Bradshaw, M., and N. Eyre, 2013. 'Geographies of energy transition: Space, place and the low-carbon economy.' *Energy Policy* 53: 331–340.
Coenen, L., Benneworth, P. and B. Truffer, 2012. 'Toward a spatial perspective on sustainability transitions.' *Research Policy* 41(6): 968–679.
Compean, R. and K. Polenske. 2011. 'Antagonistic bioenergies: Technological divergence of the ethanol industry in Brazil.' *Energy Policy* 39: 6951–6951.
Cooke, P., 2002. 'Regional innovation systems: General findings and some evidence from biotechnology clusters.' *Journal of Technology Transfer* 27: 133–145.
Cooke, P., 2004. 'The regional innovation system in Wales.' In P. Cooke, M. Heidenreich and H. C. Braczyk (eds), *Regional Innovation Systems: The Role of Governance in a Globalized World*, Second Edition. London: Routledge.
Cooke, P., 2009. 'Jacobian cluster emergence: Wider insights from "green innovation" convergence on a Shumpeterian "failure"'. In D. Fornahl, S. Henn and M. P. Menzel (eds) *Emerging Clusters: Theoretical, Empirical, and Political Perspectives on the Initial Stage of Cluster Evolution*. Cheltenham: Edward Elgar.
Corsatea, T. D., 2014. 'Increasing synergies between institutions and technology developers: Lessons from marine energy.' *Energy Policy* 74: 682–696.
Essletzbichler, J., 2012. 'Renewable energy technology and path creation: A multi-scalar approach to energy transition in the UK.' *European Planning Studies* 20(5): 791–816.
Gruenspecht, H., 2013. 'Biofuels in the United States: Context and outlook.' Biofuels Workshop, Institute of Medicine, National Academy of Sciences.
Hayter, R., 2004. 'Economic geography as dissenting institutionalism: The embeddedness, evolution, and differentiation of regions.' *Geografiska Annaler B* 40: 1–21.
Hayter, R., 2008. 'Environmental economic geography.' *Geography Compass* 2: 831–850.

Hayter, R. and R. Le Heron, 2002. 'Industrialization, techno-economic paradigms and the environment.' In R. Hayter and R. Le Heron (eds), *Knowledge, Industry, and the Environment: Institutions and Innovation in Territorial Perspective*. Aldershot, UK: Ashgate.

Hellsmark, H. and S. Jacobsson, 2009. 'Opportunities for and limits to academics as system builders – The case of realizing the potential of gasified biomass in Austria.' *Energy Policy* 37: 5597–5611.

Huber, M., 2009. 'Energizing historical materialism: Fossil fuels, space and the capitalist mode of production.' *Geoforum* 40(1): 105–115.

Huber, M., 2011. 'Enforcing scarcity: Oil, violence, and the making of the market.' *Annals of the Association of American Geographers* 101(4): 816–826.

Jacobsson, S. and V. Lauber, 2006. 'The politics of energy system transformation – explaining the German diffusion of renewable energy technology.' *Energy Policy* 34: 256–276.

Kedron, P., 2015. 'Environmental governance and shifts in Canadian biofuel production and innovation.' *The Professional Geographer* 67(3): 385–395.

Kedron, P., and S. Bagchi-Sen, 2011. 'A study of the emerging renewable energy sector within Iowa.' *Annals of the Association of American Geographers* 101(4): 882–896.

Lehmann, P., Creutzig, F., Ehlers, M., Friedrichsen, N., Heuson, C., Hirth, L. and R. Pietzcker, 2012. 'Carbon lock-out: Advancing renewable energy policy in Europe.' *Energies* 5(2): 323–354.

Martin, R. and P. Sunley, 2007. 'Complexity thinking in evolutionary economic geography.' *Journal of Economic Geography* 7: 573–601.

McGlashen, A., 2013. 'As a key partner departs, future dims for Michigan cellulosic biofuel plant.' *Midwestern Energy News*. www.midwesternergynews.com/2013/08/06/as-key-partner-departs-future-dims-for-michigan-cellulosic-biofuel-plant/.

Moreno, R., Paci, R. and S. Usai, 2005. 'Spatial spillovers and innovation activity in European regions.' *Area* 34: 82–91.

Parker, M., 2011. Range Fuels cellulosic ethanol plant fails, U.S. pulls plug. *Bloomberg Sustainability*.

Patchell, J. and R. Hayter, 2013. 'Environmental and evolutionary economic geography: Time for EEG²?' *Geografiska Annaler: Series B, Human Geography* 95(2): 111–130.

Penrose, E., 1959. *The Theory of the Growth of the Firm*. New York: Wiley.

RFA, 2015. 'Ethanol Industry Outlook 2003.' *Renewable Fuels Association*. www.ethanolrfa.org.

Simmie, J., 2012. 'Path dependence and new path creation in renewable energy technologies.' *European Planning Studies* 20(5): 729–731.

Sims, R., Mabee, W., Saddler, J. and M. Taylor, 2010. An overview of second generation biofuel technologies. *Bioresource Technology* 101(6): 1570–1580.

Smil, V., 2005. *Energy at the Crossroads: Global Perspectives and Uncertainties*. Cambridge, MA: MIT Press.

Smil, V., 2010. *Energy Transitions: History, Requirements, Prospects*. Santa Barbara, CA: Praeger.

Stenzel, T. and A. Frenzen, 2008. 'Regulating technological change – The strategic reactions of utility companies towards subsidy policies in the German, Spanish, and UK electricity markets.' *Energy Policy* 2645–2657.

Suurs, R. and M. Hekkert, 2009. 'Competition between first and second generation technologies: Lessons from the formation of a biofuels innovation system in the Netherlands.' *Energy* 34: 669–679.

Taylor, M. and P. Oinas, 2006. *Understanding the Firm*. Oxford: Oxford University Press.

United States Department of Energy. 2013a. Plant feedstock genomics for bioenergy: Joint Awards 2006–2012. *U.S. Department of Energy Office of Science*. http://GenomicScience.energy.gov/research/DOEUSDA/.

United States Department of Energy. 2013b. Follow-up audit of the Department of Energy's financial assistance for integrated biorefinery projects. U.S. Department of Energy Office of Inspector General – Office of Audits and Inspections. DOE/IG-0893. www.usdoe.gov.

United States Energy Information Administration. 2015a. 'Renewable energy production and consumption by source, Table 10.' United States Energy Information Administration. www.eia.gov/totalenergy/data/monthly/#renewable/.

United States Energy Information Administration. 2015b. 'Transportation sector energy consumption.' United States Energy Information Administration. www.eia.gov/totalenergy/data/monthly/.

United States Patent and Trademark Office (2015) US patent full-text and image database United States Patent and Trademark Office. www.uspto.gov/.

3

New perspectives on an ancient energy resource

Biomass, bioenergy, and emerging bio-economies

Kirby Calvert, Kean Birch and Warren Mabee

Introduction

The fossil fuel era – marked by massive industry, on-demand heating and cooling, and rapid mobility – represents less than one tenth of one percent of humanity's 'terrestrial tenure' (Cook, 1976). The rest of our existence was largely powered by the use of biomass. As the fuel for muscle power, edible biomass provided the energetic basis for the daily activities of hunter-gatherer societies. Later, the control and use of inedible biomass allowed a wider range of services in larger and more complex agricultural societies – e.g., wood to supply fire for warmth and cooking; grass to feed animals who were put to work for transportation or labor. In fact, much of the world's current population continues to use biomass as their primary source of energy, as approximately 12 per cent of global primary energy supply is derived from biomass. And yet, biomass is often discounted as an 'energy resource'. Nutritional energy is typically omitted in energy balance estimations of societies (Haberl, 2002) and the shift away from the use of biomass for energy is widely considered to be necessary in order to achieve modern living standards (Leach, 1992).

The ecological and geological limits of the fossil fuel economy are increasingly clear, however, and alternative energy worlds are being pursued. One of these alternatives is the 'bio-economy'. Generally speaking, the term bio-economy is used as a short-hand for a vision of a world in which institutions, infrastructures, and technologies metabolise various forms of biomass through advanced processes in order to displace fossil fuels as carriers of energy and non-energy products (e.g., biomaterials, biochemicals) (see OECD, 2006; EU Presidency, 2007; CEC, 2012; White House, 2012). Can we produce and sustainably harvest enough biomass to realize the full extent of this vision? A meta-analysis of seventeen academic studies suggests that biomass resources might provide between 50 and 450 exajoules (EJ) of non-nutritional energy by the year 2100; the range depending on assumptions about land-use, yields, harvest rates, economic competition over biomass (e.g., food versus fuel) and ecological sustainability (e.g., how direct and indirect land-use change for energy crop production influences carbon budgets) (Berndes et al., 2003; see also Deng et al., 2015). Currently, almost 500 EJ of non-nutritional energy, mainly drawn from fossil sources, are consumed globally; this figure is on the rise in response to the energy demands of emerging 'modern lifestyles' in developing nations such as India and China. In other words, even an optimistic estimate suggests that biomass, while perhaps suitable for certain sectors or end-use applications, is insufficient to power global society while also providing food and fibre needs. Thus, the bio-economy should be viewed as an important component of a clean economy which draws energy from a range of renewable resources.

In what ways might we use biomass for energy, and what are the implications on people and places of choosing among these options? How might we reconcile the increasing global need for food and fibre with our increasing tendency to utilize biomass for energy production? Is energy production the best use of biomass resources? The purpose of this chapter is to bring a geographical perspective to these questions, and to identify major trends and issues that geographers are best equipped to engage. In the next section, we situate our chapter in the broad themes of this book through an exploration of how emerging bio-economies are (re)connecting and (re)configuring sociospatial relationships and landscapes. In particular, we conceptualize and investigate the interdependencies between policies, resources, technologies, and regional geographies that shape how the bio-economy is being articulated in particular places. In the third section, we explore the various discursive frames in which bio-economy is being considered, with emphasis on the role of spatial identities in the varied and variegated visions of what a bio-economy is, and what a bio-based future might look like. We conclude with thoughts on where bioenergy and biofuels are situated within the broad constellation of resources and technologies that might contribute to a sustainable energy future.

Bio-economies and the changing geographies of energy

The changing role of energy in society is shaped by complex interactions between social and technological innovations. Through these interactions, physical entities that have certain energetic become energy resources, and the material nature of those resources play an active, if unintentional, role in shaping human-environment encounters and social relations (see Bakker and Bridge, 2006). Mitchell (2011), for example, suggests that modern democracy was both enabled and disabled by the particular biophysical characteristics of different energy regimes that have developed over the last 200 years. On the one hand, he argues that the biophysical characteristics of the coal regime (e.g. mined, transported by rail and ships, lots of workers needed, etc.) helped to create the conditions for development of large labour movements who could block key transit points until their demands were met (e.g. wages, work conditions, suffrage, etc.). On the other side, Mitchell claims that these labour pressures facilitated governments, especially in the US, to envision and implement a future based on oil resources, which have very different biophysical features that limit the ability of labor to choke supply (e.g. extracted from wells, transported mostly by underground pipeline, fewer workers, etc.). In other words, energy transitions are best understood as a dynamic interplay between social institutions, technologies/infrastructures, and the form and spatial distribution of the resource base (see also Jiusto, 2009; Birch and Calvert, 2015).

Geographers have increasingly weighed in on the discussion of energy transitions, in order to better understand how energy transitions vary across space, (re)connect places, and are implicated in our relationship with our surroundings (Bridge et al., 2013). One prevalent contribution has been to 'energize' the concept of landscape – defined simply as a geographic area perceived by people (see Howard, 2011 for elaboration) – as in the concept of 'energy landscape'. The concept of energy landscape attends to the role of energy production, distribution, and use in shaping landscape form and livelihood arrangements (material spatial expressions of energy regimes) as well as how energy is implicated in representations of, and emotional connections to, landscape form and livelihood arrangements (immaterial spatial expressions of energy regimes) (Pasqualetti et al., 2002; Pasqualetti, 2011; Nadaï and van der Horst, 2010). According to Pasqualetti (2013), energy landscapes can be understood in terms of the energy system constructs (i.e., resource types; infrastructures that access, process, and use resources; geophysical context; and socio-political institutions) that become layered as direct or indirect landscape impacts (e.g., transmission wires; hydroelectric reservoir; a remediation site). In this way, the material nature of the resource base is considered in combination with social, institutional, and geographical structures and processes when trying to understand the role of energy in changing sociospatial and society–environment relationships.

Recent scholarship in energy geographies has suggested that the 'energy landscape' concept might be extended through the concept of 'production of space' (Lefebvre, 1991; see also Calvert, 2015; Huber, 2015). Karplus and Meir (2013) interpret and describe three facets of the production of space framework; conceived space, perceived space, and lived space. In the context of energy studies, *perceived space* refers to the material landscape and the landscape aesthetics that are shaped by different systems of energy provision. *Conceived space* refers to the social value systems that establish the legal and extra-legal institutions through which resource access and spatial planning are enforced. *Lived space* refers to 'acceptable' or 'normal' spaces of everyday experience (e.g., our ideas about 'home' or 'comfort'; preferred landscape aesthetics). This tripartite framework attends to interdependence within and between the 'constructs' listed above, helping to unpack the nexus of landscape form, landscape representations, and landscape practices when thinking about how energy landscapes take shape and are stabilized across socio-spatial scales (from the home to the nation). In the following subsections, we apply these insights in order to investigate established bio-economies and to consider the geographical implications of emerging bio-economies.

From traditional to advanced bio-economies

In the context of a fossil-fuel intensive economy, it is easy to forget that bio-economies already exist. We can find them in regions which are often referred to as 'under-developed', where people derive as much as 80–100 percent of the energy they use from biomass.[1] In these regions, consumption profiles are composed almost exclusively of energy services that can be met directly with raw biomass (e.g., straw or wood) such as cooking and heating. In other words, the dominant role for biomass in these regions reflects established practices and energy needs. This helps to explain why some state-led proposals to develop hydroelectricity have failed; although an 'advanced' fuel, electricity requires not only new appliances but also new social/cultural routines and human–environment encounters.

Energy consumption patterns in traditional biomass-based economies are associated with particular social–ecological interactions and livelihood arrangements which differ in sharp ways from a fossil-fuel based system. In developing countries, women and children of the household are more likely to undertake daily fuelwood collection activities. The quantity and form of biomass that is locally available shapes the length of their trips and the ways in which biomass is managed for distribution and use. From the perspective of an urbanite who uses a high efficiency gas furnace, the home might be considered 'dirty' due to the presence of soot, ash, and smoke. The ongoing shift toward more advanced forms of energy provision, such as natural gas or liquid petroleum, opens up new living arrangements and lifestyle opportunities. For starters, they eliminate the need to collect wood on a daily basis, dramatically reducing the labor burden on women and children, and also eliminate the presence of combustion residues (soot and smoke) in the home (Leach, 1992).

In contrast to these 'traditional' bio-economies, the idea behind the 'advanced bio-economy' envisioned in developed countries is to produce and consume biomass at a scale that enables the same range and intensity of energy services currently offered by advanced fossil fuel systems, with all of the same conveniences that fossil fuel systems provide. In addition to cooking and home heating, biomass would provide energy for transport and electricity generation, along with platform biochemicals that form the material basis of a 'modern' existence (e.g. bio-based ethylene, which might be used in high-density plastic production). Central to the notion of an advanced bio-economy is the biorefinery; a technological innovation implicated in social and geographic relationships across multiple scales.

Landscapes of biorefining: the resource–technology–policy nexus

Rather than using biomass in its raw form – e.g., a wooden log in a cooking stove – a biorefinery, much like a conventional petroleum refinery, converts raw feedstocks into a range of more useful energy, fuel, and chemical products. In the context of an energy system, the term biomass refers to various forms of biological material which can be converted directly into energy (e.g. bioheat, bioelectricity), a solid or liquid biofuel (e.g. pellets, bioethanol, biodiesel), or other products (intermediates such as sugars, lignin, or biochemicals as described earlier; see Figure 3.1). Biomass can be derived directly from ecological systems which are subject to various degrees

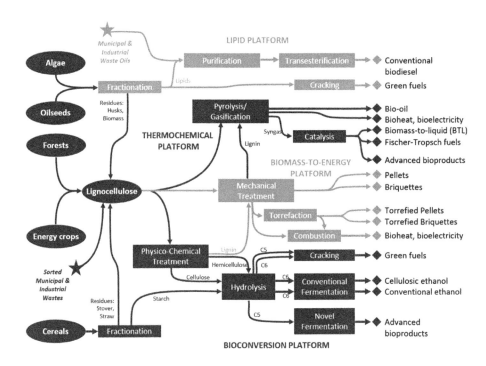

Figure 3.1 Technological pathways for the conversion of solid biomass into advanced solid fuels, liquid fuels, or other bioproducts (e.g., biochemicals)

of management (e.g., forests, agriculture, or energy crops) or from municipal and industrial waste streams (see Figure 3.2).

The range of products delivered via the biorefinery is dependent on the nature of the feedstock as well as the processing technology (see Figure 3.1). The simplest of these technologies perform physical manipulations to improve bulk density (e.g., a pelletizer) thereby improving the efficiency of transport, on-site storage and processing, and combustion. More advanced technologies aim to reduce biomass to more basic components in order to develop platform chemicals that can be formulated into something reminiscent of petroleum products (e.g., bioethanol, or more advanced 'drop-in' biofuels which we discuss in a later section).

As shown in Figure 3.1, biomass processing techniques may be divided into a number of platforms. The exact composition and titling of different platforms varies depending upon the author or agency, and so the differentiations presented in Figure 3.1 should not be considered definitive. Rather, we use this conceptual model to present four basic platforms – biomass-to-energy, bioconversion, thermochemical, and lipid-based technologies – which are distinguished by the technology that is used to deliver combinations of energy, fuels, and products. Note that these platforms are not mutually exclusive; hybrid platforms might combine elements of two or three of the approaches described in Figure 3.1.

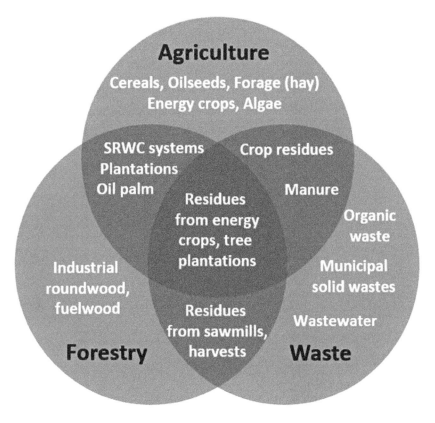

Figure 3.2 Common sources of biomass for energy and fuel production. *Trees which are not commonly used in lumber production due to species type or stem diameter, and would otherwise supply pulp production. **SRWC = Short rotation woody coppice systems.

Biorefineries are embedded within prevailing global political-economic circuits. Global commodity markets, for instance, shape a farmer's planting decisions and therefore local land-use patterns and biomass availability, while biorefining technologies are commercialized using international financing. At the same time, biorefineries are strongly localized and context-dependent. The low energy density by weight and volume of raw biomass means that it is not usually worthwhile in monetary or energetic returns to transport unprocessed biomass resources long distances from cultivation area to processing plant. In other words, biomass resource extraction and pre-processing activities generally occur in close proximity to where they will be processed and converted; typically within 100km of the processing facility. This means that biorefining technologies shape, and are shaped by, the types and quantities of biomass which are locally available. The introduction of starch-to-ethanol facilities in Ontario, Canada, for instance, led to an unprecedented spike in local year-over-year corn production (Calvert and Simandan, 2015).

The low energy density of biomass also means that biomass processing facilities (from local generating stations to larger biorefineries) will necessarily be more decentralized than a fossil-fuel based economy. Even the largest biomass processing facility in the world is approximately seven times smaller than the average petroleum processing facility in the US (see Figure 3.3). On average, approximately 1.5 tonnes of biomass, grown aboveground, are required in order to replace the energy equivalent of one tonne of coal, which is recovered from subterranean deposits. Replacing one unit of petroleum and natural gas requires as much as 2–4 tonnes of biomass which must be transported above ground rather than below ground. In other words,

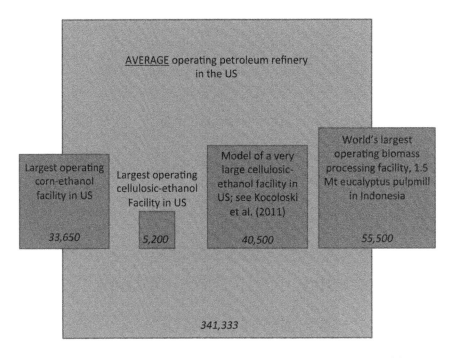

Figure 3.3 Comparison of existing and anticipated biofuel production systems with an average petroleum refinery in the US. Sizes are proportional based on total feedstock throughput on a GJ basis, and numbers represent a hypothetical daily production of 60 litre gas tanks.

each petroleum refinery will need to be replaced by no fewer than seven biorefineries, likely scattered across a region rather than centralized in a single location in order to minimize transport distances between extraction and processing. This means a higher number of communities hosting industrial-scale biomass processing and a re-surfacing of infrastructure and activity associated with biomass distribution and conversion. Unfamiliar smells and economic activity, along with increasing traffic activity associated with biomass transport and processing, are all sources of local resistance to project development (Sampson et al., 2012).

The quantity of energy that can be derived from waste streams is limited and woefully insufficient to displace a meaningful amount of fossil fuel use. A meaningful transition from a fossil to a bio-economy is therefore dependent on what are known as 'closed-loop systems', wherein biomass is derived from agricultural production activities that are conducted exclusively for energy production. Pellets or ethanol from fields of switchgrass and diesel from jatropha seeds or algae are examples of such systems. One major challenge with implementing closed-loop systems is the disruptions to local and global socioecological systems required for their establishment. Dedicated bioenergy plantations require large areas of arable land upon which to establish monoculture practices, thereby imposing new value systems on land and landscapes. This has led to concerns that natural ecosystems are being plowed under for the sake of producing bioenergy crops (direct land-use change), or to replace the land-area that has switched from food production to fuel production (indirect land-use change). Public policy makers have responded in ways to minimize these negative consequences. In the US, limits have been placed on the production of corn-based biofuels that are eligible for public support, as a way to help ensure that future investments shift toward inedible and less intensive sources of biomass (e.g., lignocellulosic material from wood and grass, or algae). In addition, the US government will not issue credits for biofuels that have been produced from crops grown on newly converted grasslands.

Another strategy to limit the land-use and land-cover impacts of closed-loop biorefineries is to encourage or restrict the production of energy crops onto land that is 'marginal' and/or abandoned for agricultural purposes. This strategy has been proven problematic and in some cases socially regressive. Land considered 'marginal' by one set of economic or ecological criteria might be considered adequate by another, while land considered 'abandoned' for commercial or industrial purposes might in fact be supporting an undocumented subsistence economy. Indeed, there are clear signs that corporate stakeholders, with the co-operation of state governments, are actively imposing categories such as 'marginal' and 'abandoned' onto landscapes in order to establish a moral and legal context in which plantations are considered an 'acceptable' use of this land. In India, for example, the growing demand for diesel as a transport fuel combined with state policy and regulations is legitimizing the dispossession of land and the displacement of heating and cooking fuels on so-called 'marginal lands' in order to grow Jatropha for its oil-rich seeds (cf. Baka and Bailis, 2014). Uncritical assessments of, and policy interventions toward, 'abandoned' or 'marginal' land has created potential for local energy and food shortages among communities of precarious land tenure situations and low socioeconomic status, as the land upon which they once relied is now being asked to serve the material needs of a rising middle class (i.e., liquid fuel for transport rather than solid fuel for cooking). In this process, generally referred to as 'land grabbing', we find evidence of social and environmental injustices that are reminiscent of the regressive organizational logics and outcomes that are typically associated with 'business as usual' petro-capitalism.

There are other implications on people and places from the development of closed-loop biomass systems. First, a range of local upstream actors (e.g. growers, land managers) must be coordinated in order to secure the resources that are necessary to keep a biorefinery operational.

Table 3.1 Estimating land-use requirements and number of farms required in order to service a modest biofuel facility

	Base-case scenario	Optimistic scenario
Average farm size in Pennsylvania (ha)	52	52
Biomass yield (ODT/ha)	10	25
Fuel Yield (l/ODT)	210	405
Farms required per day	**75**	**15**

By way of example, a petroleum refinery producing approximately 52,000 barrels of jet fuel per day at the Philadelphia airport was recently purchased by a major airline with intentions of generating bio-based jet fuel in the future. Table 3.1 summarizes the implications of this conversion in terms of the number of individual properties that would be required to keep the facility operational under two different scenarios. By contrast to the *15–75 properties that would be required on a daily basis*, petroleum refineries or petroleum aggregators can procure feedstock by way of contracts with only a *handful of suppliers on an annual basis*. Second, replacing fossil fuels with biomass represents a transfer of ownership of energy resources from public to private hands (i.e., landowners). Fossil energy resources in most states (with the exception of a few countries such as the US) are by constitutional law publicly owned. Much of the land from which biomass will be procured for bioenergy production however (e.g. agricultural land and woodlots) is privately owned.

Bio-economy as an alternative but familiar energy world?

As the building block for fossil fuel, biomass has a set of characteristics that come closer to matching conventional fossil fuels than any other renewable resource. As such, advanced bio-energy and biofuel technologies have potential to maintain a familiar interaction between an energy user and an energy conversion system. With this in mind, bio-economies are often espoused as a familiar alternative. In the terms of energy landscapes and production of space, advanced bio-economies can, so the argument goes, maintain prevailing sociospatial networks and social relationships to land.

Nowhere is this idea more explicit than in the notion of a 'drop-in biofuel'. Liquid biofuels currently represent more than half of all renewable energy production in most industrialized nations (see Chapter 4, this volume). Although the liquid biofuel industry is more than 30 years in the making, public policies and industry efforts over the last decade have focused on advanced products which are known as 'drop-in' biofuels. The name implies that said fuels are similar in terms of chemistry and energy content to petroleum fuels (gas or diesel) and thus can be incorporated ('dropped') directly into the fuel in any ratio, without being constrained by blend limits or the need for infrastructure change. The advantage of these fuels is that they can be blended with petro-based fuels at any point, and used with existing infrastructure such as pipelines, filling stations, and internal combustion engines, without negative impacts. Thus, future bio-based fuels could be transported, sold, and used using the same infrastructure as our existing fuel system, avoiding costly and dramatic modifications.

Another option to maintain familiar energy worlds using biomass is to establish district heating systems (DHS). As part of a DHS, buildings can utilize a renewable resource for their heat without increasing the personal space and maintenance requirements that might be associated with other renewable building-level heating systems (e.g., pellet stoves; rooftop solar). The DHS model is common in many Scandinavian countries, but the capacity to implement these systems elsewhere is strongly context dependent. The cost of implementation can be prohibitive if suitable infrastructure is not in place; i.e., pipes to carry steam or hot water, consuming units with similar heating systems that can be reconfigured for DHS. For these reasons, DHS are most viable and accepted in areas that have a legacy of such systems, or in campus environments with well-organized infrastructure and renter-occupied buildings.

Energy resources represent the single largest commodity traded globally, in terms of weight and wealth. There are efforts to maintain this key driver of economic globalization through biofuel trade. These long-distance flows of bioenergy are entirely dependent on strong economic pulls or willed markets created by subsidies or carbon taxes in consumer jurisdictions. That said, it should be noted that global flows of bio-based fuels (both solid and liquid) remain limited. Wood pellets are the most widely traded solid biofuel; key sources of supply include Canada, the southeastern USA, Brazil, Australia, Indonesia, Poland, and western Russia, while the primary importers are the USA, Germany, the UK, the Netherlands, Denmark, Japan, Korea, and China. Germany and the UK are predicted to have the greatest increase in biomass demand by 2020 (see Lamers et al., 2015). In the liquid biofuel market, trade of ethanol is dominated by exports from Brazil to the EU and, increasingly, the USA, which recently removed tariffs on imported ethanol (Lamers et al., 2011). In the biodiesel market, trade of finished products is relatively limited and found between EU trade partners or between neighbouring jurisdictions (e.g. Canada and the USA), although there is significantly more trade of vegetable oils which are in part used for biodiesel production. In the latter category, key sources of oils include Malaysia and Indonesia (palm oil), Argentina and Brazil (soybean oil), and Canada (canola oil) (ibid.).

Although drop-in biofuels and DHS might maintain many familiar sociotechnical relationships at the consumption end of the system, they still require massive changes upstream in terms of land-use, sociospatial and territorial relations, transport, and system siting as described above. Indeed, the notion of a biofuel that can simply be 'dropped-in' to prevailing sociotechnical systems is shown to be highly problematic and fraught when upstream implications are considered. Moreover, the global trade of biofuels is currently subsidized by fossil fuel transport and based on willed markets, so that trade patterns are unlikely to hold in the more distant future. Life-cycle analyses of long-distance transport of pellets between British Columbia and Europe, for example, reveal that ocean transport increases the greenhouse gas emissions and embodied energy associated with wood pellets by 53 per cent and 54 per cent, respectively, lowering the net energy recovered well below the energy return on investment of locally consumed pellets (Magelli et al., 2009; Pa et al., 2012). Perhaps more importantly, optimistic outlooks on biomass-based energy worlds provide examples of the ways in which energy futures are socially constructed on the basis of the conceived, perceived, and lived spaces into which they are imagined as being integrated. In the next section, we focus on these social constructions.

Visions of future bio-economies

The emergence of new energy worlds is not automatic or natural. They emerge from the creation and promotion of visions of the future, and become manifest in the development of new institutions that reconfigure socio-political expectations, demands and choices according to these visions. Visions of future bio-economies are important because they establish key parameters and define what can be done in the future (and what cannot be done); they shape current policy priorities and funding; and they enroll a range of stakeholders in the pursuit of particular policy strategies (Ponte and Birch, 2014). Consequently, they are implicated in what sort of energy worlds we will end up with in the future. A number of scholars are critical of how dominant policy visions of the bio-economy have centred on techno-scientific solutions to societal problems, rather than on rethinking the social, political, economic and ecological contexts in which agriculture and forestry are pursued (e.g. Kitchen and Marsden 2011; Schmid et al. 2012). Other scholars have noted that dominant bio-economy visions tend to be so broadly conceived that they can represent 'something for everyone' – e.g. sustainability, energy security *and* rural economic growth (Frow et al., 2009; Birch et al., 2010; Richardson, 2012).

Following Mazzucato (2013) and many other innovation scholars, we want to stress that the state is central to the creation and development of the bio-economy – as with any new energy regime. Indeed, as McCormick (2014) demonstrates, the work of the OECD has been particularly important in spreading the bio-economy as a policy strategy that countries could pursue to promote future sustainability *and* ensure future economic growth (cf. Pfau et al., 2014). Subsequently, a number of governments have developed or started to develop bio-economy strategies in the last decade, including the EU, USA, UK, Germany, Finland, Sweden, and Australia (German Bioeconomy Council, 2015). The range of stakeholders involved in state-led initiatives, and the range of future visions espoused and pursued, is exemplified by the diverse 'market development' policies pursued in various countries to create and develop the bio-economy. As a report by Consilium (Pugatch Consilium, 2014) illustrates, most market development policies focus on things like human capital; R&D funding; intellectual property rights; regulatory and legal environment; and market incentives. Research on biofuels emphasizes, in particular, the importance of crafting an economically viable energy market in which biofuels can compete with petroleum. A report by the German Bio-economy Council (2015) highlights a range of these market development policies across the G7 countries, including: product labelling and certification so that consumers can choose to support 'sustainable' products (e.g. Germany, France); tax relief for investment in (bio-)refineries (e.g. UK, Canada); feed-in tariffs to support bioenergy production (e.g. Japan, Canada); and public procurement to stimulate bio-product and bioenergy demand (e.g. USA, France).

To complicate matters, bio-economy visions and strategies are underpinned by geographically specific and territorially specific imaginaries, in that different countries have different visions of what a bio-economy is and what it means for policy-making. For example, visions of the bio-economy in Canada have been driven by imagining a radical shift in the Canadian economy away from the traditional characterization of being 'hewers of wood' to value-added processing of biological material (e.g. CRFA, 2014). Frequently these geographical imaginaries mean that specific policy strategies frame future energy worlds in very different ways. For example, there is some difference in the use of terms like 'bio-economy' versus 'bio-based economy' in certain national strategies; that is, the former is used to refer to the conversion processes (i.e. technoscience) and the latter to the raw materials (i.e. natural resources) (Staffas et al., 2013). It is important to analyze the different ways that various governments and jurisdictions frame and define the bio-economy to better understand the underlying assumptions driving policy strategies and the implications this has for the emergence of new energy worlds. McCormick (2014: 192) highlight the differences between the OECD, EC and USA as follows:

- OECD (2009): focused on the development of 'sustainable, eco-efficient and competitive products.'
- EC (2012): focused on the development of new products and integration of biological resources into the economy.
- USA (2012): focused on use of (biological) science and innovation to create economic growth.

Such definitions end up framing potential policy strategies; for example, the emphasis on consumers and markets as drivers of change precludes the use of certain measures (e.g. government incentives) to promote the bio-economy. At the same time, there are discussions, debates and differences within and between states about the 'best' model for development in terms of degree of decentralization in infrastructure development and institutional control. In many Scandinavian countries, the favoured model is a highly decentralized bio-economy in terms of infrastructure but with some centralization in terms of institutional control (e.g., DHS) while in North America, the favoured model seems to be large, centralized facilities (albeit less-so than fossil fueled facilities due to physical limits discussed above), with decentralized control (private ownership). Generally, in this regard, how the future is defined and framed ends up shaping potential policy strategies and policy-making, especially when it comes to science, innovation and market policies, and the various institutions and infrastructures on which these policies depend (Birch et al., 2010, 2014).

Whether biomass should be used for energy purposes beyond nutrition remains a normative question, as evidenced by the clash of opposing value systems and visions of a sustainable future within the food versus fuel debate. But aside from this, we can imagine bioenergy and biofuel production systems as 'transition technologies' or as a 'bridge' to higher value products. This vision is emerging in policy environments and may have implications on the future development of the bio-economy. The U.S. Agricultural Act of 2014, for example, extended the biorefinery assistance program within the Food, Conservation and Energy Act of 2008, but has expanded the scope of assistance to include systems that are using biomass to produce biochemicals.

Conclusion: geographies of bio-economy

The notion of a bio-economy refers to an ancient, a present, and a future energy world. Across space and time, these worlds differ in terms of the scale and complexity of energy and material services that are demanded; the resources and technologies that are available and used; and the social value systems that guide environmental and resource management. Central to the notion of an 'advanced' bio-economy is the biorefinery. As described above, the implementation of biorefineries and development of biomass resources for energy purposes will surely prove disruptive (or, if you prefer, transformative) to prevailing energy landscapes given the vastly different materiality between the two energy resources and fundamental differences in the territorial and social relations surrounding surface versus subsurface (or subterranean) resources.

What role can (should) geographers play in helping to understand the barriers, opportunities, and implications of bio-economy? Here we conclude with suggestions based on our discussion above. Analytically, geographers are uniquely positioned to identify the conditions under which bio-economies are 'sustainable' in social and ecological terms. Already we have done well to note the political-economic teleconnections that compromise the sustainability of biofuels, be it through land-grabbing and/or indirect land-use changes. Tracing these teleconnections, and possible governance models to best manage them, remains a key area for research, especially as nations continue to support domestic biofuel industries that reverberate through land and commodity markets across scales. Furthermore, it is important to assess the extent to which new energy regimes are co-constitutive of new socioeconomic regimes. For instance, are the 'land grabs' for biofuel production different in kind or degree from the forms of enclosure and exclusion that are typical of fossil fuel resource extraction? If not, how can they be made different through collaborative, stakeholder-based consultation and ownership models?

There is also a strong need for integrated regional analyses. Given that biomass is not a global solution due to limited resource availability, it is important to map and assess resource availability at local and regional scales, and to integrate these analyses with assessments of the ways in which local and regional spatial identities, land-use practices and land tenure relations are implicated in the strategic visioning and resource development process. Integrated studies of the sort will lead to an improved understanding of interdependencies at the resource-institution-technology interface in particular places and, in turn, the opportunities, barriers, and implications of an energy regime that includes advanced bio-economies.

Conceptually, geographers should strive for systematic frameworks by which to understand linkages and interdependencies at the resource-institution-technology interface as it relates to the (changing) geographies of energy. Energy cannot simply be something we study; it needs to be an underlying concept (Huber, 2015). One of the biggest challenges here is to take this seriously without slipping into energy determinism. By tracing the ways in which new biomass conversion technologies are implicated in and co-constitute political-economic and socioecological relationships, hopefully this chapter serves as a modest step toward meeting that challenge.

Note

1   In contrast, most advanced economies procure only 1–3 percent of their primary energy from biomass. Here, we are speaking only of non-nutritional energy needs. When considering throughputs of nutritional energy needs, proportions change. Even in societies that have transitioned to fossil fuels, for example, edible plant matter and domesticated animals still represents at least 30 per cent of the total energy throughputs (Haberl, 2002).

References

Asomaning, J., Mussone, P., Bressler, D.C. 2014. Two-stage thermal conversion of inedible lipid feedstocks to renewable chemicals and fuels. *Bioresource Technology* 158: 55–62.

Bakker, K., Bridge, G. 2006. Material worlds? Resource geographies and the 'matter of nature'. *Progress in Human Geography* 30: 5–27.

Berndes, G., Hoogwijk, M., van den Broek, R. 2003. The contribution of biomass in the future global energy supply: a review of 17 studies. *Biomass and Bioenergy* 25: 1–28.

Birch, K. and Calvert, K. 2015. Rethinking 'drop-in' biofuels: on the political materialities of bioenergy. *Science and Technology Studies* 28: 52–72.

Birch, K., Levidow, L., Papaioannou, T. 2010. Sustainable Capital? The neoliberalization of nature and knowledge in the European knowledge-based bio-economy. *Sustainability* 2: 2898–2918.

Birch, K., Levidow, L., Papaioannou, T. 2014. Self-fulfilling prophecies of the European knowledge-based bio-economy: the discursive shaping of institutional and policy frameworks in the bio-pharmaceuticals sector. *Journal of the Knowledge Economy* 5: 1–18.

Bridge, G., Bouzarovski, S., Bradshaw, M., Eyre, N. 2013. Geographies of energy transition: space, place and the low carbon economy. *Energy Policy* 53: 331–340.

Calvert, K. 2015. From 'energy geography' to 'energy geographies': perspectives on a fertile academic borderland. *Progress in Human Geography* 40: 105–125.

Calvert K., Simandan D. 2015. A polymorphic approach to environmental policy analysis: the case of the Ethanol-in-Gasoline Regulation in Ontario, Canada. *Geografiska Annaler: Series B, Human Geography* 97(1): 31–45.

Cao, L., Yuan, X., Li, H., Li., C., Xiao, Z., Jiang, L., Huang, B., Xiao, Z., Chen, X., Wang, H., Zeng, G. 2015. Complementary effects of torrefaction and co-pelletization: energy consumption and characteristics of pellets. *Bioresource Technology* 185: 254–262.

CEC. 2012. *Innovating for Sustainable Growth: A Bioeconomy for Europe* [COM (2012) 60 Final]. Brussels: Commission of the European Communities.

Cook, E.F. 1976. *Man, Energy, Society*. San Francisco, CA: Freeman.

CRFA. 2014. *Evolution and Growth: From Biofuels to Bioeconomy*. Ottawa: Canadian Renewable Fuels Association.

Deng, Y.Y., Koper, M., Haigh, M., Dornburg, V. 2015. Country-level assessment of long-term global bioenergy potential. *Biomass and Bioenergy* 74: 253–267.

EU Presidency. 2007. *En Route to the Knowledge-based Bio-economy*. Cologne: Cologne Summit of the German Presidency.

Floudas, C.A., Elia, J.A., Baliban, R.C. 2012. Hybrid and single feedstock energy processes for liquid transportation fuels: a critical review. *Computers & Chemical Engineering* 41: 24–51.

Frow, E.K., Ingram, D., Powell, W., Steer, D., Vogel, J., Yearley, S. 2009. The politics of plants. *Food Security* 1(1): 17–23.

German Bioeconomy Council. 2015. *Bioeconomy Policy: Synopsis and Analysis of Strategies in the G7*. Berlin: Office of the German Bioeconomy Council.

Haberl, H. 2002. The energetic metabolism of societies, Part 2: empirical examples. *Journal of Industrial Ecology* 5: 71–88.

Howard, P.J. 2011. *An Introduction to Landscape*. Farnham, UK: Ashgate:.

Huber, M. 2015. Theorizing energy geographies. *Geography Compass* 9(6): 327–338.

Jiusto, S. 2009. Energy transformations and geographical research. In *A Companion to Environmental Geography*, ed. N. Castree, D. Demeritt, D. Liverman, and B. Rhoads. Hoboken, NJ: Wiley-Blackwell, 533–551.

Karplus, Y., Meir, A. 2013. The production of space: a neglected perspective in pastoral research. *Environment and Planning D: Society and Space* 31: 23–42.

Kitchen, L., Marsden, T. 2011. Constructing sustainable communities: a theoretical exploration of the bio-economy and eco-economy paradigms. *Local Environment* 16: 753–769.

Kocoloski, M., Griffin, W.M., Matthews, H.S. 2011. Impacts of facility size and location decisions on ethanol production cost. *Energy Policy* 39: 47–56.

Lamers, P., Hamelinck, C., Junginger, M., Faaij, A. 2011. International bioenergy trade – a review of past developments in the liquid biofuel market. *Renewable & Sustainable Energy Reviews* 15(6): 2655–2676.

Lamers, P., Hoefnagels, R., Junginger, M., Hamelinck, C., Faaij, A. 2015. Global solid biomass trade for energy by 2020: an assessment of potential import streams and supply costs to North-West Europe under different sustainability constraints. *GCB Bioenergy* 7: 618–634.

Leach, G. 1992. The energy transition. *Energy Policy* 20: 116–123.

Lefebvre, Henri. 1991. *The Production of Space*. Vol. 142. Blackwell: Oxford.

Lynd, L.R., Laser, M.S., Mcbride, J., Podkaminer, K., Hannon, J. 2007. Energy myth three – high land requirements and an unfavourable energy balance preclude biomass ethanol from playing a large role in providing energy services. In *Energy and American society – thirteen myths*, ed. B.K. Sovacool and M.A. Brown, Dordrecht, The Netherlands: Springer, 75–102.

McCormick, K. 2014. The bioeconomy and beyond: Visions and strategies. *Biofuels* 5: 191–193.

Magelli, F., Boucher, K., Hsiaotao, T.B., Melin, S., Bonoli, A. 2009. An environmental impact assessment of exported wood pellets from Canada to Europe. *Biomass and Bioenergy* 33: 434–441.

Mazzucato, M. 2013. *The Entrepreneurial State*. London: Anthem Press.

Nadaï, A., van der Horst, D. 2010. Landscapes of energies. *Landscape Research* 35: 1–11.

OECD. 2006. *The Bioeconomy to 2030: Designing a Policy Agenda*. Paris: Organisation for Economic Co-operation and Development.

Pa, A., Craven, J.S., Bi, X.T., Melin, S., Sokhansanj, S. 2012. Environmental footprints of British Columbia wood pellets from a simplified life cycle analysis. *International Journal of Life Cycle Assessment* 17: 220–231.

Pasqualetti, M.J. 2011. Social barriers to renewable energy landscapes. *Geographical Review* 101: 201–223.

Pasqualetti, M.J. 2013. Reading the changing energy landscape. In *Sustainable Energy Landscapes: Designing, Planning and Development*, ed. Stremke, S., van den Dobbelsteen, A. Baton Rouge, FL: CRC Press, 11–44.

Pasqualetti, M.J., Gipe, P., Righter, G.W. 2002. *Wind Power in View: Energy Landscapes in a Crowded World*. San Diego, CA: Academic Press.

Pinzi, S., Leiva-Candia, D., Lopez-Garcia, I., Redel-Macias, M.D., Dorado, M.P. 2014. Latest trends in feedstocks for biodiesel production. *Biofuels, Bioproducts & Biorefining* 8(1): 126–143.

Ponte, S., Birch, K. 2014. Introduction: Imaginaries and governance of biofueled futures. *Environment and Planning A* 46(2): 271–279.

Pugatch Consilium. 2014. *Building the Bioeconomy*. Bicester, Oxon: Pugatch Consilium.

Sampson, C., Agnew, J., Wassermann, J. 2012. *Logistics of agricultural-based biomass feedstock for Saskatchewan*. Report prepared for ABC Steering Committee, SaskPower, NRCan. Project No. E7810

Schmid, O., Padel, S. and Levidow, L. 2012. The bio-economy concept and knowledge base in a public goods and farmer perspective. *Bio-based and Applied Economics* 1: 47–63.

Sims, R.E., Mabee, W.E., Saddler, J.N., Taylor, M. 2009. An overview of second generation biofuel technologies. *Bioresource Technology* 101: 1570–1580.

Snell, K.D., Peoples, O.P. 2009. PHA bioplastic: a value-added coproduct for biomass biorefineries. *Biofuels, Bioproducts & Biorefining* 3(4): 456–467.

Staffas, L., Gustavsson, M., McCormick, K. 2013. Strategies and policies for the bioeconomy and bio-based economy: an analysis of official national approaches. *Sustainability* 5: 2751–2769.

White House. 2012. *National Bioeconomy Blueprint.* Washington, DC: The White House.

4

The changing landscape of biofuels

A global review

Barry D. Solomon and J. Brad Barnett

Introduction

Global production and use of biofuels has greatly expanded in the last decade. The leading producers and consumers since the 1970s have been Brazil and the United States, but more recently several other countries have become major producers, e.g. China, Argentina, Germany, France and Indonesia. The largest production growth has occurred in the U.S., Argentina, Indonesia, China and Thailand (EIA 2015). Most of this growth is attributable to the desire to lower petroleum-based fuel use and oil imports, and to reduce greenhouse gas (GHG) emissions. In addition, biofuel expansion is partly encouraged and modeled after the successes of Brazil and the U.S. in their biofuels industries, which increase economic returns to farmers (Solomon et al. 2014). This chapter will provide a global overview of the geography of major biofuel feedstock landscapes, with a focus on landscape change, several conflicts and their resolutions.

There are many concerns over the sustainability of biofuels (Solomon 2010). These include the practice of manufacturing biofuels by diverting edible feedstock such as corn into fuel. In addition, rising demand for biofuels often directly and indirectly leads to land clearing or displacement and thus increased emissions, loss of rainforests, peatlands, savannas and grasslands in Latin America and Southeast Asia (Richards 2015; Koh and Wilcove 2008). However, estimates of indirect land use change vary widely and are contested, and in some cases impacts can be positive such as when biofuel feedstock is grown on degraded pasturelands (e.g. Fargione et al. 2008; Lapola et al. 2010; Solomon and Bailis 2014). Many other aspects of the sustainability of biofuels should be considered: emissions, soil fertility and loss, use of marginal lands, water use and access, biodiversity, land tenure, human rights, government subsidies, economic development trade, health, gender issues and ethics (Solomon 2010).

Three biofuels are currently produced commercially: ethanol, which is blended with gasoline; biodiesel, which is blended with diesel; and methanol. Ethanol is produced from starch and sugar crops such as corn, sugarcane and sugar beets, and in a few cases cellulosic residues such as corn stover (leaves, stalks and cobs), wheat straw and switchgrass. Biodiesel is produced from oilseed crops such as soy and rapeseed oils, animal fats, and waste vegetable oil. Methanol is today made primarily from natural gas rather than wood, and most of its use is for non-energy purposes.

Ethanol was commercialized earlier than biodiesel and is produced in much larger volumes, although in regions such as the European Union (EU), Southeast Asia and South America, biodiesel is more common. Since the mid-1970s, Brazil and the U.S. have dominated global ethanol production. Brazil began to develop its fuel-ethanol industry in the mid-1970s soon after the oil embargo by Organization of the Petroleum Exporting Countries (OPEC) exposed the vulnerabilities of importing countries. Brazil led in global production until the U.S. passed it in 2006 (Solomon and Bailis 2014). In order to allow their industries to mature, governments in the U.S. and Brazil implemented numerous policies including subsidies, tax incentives and tariffs (Solomon et al. 2007; Hira and de Oliveira 2009). In 2012, the U.S. and Brazil accounted for 87 percent of global ethanol production (Table 4.1).

Biodiesel production is geographically more dispersed (Table 4.2). North and South America account for 40 percent of global production. The U.S., Brazil and Argentina have been the three largest producers for many years. In the U.S., biodiesel production grew dramatically from

Table 4.1 Top 10 ethanol-producing countries, 2007–2012 (million liters/year)

	2007	2008	2009	2010	2011	2012	Ethanol incorporation rate in road fuel in 2012*
United States	24,685	35,141	41,404	50,338	52,728	50,809	10
Brazil	22,557	27,059	26,103	28,203	22,748	23,357	25
China	1,665	1,996	2,179	2,128	2,255	2,509	2
Canada	801	870	1,161	1,393	1,741	1,898	4.7
France	540	928	906	1,208	1,208	1,241	7
Germany	395	580	752	765	730	776	5.7
Thailand	174	331	419	451	486	471	6
Guatemala	168	168	174	174	232	464	1.5
Belgium	0	25	154	308	400	450	7
Netherlands	14	15	0	140	275	450	5

Note: * Many of these countries have large imports and exports of ethanol.
Source: EIA (2015).

Table 4.2 Top 10 biodiesel producing countries, 2007–2012 (million liters/year)

	2007	2008	2009	2010	2011	2012
United States	1,854	2,560	1,973	1,277	3,656	3,714
Germany	3,308	3,192	2,611	2,843	3,321	3,174
Argentina	209	807	1,341	2,089	2,747	2,780
Brazil	404	1,164	1,608	2,386	2,673	2,710
Indonesia	58	116	330	740	1,800	2,200
France	1,085	1,996	2,379	2,147	1,973	1,898
China	116	290	591	568	852	909
Thailand	70	447	610	660	630	900
Italy	534	760	905	841	650	569
Poland	52	290	348	406	395	563

Source: EIA (2015).

2005–8, but has been somewhat volatile in the years since, dropping in 2009–10 and rebounding since. The balance of global biodiesel production is divided between the EU and Southeast Asia. The EU accounts for another 40 percent of global production (EIA 2015). However, the popularity of diesel-powered passenger cars and the region's renewable energy policies make it impossible for the EU to meet its biodiesel demand. Thus, the EU is an attractive market for Southeast Asian and Latin American nations developing export-oriented biodiesel industries.

This chapter is organized as follows. The next section provides a world regional review of biofuel landscapes, with attention to crop choice and characteristics, industrial location, and whether the biorefineries are export oriented or focused on meeting domestic fuel demand. The subsequent section provides five case studies of conflicts associated with biofuel development, such as feedstock selection, environmental issues and sustainability. These energy landscape issues are socioeconomic as well as environmental, and have been concentrated in Latin America and Southeast Asia. Finally, we will close with some conclusions and policy recommendations.

Regional patterns of biofuel landscapes

While many nations produce biofuels, a few countries and world regions dominate. In this section, we concentrate on the status of biofuels in the principal feedstock growing regions (such as the U.S. and Brazil), while ignoring regions of minor biofuel production. Dominant feedstock and geographic patterns in each major region are highlighted.

North America

In North America there is large overlap in the geography of corn and soybean production, by far the two major biofuel feedstocks. Food crop based biofuels are often called first generation biofuels. Corn accounts for over 95 percent of ethanol production in the U.S., with a slightly lower share in Canada (Canada uses wheat to make roughly 20 percent of its ethanol). Biofuels are simply another product market for the well-established agricultural sector concentrated in the North Central States and thus any landscape change from biofuel feedstocks is inseparable from food production. U.S. corn production acreage expanded 10 percent from 2000–9, a period of rapid expansion in ethanol production, but most of this was growth concentrated in existing farm belt states (USDA 2011). Nevertheless, with rapid growth over the last decade, corn ethanol in the U.S. today accounts for around 40 percent of the total corn crop (and 30 percent in Canada), nearly as much as is dedicated to feed livestock, though much more than for direct human consumption (Iowa State University 2015). The largest corn-ethanol production by volume occurs in Iowa, Nebraska and Illinois (U.S.), also the largest corn producers (Figures 4.1–4.3), and Ontario, Canada.

U.S. ethanol use dates back to the start of the automobile industry. It then fell, only to be revived in the late 1970s in response to the import vulnerabilities exposed by the OPEC oil embargo (Solomon et al. 2014). Initially U.S. farmers concentrated production in small-scale distilleries to produce ethanol, though after 1978 the industrial giant Archer Daniels Midland (ADM) dominated the industry for the next two decades (Federal Trade Commission 2006). As U.S. ethanol output expanded, the nation also imported some Brazilian ethanol to meet domestic demand, typically after final dehydration in Caribbean or Central America countries. This pattern changed in 2011 as Brazilian output fell and the nation started importing ethanol from the U.S. Today a large volume of ethanol is moving in both directions, and the U.S. also exports a significant volume to Canada to help meet its domestic biofuel production mandate (Lamers et al. 2014). Until the 1980s, U.S. corn ethanol production came exclusively from

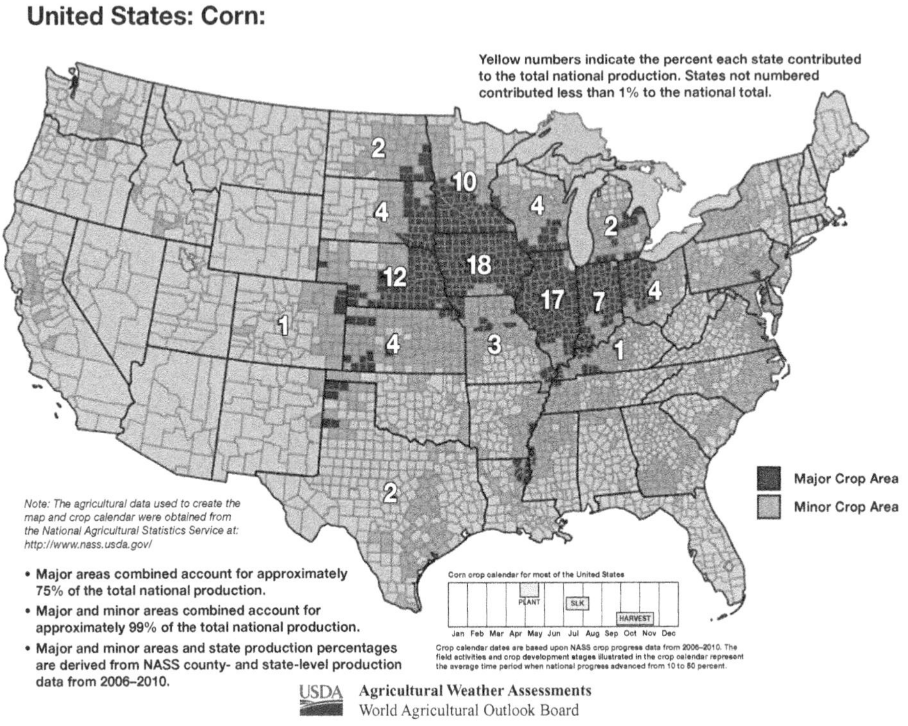

United States: Corn:

Yellow numbers indicate the percent each state contributed to the total national production. States not numbered contributed less than 1% to the national total.

Major Crop Area

Minor Crop Area

Note: The agricultural data used to create the map and crop calendar were obtained from the National Agricultural Statistics Service at: http://www.nass.usda.gov/

- Major areas combined account for approximately 75% of the total national production.
- Major and minor areas combined account for approximately 99% of the total national production.
- Major and minor areas and state production percentages are derived from NASS county- and state-level production data from 2006–2010.

Corn crop calendar for most of the United States

PLANT SLK HARVEST

Jan Feb Mar Apr May Jun Jul Aug Sep Oct Nov Dec

Crop calendar dates are based upon NASS crop progress data from 2006–2010. The field activities and crop development stages illustrated in the crop calendar represent the average time period when national progress advanced from 10 to 80 percent.

USDA **Agricultural Weather Assessments**
World Agricultural Outlook Board

Figure 4.1 U.S. corn production map

Source: USDA.

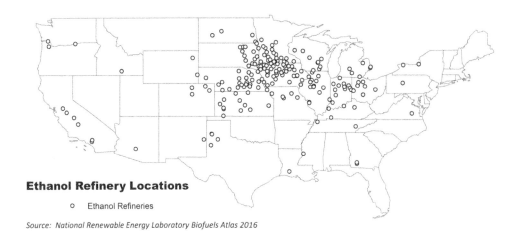

Ethanol Refinery Locations

○ Ethanol Refineries

Source: National Renewable Energy Laboratory Biofuels Atlas 2016

Figure 4.2 Ethanol refinery locations in the U.S.

Source: RFA (2016), figure by Brad Barnett.

Figure 4.3 A Springfield, Nebraska cornfield ready for harvest
Source: Nati Harnik, AP, reprinted with permission.

hydrous (wet grind) mills. As the industry has grown to over 210 distilleries more recently most production has shifted to anhydrous (dry grind) mills, which are cheaper and more efficient (RFA 2016; Solomon et al. 2014; Figure 4.2).

The U.S. soy biodiesel industry is much smaller than the ethanol sector, and very little production occurred before 2005. Today it is the world leader, though energy landscape change has been minimal (Table 4.2). Canadian biodiesel production is negligible. The North American demand for diesel fuel is smaller because of a lower rate of use of diesel cars and trucks than in Europe. The region's soy crop is produced on many of the same lands that produce corn as part of regular crop rotation, and thus there is a tradeoff in crop selection. Some producers also use canola and corn oils, tallow, and poultry fats for feedstock in addition to soy oil. The largest biodiesel producing states are Texas, Missouri, Iowa, Illinois and Pennsylvania.

Latin America

The Latin American ethanol market is dominated by sugarcane production in Brazil, where production is more efficient than making it from corn (Figure 4.4), with much smaller sugarcane output elsewhere such as Colombia, Argentina and Guatemala. In addition, a small amount of ethanol is produced from grain. Although Brazil (like the U.S.) experimented with ethanol production for many decades, it began to develop its fuel-ethanol industry in earnest with fuel blending mandates in the mid-1970s, soon after the OPEC oil embargo. It led global production until the U.S. surpassed it in 2006. Almost 90 percent of sugarcane production in Brazil occurs in the South-Central states, with the majority being harvested in Sao Paulo (Figure 4.5). Given the close association between ethanol and sugarcane outputs in Brazil, and similar production levels, the two sectors have co-evolved, as have any energy landscape impacts. When market conditions improve for one the output can be quickly adjusted in response between the two products (Hira and de Oliveira 2009). Thus, in years when sugarcane ethanol production is

Figure 4.4 Sugarcane stretches to the horizon around Ribeirão Preto, Brazil
Credit: Mario Osava/IPS.

Figure 4.5 Brazil map of sugarcane production areas vs. forest locations
Source: Goldemberg (2008). Reprinted with permission of Jose Goldemberg.

high in Brazil it meets domestic needs along with significant exports, and in down years it imports more from the U.S.

As was the case in North America, biodiesel production in Latin America started more recently than did production of fuel ethanol and it is less spatially concentrated. Latin America accounts for roughly one quarter of global output. Argentina and Brazil are the largest producers, with each supplying around 45 percent of Latin American output, having lagged behind only the U.S. and Germany in total production since 2011 (Solomon and Bailis 2014). Colombia contributes most of the region's remaining supply. The region's biodiesel feedstock is dominated by soy oil, which accounts for around 73 percent of the total in Brazil and virtually all of Argentina's output. The vast growing areas extend from northeast Brazil south through southern Paraguay to northeast Argentina. However, in Brazil, the west-central State of Mato Grosso is the leading producer, and the proximity of growing areas to the Amazon and Cerrado is a major concern (Castanheira et al. 2014). Columbia, in contrast, uses African palm. In all cases biodiesel oil comprises a very small part of the agricultural output in each country and thus any energy landscape change. Argentina had been the world's largest exporter of biodiesel since 2009, mostly to Europe, but because of a trade dispute with the European Commission it has focused more on the domestic market since 2013 (Evans 2015).

Europe

Lack of available land limits biofuel production in Europe. Thus, energy landscape change due to biofuel production has been negligible. Unlike the U.S. and Brazil, ethanol distilleries in Europe are usually closer to the product market and designed to use multiple feedstocks, with sugar beets, corn, and wheat crops dominating (though ethanol accounts for just a small portion of output). The major producers are Germany, France, the Benelux countries and Spain (Flach et al. 2015). Because of land constraints and renewable energy mandates, Europe has some ethanol import reliance (especially the UK), with the exception of Belgium and the Netherlands. Major import sources have included Brazil and the U.S. Recent EU policy has lowered imports from the U.S. through an anti-dumping duty, and imports have fallen since 2011 (ibid.).

The biodiesel markets and feedstock production in Europe are larger than for ethanol. The continent constitutes the world's largest output since biodiesel was first promoted in the 1990s. This is because of a higher reliance on diesel cars and trucks. The main feedstock is rapeseed oil, accounting for about two thirds of the total. For biodiesel production, Germany and France again dominate. Germany, Belgium and the Netherlands are self-sufficient in supply and can export, with most others being import dependent. Until 2013, Argentina and Indonesia were the main import sources. Due to trade disputes, most European countries have greatly lowered biodiesel imports from these two countries since 2012 (ibid.).

Southeast Asia

Biofuel production in Southeast Asia is dominated by biodiesel from oil palm plantations in Indonesia, with major landscape change, and to a lesser extent Thailand and Malaysia. Ethanol production is smaller and mostly comes from molasses and sugarcane feedstock in Thailand, with a growing contribution from cassava (Wright and Wiyono 2014; Preechajarn and Prasertsri 2014; Wahab 2014). Much biodiesel production in Southeast Asia has been destined for profitable export markets in Europe (a small amount is shipped elsewhere in Asia), with a significant amount of palm oil also exported unrefined or for food oil. Indonesia produces much more biodiesel than does Malaysia and Thailand. Significant attention has been given to

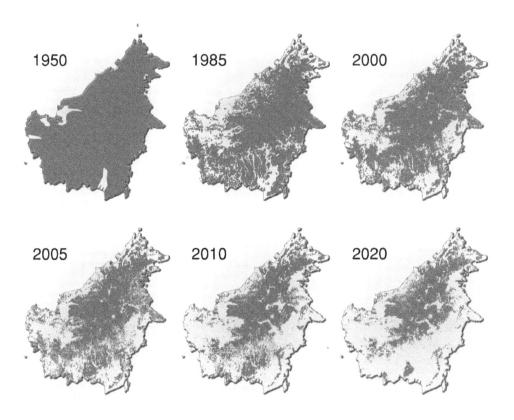

Figure 4.6 Extent of deforestation in Borneo, Indonesia 1950–2005 and projections towards 2020

Source: Hugo Ahlenius, UNEP/GRID-Arendal, further information available at www.grida.no/graphicslib/detail/extent-of-deforestation-in-borneo-1950–2005-and-projection-towards-2020_119c.

the deforestation of original forest cover coincident with growth of oil palm monocultures in the region (Figure 4.6), with adverse effects on biodiversity, especially orangutan and tiger habitats, and GHG emissions (Koh and Wilcove 2008). Data from Indonesia indicate that 100 percent of the extent of the oil palm plantations was created from the conversion of primary, secondary, or plantation forests from 1990–2005. In the case of Malaysia, the forest loss rate was between 41 percent and 45 percent (ibid.), though most of its palm oil is destined for food, not biofuel.

Conflicts and resolutions

Deforestation of the Brazilian Amazon

Deforestation in Amazonia, as is the case in most of the world, has been going on for many centuries (Denevan 1992). Such landscape changes result from land clearing for wood, agriculture, industry and urban development. As a result, it is important to distinguish between continuing long-term patterns of deforestation and those that result from biofuel crop growth (e.g. Fargione et al. 2008; Lapola et al. 2010).

The cultivation of sugarcane occurs on approximately 9 million hectares in Brazil, and roughly 4 million hectares were added from 2000–10, for a 79 percent increase (IBGE 2013; Figure 4.5). Even so, the land area dedicated to food crops was not reduced. Sparovek et al. (2009) found that the expansion of sugarcane that occurred from 1996–2006 in the states of São Paulo, Minas Gerais, Paraná, Goiás and Mato Grosso do Sul came at the expense of cattle pastures, without causing deforestation in these states. This expansion of sugarcane production contributed to economic growth that exceeded land areas that did not experience sugarcane expansion (Coelho and Guardabassi 2014).

The use of soy oil for biofuel in Brazil is more problematic, though the effects on deforestation are complex (see Bailis 2014, upon which the rest of this section is derived). It should also be noted that less than 10 percent of total soy oil production is used for biodiesel. While Brazil's early soy expansion occurred primarily in southern Brazil, on previously farmed lands, the second phase rapidly encroached upon the Amazon Forest and Cerrado savanna grassland biomes (Figure 4.7; Sawyer 2008). From 2000–5, when Brazil's cultivated area of soybeans peaked, the area of soy cultivated in the Legal Amazon Region (LAR) increased by 120 percent, accounting for 40 percent of nationwide soybean expansion (IBGE 2012). This expansion was divided between land that had been deforested prior to 2000 and recently deforested land, but used for pasture prior to being converted to soy crop (Macedo et al. 2012). In the State of Mato Grosso, which accounted for one third of the country's soybean expansion in this period, one fourth

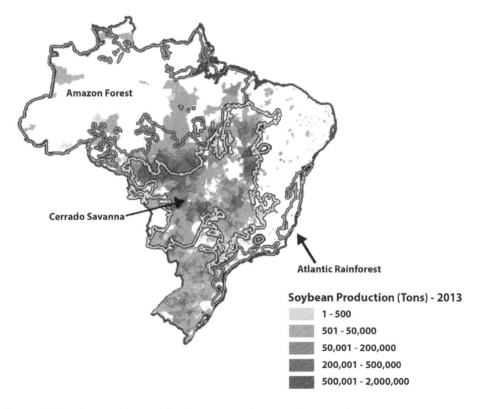

Figure 4.7 Brazil map of soy production areas vs. forest and savanna locations
Source: authors.

of new soybean production occurred as direct expansion into forests, while the rest occurred on former pastureland, the majority of which was deforested before 2000 (Macedo et al. 2012).

Although over 90 percent of soybeans were planted on former pastureland (and thus the direct effect of soy crop production on deforestation has been minimal), many analysts think that there is still a negative impact on forest cover (Lapola et al. 2010). Changes in these types of energy landscape happen *indirectly* through regional land markets as follows: when soybean prices are high, as they were for most of the last decade or so (CEPEA 2012), the price of land suitable for soybean production increases. This incentivizes the sell off of pasturelands and helps spread cattle deeper in the Amazon Forest region, where cheaper land exists (Lapola et al. 2010; Richards 2015). Differences in the value of former pasturelands targeted for soybean production during this time and land in undeveloped Amazon areas allowed ranchers to purchase and clear much larger areas of land than they sold. This gave rise to large indirect effects of pasture-soybean transitions (Arima et al. 2011).

In mid-2006, only one year after Brazil began producing biodiesel, the nation began a moratorium on soy production in 'recently deforested' areas of the LAR, which has been extended ever since. The area under soybean cultivation expanded until 2006 and then fell for two years, but has since recovered to 2005 levels in Mato Grosso and elsewhere. While some analysts claim that the moratorium has been effective at inhibiting further geographical expansion of soybean cultivation into the LAR (Rudorff et al. 2011), the contribution of the moratorium to any slowing of soy expansion is unclear. Soy prices are closely correlated with agriculturally induced deforestation (Morton et al. 2006), and have also played a role. Between March 2004 and March 2006, real soy prices declined 54 percent and remained erratic until late 2011 (CEPEA 2012). Other analysts have noted advancement of the agricultural frontier into the LAR has continued since the moratorium began, driven in part by the indirect effects of soybean production on pasture expansion noted above, which the moratorium does little to address (Arima et al. 2011).

Protection of the Brazilian Cerrado savanna and Atlantic forest

Indirect land use change from biofuel expansion in Brazil can also encroach upon regions besides the Amazon Forest. For example, modeling by Lapola et al. (2010) has raised concerns about reduced biodiversity in the Atlantic Forest, which has much more limited geography than the Amazon Forest, and especially on the Cerrado tropical savanna grassland in central to southeastern Brazil (Figure 4.7). The Cerrado has more than 7000 plant species, with a very high rate of endemism (Klink and Machado 2005). Expansion of the soy crop is the main culprit for lost diversity in the Cerrado, because of likely cultivation in the central state of Goiás and the southeastern state of Minas Gerais; increased sugarcane plantations would also occur in Minas Gerais. Sugarcane expansion could also lead to indirect deforestation of the Atlantic rainforest in southeastern Brazil (Figure 4.5), a region with biodiversity similar to the Amazon (Carnaval and Moritz 2008).

A partial solution to prevent further destruction of the Cerrado savanna has taken the form of agro-ecological zoning, which is aimed at protecting existing biomes. Indeed, the state of Minas Gerais pioneered this process in 2007, and a sugarcane zone was created in 2009 (Coelho and Guardabassi 2014: 82–3). While an oil palm zone was created in 2010, no such zone has been established in Brazil for soy crops. Moreover, the Atlantic Forest lacks such protection. In the case of Minas Gerais, social, economic and environmental data are used to determine regional areal characteristics, potential for sugarcane expansion, and vulnerabilities, including soil, water, climate and biodiversity data. However, enforcement is critical in ensuring that zoning works as planned.

Figure 4.8 Tiger habitat deforested for oil palm plantations in Longkib, Sumatra, Indonesia

Source: Steve Winters, National Geographic Creative. Reprinted with permission.

Oil palm monocultures, Southeastern Asia deforestation, and the Roundtable on Sustainable Palm Oil

The growth of oil palm, a feedstock option for biodiesel as well as food oil, has been blamed for large-scale deforestation of Southeast Asian tropical rainforests in countries like Malaysia and Indonesia (Figure 4.8). Indonesia and Malaysia have dedicated ~ 13.3 million hectares to the commodity, accounting for approximately 86 percent of the world's oil palm production (USDA 2014). The vast majority of the plantations, particularly in Indonesia, are on cleared primary and secondary forestland (Danielsen *et al.* 2009). This has been a key contributor to the loss of nearly 40 million hectares of forestland in Indonesia and 5 million hectares of forestland in Malaysia since 1990 (Wicke et al. 2011).

Deforestation from oil palm development has led to several pressing sustainability issues. As the top feedstock for biodiesel in Southeast Asia, along with oil production for food, palm oil production growth has been championed as a mechanism to combat climate change by displacing fossil fuels. However, conversion of tropical rainforests to oil palm plantations can result in a loss of carbon sequestration capacity requiring almost 100 years of fossil fuel displacement to make a positive net reduction in carbon emissions. When grown on degraded lands or lands already converted for other agricultural commodities, the carbon payback period for oil palm-based biodiesel can be as little as ten years (Danielsen et al. 2009). In addition, oil palm plantations represent a monoculture system where only one crop is planted and competing plants are removed or prevented from growing. Studies on oil palm plantations indicate that the conversion of primary and secondary forests leads to a reduction in the overall number of forest plants and animal biodiversity because of alteration of forest ecosystem. Lost habitats for the Bornean orangutan and the Sumatran orangutan and tiger are especially worrisome (Danielsen et al. 2009; Koh and Wilcove 2008).

In 2004, the World Wildlife Fund (WWF) led a coalition of 47 organizations to establish the Roundtable on Sustainable Palm Oil (RSPO) in response to biodiversity losses and deforestation linked to the oil palm industry in Southeast Asia (Figure 4.8) (Tan et al. 2009). RSPO represents a non-state market driven governance system consisting of businesses and non-governmental organizations (NGOs) to provide a mechanism to enhance the sustainability of commodities like biofuel, coffee, forest products, etc. (Cashore 2002). In 2007, RSPO created a sustainability certification system for oil palm, with more than 2.65 million hectares or 11.75 million tonnes now certified globally (RSPO 2015).

Roundtables such as RSPO can influence a commodity's production chain by creating 'standards' that outline best practices for feedstock producers and processers. Producers and processers that meet these standards can have their products 'certified' by a third-party organization such as the RSPO. Certifying agencies work to create an environment that leverages partnerships with processors, retailers and governments to encourage and/or require feedstock growers to meet these standards. RSPO's oil palm certification system consists of environmental and social criteria that must be met in order for production to be labeled as 'Certified Sustainable Palm Oil'. Among other benefits, certification gains access to major markets in Europe where many industries in countries such as the UK, France and Germany require 100 percent RSPO certified palm oil (RSPO 2015). Access to the European markets and a desire to counter environmental groups' assertions of unsustainable production practices led the Malaysia Palm Oil Association to join RSPO. This in turn led the Indonesian Palm Oil Association to become an RSPO member as well (Nikoloyuk 2009).

Currently, RSPO certifies around 20 percent of global palm oil production (RSPO 2015); however, it has been criticized for many reasons. First, small-scale oil palm growers have found it difficult to be represented in RSPO's governance system. Second, RSPO lacks effective means for verifying that its members actually improve oil palm production sustainability. Third, RSPO has not achieved a monopoly on the palm oil market, which allows palm oil producers to seek alternative destinations for their product. Fourth, critics assert that the criteria for sustainability certification are too weak and favor large-scale production systems, which are argued to be unsustainable. Fifth, the use of non-state market driven mechanisms like RSPO allow governments to delay, soften or neglect direct regulation of the industry (Schouten and Glasbergen 2011).

EU demand for sustainable biofuels and the rejection of 'subsidized' biodiesel from Indonesia and Argentina

For many biofuel and feedstock producing countries, the EU represents a critical source of demand for their products, particularly biodiesel (Lamers et al. 2014). In 2015, EU member states consumed approximately 5.7 billion liters of ethanol and nearly 12.3 billion liters of biodiesel. While most of these fuels are domestically produced, 9 percent of ethanol and 11 percent of biodiesel supply were imported from outside the EU. In 2011, ethanol imports to the EU peaked at 1.3 billion liters; however, domestic production has steadily increased while overall demand has remained relatively constant. As shown in Figure 4.9, biodiesel imports may have peaked in 2012 at ~ 3.3 billion liters; however, biodiesel consumption has declined from its peak of 14.1 billion liters in 2011 while overall production is expected to remain relatively flat at 11 billion liters (Figure 4.9). For both biofuels the gap between domestic production and demand has narrowed considerably, yet it still represents a critical market for exporting countries (Flach et al. 2015).

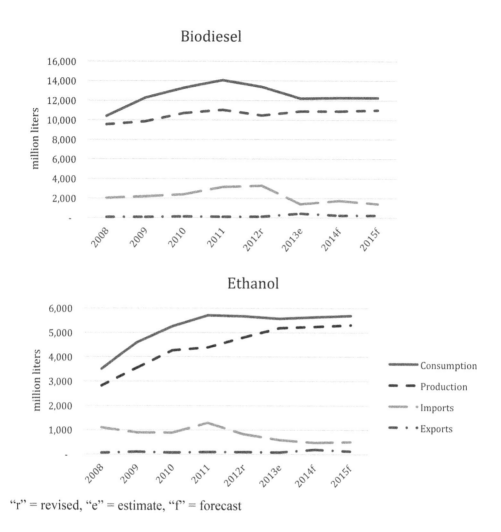

"r" = revised, "e" = estimate, "f" = forecast

Figure 4.9 EU liquid biofuel production, supply and demand 2008–15
Source: Flach et al. (2015).

The EU's demand for biofuel is driven by several policies intended to increase the use of renewable fuels; the main mechanism is the EU's Renewable Energy Directive (RED) of 2009. RED requires 10 percent of member states' final transportation fuel to come from renewable sources by 2020. It also contains a myriad of sustainability requirements focusing on climate change, land use change, and biodiversity, which are met by NGO certification standards like the RSPO among others (European Parliament 2009; Solomon and Bailis 2014). While RED helps to influence biofuel production and trade within the EU, it also generates demand for imports from non-EU states.

In 2013, Indonesia and Argentina accounted for around 90 percent of biodiesel exports to the EU (ICTSD 2013). However, in 2012 the two countries were accused by EU member states of fuel 'dumping' or selling artificially low-priced biodiesel below the cost of producing the feedstock (e.g. soy beans and palm oil). This was caused by these countries using differential export taxes (DET), where the export tax rates for raw feedstock materials are higher than the taxes placed on exported biodiesel. This was done to make importing biodiesel more attractive than importing the raw materials. From the perspective of Argentina and Indonesia, this helped grow demand for value added products, but to EU countries like Spain, the DET resulted in artificially low cost biodiesel. However, the 'dumping' allegation led the EU to levy import duties on biodiesel imported from Argentina and Indonesia at the rates of €217–246 per metric tonne and €122–149 per metric tonne, respectively. In turn, Argentine and Indonesian trade officials submitted a challenge to the World Trade Organization (WTO) in June 2014 to dispute the tariffs (ICTSD 2013; WTO 2014). In April of 2015, the EU decided to limit the use of fuels generated from food crops like soybeans and oil palm because of concerns that these fuels were contributing to deforestation, rising food prices, and overall GHG emissions. This new agreement restricts crop-based biofuels to 7 percent of consumed transportation fuels and signals a move to pursue second generation biofuels derived from crop residues and non-food crops like seaweed (European Parliament 2015).

Corn for food vs. fuel

Conventional biofuel production and use has also been criticized on the grounds that food is used to make fuel. This is a direct result of the dramatic growth in corn use for ethanol in the United States, the leading corn producing country (FAO 2015). Growth in corn production for biofuel has indirect impacts on global food markets. Increased demand for biofuels can increase global corn prices leading to concerns that biofuels production may be starving the poor (Runge and Senauer 2007). This potential conflict between corn used for food vs. fuel has led to policy responses in Mexico, the U.S. and China.

With respect to food prices, the U.S. National Academy of Sciences among others found that the expansion of biofuel landscapes accounted for 20–40 percent of the food price increased experienced in 2007–8, when the price of many food items doubled (ActionAid International USA 2012). However, several dimensions of food security and prices need to be considered. First, as corn use for ethanol has grown over time, so has total corn production. Second, while corn use for ethanol is substantial, accounting for around 40 percent of the total U.S. corn output in the past several years, its use for human food production is much smaller than its use for both ethanol and livestock feed grain. The U.S. is the world's largest corn exporter, with around 20 percent of output leaving the country (FAO 2015). Third, while growth in biofuel production can lead to increased food prices, the prices of other staple foods that are not used for biofuel such as rice have often risen in the same period. This is because multiple factors are at work, including rising energy costs, increased demand, market speculation and adverse weather (de Gorter et al. 2013). Finally, food prices can sometimes increase in one world region while falling in others.

The largest effects of the diversion of U.S. corn into ethanol production was experienced in Mexico, especially during the 2007 'tortilla crisis' (Keleman and Raño 2011). These concerns subsided as food prices fell during the Great Recession and into 2010, but have grown again since then. As a result, Mexico has added a specific provision in its biofuel law banning or restricting the use of corn for ethanol production (Eastmond et al. 2014). The U.S. addressed this concern by capping the use of corn for biofuel production at 15 billion gallons per year, as part of its Renewable Fuel Standard (RFS) (Solomon and Bailis 2014). This limit was lowered to 14 billion gallons in 2016 as policy seeks to stimulate investment in 2nd generation biofuels such as ethanol from crop residues, wood chips or switchgrass (Drajem 2015).

China, the second largest corn producer after the U.S., in 2007–8 banned the use of corn/grain in new processing projects such as ethanol plants. State policy prescribed that biofuel development should not compete for arable land with cropland dedicated to human food consumption. This reflected a growing concern with staple food security in China. However, in 2014 corn remained the major feedstock for ethanol production in China (Anderson-Sprecher and Junyang 2014).

Conclusions

Over the last decade, the global consumption and production of biofuels has dramatically increased. These increases have been concentrated in regions such as North and South America, Europe and Southeast Asia. Significant landscape changes from deforestation have occurred in Brazil and Indonesia, but the root causes of the changes are complex and much of it may be due to food crop production or ranching rather than biofuel crops. The U.S. and Brazil dominate the ethanol market, relying on corn and sugarcane feedstocks respectively. The international biodiesel market is propelled by European biodiesel consumption, which has stimulated export opportunities for nations like Argentina and Indonesia. However, EU protectionist policies to shelter domestic biodiesel producers from low cost imports and low oil prices have slowed the international biodiesel trade.

The recent growth in biofuel production and consumption has been driven by government policies intent on increasing the use of biofuels to combat climate change, increase rural development and foster energy security. Policies such as the U.S. RFS, Brazilian requirements, and the EU RED have created mandates for renewable transportation fuels, stimulating the growth of first generation (food crop based) biofuels. Even so, concerns over biofuels' pressure on food prices, land conversions leading to deforestation, and subsequent effects on GHG emissions and biodiversity have caused governments to begin the process of transitioning to second generation biofuels created from crop residues or non-edible crops grown on marginal lands.

As this chapter demonstrates, the potential effects of biofuel production can be challenging to measure. The largest landscape effects can result from indirect land use change as demonstrated in the case of Brazilian soybean production affecting the Amazonian rainforest and the Cerrado savanna, and the potential indirect influence of sugarcane production on the Atlantic forest. However, only a small portion of soy oil production is used to make biodiesel, and thus the indirect effect on deforestation of the Amazon is similarly modest. The Brazil case also highlights that some biofuel feedstock production does not result in deforestation (i.e. sugarcane production and the Amazon rainforest). Even in cases of clear landscape change, such as with Indonesian deforestation caused by oil palm plantations, the interplay between food and fuel market end uses make it difficult to directly attribute the effects to biofuel feedstock production.

In order to address challenges such as lessening undesired energy landscape changes, many governments have signaled a desire to shift toward second-generation biofuels. For example, the EU has recently capped the use of first generation biofuels at 7 percent of its overall transportation fuel energy mix. In order to meet its goal of renewables accounting for 10 percent of all liquid transportation fuels by 2020, the use of second-generation biofuels will need to dramatically increase along with other renewable energy transportation technologies. The U.S. has also indicated a desire to increase the use of second-generation biofuels by capping corn ethanol. This shift in policy focus by major biofuel producing and consuming countries, combined with advances in biofuel conversion technologies, creates an environment conducive to transitioning away from food crop based biofuels.

References

ActionAid International USA. 'Fueling the Food Crisis: The Cost to Developing Countries of US Corn Ethanol Expansion'. 2012, at: www.ase.tufts.edu/gdae/Pubs/rp/ActionAid_Fueling_Food_Crisis.pdf (accessed 19 January 2016).

Anderson-Sprecher, A., and J. Junyang. 'People's Republic of China Biofuel Annual: China's 2014 Fuel Ethanol Production is Forecast to Increase Six Percent'. GAIN Report Number: CH14038. Beijing: USDA Foreign Agricultural Service, 2014.

Arima, E. Y., P. Richards, R. Walker, and M. M. Caldas. 'Statistical Confirmation of Indirect Land Use Change in the Brazilian Amazon'. *Environmental Research Letters* 6, no. 2 (2011): 024010.

Bailis, R. 'Brazil: Biodiesel'. In *Sustainable Development of Biofuels in Latin America and the Caribbean*, eds. B. D. Solomon and R. Bailis, 103–26. New York: Springer, 2014.

Carnaval, A. C., and C. Moritz. 'Historical Climate Modeling Predicts Patterns of Current Biodiversity in the Brazilian Atlantic Forest'. *Journal of Biogeography* 35, no. 7 (2008): 1187–1201.

Cashore, B. 'Legitimacy and the Privatization of Environmental Governance: How Non-State Market-Driven (NSMD) Governance Systems Gain Rule-Making Authority'. *Governance* 15 (2002): 503–529

Castanheira, E. G., R. Grisoli, F. Freire, V. Pecora, and S. T. Coelho. 'Environmental Sustainability of Biodiesel in Brazil'. *Energy Policy* 65 (2014): 680–691

CEPEA (Centro de Estudos Avancados Em Economia Aplicada). 'Indicador Soja CEPEA/ESALQ – Paraná'. 2012, at: http://cepea.esalq.usp.br/soja/?page=351&Dias=15#> (accessed 19 January 2016).

Coelho, S. T., and P. Guardabassi. 'Brazil: Ethanol'. In *Sustainable Development of Biofuels in Latin America and the Caribbean*, eds. B. D. Solomon and R. Bailis, 71–101. New York: Springer, 2014.

Danielsen, F., Beukema, H., Burgess, N. D., Parish, F., Bruehl, C. A., Donald, P. F., Murdiyarso, D., Phalan, B., Reijnders, L., Struebig, M., and E. B. Fitzherbert. 'Biofuel Plantations on Forested Lands: Double Jeopardy for Biodiversity and Climate'. *Conservation Biology* 23, no. 2 (2009): 348–358.

de Gorter, H., D. Drabik, and D. R. Just. 'How Biofuels Policies Affect the Level of Grains and Oilseed Prices: Theory, Models and Evidence'. *Global Food Security* 2, no. 2 (2013): 82–88.

Denevan, W. 'The Pristine Myth: The Landscape of the Americas in 1492'. *Annals of the Association of American Geographers* 82, no. 3 (1992): 369–385.

Drajem, M. 'EPA Relents on Ethanol Mandate with Overdue Renewables Quota'. 2015, at: Bloomberg Politics. 29 May www.bloomberg.com/politics/articles/2015-05-29/epa-cuts-mandates-for-corn-ethanol-with-overdue-renewables-quota-ia9q06do (accessed 19 January 2016).

Eastmond, A., C. Garciá, A. Fuentes, and A. Becerril-Garciá. 'Mexico'. In *Sustainable Development of Biofuels in Latin America and the Caribbean*, eds. B. D. Solomon and R. Bailis, 203–222. New York: Springer, 2014.

EIA (Energy Information Administration). 'International Energy Statistics'. 2015, at: www.eia.gov/cfapps/ipdbproject/iedindex3.cfm?tid=79&pid=81&aid=1&cid=r1,r2,r3,r4,r5,r6,r7,&syid=2000&eyid=2012&unit=TBPD (accessed 19 January 2016).

EPA (U.S. Environmental Protection Agency). 'EPA Proposes Renewable Fuel Standards for 2014, 2015, and 2016, and the Biomass-Based Diesel Volume for 2017'. 2015, at: www.epa.gov/otaq/fuels/renewablefuels/documents/420f15028.pdf (accessed 19 January 2016).

European Parliament. 'Directive 2009/28/EC'. 2009, at: http://eur-lex.europa.eu/legal-content/EN/ALL/?uri=CELEX%3A32009L0028 (accessed 19 January 2016).

European Parliament. 'Parliament Supports Shift Towards Advanced Biofuels'. 2015, at: www.europarl.europa.eu/news/en/news-room/content/20150424IPR45730/html/Parliament-supports-shift-towards-advanced-biofuels (accessed 19 January 2016).

Evans, B. 'Biodiesel Industry Seeks to Reverse Argentina Import Decision'. 2015, at: Biodiesel.org, http://biodiesel.org/news/news-display/2015/03/31/biodiesel-industry-seeks-to-reverse-argentina-import-decision (accessed 19 January 2016).

FAO (Food and Agriculture Organization of the United Nations). FAOSTAT, 2015, at: http://faostat.fao.org/site/339/default.aspx (accessed 19 January 2016).

Fargione, J. J., Hill, D. Tilman, S. Polasky, and P. Hawthorne. 'Land Clearing and the Biofuel Carbon Debt'. *Science* 319, no. 5867 (2008): 1235–1238.

Federal Trade Commission. 2006 Report on Ethanol Market Concentration. Report to the U.S. Congress. Washington, DC: Federal Trade Commission, 2006.

Flach, B., K. Bendz, and S. Lieberz. 'EU Biofuels Annual 2015'. GAIN Report Number: NL5028. The Hague: USDA Foreign Agricultural Service, 2015.

Goldemberg, J. 'The Brazilian Biofuels Industry'. *Biotechnology for Biofuels* 1, no. 6 (2008): 1–7.

Hira A., and L. G. de Oliveira. 'No Substitute for Oil? How Brazil Developed its Ethanol Industry'. *Energy Policy* 37, no. 6 (2009): 2450–2456.

IBGE (Instituto Brasileiro de Geografia e Estatística). 'Área Plantada/Colhida'. Instituto Brasileiro de Geografia e Estatística, 2012, at: www.sidra.ibge.gov.br/bda/agric/default.asp?t=4&z=t&o=11&u1=1&u2=1&u3=1&u4=1&u5=1&u6=1 (accessed 19 January 2016).

IBGE (Instituto Brasileiro de Geografia e Estatística). 'Pesquisa Agrícola Municipal'. 2013, at: www.sidra.ibge.gov.br/bda/pesquisas/pam/default.asp (accessed 19 January 2016).

ICTSD (International Centre for Trade and Sustainable Development). 'EU Confirms Duties on Argentine, Indonesian Biodiesel Imports'. 2013, at: www.ictsd.org/bridges-news/bridges/news/eu-confirms-duties-on-argentine-indonesian-biodiesel-imports (accessed 19 January 2016).

Iowa State University. 'Ethanol Usage Projections & Corn Balance Sheet (mil. bu.)'. Agricultural Marketing Resource Center, 22 July 2015, at: www.extension.iastate.edu/agdm/crops/outlook/cornbalancesheet.pdf (accessed 19 January 2016).

Keleman A., and H. G. Raño. 'The Mexican Tortilla Crisis of 2007: The Impacts of Grain-Price Increases on Food-Production Chains'. *Development Practice* 21, nos. 4–5 (2011): 550–565.

Klink, C. A., and R. B. Machado. 'Conservation of the Brazilian Cerrado'. *Conservation Biology* 19, no. 3 (2005): 705–713.

Koh, L. P., and D. S. Wilcove. 'Is Oil Palm Agriculture Really Destroying Tropical Biodiversity?'. *Conservation Letters* 1, no. 2 (2008): 60–64

Lamers, P., Rosillo-Calle, F., Pelkmans, L., and C. Hamelinck. Developments in International Liquid Biofuel Trade. In *International Bioenergy Trade*, eds. M. Junginger and C.S. Goh, 17–40. Dordrecht: Springer, 2014.

Lapola, D. M., R. Schaldach, J. Alcamo, A. Bondeau, J. Koch, C. Koelking, and J. A. Priess. 'Indirect Land-Use Changes Can Overcome Carbon Savings from Biofuels in Brazil'. *Proceedings of the National Academy of Sciences of the United States of America* 107, no. 8 (2010): 3388–3393.

Macedo, M. N., R. S. DeFries, D. C. Morton, C. M. Stickler C. M., G. L. Galford, and Y. E. Shimabukuro. 'Decoupling of Deforestation and Soy Production in the Southern Amazon During the Late 2000s'. *Proceedings of the National Academy of Sciences of the United States of America* 109, no. 4 (2012): 1341–1346

Morton, D. C., R. S. DeFries, Y. E. Shimabukuro, L. O. Anderson, E. Arai, F. del Bon Espirito-Santo, R. Freitas, and J. Morisette. 'Cropland Expansion Changes Deforestation Dynamics in the Southern Brazilian Amazon'. *Proceedings of the National Academy of Sciences of the United States of America* 103, no. 39 (2006):14637–14641.

Nikoloyuk, J. 'Sustainability Partnerships in Agro-Commodity Chains: A Model of Partnership Development in the Tea, Palm Oil and Soy Sectors'. Utrecht-Nijmegen Programme on Partnerships, 2009.

Preechajarn, S., and P. Prasertsri. 'Thailand Biofuels Annual 2014'. GAIN Report Number: TH4057. Bangkok: USDA Foreign Agricultural Service, 2014

RFA (Renewable Fuels Association). 'Biorefinery Locations'. 2016, at: www.ethanolrfa.org/bio-refinery-locations/ (last accessed 19 January 2016).

Richards, P. 'What Drives Indirect Land Use Change? How Brazil's Agricultural Sector Influences Frontier Deforestation'. *Annals of the Association of American Geographers* 105, no. 5 (2015): 1026–1940.

RSPO (Roundtable for Sustainable Palm Oil). 'About'. 2015, at: www.rspo.org/about (accesed 19 January 2016).

Rudorff, B. F. T., M. Adami, D. A. Aguiar, M. A. Moreira, M. P. Mello, L. Fabiani, D. F. Amaral, and B. M. Pires. 'The Soy Moratorium in the Amazon Biome Monitored by Remote Sensing Images.' *Remote Sensing* 3, no. 1 (2011): 85–202.

Runge C. F., and B. Senauer B. 'How Biofuels Could Starve the Poor'. *Foreign Affairs* 86 (May/June 2007): 41–53.

Sawyer, D. 'Climate Change, Biofuels and Eco-Social Impacts in the Brazilian Amazon and Cerrado'. *Philosophical Transactions of the Royal Society of London* B 363, no. 1498 (2008): 1747–1752.

Schouten, G., and P. Glasbergen. 'Creating Legitimacy in Global Private Governance: The Case of the Roundtable on Sustainable Palm Oil'. *Ecological Economics* 70, no. 11 (2011): 1891–99.

Solomon, B. D. 'Biofuels and Sustainability'. *Annals of the New York Academy of Sciences* 1185 (2010): 119–134.

Solomon B. D., and R. Bailis. 'Introduction'. In *Sustainable Development of Biofuels in Latin America and the Caribbean*, eds. B. D. Solomon and R. Bailis, 1–26. New York: Springer, 2014.

Solomon, B. D., J. Barnes, and K. E. Halvorsen. 'Grain and Cellulosic Ethanol: History, Economics, and Energy Policy'. *Biomass & Bioenergy* 31, no. 6 (2007): 416–425.

Solomon, B. D., J. Birchler, S. L. Goldman, and Q. Zhang. 'Basic Information on Maize'. In *Compendium of Bioenergy Plants: Corn*, eds. S. L. Goldman and C. Kole, 1–32. Boca Raton: CRC Press, 2014.

Sparovek, G., A. Barretto, G. Berndes, S. Martins, and R. Maule. 'Environmental, Land-Use and Economic Implications of Brazilian Sugarcane Expansion 1996–2006'. *Mitigation and Adaptation Strategies for Global Change* 14, no. 3 (2009): 285–298.

Tan, K. T., K. T. Lee, A. R. Mohamed, and S. Bhatia. 'Palm Oil: Addressing Issues and Towards Sustainable Development'. *Renewable and Sustainable Energy Reviews* 13, no. 2 (2009): 420–427.

USDA (United States Department of Agriculture). 'The Ethanol Decade: An Expansion of U.S. Corn Production, August 2011, at: www.ers.usda.gov/media/121200/eib79_reportsummary.pdf (accessed 19 January 2016).

USDA (United States Department of Agriculture).'Southeast Asia: Post-2020 Palm Oil Outlook Questionable', 22 Sept 2014, at: http://pecad.fas.usda.gov/highlights/2014/09/SEAsia/index.htm (accessed 19 January 2016).

Wahab, A. G. 'Malaysia Biofuels Annual 2014'. GAIN Report Number: MY4011. Kuala Lumpur: USDA Foreign Agricultural Service, 2014.

Wicke, B., Sikkema, R., Dornburg, V., and A. Faaij. 'Exploring Land Use Changes and the role of Palm Oil Production in Indonesia and Malaysia'. *Land Use Policy* 28, no. 1 (2011): 193–206.

Wright, R. T., and I. Edy Wiyono. 'Indonesia Biofuels Annual 2014'. GAIN Report Number: ID41420. Jakarta: USDA Foreign Agricultural Service, 2014.

WTO (World Trade Organization). 'Indonesia Files Dispute Against EU on Biodiesel', 11 June 2014, at: www.wto.org/english/news_e/news14_e/ds480rfc_11jun14_e.htm (accessed 19 January 2016).

5

Geographical pivots and 21st-century Eurasian energy flows

An energy heartland from the Arctic to Central Asia

Jessica K. Graybill

Enduring geographical notions: pivots and their heartlands

The concept of a geographical pivot located in Eurasia has been around for a little over a century and interest in it has recurred periodically over the last century by geographers, foreign policy experts, and international relations scholars. The ideas of a geographical pivot and a heartland located in Eurasia around which international politics and geopolitical strategizing were spatially centered were proposed by Mackinder in 1904 and 1919, respectively. These maintain a following today, especially related to the imagined spatial relationships in global politics (see, for example, Kearns, 2009; Sengupta, 2009). Specialists attempting to interpret or forecast patterns of socioeconomic and demographic growth and political interest in this region have all invoked Mackinder's notion of a crucial pivot around which urban centers are established, industrial complexes are built, resources are extracted and mobilized in the name of establishing, and expanding the power and prestige of the nation state (Mackinder, 1904, 1919). As Mackinder recognized just over a century ago, the potential for resource development in Eurasia is large, and whoever controls the territory holds a key to geopolitical power because of the importance of resources—especially energy-producing resources such as coal, oil, natural gas and uranium—to our modern world.

The territory considered Eurasia has undergone great change in just one hundred years. Indeed, scholars of this vast region do not agree where, exactly, this region lies and which countries it encompasses. This could be considered attributable to the rapid pace of geopolitical and socioeconomic change in this region, and as ethnic homelands have been colonized by multiple large powers, developed under Communism and now under a mixture of socialism and free market capitalism. As the geopolitical considerations of Eurasia have changed over the course of the last century, so has the imagined location of the pivot, and associated heartland, also changed. Here, I focus attention on Russia and Kazakhstan. While these countries comprise only part of the European and Asian landmass, they were chosen for two reasons: to focus the discussion of pivots and heartlands to better understand their (re)conceptualizations, functions and possible futures; and to explore the contribution of energy resources in the creation of a pivot and heartland region.

In this chapter, I provide a spatially informed chronology of the main contributing theoretical and empirical notions of a Eurasian pivot and heartland. This provides a foundation for understanding how these concepts have informed discussions of what and where a Eurasian heartland might be, their meaning for people and places of Eurasia, and the future of extractive energy resources amidst transforming socioeconomic and biophysical worlds.

Mackinder's Eurasian pivot and heartland

Sir Halford Mackinder has been called a "wise and vigorous founding father" (Hooson, 1962) of geography who emphasized the need for "academic synthesis" (Sengupta, 2009) that would supplant the "old" geography (Mackinder, 1904) that had been predicated on discovery and exploration. Especially, he advocated the need for broad regional understanding, within geography, as we attempt to make sense of physical and cultural regions largely for geopolitical reasons. As such, his interests in geography were broadly based and his contributions attempted to make sense of how physical geographic regions were related to societal and geopolitical concerns. Writing at the turn of the last century, his lasting contributions, especially to political geography and international relations, are the concepts of the geographic pivot and the geographic heartland.

Originally, Mackinder conceptualized the pivot area as a hydrologic-strategic geographic concept to explain the importance of the vast Eurasian steppe region in the mobilization of populations from Asia into Europe. In this way, he related the biophysical to societal and to economic development. The area of the Eurasian pivot, as envisioned by Mackinder, was spatially expansive, encompassing watersheds that began in the central Asian steppe, flowed laterally and northward across the Eurasian continent and drained to the Arctic (Figure 5.1). His major interest was in understanding how waterways historically acted as corridors for mobilization of people. His reason for envisioning a vast pivot region could be understood as his geopolitical concern that a powerful, united heartland, with access to multiple resources (human, industrial, energy) could threaten Western interests across southern Asia and into Central Europe.

Mackinder, in describing the relationship of space and time in the Old World at an historic moment on the edge of great change, divided this region into three zones. From north to south, he inscribed a Eurasian heartland, the inner (marginal) crescent and the outer (insular) crescent. He envisioned Eurasia as:

> A continuous land, ice-girt in the north, water-girt elsewhere, measuring 21 million square miles, or more than twice the area of Europe, have not water-ways to the ocean, but, on the other hand, except in the subarctic forest, are very generally favourable to the mobility of horsemen and camelmen. To east, south, and west of this heart-land are marginal regions, ranges in a vast crescent, accessible to shipmen.
>
> (Mackinder, 1904, p. 431)

This geographic imagination continues to provide a theoretical lens through which much foreign policy has been conceptualized at both regional and global scales Indeed, Mackinder's famous words from *Democratic Ideals and Reality* still resound to inform some geopolitical strategy today: "Who rules East Europe commands the Heartland: Who rules the Heartland commands the World-Island: Who rules the World-Island commands the World" (Mackinder, 1919, p. 150). He noted that the "Soviet heartland is the greatest natural fortress on earth, for the first time in history it is manned by a garrison sufficient in number and quality" (ibid., p. 273).

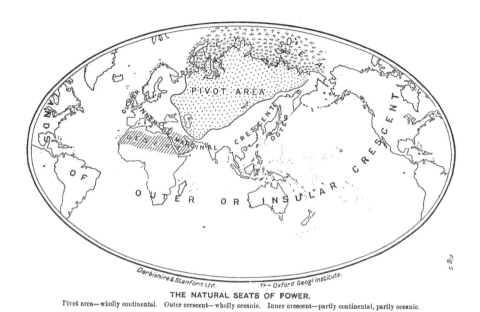

THE NATURAL SEATS OF POWER.
Pivot area—wholly continental. Outer crescent—wholly oceanic. Inner crescent—partly continental, partly oceanic.

Figure 5.1 The geographic pivot and heartland conceptualized by Mackinder (1904)

It seems clear that Mackinder viewed the burgeoning geopolitical power of the nascent Soviet Union with awe and also perhaps with trepidation for what its development could mean for European and Asian relations and the world balance of power. Since that time, the concepts of a geographical pivot and a Eurasian heartland have been seemingly imprinted, almost indelibly, on the geographic imaginary of this massive region from regional and global viewpoints (Smolka, 1938; Brzezinski, 1997). The main geopolitical change that Mackinder that anticipated, in the early twentieth century, was the development of the Eurasian heartland by a powerful nation state that could disturb the existing balance of power. The key was, and remained for several decades, the construction of a cross-continental railroad for the exploration, extraction, refining, and mobilization of Eurasia's virtually untapped industrial and energy resources.

Mackinder's pivot and heartland, geographically located in Central Asia, endure for scholars of Central Asia. Just recently, two entire book-length treatises were published that explore the meaning of these concepts for the Central Asian states in the twenty-first century (Kearns, 2009; Sengupta, 2009). For these authors, the heartland concept informs global geopolitics and the "strategic partnership" (Sengupta, 2009) that the West, especially the US, has made with many Central Asian states in the post-Soviet world. After all, Mackinder's heartland was largely imagined to be located in this part of Eurasia, and to be predicated on the ability of a great power to control resources across this territory. Below, the focus is on understanding Kazakhstan in the heartland, as it is the largest Central Asian state with the most developed energy industry and perhaps indicative of the kinds of geopolitical relationships burgeoning in the entire Central Asian region regarding the development of energy resources.

Three pipelines connect Kazakh oil to Russia: one enters in the southern Volga region and two enter into different parts of Siberia. Built during the Soviet era, these pipelines fueled the eastward expansion of urban-industrial complexes across the Russian part of the Soviet Union. Today, pipelines across Kazakhstan continue to mobilize some of the most abundant energy resources in Eurasia, but the three lines that enter Russia are joined by others that carry oil laterally across Kazakhstan to European markets in the west and Chinese markets in the east. In this sense, the Central Asian heartland endures as a resource-rich region that produces energy in a relatively isolated place and transports it for global markets.

Hooson's post-WWII Soviet heartland

David Hooson, a geographer and Soviet area studies specialist, re-engaged with Mackinder's concept of the Eurasian heartland during the period of post-WWII urban and industrial growth. Hooson's thesis was that a new Soviet heartland was developing along what he called the Volga-Baykal zone, where the Soviet Union's "principal industrial and urban growth poles" were now located (Johnston, 2009, pp. 184). Thus, Hooson reconceptualized the heartland as coinciding not with watersheds as the most crucial corridors of human activity but instead with railroad networks and the urban nodes developing along them. He relocated the heartland, thus, to southern Russia, connecting the European and Asian parts in this imagination. This new, industrial heartland was conceptualized as spreading eastward along the east-west axis of Russia's southern railroad networks (Figure 5.2). Hooson mentions Novosibirsk as a northern point of importance, but does not address Central Asia in this heartland.

Hooson recognized the importance of the urban-industrial complex to the continued economic expansion of the Soviet Union, especially as it advanced away from the traditionally populated core of western Russia into the Ural region, Siberia and the Far East. To contextualize

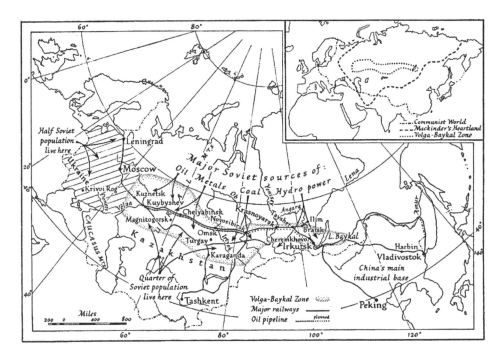

Figure 5.2 The Soviet-era Eurasian heartland as reimagined by Hooson (1964) in which growth across the southern Siberian belt was due to development of urban-industrial complexes and the resettlement of Soviet citizens into newly opened territories

the relative importance of this spatial expansion, he developed the concept of the Effective National Territory, or "that major part of the country which consistently produces a surplus in relation to its population and which, by implication, is therefore supporting the rest of the country in a real sense" (Hooson, 1966, pp. 342–343). He provided six criteria for judging how individual regions contributed to the economic, military and political power of the USSR: scale of a region's contribution to the overall economy; population growth rate, especially in urban areas; accessibility of valuable resources; economic specialization; historical community; and regional ethnic distinctiveness. With this multivariate type of analysis, Hooson accounted for the roles of resources, culture, economy and population size in assessing the Soviet Union's development at a time when most analyses were devoid of detail (Johnston 2009). Using his criteria, he divided the Soviet Union into three regions: the Established European Core (in westernmost Russia), the Expanding Volga-Baykal Zone (from the middle Volga region into the eastern Siberia) and the Marginal Zone (the southernmost USSR, the extreme Far East and the North). In his understanding, the Marginal Zone was comprised of a heterogeneous group of regions that were located across two-thirds of the Soviet area but remained of lesser importance due to low population (less than one-third), lesser resources, and smaller overall productivity (Hooson 1964, pp. 16–17). While his major focus on imagining a Soviet heartland was on the growth of urban-industrial complexes and the resettlement of Soviet citizens into newly opened urban territories, extractive resources—coal, oil, gas—provided the fuel for this expansion and contributed to the industrialization of national economy, and the mobilization of its extractive resources.

In reimagining the heartland in the established Soviet era, there is both continuity and change with Mackinder's original conceptualization. Hooson maintained focus on the importance of railways for mobilizing people, accessing natural resources, and increasing industrial development. Hooson also remained greatly influenced by Mackinder's notion of the importance of locating geopolitical strategy in actual territories. However, he narrowed both the conceptual and spatial understandings of a Soviet heartland as one predicated on urban and industrial growth and located in a specific east-west zone in southern Russia. Hooson's heartland, thus, was not predicated on prior mobilities or colonizations; rather, it re-imagined the Soviet heartland as the spatial expression of an industrially fueled and expansionist eastward gaze. Mobility pathways and resources still played an important role, but the development of territory and access to resources were central to the nation's present and future. Perhaps because the vast territories of the Soviet Union had already been consolidated, he could reflect further on what Mackinder also found important in control of the World-Island:

> The most fundamental of all the elements of national power is sheer location on the globe. Although this is often regarded as absolute and immutable, it is in reality relative to the other parts of the world and its significance changes as they change, and therefore has to be constantly reassessed.
>
> (Hooson 1964, p. 117)

Thus, while the location of Hooson's heartland along the Volga-Baykal corridor re-imagined and re-placed the original heartland, it seemingly still served to connect northern and southern Eurasia.

Bradshaw and Prendergrast's Russian heartland

Revisiting the heartland concept, Bradshaw and Prendergrast (2004) update it based on the new economic, demographic and political context of the early post-Soviet era. Following Hooson

(1964), they analyze only Russia, thereby removing other former Soviet states from their considerations of the locations for and conditions of the heartland. And, just as Hooson's analysis was an exercise in contextualizing development during the Soviet era, so Bradshaw and Prendergrast (2004) also attempted to create a new suite of variables to demarcate a "New Russian Heartland." They did not, however, specifying a physical location, "given the complexity of the current situation" (ibid., p. 87). Instead, they focus on the transformation of the economic sector from a relatively advanced stage of development—especially in Russia's urban- and military-industrial complexes—to one of fragmentation across Russian territory and by type of industry. They suggest that the dissolution of the Soviet Union and subsequent political, economic and social upheaval in Russia were creating even greater conditions for "uneven development, increasing disparities and territorial fragmentation" (ibid.). And, rather than delineate a physical heartland, they note important changes that could herald trends related to urban, industrial and military development in the post-Soviet era.

For example, changes in the industrial sector after the collapse of the Soviet Union did not result in a complete dismantling of industrial activity, as the term deindustrialization would signify. Instead, their more nuanced discussion of the *primitivization* insightfully noted that the partial collapse of industrial activity across the Soviet Union resulted in the "retreat" of the industrial sector into the raw resource base. What this meant was that "the manufacturing economy has experienced significant decline and limited recovery, while the resource economy experienced less significant decline and recently has exhibited appreciable growth" (ibid., p. 91). Industrial primitivization, thus, has had geographical effects, which were largely to increase the economic importance of regions that actively produce and export raw resources. Indeed, it was the peripheral regions that contributed greatly to the early post-Soviet Russian national economy (in Hooson's terminology, the Effective National Territory) and gave Bradshaw and Prendergrast pause in (re)composing a new Russian heartland, both in terms of determining variables and location. The lack of a new map of the Russian heartland for the early post-Soviet period bespeaks a period of upheaval where nothing could be permanently fixed in space given the complexities of socioeconomic and political transformations, including further economic collapse in 1998 and political reorganization of federal territories during the period 1989–2002.

From the viewpoint of the early post-Soviet era, Hooson's Volga-Baykal Soviet heartland did not remain economically important enough to still be considered as a heartland. Bradshaw and Prendergrast's comparison of a population density map of Russia in 2002 (Figure 5.3a) with the outline of Hooson's heartland does indicate the lasting spatial imprint of population expansion eastward, yet also notes the primacy of the Moscow region in western Russia. However, Bradshaw and Prendergrast's spatial analysis of Russia's Gross Regional Product (Figure 5.3b) for 2001 indicates that while population may be concentrated in a western and southern belt across Russia, the economic power now lies to the north and east. That power is in raw resources, especially mineral and energy resources. They conclude that the

> New Russian Heartland . . . is structurally and geographically quite different from Hooson's New Soviet Heartland of the 1960s. It is a product of Russia's natural resource wealth and export potential rather than its domestic industrial capacity. Consequently there is a clear mismatch between those regions that carry the greatest economic weight and those that are the most populous
>
> (Bradshaw and Prendergrast, 2004, p. 96)

Indeed, examination of the highest amounts of Foreign Direct Investment (FDI) in Russia's regions indicate that the highest rates (over 1 bln USD) from 2006–2009 were invested in only

(a)

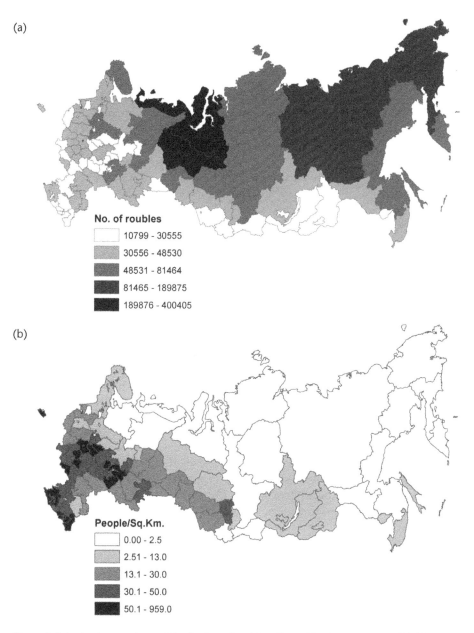

No. of roubles

- 10799 - 30555
- 30556 - 48530
- 48531 - 81464
- 81465 - 189875
- 189876 - 400405

(b)

People/Sq.Km.

- 0.00 - 2.5
- 2.51 - 13.0
- 13.1 - 30.0
- 30.1 - 50.0
- 50.1 - 959.0

Figure 5.3 Lack of an early post-Soviet heartland is reflected in the maps provided by Bradshaw and Prendergrast (2004) that indicate (a) disparate concentrations of rubles (and thus economic investment) in Russia in 2001 and (b) population across Russia's regions in 2002

four highly disparate geographic regions (oblasts): Moscow, St. Petersburg, Archangelsk (along the western Arctic White Sea) and Sakhalin (in the Far Eastern sub-Arctic Sea of Okhotsk), where "the majority of FDI is geographically confined to a handful of regions that are attractive through their market size and the wealth of their natural resources" (KPMG, 2013, p. 17).

Thus, foreign entities investing in Russia saw potential in Russia's two largest cities, in the largest existing Arctic port and in a site of offshore hydrocarbon production. While FDI is only one measure of a specific type of economic performance, a spatial understanding of where it is occurring today provides context for envisioning its currently fragmented spatial and resource-dependent heartland.

Bradshaw and Prendergrast's suggest that three critical factors—demographics, economic and internal political realignment—will continue to influence the ongoing transformation of the Russian Federation. They identified four challenges to Russia's future development: the Moscow effect, the missing millionaires, the Siberian dilemma and fragile borderlands. The first two points are closely related: as Moscow continues to act like a primate city, it draws labor and creativity away from other large urban centers (with a million-plus inhabitants), effectively rendering their continued or re-emergent development less effective and relegating them to the status of less significant regional centers unable to pull out of the orbit of existing large cities. The third and fourth points are also important for contextualizing the primitivization of industry because many of the border regions are "demographically stressed and economically underdeveloped" and "include the majority of Russia's ethnic republics" (Bradshaw and Prendergrast, 2004, p. 118). While many of the fragile borderlands are located in the southernmost regions of the country, some are in northern and eastern Russia precisely where reliance on continued and possibly increasing export of energy resources are anticipated for national-level development.

Antrim's Artic pivot

With growing global awareness of the transformations wrought by climate change across the Arctic, comes increasing attention to this region's extractive resource potential, particularly in offshore oil and gas, and possibilities for new marine shipping routes. While the potential for development of resources and places exists in multiple Arctic regions, it is most latent in the Russian North, as it is the least developed and most spatially extensive Arctic and sub-Arctic territory. Antrim turns to Mackinder's original concept of the pivot to explain why the physically increasing access to the Arctic region and its resources in the twenty-first century is of paramount importance for Russia's development. She writes that the "Arctic has played an essential, yet unrecognized, role as the northern wall in the Western strategy to enclose and contain the world's largest land power" and that Mackinder's proposed "natural conflict between the landlocked Eurasian heartland and the Western maritime nations" is at a historical turning point as the Arctic is transformed by new mobilities of people and resources via human labor, energy resource networks and nodes that are open to the northern seas (Antrim, 2010, p. 15).

Thus, as a marine political analyst, Antrim's geographical pivot for the coming twenty-first century is located in the Russian Arctic, where largely unexplored on- and offshore energy and mineral resources are abundant and the possibility of developing navigation along Eurasia's Arctic coastline and extensive river networks is increasing due to socioeconomic and biophysical transformations. Antrim locates a heartland in northern Eurasia (i.e., Siberia) that is of both economic (for extractive and agricultural potential) and geostrategic interest (Figure 5.4). Antrim proposes that geopolitical interests of the twenty-first century will differ from those of Mackinder's and Hooson's eras, in that Russia will no longer be considered a landlocked nation struggling with the concerns of distance, mobility and accessibility that encumbered its development through the last century. Instead, the technological capabilities of the modern era will unlock the again re-located Eurasian heartland, for which mineral and energy resources will become the major driver of socioeconomic and infrastructural development.

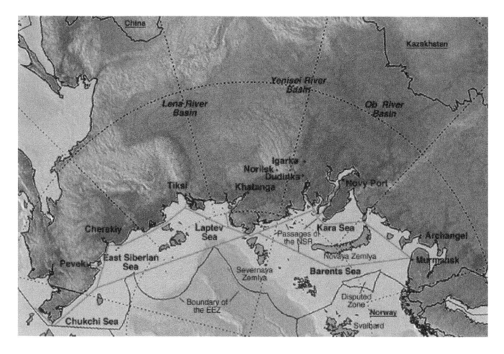

Figure 5.4 The coming Arctic heartland as conceptualized by Antrim (2010)

In this geographic imagination, the development of Russia's Arctic coastline in Eurasia becomes a major export route for resources extracted in northern Eurasia. Largely due to climate change and predicated on continued warming, transformations of the Eurasian Arctic will make this new heartland no longer a "'black hole' in the center of Eurasia" where "the geopoliticians' 'heartland' had been suddenly yanked from the global map" (Brzezinski, 1997, p. 87). Instead, Antrim proposes that in a "new geostrategy, Russia assumes a role as one of the maritime powers of the 'rimland,' and the Russian Arctic becomes a new geographical pivot among the great powers" (Antrim, 2010, p. 16). While this status has not yet been realized, the adoption of new policies and economic directives from the political center (Moscow) suggest that Russia is carefully considering the Arctic opportunities afforded them by climatic and geopolitical shifts in the twenty-first century.

Part of that shift is a re-imagination of Russia as strong a maritime state. This shift, Antrim writes, would reduce Russia's geographic isolation and also "require Russia to become more closely integrated into global commercial and financial networks, to welcome international business involvement, and to participate in international bodies that harmonize international shipping, safety, security, and environmental regulations" (Antrim, 2010, p. 16). Thus, potentially the "Great Game" of the twentieth century is pivotally refocused on extractive resource potentials and re-located to the area surrounding Mackinder's "Icy Sea."

Locations, functions and futures: energizing the pivot and heartland

From approximately a century of engagement with the concepts of the geographical pivot and the heartland, it is clear that its importance to scholars concerned with the Eurasian landmass, from Central Asia to the Arctic, is enduring. However, while there has been a continuous interest in this concept regarding its implications for geopolitical relations and regional-national scale development, its functional meaning and imagined locations have changed. In other words, the pivot and the heartland have been reimagined conceptually and spatially by their functions and futures rather than by a static, physical location. Essentially, the focus has increasingly become on identifying *what* is functionally crucial (e.g., territorial expansion and consolidation, connective corridors, urban growth, resource export) to the ongoing development of Eurasia during particular development regimes (e.g., Tsarist era, Soviet or post-Soviet) and how that intertwines with international relations and geopolitical strategizing.

For example, Mackinder noted in his 1904 essay that certain geopolitical ideas act to form a nation at certain moments in history. He noted that "the idea of England was beaten into the Heptarchy by Danish and Norman conquerors; the idea of France was forced upon competing Franks, Goths and Romans by the Huns at Chalons, and in the Hundred Years' War with England . . ." (ibid., p. 423). He suggests that, when attention is attuned to the outcome of a civilization, we lose sight of the "more elemental movements" of people over space and time. While Mackinder's argument was developed to understand how physical geographic features (watersheds) connected multiple ethnicities over time and Eurasian space, the point remains the same in the modern era: the mapping of the location and territorial belonging to social, cultural or economic groups is something we imagine. How people and resources belong and have been connected in the "geographically charmed sanctuary in the Eurasian interior" (Hooson, 1962, p. 20), then, are continually (re)constructed as in the attempt to make sense of socioeconomic, political and biophysical transformations and their spatial expressions across a given landscape.

Reconceptualizations of the Eurasian heartland's location, function and future indicate continued interest in the ongoing and sometimes precipitous changes in socioeconomic and political developments within Eurasia over the past century. Below, I discuss three important contextual concerns for understanding why the location, functions and potential futures of the heartland have continued to evolve: socioeconomic development, resource mobilization, and ethnic homelands. At the core of each lies a preoccupation with energy resources.

Socioeconomic development

Less than a century ago, the Tsarist Empire still existed. During that era, as Mackinder and many others have noted, the dominant mode of development in lesser-developed places in Russian-controlled Eurasian territories was geologic and geographic exploration and discovery. Exploration was often as elemental as physically mapping landscapes and natural resources in lesser-known territories to understand the potential for human colonization and resource extraction, both for the benefit of the dominant Russian state. While explorers had already discovered and indeed even colonized parts of what we now call Central Asia, the vast territories of Siberia, the Arctic, and the Far East remained less known, especially away from coastal regions. Just a few decades later, that mode of knowing places and peoples had been drastically altered due to the expansion of a federated Union of Soviet Socialist Republics across Eurasian space. In the Soviet incursion across Eurasian space, massive urbanization and industrialization moved like a tsunami wave rolling eastward, southward and northward from the western core. The Trans-Siberian railway corridor and its branches connected west to east and served as a transcontinental lifeline for moving people, goods and raw materials, thereby consolidating the Soviet Union's economic power and expanding an empire based on industrial growth and resource extraction.

For Mackinder, writing before the completion of the Trans-Siberian railway, this development was embryonic but he recognized the potential for changing Eurasia's development trajectory. Half a century later, eastward expansion of cities and transportation corridors—all built on the promise of natural resource extraction—is the trend that Hooson found to be most concentrated along the Volga-Baykal urban-industrial corridor. In the post-Soviet era, the development potential of much of Eurasia has again become latent rather than active due to systemic collapse of the Soviet Union and ongoing socioeconomic and political transformations in Russia and the newly independent states. Interest in and knowledge about the spatial extent of natural resources, especially energy resources, is the greatest it has ever been for this region but extraction and development of them is emerging only slowly across Eurasian territories.

At its core, colonization of and control over Eurasian space by Tsarist Russian, Soviet and post-Soviet nation states have been predicated on the discovery and exploitation of mineral and energy resources. Extraction of natural resources has remained at the core of Eurasia's economy even with the cessation of eastward urbanization and industrialization and the ensuing demographic transformation, largely characterized by overall migration from the east and concentration of population in large urban centers everywhere (Heleniak, 1997). Due to primitivization, the focus turns away from manufacturing of goods to the extraction and export of unprocessed raw resources. Well documented in the resource peripheries of eastern Russia for energy resources (Bradshaw, 2006; Graybill, 2013), these processes continue across many regions of the former Soviet Union and include the removal and export of a wider range of resources, including forest, fish and wildlife (Newell, 2004; Holzlehner, 2006). Some urban and industrial places have managed to rise out of economic collapse, especially where the transition to profit-oriented economic structures have been assisted by Western or Asian backing. Commonly, this occurs as urban or industrial infrastructure is (re)developed, as in the peripheral Archangelsk and Sakhalin regions in Russia or in the case of the building of Astana, the new capital of Kazakhstan. Established in 1997, a new location and architectural style for the new capital was undertaken ostensibly—and ostentatiously—to attract global investors into the very "heart" of the heartland (Koch, 2013).

Russia and Kazakhstan increase their power on the global economic stage as they seek to modernize infrastructure in existing oil and gas fields, such as the Tengiz in Kazakhstan, Yamal in the Arctic and across multiple locations in Siberia. There has also been entrance into agreements with multinational oil and gas corporations to develop newer sites, such as Kazakhstan's Karachaganak and Kashagan fields, Kovykta and other eastern Siberian fields, as well as offshore fields near sub-Arctic Sakhalin Island or Arctic Yamal Peninsula. Despite the drop in price per barrel of oil in 2014, the extractive resource potential of these regions is still considered as a mainstay of the regional and national economies. Partially, this is because of a century of investment in these resources and the lingering notion that heavy industry remains a viable path forward for the national state (Alekperov, 2012; Gustafson, 2012). Another lingering notion is the idea of authoritarian control over land and resources: President Putin is characterized as a leader seeking total state power, which is "the precondition for stability and economic growth, and control of natural resources is the precondition of state power" (Gustafson, 2012, p. 249). Additionally, the culture around the development of oil and gas is deeply ingrained in Russian and Kazakh mythology about the source of power and wealth in these nation states. After all, "it is oil, then as now, that makes the weather in the Russian economy" (ibid., p. 185). Indeed, younger generations on Sakhalin Island consider themselves to be the *zolotie deti*, golden children, of that region due to the potential for continued hydrocarbon extraction in this place for at least their lifetimes (Graybill, 2009).

In both countries, but especially in Kazakhstan, development of energy resources includes mining of uranium for nuclear power (EIA, 2015). Additionally, the development of renewable energy resource infrastructure is nascent but beginning to take hold in some regions. For examples, solar energy technology is garnering more interest in Kazakhstan (Karatayev and Clarke, 2014) and former President Medvedev promoted renewable energy development beginning in 2008. In both countries, part of the Soviet legacy of urban-industrial development was the creation of large and widespread dams for hydroelectric power production. Generally, due to the abundance of hydrocarbon resources—coal, oil and gas—and the presence of hydroelectric power stations near large urban centers in both countries has caused sluggish development of other non-fossil fuel energy sources.

Mobilization of energy resources

Current economic growth in Russia and Kazakhstan depends on the continued export of raw or only lightly manufactured resources for their participation in the global economy. Without the means to export resources, these Eurasian countries will stagnate, due to the lack of sufficient internal markets for energy resources and the lack of modern factories for refining them prior to export. This is precisely the reason for international investment in Russia's urban and industrial infrastructure in resource regions, such as Archangelsk and Sakhalin. In these regions, European and Asian companies have gambled in their investments, but are betting on the world's continued reliance on hydrocarbons to fuel our global futures. While some resource-rich locations in Eurasia are beginning to be developed via fly-in fly-out (FIFO) operations, most remain tethered to existing urban-industrial nodes in the transportation networks that exist, due to multiple and intertwined social, economic, and political barriers. These relate strongly to the Soviet-era command economic system and its associated "economy of favors," which revolved largely around the social-political networks in specific places (Ledeneva, 1998). For resource development, these places were often precisely the urban-industrial centers that remain important today (e.g., Yuzhno-Sakhalinsk in Sakhalin region) within the oil and gas industry. From these nodes of social-political power transportation networks, such as pipelines, are also controlled, especially by Transneft, Russia's monopolistic pipeline operator (Alekperov, 2012).

Thus, the most developed places in the Russian Arctic, Siberia and Kazakhstan remain the urban-industrial complexes that were promulgated during the Soviet era. The ones that have had the greatest possibilities for continued growth into the post-Soviet are those located nearest to transportation networks that have international reach. This includes railways and pipelines across southern parts of Eurasia and, increasingly, marine ports that can assist the movement of raw resources globally. Infrastructure development remains key to resource mobility along the entire pathway from extraction to export. Thus, imagining a Eurasian heartland comprised of urban and industrial nodes that are connected by transportation networks is useful for understanding how all of the different kinds of resources—human, manufactured, raw—are mobilized along railways, in pipelines and via seaports for the purposes of regional and national development. This imagination emphasizes that any place that is exporting raw natural resources is part of the effective national territory of Russia or Kazakhstan, to use Hooson's terms. The dilemma, extending Bradshaw and Prendergrast's arguments about Russia, to also include Kazakhstan, is the need "to find a new way of developing those resources, a way that is economically viable and ecologically sustainable" (Bradshaw and Prendergrast, 2004, p. 116) to avoid future mis-steps for the long-term future of these regions.

Largely because of the need for modernized transportation networks, the new model for Eurasian development—especially in the Russian Arctic and Siberia—is assumed to be "capital and technology-rather than labor-intensive". This is the Siberian dilemma: the cost of development will only "amplify the imbalance between those regions that contribute to the national economy and those regions that house the bulk of Russia's population" (ibid., p. 116).

In Kazakhstan, President Nazarbayev has attempted to side-step regional inequality by redistributing wealth derived from western Kazakhstan's hydrocarbon resources nationwide (Ostrowski, 2010).

Ethnic homelands

The Arctic, Siberia and Central Asia contain numerous ethnically diverse populations for many of whom entire territories are intricately woven together with concepts of ethnic and/or national identity. For example, Bradshaw and Prendergrast (2004, p. 98) note:

> Identity conceptions in Russia operate at a variety of spatial scales; they can be primarily civic, ethnic, or regional but all have an integral territorial dimension. Some of those identity arenas . . . inhabit a space larger than the state, be they anti-national (for instance, global), supra-national (Orthodox, Islamic or Soviet), or trans-national (Eastern-Slavic, Pan-Mongolian, or Eurasian). Others nestle within the boundaries of the state (sub-national), and may be intra-regional (Siberian of North Caucasian), singularly regional (Tatarstani or Muscovite), or subregional (Cherkess or Lezgin). Many of these identity conceptions crosscut one another and, to varying degrees the boundaries of Russia civic conception.

This excerpt suggests that, in Russia, ethnicity has long been intricately intertwined with specific territories, and the creation of somewhat autonomous ethnic homelands at the beginning of the Soviet era. The most volatile ethnic regions lie in the southwestern part of Russia, such as in Chechnya and Dagestan, where ethnic-based conflict has existed for centuries and has largely been related to devastating colonization and other territorial conflicts.

However, indigenous residents in Siberia and the Arctic are also no longer passively accepting extractive industrial incursions into their ethnic homelands. In the early 1990s, anthropologists and indigenous activists worked to raise awareness about the lack of protected rights to ancestral territories and culturally important natural resources (e.g., caribou, salmon, whales) the indigenous peoples of the Russian Far East and North (e.g., Pika and Grant, 1999). Attention to their plight and incursions in their ancestral landscapes by extractive industries continued throughout the 1990s and into the twenty-first century as Western and Asian scholars were granted increasingly open access to formerly closed territories in the far eastern and northern border zones. Perhaps the most publicized and researched case study of the contested landscape of indigenous rights to land and resources in the midst of energy resource development is the transnational development of the offshore oil and gas fields in the Sea of Okhotsk associated with the Sakhalin-2 project (e.g., Wilson, 2003; Bradshaw, 2005; Graybill, 2008). In this location, until the 2006 takeover of majority stakeholder shares by Russian national companies, environmental and cultural activists strongly contested the development of onshore infrastructure, including an onshore pipeline through salmon watersheds and a liquefied natural gas export terminal (Rutledge, 2004), and offshore infrastructure, such as the platform for the rig designed for the Sakhalin-2 project sited in the breeding grounds for the Korean Grey Whale (Sakhalin Environmental Watch, 2008). At the same time that international communication of environmental and cultural conditions for ethnic minorities and indigenous groups became more widely transmitted, representatives of indigenous groups were more easily able to travel abroad to participate in meetings with native peoples, cultural activists or environmentalists from other places who encouraged them to resist territorial exploitation without receiving benefits from transnational corporations or the Russian state. Such interactions continue to provide fresh energy for communicating the importance of ethnic or cultural territories and fighting for rights to resources and landscapes (i.e., personal communications on Sakhalin Island and Kamchatka Peninsula with multiple cultural activists).

Before being delineated by the Soviet Union as the Kazakh Soviet Socialist Republic in 1936, the territory in Kazakhstan was home to multiple ethnic groups, mostly Turkic-speaking nomadic and semi-nomadic tribes. Today, over 100 ethnic groups comprise the population of Kazakhstan. And, for the majority ethnicity Kazakh, there is also historic kinship division among three major tribes, or *zhuz*, that comprise Kazakh culture. Geographically, each subethnic group had its own territory: the Small Zhuz in the west, the Middle Zhuz across the middle and east and the Great Zhuz in the southeast. These kinship ties and their geographical imprint on the Kazakh cultural landscape remain strong and pervade everyday social, economic and political realities of today. Even seventy years of Soviet rule did not eliminate the kinship-based ties among ethnic Kazakhs, and the subethnic clans (re)emerged in the post-Soviet period. Beginning "in the 1990s, the subethnic pedigree of political actors in Kazakhstan became a critical factor in the allocation of political resources (especially posts in the state bureaucracy) and economic resources (especially access to revenues from extractive industries)" (Schatz, 2005, p. xviii). Control over the possibility of infighting among the *zhuz* has been theorized as important for control over the extractive resource economy from local-national scales in Kazakhstan. Indeed, some experts suggest that much of President Nazarbayev's internal geopolitical strategy has been created to minimize the possibility of dissent among any one *zhuz* (Schatz, 2005; Aitken, 2012). Rather, he has worked, especially within Kazakh industries, to form ties that bind the three major clans together economically and politically (Schatz, 2005; Ipek, 2007).

Also, of concern, are the relations between Russia and Kazakhstan, who share a lengthy border and, in some cases, uneasy relations due to Soviet-era "progressive" development. For example, the human health and environmental catastrophes experienced because of Soviet agricultural practices in the Aral Sea region, space flights from the Baykonur Cosmodrome, and nuclear testing in Semipalatinsk continue to impact local and regional populations and landscapes. Perhaps, the rupture in political relations between Russia and Kazakhstan means that the Eurasian heartland is no longer unified or isolated. Instead, it has been incised by a new border and accompanied by the possibility of new geopolitical relations related to energy development. Thus far, Kazakhstan and Russia have maintained good relations, largely for economic reasons (Aitken, 2012). New authoritarian myth-making and architectural monument-building from the new, post-independence capital would have the world imagine this nation as a peace-promoting entity that desires cooperative multilateral international relations. In the current scenario, Russia is part, but not sole, beneficiary of Kazakh economic development.

An energy resource heartland?

Is it possible to envisage an energy heartland from Central Asia to the Arctic via eastern Russia (through Siberia)? If such a territory could be circumscribed, where exactly would it exist? Would it again be imagined to encompass the vast territory of Mackinder's imagination? Or would it be more confined, perhaps to a north-south path from Central Asia through Siberia to the Russian Arctic? What would be its meaning(s), its function(s) and possible future(s)? Would the heartland relate to population concentrations, industrial development, or the export of natural resources? These questions are, of course, rhetorical. Asking them, however, stresses the import-ance of understanding the impact of different historic moments, and modes, of development on space, place and peoples. In the last century, Eurasian places have moved quickly through multiple stages of development in both market and command economic systems: conquest and colonization, industrialization and urbanization, primitivization and (partial) collapse, and finally, in some cases, redevelopment. This has occurred not only in Russia, but in all of the former Soviet states. As the second largest territory of the former Soviet nations, and coming in second in Gross Domestic Product in 2013, Kazakhstan remains an important part of the Eurasian heartland and an important pivot, especially for other Central Asian states as they continue to develop multiple kinds of resources (e.g., energy, agricultural and human).

One thing that is glaringly obvious from the exploration of the Eurasian heartland is that scholarship about and understandings of Russia and Central Asia have become largely divorced from each other. With the dissolution of the Soviet Union and the reinvention of this region of area studies inquiry, Russia and Central Asia became separate paths of inquiry for scholars. Thus, understanding of how the originally imagined Eurasian heartland, all of which experienced the same type of twentieth century development during their greatest urban-industrial development leap forward, diverged greatly. What this means is that understandings about cross-border flows, especially regarding energy resources, has lessened. Knowing that Central Asia, like much of eastern Russia, is rich in industrial and energy resources, and knowing, from looking at maps, that Kazakhstan is poised as the largest, northernmost Central Asian country with three operational oil pipelines into Russia, our geographic imaginaries have become limited in terms of what Central Asian energy resources do, and could, mean for a Eurasian energy heartland. The increasingly northern imagination of the Eurasian heartland neglects the important role that Central Asia has played in Russian development and vice versa. The important and shared development of these regions for the last century continues today in the form of international relations between the two countries, largely over the future of extractive industrial resource mobility. And, while perhaps not as long lived as the ancient cultures of the region, the legacy of Soviet development in Central Asia is lasting as a physical imprint on the built, urban, rural and development landscapes of the region.

It would serve scholars interested in the concepts of geographical pivots and heartlands well to "re-assert the importance of 'thick' regional geographical knowledge in the face of 'thin' universal theorizing about world affairs" (Toal, 2003, p. 655). In this way, our understanding of what, and where, the Eurasian heartland may be becomes grounded in geographical knowledge of our human and biophysical surroundings, all of which inform, transform and energize the worlds in which we live.

References

Aitken, J. *Kazakhstan: Surprises and Stereotypes After Twenty Years of Independence.* New York, NY: Continuum, 2012.

Alekperov, V. *Oil of Russia: Past, Present and Future.* Minneapolis, MN: East View Press, 2012.

Antrim, C. 'The Next Geographical Pivot: The Russian Arctic in the Twenty-First Century.' *Naval War College Review* 63, no. 3 (2010): 15–38.

Bradshaw, M. J. 'Environmental Groups Campaign against Sakhalin-2 Project Financing. Pacific Russia Oil and Gas.' *Report* 3 (2005): 13–18.

Bradshaw, M. J. 'Russia's Oil and Gas: State Control, the Environment, and Foreign Investment.' *World Today* (2006): 18–19.

Bradshaw, M. J. and Prendergrast, J. 'The Russian Heartland Revisited: An Assessment of Russia's Transformation.' *Eurasian Geography and Economics* 46, no. 2 (2004): 83–122.

Brzezinski, Z. *The Grand Chessboard.* New York, NY: Basic Books, 1997.

EIA. 'U.S. Reactor Operators Increase Uranium Purchases from Kazakhstan.' 2015, at www.eia.gov/todayinenergy/detail.cfm?id=23212 (accessed 27 August 2016).

Graybill, J. K. 'Regional Dimensions of Russia's Transformation: Hydrocarbon Development and Emerging Socio-Environmental Landscapes of Sakhalin Island. Kring Beringia: Expeditioner och folk.' *Geography and Anthropology Yearbook of the Swedish Society for Anthropology and Geography* No. 127 (2008): 66–97.

Graybill, J. K. 'Places and Identities on Sakhalin Island: Situating the Emerging Movements for "Sustainable Sakhalin".' *In Environmental Justice of the Former Soviet Union.* Agyeman, J., Ogneva-Himmelberger, E., eds. Boston, MA: MIT Press, 2009.

Graybill, J. K. 'Mapping an Emotional Topography of an Ecological Homeland: The Case of Sakhalin Island, Russia.' *Emotion, Space and Society* 8, (2013): 39–50.

Gustafson, T. *Wheel of Fortune: The Battle for Oil and Power in Russia.* Cambridge, MA: Belknap Press, 2012.

Heleniak, T. 'Internal Migration in Russia During the Economic Transition.' *Post-Soviet Geography and Economics* 38, no. 2 (1997): 81–104.

Holzlehner, T. *Shadow Networks: Border Economies, Informal Markets, and Organized Crime in Vladivostok and the Russian Far East.* Fairbanks, AK: University of Alaska Fairbanks, 2006.

Hooson, D. M. 'A New Soviet Heartland?' *The Geographical Journal* 128, no.1 (1962): 19–29.

Hooson, D. M. *The Soviet Union: A Systematic Regional Geography.* London, UK: University of London Press, 1966.

Ipek, P. 'The Role of Oil and Gas in Kazakhstan's Foreign Policy: Looking East or West?' *Europe-Asia Studies* 59, no. 7 (2007): 1179–1199.

Johnston, R. 'David Hooson: Political Geography, Mackinder and Russian Geopolitics.' *Geopolitics* 14, (2009):182–189.

Karatayev, M. and Clarke, M. L. 'Current Energy Resources in Kazakhstan and the Future Potential of Renewables: A Review.' *Energy Procedia* 59, (2014): 97–104.

Kearns, G. *Geopolitics and Empire: The Legacy of Halford Mackinder.* Oxford, UK: Oxford University Press, 2009.

Koch, N. 'The "Heart" of Eurasia? Kazakhstan's Centrally Located Capital City.' *Central Asian Survey* 32, no. 2 (2013): 134–147.

KPMG. 'Investing In Russia: An Overview of the Current Investment Climate in Russia. KPMG Miejburg & Company.' 2013, at: www.kpmg.com/NL/nl/IssuesAndInsights/ArticlesPublications/Documents/PDF/High-Growth-Markets/Investing-in-Russia1.pdf (accessed 14 January 2016).

Ledeneva, A. *Russia's Economy of Favours: Blat, Networking and Informal Exchange.* Cambridge, UK: Cambridge University Press, 1998.

Mackinder, H. J. 'The Geographical Pivot of History.' *Geographical Journal* 21, no. 21 (1904): 421–437.

Mackinder, H. J. *Democratic Ideals and Reality.* New York, NY: W.W. Norton, 1962. (Originally published in 1919.)

Newell, J. *The Russian Far East: A Reference Guide for Conservation and Development.* McKinleyville, CA: Daniel and Daniel, 2004.

Ostrowski, W. *Politics and Oil in Kazakhstan.* New York, NY: Routledge, 2010.

Pika, A. and Grant, B. (eds.) *Neotraditionalism in the Russian North: Indigenous Peoples and the Legacy of Perestroika.* Seattle, WA: Edmonton Canadian Circumpolar Institute, 1999.

Rutledge, I. *The Sakhalin II PSA—A Production 'Non-Sharing' Agreement: Analysis of Revenue Distribution.* Prague, CZ: CEE Bankwatch Network, 2004.

Sakhalin Environmental Watch. 'Sakhalin II Victory—Who is Now Prepared to Touch Beleaguered Project?'. 2008, at: www.sakhalin.environment.ru (accessed 20 December 2015).

Schatz, E. *Modern Clan Politics: The Power of 'Blood' in Kazakhstan and Beyond.* Seattle, WA: University of Washington Press, 2005.

Sengupta, A. *Heartlands of Eurasia: The Geopolitics of Political Space.* Plymouth, UK: Lexington, 2009.

Smolka, H. P. 'Soviet Strategy in the Arctic.' *Foreign Affairs* 16, no. 2 (1938): 272–278.

Toal, G. 'Re-asserting the Regional: Political Geography and Geopolitics in a World "Thinly-Known"'. *Political Geography* 22, (2003): 653–655.

Wilson, E. 'Freedom and Loss in a Human Landscape: Multinational Oil Exploitation and Survival of Reindeer Herding in North-Eastern Sakhalin, the Russian Far East.' *Sibirica* 3, no. 1 (2003): 21–48.

6

Mobile energy and obdurate infrastructure

Distant carbon and the making of modern Europe

Corey Johnson

Introduction

This chapter considers the oft-used trope of Europe's "dependence" on Russia for energy, particularly natural gas, in historical and geographical contexts. Rather than focusing on Russia as a supplier, as many analyses do, my entry points are in Europe itself as a massive consumer of hydrocarbon energy. Rather than placing states at the center of the analysis, it considers multiple geographic scales to understand linkages: the household, the urban, the regional and the planetary. In short, I argue that energy dependence cannot be understood without understanding energy's mobility, and thus energy and transport—especially, in the case of gas, via pipeline— must be considered in tandem (Seow 2014). Furthermore, Europe's reliance on distant carbon is part and parcel of planetary urbanization (Merrifield 2013) and an outcome of processes of concentration and extension (Brenner 2013) at multiple temporal and spatial scales. This has many implications. For example, the oft-heard aspiration of turning on a dime—kicking out Russian carbon in favor of better, less politically tainted carbon from Qatar, domestic shale rock, or the US—runs up against the obduracy of the networked infrastructure that constitutes the system. Seeking relief in renewable energy—certainly a worthwhile goal—similarly is only a partial fix in the medium term due to the sheer complexity of an energy system that has developed over centuries.

Curiously, the history and geography of the socio-technical system of natural gas in the US is much better understood than the European (see, e.g., Nye 1998; Herbert 1992; Makholm 2012), in spite of the fact that gas is far more politicized in Europe. The EU is by its own reckoning the "world's largest energy importer" (European Commission 2015), but discussions of energy dependence in popular media consistently end in hackneyed geopolitical discussions of individual states being reliable or unreliable suppliers, thereby falling into the "territorial trap" (Agnew 1994) that many scholars of the geography of energy have been critiquing in recent years (see, e.g., Bouzarovski et al. 2015). In particular, there is a need to consider the imbrication of energy infrastructure in territory and consider the "socio-technical assemblage" that complicates overly simplified "states vs. markets" duality common in discussions of energy governance (ibid.).

As this book makes abundantly clear, energy infrastructures comprise the largest economic undertaking in world history. The transitions described in this chapter, however, while politically sensitive at times, and highly visible in certain moments and in particular places, have largely occurred without the amount of attention one might expect for such a massive undertaking. The growing spatial ambit of energy provision, particularly as it relates to building natural gas into the energy system of Europe, has entailed geographical questions at the heart of this volume: from local to regional and in some cases even global webs and networks that move hydrocarbon molecules from source to flame with remarkable expediency. Although largely invisible, the mere fact that very few Europeans actually awaken wondering if the supply of energy to fire their boilers, cook their food, or heat their water will suffice for that day. This tends to obscure the technological marvel that emerged in a relatively short time to enable those molecules to move through these spaces, and it obscures the political geographies such infrastructures transcended—not always smoothly—to bring us where we are today.

In that vein, the first half of the chapter covers the energy transitions that eventually led to natural gas—a fuel once considered more or less useless because it was more tethered to geography than other energy sources—becoming now the second most important source of primary energy in the EU, surpassed only by petroleum in the energy mix. The second half of the chapter draws on geography, urban studies and science and technology studies to argue for how scholars might better conceptualize the material geographies of interdependence that arise out of the development of networked architecture of natural gas that built a virtually uninterrupted conduit from a hot water spigot in Berlin to a well in West Siberia.

Overcoming energy's limits

The relationship between European energy consumers and distant carbon was one born of necessity: large scale industrial capitalism, the mass movement of people to dense cities and the energy intensive lifestyles that cities engendered, and internal combustion engines rapidly depleted local energy sources. As Wrigley (2013) has argued, the industrial revolution was in large part an energy transition. Prior dependence on local forests for charcoal and firewood as supplements to animate energy sources (an *organic energy regime*) would have to give way to make possible the concentration and up-scaling of production. This was both a time-scale and geographic-scale problem: reliance on biomass outputs made possible by photosynthesis on a limited area of land over the course of a year simply could not sustain the geographically concentrated, energy-intensive industries and lifestyles that industrial capitalism created (Huber 2009). The *mineral energy regime* that took form in geographically specific contexts, such as the British Midlands and later the Ruhr River region of Germany, freed energy consumers from the temporal limitations of the organic regime because it could depend on millions of years of stored photosynthesis in the form of coal, then later oil and methane (Jones 2014). During the early part of the Industrial Revolution, intensification of production and urbanization made possible by the mineral energy regime were geographically coincident with the sources of energy, such as the coal fields of northern England, the Ruhr basin, or eastern Pennsylvania. Freed from Jevons' "laborious poverty" of the organic energy regime, productivity shot up, living standards increased, and economic growth could increase vastly because the system was freed, for a time at least, from natural constraints on energy production that yearly cycles of photosynthesis imposed (Wrigley 2013). As the title of historian David Landes' classic book on the Industrial Revolution suggests, the "unbound Prometheus" radically transformed life in much of Europe (Landes 1969).

This energy transition created "landscapes of intensification" – cities that were home to factories and populations living ever more modern and consumptive lifestyles; places of energy and raw material extraction; and the infrastructure (roads, railroads, canals, and later pipelines and electricity lines) that tethered the system together (Jones 2014; Hughes 1983). Importantly, it was this last piece–the transportation networks–that freed the sites of energy consumption from the sites of energy production (mines, wells, etc.). While energy intensity was increasingly dramatically, local sources of carbon energy became ever scarcer, requiring that if growth were to be sustained, energy would need to be brought in from farther away. It was much easier to move energy from source to consumer than it was to move the factories, cities, and labor pool to the sources of energy. This scaling up of the energy catchment area had two consequences. First, it allowed intensive and extensive urbanization to continue *in situ*, in spite of the lack of locally available energy sources. Second, it allowed the industrial and energy revolutions to spread to places that did not have local stocks of coal or oil. While the first places in Europe to develop heavy industries were those close to coal seams and orebodies, eventually nodes of economic intensification could be found farther and farther afield from necessary raw materials.

At this point it is important to note what many scholars of mobile energies and large technological systems have already noted. The energy transitions of the past two hundred years were not simply about technological innovations exploiting natural endowments at the service of economic needs. Rather, these were highly socially mediated transitions (Hughes 1983; Coutard 1999; Graham and Marvin 2001; Bijker et al. 1987). Development and innovation, growth and competition, and the momentum of a large system once in place are not the byproducts of self-organization but rather of actor networks and human decision-making that, for a variety of complex reasons, maintain the momentum of a particular system once in place (see chapters by Hughes and Callon in Bijker et al. 1987) and lead over time—slowly and with much resistance—to the introduction of new technologies and energy sources. Factors such as vested interests of actors and sunk costs also contribute to momentum (ibid.). The choice of preferred fuel, followed by the investment of large sums of capital in building up that system, create a form of path dependency not entirely unlike what the seemingly arbitrary choice of a railroad gauge created (Puffert 2009). This is certainly the case for natural gas, as will be explored below. From a geographical perspective, it merits mention that this entire complex, expensive, environmentally problematic system exists entirely to move vast stores of subterranean carbon from one part of the Eurasian landmass to another. The spatial dislocation of carbon is an aspect of the energy transition that merits further critical attention.

Urban lifestyles and socio-technical systems

The massive growth of cities across Europe in the 19th century geographically involved both concentration and extension. In considering networked infrastructure, re-working centuries of built environment in historic city cores around the possibilities of mass transit, mass consumption, and "modern" amenities could only happen incrementally, at great cost, and only by overcoming built-in resistance to changing lifestyles and consumption habits. So it was in the extended city where the earliest and clearest evidence of the energy transition described above is to be found (to accompany this brief summary, see Osterhammel 2014). Streetcars, commuter railways, and, eventually, automobiles substituted for the pedestrian life of the old urban core. Energy-intensive iron and brick could be turned into ever larger housing, and larger housing units could be warmed by piped steam from boilers heated by fossil fuels. Water, in turn, could be piped into homes and sewage taken away by a different pipe. Mechanical pumps did much of the work, aided by gravity. Gas lighting, which had first been used to lengthen the work day in textile factories, saw increasing use in street lamps, theaters, and starting in the 1880s in Britain, in home heating, cooking, and lighting (ibid.).

Although gas was making inroads into the urban energy system in European cities at the end of the 19th century, it was electricity that would come to dominate household lighting. Early gas was manufactured locally, often from coal, and this manufactured gas, or "light gas" or "town gas," was costlier than the natural gas from wells that tapped subterranean stores of methane (*Erdgas*, or earth gas, in German) that became predominant in the second half of the 20th century in Europe. Although electricity was not without its own risks, it came to be viewed as safer than gas for domestic uses such as lighting since gas could explode and poison. Electric lights could be turned on and off in an instant. Many households in cities across Europe came to depend on both electricity and town gas to meet the various household needs of lighting, heating, cooking, etc. (Leuschner 2009). The provision of gas only made economic sense in the eyes of the private gas companies in densely populated cities, since not only did town gas need to be manufactured but it could not be transported (yet) over long distances. Town gas, as the name suggests, was only suitable in urban areas where a profit could be made producing and distributing it, and thus the distinctively urban consumption patterns that developed in northwest Europe in the early twentieth century were at least in part shaped by the energy sources that were available to household consumers.

In his book *Cities of Light and Heat*, historian Mark Rose chronicles the adoption of gas and electricity in Denver and Kansas City during the late 19th and early 20th century (Rose 1995). He calls the boosters of technologies that used electricity and gas "agents of diffusion"; these agents, which included power company owners, appliance salespersons, real estate developers, and others, were instrumental in making certain types of consumption indispensable to the urban household. Highly gendered marketing campaigns implored housewives to "cook with gas!" while others attempted to alleviate commonly held fears about electric clothes irons by proclaiming that a new iron would "remove the feeling which tangles nerves and tires bodies" (ibid.: 86). These agents of diffusion made the non-vital seem essential: instant hot water at the turn of a knob, uniform heat that did not require constant attention, irons, automatic washing machines, "ice boxes" that did not require delivered ice. Without these conveniences, modern life was not possible. Similar marketing was happening in European cities, as electric and gas companies competed for customers in an environment in which the entire pie of energy consumption was growing, meaning that a transition from gas lighting to incandescent lightbulbs only meant that consumers would need to be persuaded that everyone needed a home hot water heater that burned gas.

By the early 20th century, gas was firmly woven into the urban metabolism of most northern European cities. At the household scale, piped town gas that had once provided light was increasingly used for heating and cooking. At the urban scale, an extensive and expensive infrastructure of gas plants, gasometers for storage and pressure maintenance, and a pipeline network to move it to consumers was now largely in place. As cities continued to grow, there was little question that new homes would be connected to gas and electricity service because that is what it meant to be urban. At some point, however, the costs of locally manufactured gas would become too high, just as the requirements of firewood and charcoal to sustain urban growth had been outstripped in the late 18th century (Kim and Barles 2012). The sheer size and consumptive appetite of the modern European city met the natural limits imposed by a highly localized or regionalized regime of energy provision.

It was actually France, not typically thought of as a gas innovator, where long distance transport of gas first entered the picture in Europe. In the 1950s, a 312 km long pipeline, the so-called "eastern artery," was constructed to supply the Paris region with gas manufactured in the coking plants of industrial Lorraine (Beltran 1992). Attention then turned to the Lacq region in southwest France, where oil exploration had yielded the discovery of a large deposit of natural gas. For

France, the construction of new pipelines to move natural gas to markets in Paris, Lyon, and Nantes marked an important milestone in several respects. First, this was the first time France had essentially a national network of gas distribution, instead of the polycentric town gas model. Here "national" must be qualified, since the provision of gas was still focused on larger cities. Second, it marked the transition away from manufactured gas to natural gas (ibid.). This required household energy transitions as well, since natural gas had different properties to the then customary town gas, including approximately double the heat content (Heymann 2012). Appliances would need to be replaced to accommodate the more potent fuel. But gas was by now a widely accepted energy source, and households were more than willing to assume the expense of transition given the benefits of gas over other energy sources in household applications. Households and industry together were the largest consumers, but the French electricity monopoly burned around one-third of the gas in generating plants. By late 1960s France still had no nuclear generating capacity.

While not matching the sheer profligacy of American energy consumption, post-World War II growth in Europe was fed by energy: the energy consumption of European member states of the OECD as measured in metric tonnes of oil equivalent (mtoe) roughly doubled from 1960 to 1973 (Clark 1991). As for natural gas as an important part of the energy mix, the discovery of the supergiant gas field near Groningen, Netherlands, in the early 1960s and the increasing appreciation of gas as a clean, efficient source of energy set into motion events that would create an increasingly cross-border, Europe-wide transmission system (Bouzarovski et al. 2015). These dynamics would also fairly quickly necessitate looking beyond domestic sources to meet increasing demand as the following section explores.

Before turning to Russia's role in the European energy system, it may be useful here to provide a bit of a conceptual mop-up. The exponential growth of energy demand in European cities encompassed nearly all aspects of urban life: brick and steel for buildings, concrete and asphalt for roadways, lights, elevators, refrigerators, space heating, factories—all of these were energy intensive, and none of them was considered optional. Despite economic analysis that treats some modern energy uses as intensive and others as not, when compared to the organic energy regime *all* modern energy uses are intensive, relying on million-year time scale processes of subterranean carbon concentration and storage to power consumptive lifestyles. Part of the "landscapes of intensification" in urbanized Europe was an increasingly complex set of infrastructures at the household, urban, regional, and increasingly supra-regional and planetary scales, that moved hydrocarbons from source to flame. Importantly, as the scale of energy provision increased, more and more capital was sunk in long distance infrastructure to move those carbon molecules over longer distances meaning that "distant carbon," as I have called it here, resulted in a form of path dependency in the system and the *obduracy* of the system made wholesale shifts in energy provision more and more difficult (Johnson 2014; Hommels 2005).

This "background of technology" (Verbeek 2005) is not completely invisible, and the 20th-century consumer recognized that the massive, unsightly gasometer was related to her having a warm living room. But the background of technology certainly became taken-for-granted, which in itself acts as a form of obduracy in the system. The modern energy regime requires little or no labor for the consumer, is seldom interrupted, and costs a small fraction of the middle-class household's income to operate. With practically no one taking note, the supply area of natural gas for European cities grew and grew. Pipelines were built over the 20th century that extended the network over an ever-larger territorial extent, precisely to maintain that dependability and low cost at the core of what it means to be urban, Western, cosmopolitan, etc. As energy systems extended, inevitably they would—and did—run up against geopolitical realities, whether the fraught politics of the Mideast or the Iron Curtain.

Carbon mobility and the role of Russian gas

Part of the common narrative that I want to disrupt with this work is that "country A is dependent on country B." The point of departure of neatly contained territorial units is alluring, since political decisions that impact energy use often do happen in state institutions and much of the data available is aggregated at the level of the state. As I hope the first part of this chapter has shown, though, energy systems operate at multiple scales, and the "territorial trap" (Agnew 1994) and accompanying state fetish impoverishes our ability to constructively engage with the topic of our dependency on distant carbon.

Important recent work by Högselius (2013) has greatly enriched our understanding of the development of energy infrastructures in Eurasia. He upends the metaphor of a politically motivated "energy weapon" being wielded against innocent, helpless Europeans by chronicling the "hidden integration" of Europe that was occurring during the height of the cold war through networks of natural gas infrastructure. For the purposes of this chapter, the essential point is that the nature of post-war European economic systems—on both sides of the Iron Curtain, but especially in the West—required the scaling up of energy provision that resulted in the current system of long-distance carbon mobility.

The 1960s was a key decade for what I will refer to as the *natural gas transition* in Central Europe. Most of West Germany depended on coal, petroleum, and town gas for home heating and industrial applications. Coal was abundant in the industrial northwest of the Federal Republic of Germany (FRG), but the coal industry was suffering from competition from imported petroleum, while areas without coal but with growing political and economic weight—mainly Bavaria and Baden-Württemberg in the south—were loath to depend on "imported" coal from northwest Germany to fuel their growth. At the household level, consumers in the coal-rich industrial region around the Ruhr River were already well accustomed to gas, which was supplied to them by coal syndicates. In a sense, coking plants contributed to the demise of the residential market for coal by building the infrastructure to find a market for methane, which was a byproduct of the coking process that turned mined anthracite coal into a product more usable in the steel industry (Leuschner 2009). Ruhrgas, founded in 1926 by coking conglomerates and now part of the energy company E.ON, was a pioneer in bringing the urban gas infrastructure to households. That company, headquartered amidst the most productive coal-producing region in Europe, also eventually became the largest purchaser of Russian natural gas.

The scaling up of gas provision in the 1960s involved political and economic decisions at several levels. With the discovery of giant natural gas fields in the Netherlands (Slochteren) and Algeria (Hassi R'Mel), politicians and business executives were actively toying with the idea of how to move gas over long distances. This was not just a political or economic question but also a technological one: gas under pressure required strong steel pipes that did not leak and technologically sophisticated pumps to maintain pressure along the route of the pipe. As the decade progressed, sustained post-war economic growth continued while events such as the 6 Days War in the Middle East called into question the reliability of oil supplies from there. The Soviet Union, meanwhile, had discovered that it sat on huge reserves of natural gas; in fact, they were the largest reserves in the world. Despite early skepticism among Soviet leadership of building out a large scale gas network, and owing largely to the boisterous promoting of gas by Alexei Kortunov, director of the USSR's gas directorate Glavgaz, the Soviet's committed to a strategy of gas exports (Högselius 2013). From the very early stages, Soviets set their sights not only on their allies in the Warsaw Pact as the primary export markets, but also western European countries. Not only would capitalist countries provide much needed sources of hard currency, but also access to the steel pipe technology that would be required for long distance, high pressure movement of natural gas (ibid.).

Growing interest in natural gas was not just about political decisions being made by national governments in the context of cold war geopolitics. In the Netherlands, for example, the discovery of vast quantities of natural gas near Groningen in 1959, caused the two main shareholders in the exploration company, Exxon and Shell, to have internal discussions about how best to "create" a market for this newfound natural wealth. While Shell was primarily interested in supplying large-scale users, such as electricity generators and industry, Exxon argued for a strategy that would reshape the household market by convincing users to use gas as their *primary* energy source, instead of a mix of town gas, heating oil, and coal, which was then customary (Correljé et al. 2003). This involved some risk, since large scale investments in infrastructure would need to be made *and* consumers would need to be persuaded of the need to buy expensive new household appliances or retrofit their existing ones. But that is precisely what happened (ibid.). A high-pressure network was constructed, with the help of US engineering firm Bechtel (the US had long since made the conversion to natural gas, and most of the world's expertise was thus in American companies). Conversion workshops were set up in municipal gas plants, where existing stoves and cooktops were tweaked to allow the burning of natural gas with its higher heat content than the coal-derived town gas. Marketing campaigns touted the clean, efficient, and space-saving advantages of natural gas heat over coal and heating oil. By the end of the 1960s, 80 per cent of Dutch households had gas service and 60 per cent were heating with it, a dramatic change in less than a decade (ibid.). Similar sorts of transitions were happening in Germany even without the large domestic supply enjoyed by the Netherlands.

In Central Europe 1968 was an important year: the Prague Spring was brutally crushed by Soviet tanks in August, and a mere ten days after tanks rolled into Czechoslovakia, the Austrian minister of transport and the Soviet gas minister stood atop a pipe in Baumgarten and ceremoniously cranked open the valve that allowed the first "red gas" to enter Austria from Czechoslovakia (Högselius 2013). Baumgarten is still among Europe's most important physical gas hubs, and it is a key node in the Central European Gas Hub, through which much of the gas in Central Europe is traded (Heather 2012). (Incidentally, the ceremonious valve opening or button pushing has been repeated in various locations since, such as in 1973 when German and Soviet authorities pushed the "red button" starting gas shipments and most recently in 2011 when Angela Merkel and Dmitry Medvedev, along with the prime ministers of the Netherlands and France, were on hand to open the Nord Stream pipeline by spinning a wheel.)

The early years of gas deliveries were not without difficulty, but it was not because of politically motivated supply shutoffs. Rather, the Soviets were unable to deliver promised quantities of gas due to technical snafus and the fact that much of the gas in the first years was coming not from Siberia or the Caspian, but rather from Ukraine. Knowing that being seen as undependable would put at risk the prospects of the USSR making further inroads into western European markets, the Soviets instead cut deliveries in Ukraine by approximately the same volume of gas that had promised to Austria and Czechoslovakia (ibid.: 100–101). Ukraine was therefore subjected to rationing of Ukrainian-produced gas (from Galicia) in the late 1960s and early 70s so that Soviet contracts with Austria could be more or less fulfilled.

These early delivery problems were short lived. Long-distance pipelines from West Siberia fed copious quantities of gas into the system, while in Galicia, on Ukraine's border with Poland and Czechoslovakia, new storage capacity came online (the largest gas storage facilities in Europe, in fact, are in western Ukraine). As a result, the Soviet Union turned into what Russia continues to be to this day, at least to its west European customers: a very reliable supplier of large quantities of natural gas. By the mid-1970s, in addition to the Warsaw Pact countries, West Germany, Finland, France, and Italy were receiving natural gas from the USSR. Turbulence in global oil markets as a result of oil embargoes further solidified the role of "blue gold" in the European energy system, and an ever-larger share was coming from the large fields of the North Sea, North Africa, and the Soviet Union. Large-scale pipelines were under construction or planned, nearly all using German-built steel pipe, to transport gas from increasingly remote areas of Eurasia to consumers in urban Europe.

Even though 2014 and 2015 saw less of Russia's gas transiting Ukraine, much of the difference was simply made up by increased volumes in the new Nord Stream pipeline that links Russia and its largest EU consumer, Germany, under the Baltic Sea. The Nord Stream was completed long before Russia's annexation of Crimea and the ongoing hostilities in eastern Ukraine, but the project was explicitly about avoiding transit risks (including price disputes, conflict, theft, politically motivated shutoffs, etc.) that could interrupt smooth deliveries of hydrocarbons that have flowed practically without interruption between Russian gas fields and homes and companies in Central Europe since the 1960s (Johnson and Derrick 2012). The political elite in Russia understand that becoming unreliable to "*dickes Deutschland*" ("fat Germany," as a very senior German diplomat told me in an interview several years ago) is in no one's best interest, in spite of the ethno-nationalist war of aggression being waged by Russia in Ukraine. As many observers have noted, the EU's rather feckless response to Russia's involvement in the Ukraine crisis can be tied to Europe's dependence on Russia's natural resources. I would frame it somewhat differently: it is tied to the political calculus of how European voters would react if the heating bill for the flat in Munich suddenly doubled or tripled.

Linking scales: toward a planetary urban Europe?

Only by challenging the state-centric understandings of energy provision and consumption can we adequately come to terms with the role infrastructure plays in shaping modern life—and contemporary politics—in Europe. To tie together the histories and geographies of the *natural gas transition* and scaling up of energy provision with relevant geopolitical questions, I turn here to an emerging body of work that challenges the tidy conceptual lines between urban geography on the one hand, and increasingly global processes of resources, markets, and mobilities on the other. There is growing interest among historians of technology and some geographers in networked infrastructure in Europe, particularly in light of the six-decade European integration project. Much of this work points to increasingly transnational, long distance material infrastructures that constitute an important, if largely "hidden," form of territorial integration (Badenoch and Fickers 2010; Misa and Schot 2005). This line of inquiry is a welcome development, because it explicitly asks us to think outside of the territorial boxes when considering European integration (statist, territorial-entrapped analysis). This growing body of work at the intersection of science and technology studies, geography, and history has indeed largely informed this chapter. However, I wish to push the scalar imaginaries beyond just thinking about how "Europe"—as messy that term is—is integrated in ways that push the borders of territorial Europe as well as the borders of what integration means.

In the previous sections, I provided an examination of Europe's natural gas transition: how the requirements of industrializing, urbanizing Europe resulted over time in the creation of a geographically expanding network of infrastructure to provide especially urban spaces with energy, and how natural gas became a key constituent part of the energy equation as the spatial scale of energy provision increased. In this section I provide a charcoal sketch of how we might conceptualize the linkages between scales, drawing in particular on recent work under the broad rubric of *planetary urbanization*. This emerging set of interventions in urban theory seeks to destabilize the category of analysis and practice "urban" by suggesting that the distinction does not capture how in fact urbanization works, namely as a process and set of interactions between economy, nature, politics, and social life that transcends purported boundaries between, for example, cities and countryside (Brenner 2014). This work builds on Henri Lefebvre's argument that society had become thoroughly urbanized (Lefebvre 1970), and that capitalist urbanization is characterized by a dialectic of implosions and explosions (concentration and extension) that draw not just surrounding hinterlands into capital accumulation and the spatial division of labor, but rather every corner of the planet.

It is in this vein that planetary urbanization offers a window into relating the seemingly mundane act of hooking a new home in Leipzig, Lille, or Sofia up with gas service to the exhumation of Mesozoic-era hydrocarbon molecules in West Siberia. Brenner and Schmid observe that spaces well beyond what are traditionally thought of as urban and suburban spaces are tied into the global urban fabric through infrastructure that includes natural gas pipelines and a host of other infrastructures (Brenner 2014).

Thinking through this temporally as well as spatially, a conceptual thread can be pulled from the diminishing returns of Britain's forests in the lead up to the Industrial Revolution, to the shift from local coal to gas, nuclear, oil, and renewables (with their longer supply chains) well over a century later. At each stage in the energy transitions, the spatial ambit of the energy source grew. In a lengthier exploration of this topic, here might be a good point to turn to the field of Urban Political Ecology and consider the ecological footprints of urban metabolisms (see, e.g., Luke 2003). For now I would simply like to suggest that planetary urbanization offers some possibilities for thinking through how the political geography of energy networks has been formed over time through the outcome of processes happening at multiple spatial scales, mediated by consumption habits, political decisions, and nature's possibilities, and that there is considerable momentum built into the energy system once in place.

Thinking about the issues presented in terms of the planetary urban is also useful because it helps us to conceptualize the transitions of energy systems and infrastructure. For example, oil and coal can be readily put on a ship and transported around the world; as a consequence, more or less global markets have emerged for those commodities. Natural gas, by contrast, is still mostly transported by pipeline and therefore is largely traded in regional markets by contractual relationships rather than spot markets (Freifeld 2009). Liquefied natural gas (LNG) transported by ship is growing in importance but is still very expensive when compared to, for example, conventional piped gas from Russia. Three developments are moving gas towards a more globalized commodity: consumption of energy increasing rapidly in areas outside of the traditional big markets in North America and Europe, the diminishing returns in conventional sources of gas (North Sea, Russia, Netherlands), and the development of new, more challenging hydrocarbon sinks, such as tight gas, oil sands, shale oil, etc. (Johnson and Boersma 2015). As a more globalized market for gas emerges, a more variegated and global network of energy provision to Europe will likely also emerge.

Conclusion

This chapter examined the temporal and spatial aspects of the natural gas transition in Europe by thinking through the linkages between the geographic scales of the household, the urban, the regional, and the planetary. It argues that the geography of natural gas provision in Europe has been both a product of, and a creator of, modern, urban life as it has developed over the past century. However, as the expanding network of pipelines that link, for example, West Siberian gas fields to homes and businesses in Germany, Italy, etc., shows, the "urban" cannot be separated or considered fundamentally separate or distinct from regional or even planetary scales that increasingly make up Europe's energy infrastructure network.

The implications of this for research and policy making are numerous. Researchers must be mindful of the spatiality of energy production of consumption, and integrate microscale (household) analysis into geopolitical analysis. As this chapter shows, the two are inseparable. As I have argued elsewhere (Johnson 2014), the obduracy of the energy system makes political action designed to reduce dependence on suppliers such as Russia difficult and costly. It also illustrates, though, that political solutions require thinking through the spatiality of the system. Assuming natural gas is to remain a major part of the energy system in Europe, alternate entry points into the system, whether from the Caspian, North Africa, global LNG flows, etc., can serve to challenge the unidirectionality of energy flows that has resulted in a rather perverse politics of dependence between Europe and Russia over the last several decades. There are also very serious policy questions that must be asked about the sustainability of a model based on "distant carbon" – there is abundant evidence that the planet is ill equipped to see the European or North American scaling-up of energy provision replicated in other parts of the world, yet this is precisely what is happening when, for example, China signs megadeals with Russia, Turkmenistan, and other suppliers for natural gas, or when hydraulically fractured natural gas from the Eagle Ford shale formation is liquefied and shipped to Brazil (Grattan 2016).

It is interesting to witness the growing calls for an Energy Union for Europe then, in a sense an attempt to reterritorialize energy markets and networks at the scale of the EU. An Energy Union has been proposed many times, and in some ways is foundational to the entire European integration project since the Coal and Steel Community is commonly thought of as the precursor to the EU. A common energy policy is still elusive, but may come to pass given recent events in eastern Europe. What it will not change, however, is the underlying networked infrastructure that transcends, even spites, political boundaries as it continues to grow in extent. Nor will it change the lifestyles and livelihoods of modern, urbanized Europe that create appetites for distant carbon at practically any cost.

References

Agnew, J. A. 1994. The territorial trap: the geographical assumptions of international relations theory. *Review of International Political Economy*, 1(1), 53–80.

Badenoch, A., and A. Fickers eds. 2010. *Materializing Europe: Transnational infrastructures and the project of Europe*. Basingstoke, UK: Palgrave Macmillan.

Beltran, A. 1992. The French gas network and new technologies since 1946. *History and Technology, an International Journal*, 8(3–4), 263–273.

Bijker, W. E., T. P. Hughes, and T. J. Pinch eds. 1987. *The Social Construction Of Technological Systems: New Directions In The Sociology And History Of Technology*. Cambridge, MA: MIT Press.

Bouzarovski, S., M. Bradshaw, and A. Wochnik. 2015. Making territory through infrastructure: the governance of natural gas transit in Europe. *Geoforum*, 64, 217–228.

Brenner, N. 2013. Theses on Urbanization. *Public Culture*, 25(1), 85–114.

Brenner, N. ed. 2014. *Implosions/Explosions: Towards A Study Of Planetary Urbanization*. Berlin: Jovis.

Clark, J. G. 1991. *The Political Economy Of World Energy: A Twentieth-Century Perspective*. Chapel Hill, NC: University of North Carolina Press.

Correljé, A., C. van der Linde, and T. Westerwoudt. 2003. *Natural Gas In The Netherlands: From Cooperation To Competition?* Amsterdam: Oranje-Nassau Groep.

Coutard, O. 1999. *The Governance of Large Technical Systems*. London: Routledge.

European Commission. 2015. Framework Strategy for a Resilient Energy Union with a Forward-Looking Climate Change Policy. Brussels: European Commission. 25 February COM/2015/080 final. Available at: http://eur-lex.europa.eu/legal-content/EN/TXT/?uri=COM%3A2015%3A80%3AFIN.

Freifeld, D. 2009. The great pipeline opera. *Foreign Policy*, 174, 120–127.

Graham, S., and S. Marvin. 2001. *Splintering Urbanism: Networked Infrastructures, Technological Mobilities and the Urban Condition*. London; New York: Routledge.

Grattan, R. 2016. Cheniere's Sabine Pass to ship first cargo later today. *FuelFix*, 24 February. Available at: http://fuelfix.com/blog/2016/02/24/chenieres-sabine-pass-to-ship-first-cargo-later-today/.

Heather, P. 2012. Continental European Gas Hubs: Are they fit for purpose? Oxford, UK: The Oxford Institute of Energy Studies, Document NG 63, June. Available at: www.oxfordenergy.org/wpcms/wp-content/uploads/2012/06/NG-63.pdf [accessed: 10 October 2015].

Herbert, J. H. 1992. *Clean Cheap Heat: The Development Of Residential Markets For Natural Gas In The United States.* New York: Praeger.

Heymann, M. 2012. Natural gas. In *Berkshire Encyclopedia of Sustainability*, edited by D. E. Vasey, S. E. Fredericks, S. Lei and S. Thompson. Great Barrington, MA: Berkshire, 312–318.

Högselius, P. 2013. *Red Gas: Russia and the Origins of European Energy Dependence.* 1st. ed. New York: Palgrave Macmillan.

Hommels, A. 2005. *Unbuilding Cities: Obduracy in Urban Socio-Technical Change.* Cambridge, MA: MIT Press.

Huber, M. T. 2009. Energizing historical materialism: fossil fuels, space and the capitalist mode of production. *Geoforum*, 40(1), 105–115.

Hughes, T. P. 1983. *Networks of Power: Electrification in Western Society, 1880–1930.* Baltimore, MD: Johns Hopkins University Press.

Johnson, C. 2014. Geographies of obdurate infrastructure in Eurasia: the case of natural gas. In *Eurasian Corridors of Interconnection: From the South China to the Caspian Sea*, edited by S. M. Walcott and C. Johnson. New York: Routledge, 110–129.

Johnson, C., and T. Boersma. 2015. The politics of energy security: contrasts between the United States and the European Union. *Wiley Interdisciplinary Reviews: Energy and Environment*, 4(2), 171–177.

Johnson, C., and M. Derrick. 2012. A splintered heartland: Russia, Europe, and the geopolitics of networked energy infrastructure. *Geopolitics*, 17(3), 482–501.

Jones, C. F. 2014. *Routes of Power, Energy and Modern America.* Boston, MA: Harvard University Press.

Kim, E., and S. Barles. 2012. The energy consumption of Paris and its supply areas from the eighteenth century to the present. *Regional Environmental Change*, 12(2), 295–310.

Landes, D. S. 1969. *The Unbound Prometheus: Technological Change and Industrial Development in Western Europe from 1750 to the Present.* London: Cambridge University Press.

Lefebvre, H. 1970. *La révolution urbaine.* Paris: Gallimard.

Leuschner, U. 2009. *Die deutsche Gasversorgung von den Anfängen bis 1998.* Available at: www.udo-leuschner.de/basiswissen/SB100-002.htm.

Luke, T. W. 2003. Global Cities vs."global cities": rethinking contemporary urbanism as public ecology. *Studies in Political Economy*, 70, 11–33.

Makholm, J. D. 2012. *The Political Economy of Pipelines: A Century Of Comparative Institutional Development.* Chicago, IL: University of Chicago Press.

Merrifield, A. 2013. *The Politics of The Encounter: Urban Theory And Protest Under Planetary Urbanization.* Athens, GA: University of Georgia Press.

Misa, T. J., and J. Schot. 2005. Introduction. *History and Technology*, 21(1), 1–19.

Nye, D. E. 1998. *Consuming Power: A Social History of American Energies.* Cambridge, MA: MIT Press.

Osterhammel, J. 2014. *The Transformation of the World: A Global History of the Nineteenth Century.* Princeton, NJ: Princeton University Press.

Puffert, D. J. 2009. *Tracks Across Continents, Paths Through History: The Economic Dynamics of Standardization in Railway Gauge.* Chicago, IL: University of Chicago Press.

Rose, M. 1995. *Cities of Light and Heat: Domesticating Gas and Electricity in Urban America.* University Park, PA: Penn State University Press.

Seow, V. 2014. Fuels and flows: rethinking histories of transport and mobility through energy. *Transfers*, 4(3), 112–116.

Verbeek, P.-P. 2005. *What Things Do: Philosophical Reflections on Technology, Agency, and Design.* University Park, PA: Penn State University Press.

Wrigley, E. A. 2013. Energy and the English Industrial Revolution. *Philosophical Transactions of the Royal Society A: Mathematical, Physical and Engineering Sciences*, 371(1986).

7

Emerging countries, cities and energy

Questioning transitions

Sylvy Jaglin and Éric Verdeil

Introduction

This chapter presents the main conclusions of two comparative research programmes[1] exploring urban energy transformations, specifically energy for buildings and economic activities. Their starting point is rooted in a body of work dedicated to energy transition, a notion extensively employed in academic circles and, since the 1992 Rio Summit, increasingly associated with cities: "Cities, as entities within which an ever-larger share of energy is used, are seen as simultaneously constituting a key target of such an energy transition, as well as a key 'instrument' in delivering it" (Rutherford & Coutard 2014, p. 2). The emphasis placed on the growing role of cities in the transformation of energy systems historically dominated by national players (Thorp & Marvin 1995; Bulkeley et al. 2010; Hodson & Marvin 2010) raises questions about the specifics of urban transition pathways and their geographical distribution (Bouzarovski, Introduction to this volume). The aim of the programmes was thus to explore the factors of change observable at urban scales and the main actors holding a "vision" or a strategy for energy in order to understand if and how transformations are framed, devised and implemented in relations with urban concerns and to assess the capacity for action of urban authorities. This capacity needs to be gauged in relation, on the one hand, to the traditional energy sector operators, often maintaining close links with central governments, and on the other hand, to the private players which have been strengthened by reforms in the sector (liberalisation, unbundling, privatisation or opening up to private investors). Globally, the question is about the emergence of a territorialisation of energy issues at metropolitan scales and, in this chapter, about the specificities of this process in the context of cities in emerging countries. As pointed in the introduction of the book, such metropolises must be placed in the agenda of future research. That is urban entities where development is taking place in conditions marked by a combination of high economic growth, strong integration into globalised markets and robust institutional know-how (Sgard 2008; Piveteau & Rougier 2010). Urban energy demand here is strong, driven both by high urban growth rates and by rising consumption among the urban middle classes.

The chapter offers a synthesis of in-depth empirical analyses exploring these questions in four big cities in emerging countries (Buenos Aires, Delhi, Istanbul, Cape Town) and a number of secondary cities (Sfax in Tunisia, Turkish cities). These cases were chosen for contingent reasons, including the familiarity the researchers involved in the programmes had gained thanks to previous research. But as will be explained further, the research in these cities clearly shows that urban energy issues are subjects to very different framings strongly influenced by local and place-specific concerns. In view of this and considering the uneven capacity and effectiveness of local urban action in determining the scope and nature of the cities' commitment, the chapter argues against the hypothesis of a convergence of developments towards a model of "energy transition" as set out, for instance, in national sectorial policies (see the main characteristics of the case studies in Table 7.1). The analysis also critically reviews the assumption of a growing role for local authorities in energy governance (see also Rutherford & Jaglin 2015).[2] Although cities are not passive in response to the ongoing changes and to the tensions and contestation which materialise in urban spaces in relation with the politicisation of energy issues, the chapter suggests that energy transition is not the primary focus of urban governance in cities of emerging countries, which must address context-specific priorities pertaining to broader perspectives of urban development and social regulation.

The first section explains why the focus on urban energy issues challenges the idea of a convergent and stable energy transition and leads us to favour the notion of energy changes, which is less normative and restrictive. In the second section we stress the urbanisation of energy issues, understood as the rescaling of these issues at the urban level, and discuss why it does not result nor contribute to a greater autonomy of urban stakeholders vis-à-vis national authorities and sector firms. The third section is dedicated to the analysis of the very diverse politicisation processes that occur in the surveyed cities with respect to energy policy. We conclude by summarising the main implications of the cases findings for a research on the urban governance of energy transition sensitive to the variety of issues and contexts.

Energy transition, what transitions?

In recent literature, energy transition is often presented as an imperative imposed by climate change and growing pressure on fossil fuels. In a normative sense, it means a deliberate transformation of present sociotechnical energy systems necessary for the emergence of a more sustainable, less carbon-dependent energy model, more reliant on renewables and less energy intensive (Rojey 2008).

One of the questions for researchers therefore concerns the mechanisms and phases of sociotechnical change, as explored in transition studies addressing the transformation of large-scale socio-technical systems (Smith et al. 2005), and conceptualised as a result of the interaction between regimes, niches and landscape pressures by scholars who adopt a multilevel perspective (Geels 2002). Recent research has identified the urban level of governance as a central locus of such transition processes (Bulkeley et al. 2010), either because cities, and particularly global cities, are perceived as actors deploying coherent strategies (Hodson & Marvin 2010), or because they are places where negotiations and contestation also shape transition pathways (Späth & Rohracher 2015). Despite the emphasis placed on urban areas, there have been few systematic investigations of the specific urban approaches to energy issues and of the distinct dynamics of urban-level transition: focusing on aspatial innovation processes, most of the literature on energy transition underestimates the politics and challenges of national-local alignment while, more generally, inadequately recognizing the spatial and scale dimensions of the transition approaches (Raven et al. 2012).

Our findings, however, show that the assumption of a relatively stable preliminary consensus around broad objectives should not be taken for granted. Indeed, the question of energy transition, in the sense mentioned above, is far from omnipresent in the cases studied and therefore in fact constitutes only one of the instruments of contemporary energy change. In other words, energy systems evolve regardless of the overt rhetoric of energy transition and with varying degrees of coherence and coordination. At urban scales, these processes of change give rise to a politicisation of energy questions which, far from occurring through the simple transposition of national debates, incorporates factors and objectives specific to each place, as well as limitations on action associated with the resources and skills available. Energy issues are framed in close relation to local concerns and interests, embedded in territorial compromises constructed around disparate priorities and a multiplicity of arrangements that reflects diverging conceptions of the role of energy in urban policies and development. These urban "resistances" to normative discourses on the energy transition mirror the specificities of the sociotechnical systems of each city studied and the potential risks of destabilisation for local economic networks, labour markets, territorial organisation. They cast light on the origin and nature of the conflicts and resistances thus generated but also on the local conditions of any compromises reached to overcome them, the drivers and resources of a genuinely local trajectory of change.

A territorial perspective, focusing on the processes and their transversal dimensions, is of particular heuristic value here in investigating – beyond their strict sectorial dimension – energy changes that are unfinished, disputed and politically highly loaded, and which reflect different societal interests and preoccupations that cannot, *a priori*, be assumed to converge across spaces and scales. Thus, in many emerging countries, the failings of the electricity system are of a more immediate concern than the internationally publicised global factors of crisis (peak oil, climate change) and the politics of energy are typically related to dispute around reforms of the national electricity systems (privatisation and deregulation, cuts in energy sector subsidies), matters of development specific to societies in emerging countries (for instance the growth in energy demand, which inadequate and/or ageing production and distribution systems are stretched to meet), and advancement of urban environmental transition (Lee 2006) (through the replacement of inefficient and polluting traditional energy sources, like biomass and charcoal, by modern energy as electricity or gas). Taken together – and however significant they may be – these changes cannot immediately be classed as an energy transition, because methodological obstacles limit the evaluation of their long-term effects (Grubler 2012, p. 11), and because the policies and measures observed are not always consistent and coordinated, and indeed are riddled with contradictions, thus obfuscating the coherence of any shared vision of energy transition.

Without prejudging whether or not such a vision exists, the changes considered affect existing sociotechnical energy systems. They are a combination of material factors (infrastructures, equipment), social actors (equipment manufacturers, producers and service suppliers, public decision-makers, consumers), regulatory frameworks, standards, and also values and representations internalised by the different actors. Whatever their content, the changes hold a strong political dimension. An indispensable component of wealth production, electricity is not reducible to the technical dimensions of generation and distribution activities. First, it encompasses all the value chains, in manufacturing and services, in terms of competitiveness and capital accumulation. Second, it is used as an instrument of political regulation and social redistribution, especially in cities. Necessary to the exercise of both economic and political power, its control – disputed – brings into play the balances of power between not only public and private, but also national and local actors. So beyond the "grid object", which has its own actors and dynamics, an analysis of electricity issues – including their urban dimensions – cannot ignore the sector's political economy, which incorporates social and political transformations correlated with the changes observed at all scales, as well as the way in which they transform or threaten to transform inherited power relations.

Table 7.1 Case studies in the dossier

Case study	Common features	National energy policy framework	Version and local specificity(ies) of energy policy	Forms of local politicisation of the energy issue
Buenos Aires (Prévôt-Schapira & Velut 2013)	Sustained demographic and spatial growth. Sharp increase in energy consumption (industry, middle classes). Big social contrasts and serios poverty in a significant section of the urban population	Security of supply. Energy mix: gas (for thermal power centres and households), hydroelectricity, nuclear	Political management via consumer prices (more subsidised in BA than the rest of the country) and investments (quasi-regulated economy). Innovations reinforcing the status quo. Energy transition: no	Opposition to plans increases in energy prices, stigmatisation of "big" consumers and "cheaters"
Delhi (Zérah & Kohler 2013)		Electricity sector reforms (privatisation, energy efficiency) to move from production deficit to production surplus. Energy mix: coal, Electricity sector reforms hydroelectricity. Model of low carbon intensity redistributive growth	Energy transition: nascent. Sector of application: transport. Proliferation of tools to support clean energy but little operational impact, wait-and-see and opportunistic approach in companies, distrust among users	Urban society interested in other priorities: transparency, continued low electricity prices, quality of living conditions. Rising demand driven by a high consuming middle-class hostile to cost of living increases
Cape Town (Dubresson 2013; Jaglin & Subrémon 2015)		Security of supply in response to the crisis in the South African energy model based on massive use of cheap coal. National policy to promote renewable energy and energy savings	Ambitious energy transition strategy, notably including the promotion of solar water heaters, but implementation at a standstill. Conditions not in place for a shift to an autonomous local urban policy of energy transition	Construction of a green coalition and incorporation of energy transition into municipal policy. Very marked socio-economic contrasts and problems of municipal action in poor neighbourhoods.

continued . . .

Table 7.1 Continued

Case study	Common features	National energy policy framework	Version and local specificity(ies) of energy policy	Forms of local politicisation of the energy issue
Istanbul (Arik 2013)		Security of energy supply at an affordable price through privatisation of operators and the regulation of energy markets Encouragement for the spread of gas, linked with the country's strategic position Real but marginal development of renewables	Policy of extending the natural gas network to replace household coal use (in order to reduce air pollution) No promotion of transition to renewable energy	Energy clientelism encouraging the use of coal in working-class areas, the electoral bases of the ruling party Universalisation countered by the diversity of uses and the social economic inequalities of households
Secondary Turkish cities (Pérouse 2013)		National political priorities: energy independence and security (privatisations, private investments, state governance) Reality of prices for fossil fuels (petrol, electricity)	Local administrations confined to secondary roles Development of public transport in several big cities (subway, BRT) Experimental climate plan in Gaziantep	Opposition to energy prices and fraud prevention policies

	Energy transition: development of urban gas network Encouragement for renewable energy (solar, wind power) which remain marginal and are not a prerogative of the cities		
Sfax (Bolzon et al. 2013)	Reduction in expenditure on fossil fuel imports Cut in electricity price subsidies and substitution of less subsidised natural gas for other fossil fuels (LPG) Promotion of energy savings to the industrial, tertiary and residential sectors Maintenance of public employment	Public energy consumption control policies little territorialised and partially in competition Promotion of natural gas impeded by urban morphology and poverty Success of solar water heaters Start of a policy to control energy consumption by the municipality	Opposition to hikes in the price of electricity, gas and LPG

An urbanisation of energy questions under national control

In an increasingly urbanised world, cities and urban regions are not only places and players in fossil fuel dependence and the production of greenhouse gases; they are also powerful markets for renewable energy and new technologies, centres of political and economic powers and civil organisations, a focus for the emergence and spread of new consumption practices. They are therefore scales potentially appropriate for energy change, provided that there is commitment by the social actors. A growing number of experts therefore believe that big cities and urban regions will emerge as driving forces in the implementation of measures to bring about the necessary energy revolution. Back in 2002 in Johannesburg, Peter Droege, a member of the World Council for Renewable Energy, made the following claim: "Cities, towns and other urban communities are increasingly regarded as settings for coordinated policy implementation efforts aimed at global renewable energy technology introduction and carbon emissions reduction programs" (Droege 2002, p. 2). This is also the starting point for the work of Hodson and Marvin (2010), for whom the ecological pressures imposed, for example, by climate change or peak oil on "global" cities are forcing them to develop metropolitan strategies designed both to ensure security of supply and to achieve greater autonomy in the control of their resources. Our case studies reflect a somewhat different reality.

To employ an analytical framework familiar to geographers, the question raised is therefore that of the urban rescaling of energy systems. These are "geographically embedded" (Bridge et al. 2013), in other words their infrastructures (and the spatial distribution of sunk costs) outline a specific geography of connections and interdependencies, their control and governance arrangements structure powers and interests at defined scales, the technologies implemented reflect local cultures of consumption, which in return they help to shape. Modifying and rescaling energy systems entails technical and sociotechnical changes that affect inherited power relations and transform the socio-geography of "energy landscapes".[3] Among the range of possible restructurings, the research thus explores the nature and scope of a process of urban territorialisation of energy systems, i.e. of the growing influence of the material infrastructure and urban issues on the decisions that govern them, the enhanced role of urban local authorities in their governance, a strategic reintegration (or a de-marginalisation) of the decentralised parts of the energy system in their political economy (small-scale local decentralised production, technology designed with a view to the ultimate supply and end uses, local determinants of urban demand).

If we stick to this definition, our case studies contain few examples of urban territorialisation. In energy systems that are everywhere expanding (extension of electricity and gas networks, increasing consumption), the security of urban energy supply is a major preoccupation for both central government and urban authorities, especially in capitals, as the example of Buenos Aires resoundingly demonstrates. From this point of view, there is little room for a divergence of interests and strategies: everywhere, the national and local political priorities are to meet growing demand at an affordable price, which tends somewhat to reinforce the dependency of cities first on large national infrastructures, and second on government actors and powerful new private actors, both national and international. Cape Town is dependent on the coal-powered Gauteng electricity complex, Turkish cities on Russian or Central Asian hydrocarbon suppliers. Everywhere, renewables are introduced into the energy mix to complement (rather than avoid) the increase in energy supply, whether from conventional sources (oil, gas, hydroelectric) or from nuclear power. Therefore, processes of decentralisation (development of local systems employing innovative technologies: photovoltaic, solar thermal, wind power, small-scale hydroelectric; mechanisms to increase energy efficiency in buildings or in household and industrial equipment) and of centralisation (continued construction of big fossil fuel dependent networks based on electricity, natural gas or oil; revival of large hydroelectric and nuclear projects) work together, and at the same time, to reconfigure energy systems. This enables certain cities, for instance Cape Town and the Turkish cities, to encourage innovations and even to become the drivers of appropriate supplementary solutions, for example renewables, measures that nevertheless have a limited impact on the energy sector and, as in the case of Cape Town, come into competition with national measures (Jaglin 2017).

Nevertheless, another process may be observed, which we call the *urbanisation of energy issues*. By this we mean, first, the growing inclusion of energy questions in urban policies, and also, the growing importance of rhetoric, initiatives and conflicts relating to energy issues, which are expressed in the cities and influence energy changes, even though these are controlled at other scales. Rather than growing autonomy for urban actors, and urban energy interests, what we are seeing therefore is these interests being increasingly taken into account in energy governance at national level. This process has significant consequences: it positions cities, in particular the biggest cities, as possible interlocutors in a multilevel play of actors, it enhances their role in incubating or driving energy changes, but it also helps to import into energy systems demands, claims and resistances emanating from urban consumers.

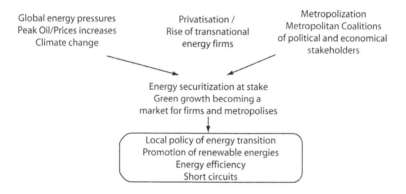

Figure 7.1 Governance of the energy transition in world cities

Source: interpreted from Marvin & Hodson (2010).

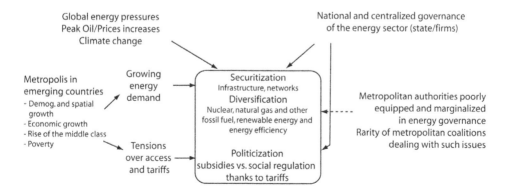

Figure 7.2 Governance of energy changes in the metropolis of emerging countries

Source: synthetic modelling by Jaglin & Verdeil.

Energy questions through the local lens: partial and situated politicisation

While highlighting the emergence of the energy issue as a "public urban issue" (Verdeil 2014), the case studies help to identify three inherent characteristics of the politicisation at work (see Table 7.1). First, they stress the importance of national political and institutional frameworks, and the overwhelming role of strategic central government choices in recent sectorial changes: power company privatisations in India and Argentina, market liberalisation in Turkey, a macroeconomic strategy of public subsidy reduction in Tunisia, a readjustment of the energy mix in response to the crisis of the low-cost coal model in South Africa, the Argentinian government's energy populism. Second, they emphasize the multiple – not to say contradictory – aims of the decentralised approaches to tackling energy questions, together with their embeddedness in specific territorial configurations. Indeed, in taking hold of energy issues, urban authorities are less interested in meeting sectorial concerns (and, among them, the place-dependent energy mix whose variety our case studies illustrate) than in reinforcing across-the-board urban policies, of which those concerns constitute only one facet.

In consequence, there are frequent divergences between the way in which the object "energy" is constructed at the urban scale, on the basis of local priorities or controversies (gas in Istanbul, subway in Delhi, environment in Cape Town, etc.), and the way in which energy policies are established at national scales, where the concerns are more strategic and sectorial (Jaglin 2014). It is therefore difficult to pinpoint where and how the big cities can engage in order to debate and construct, between themselves and at national level, a shared vision of energy changes, their consequences and their joint management. Depending on local patterns, cities may be places of receptiveness, resistance or impetus, or any combination of the three, with no assurance of an alignment between their energy priorities and those promoted by national policies.

Against this background, three primary local preoccupations can be identified in the cities studied.

The first priority of urban authorities is to secure a high-quality and cheap energy supply, seen as essential to local economic development and urban competitiveness. In circumstances where the local energy supply will continue – for a long time – to depend on large infrastructures and their associated technologies, this security entails less a transition towards low carbon energy, which is very far from constituting a dominant or even stated objective, than the inclusion of decentralised energy production systems (e.g. solar water heaters), as a supplement to the large networks (Cape Town, Sfax, Turkish Cities). The current changes are therefore less about a desire for substitution, than about the organisation of long-term coexistence between different sociotechnical systems. This coexistence itself poses considerable challenges, both of technical and regulatory nature, that cities are ill-equipped to meet. In addition, by combining heterogeneous systems, whose dynamics of development/decline can vary from one place and time to another, it creates the possibility for a diversification – deliberate or accidental – of energy systems at local scales, with impacts on urban inequality and fragmentation that are still largely unknown. These also depend on the preferences of local political and economic actors, which greatly differ according to the context: while the failings of the national system fuel the enthusiasm of business circles for alternatives (e.g. gas instead of coal-based electricity) in Cape Town, in Delhi, the local players echo aggressive national rhetoric on the "right level" of energy efficiency and consumption (in view of the historical frugal development trajectory of India) to justify a certain continuity in energy choices and practices. The emphasis on coexistence, rather than on substitution, opens up a debate about the relevance of the concept of energy transition, as already stated by Fressoz:

> To extract oneself from the transitionist imaginary is not easy because it so much structures the common perception of the history of technics, punctuated by a few great innovations that define technical ages. This vision is not only linear, it is simply wrong: it does not account for the material history of our society that fundamentally is cumulative.
>
> (Fressoz 2013)[4]

Second, all the cities also take positions in favour of measures to facilitate access to electricity and/or gas, within the framework of programmes to fight poverty (everywhere), ill-health (linked with particles and smoke from the burning of coal and wood in Istanbul) and insecurity (fire accidents in informal settlements in Cape Town). The expansion of the electricity and gas networks (Istanbul, Sfax), and the universalisation of a still rationed supply, thus remain essential components of urban policy and are reflected in huge programmes of investment in and extension of physical infrastructures. In emerging economies that possess the capacity to absorb and even promote sociotechnical innovations, these can benefit from technological leaps forward (spread of individual photovoltaic panels and solar water heaters, as in Sfax or in Turkey, development of micro-wind power) and foster off-grid alternatives. However, modern policies to increase access to energy favour – and will long continue to do so – programmes of massive investment in the construction and expansion of integrated and centralised networks, the substitution of "clean" fossil energy (gas) for other, more polluting fossil fuels (coal or oil), as in Istanbul, being, at this stage, of a much greater urgency and relevance than decarbonising the energy model.

Beyond access to the infrastructure, the question of consumption and its cost is a major political issue. On the one hand, the emergence and consolidation of the urban middle classes is a structuring factor in changes to the energy sector: these new population categories enter a consumer society characterised by the personal acquisition of numerous energy-hungry goods and aspire to levels of comfort and mobility which, in the short term, are hardly compatible with new standards of frugality. Delhi and Sfax, like Buenos Aires, reflect the impact of these new middle-class practices (in particular air-conditioning) on the sustained growth of urban energy consumption, a fact equally true of the secondary Turkish cities.

On the other hand, however, the access of large swathes of these urban societies to efficient energy continues to be hindered by poverty and inadequate housing. In research on relations between energy and poverty, numerous economic studies look at the determinants of demand and the conditions of an "urban energy transition", in which household consumption patterns progress up a scale of fuel types, from the least to the most efficient, depending on income and degree of adjustment to city life (Leach 1992). Some of these works emphasise the impact of public policies in the modalities of this transition, but also the persistence of the use of mixed energy sources for economic and/or cultural reasons (Barnes et al. 2005). This is clearly confirmed by the example of Istanbul, where the policy of universal gas supply comes into conflict with both poverty and the "coal clientelism" of the local authorities, and of Sfax, where the spread of natural gas comes up against the additional costs of installation in low-density neighbourhoods as well as competition from other systems.

Third, the cities studied have all adopted energy efficiency policies and projects (at least in their own building stock), consumer education and awareness-raising programmes and measures to support the most vulnerable households by helping them to reduce their dependency on energy resources (by exchanging energy intensive household appliances, sometimes in return for the regularisation of illegal connections[5]). Some municipalities, such as Cape Town, run workshops and forums aimed at industrial users and occupants of commercial premises. While active on the energy efficiency front, urban authorities are, by contrast, more reserved in their criticism of existing production and consumption patterns, even when they possess the tools to influence these, such as urban planning, transport systems, building standards and planning rules, etc. In general, they are content to justify energy efficiency measures on the grounds of potential economic gains (lower bills), without discussing the question of consumer practices and habits (e.g. air conditioning). In fact, any such challenge tends to yield to another priority, the right to retain a recently acquired "quality of life" (Delhi, Cape Town, Tunisian and Turkish cities). Despite almost universal energy efficiency measures, and in the absence of a more profound transformation of urban economic organisation and ways of life, which are particularly hetero-geneous in emerging countries, technical innovation and the spread of "modern" energy serve, for the moment, more to satisfy consumer demand than to achieve energy sobriety.

Altogether, these different aspects of energy policy rarely coincide with a clear and unequiv-ocal position of urban authorities on energy and climate change, and are subordinate to the very nature of urban government. The tendency is thus for city authorities – multi-sectorial, territorialised and legitimised by a relatively short electoral cycle – to embed sectoral ques-tions structurally within a transversal and multidimensional approach, justified by short-term outcomes. Cities are therefore the arena for the expression of energy concerns shaped by their territorial specificities: the nature of the economic base (dominated by heavy industry/ services), the origin of primary energy resources (coal, nuclear, hydroelectric), the local poten-tial of renewables (sunshine, wind), whether or not civil society is organised and motivated on energy issues, the impact of certain problems (poverty, industrial conversion or deindustrial-isation, political crisis, regional energy geopolitics). The nature of the pressure groups is also crucial: while coal or nuclear lobbies can be particularly influential at national level, local busi-nesses (as big energy consumers and major sources of taxes and/or jobs), environmental pressure groups or resident associations may also be key sources of local pressure. The example of Delhi thus shows the influence of middle-class consumer associations in setting priorities (transparency, living conditions, low electricity prices). Municipal energy agendas are also dependent on the need to join up with other facets of urban policy: cutting greenhouse gas emissions while transforming the city with a public transport system subject to recent restructuring (Delhi); developing renewable energy to sell the image of a "green" destination to investors (Cape Town); combating atmospheric pollution and a poor urban image (Istanbul). In cities, changing the energy system can mean altering spatial planning principles and ways of life as much as transforming the energy mix and the urban politicisation of energy issues is rarely about energy change *per se*. The capacity of local actors to create formal urban coalitions around shared objectives in this sphere – transcending real conflicts of interest between economic sectors whose primary concern is a reliable supply of cheap energy, others more preoccupied with the competitive advantages of a green economy, and yet others directly interested in the development of a renewable energies industry – is, for the moment, far from amply apparent in the cities studied. From this point of view, and subject to a more in-depth study, the participation of cities like Istanbul, Rio or Buenos Aires in a network like C40[6] seems more attributable to its organisers' wish to parade the participation of the cities of the South, than to any commitment by the representatives of those cities, who can nevertheless use it as a resource for their territorial marketing strategy.

Finally, the case studies emphasize that, whatever the ambition and the content of urban authorities' actions, it is primarily in relation to rationing and scarcity, on the one hand, and prices on the other, that energy becomes an urban political question. The price issue is parti-cularly sensitive and linked to the impact of increases in production costs (use of gas-fired power stations to meet peak demand, new environmental standards, the impact of drought on hydroelectric production, rising oil prices, etc.), or to a gradual removal of subsidies previously applicable to consumer prices before the liberalisation and privatisation policies implemented since the 1990s, as in Istanbul, Delhi or Sfax and, more partially and indecisively, in Buenos Aires. Ratcheting up the price of energy without giving consumers ways to reduce their dependency can lead to very serious social and economic consequences. For the moment, however, there has been very little close analysis of the socio-spatial distribution of the costs of potential changes in energy systems, nor systematic investigation of the consequences of these additional costs for different categories of urban energy users, as evidenced by the lack of data and analysis on the response of Cape Town's lower middle-classes to increases in electricity prices and to ill-conceived energy efficiency campaigns (Jaglin and Subrémon 2015). More generally, the shift from subsidised energy to less subsidised energy threatens the existing social balances, a threat exploited in different ways by movements of opposition to a less regulated electricity economy in Buenos Aires and of middle-class protest against discredited political and bureaucratic elites in Delhi. The tensions thus generated also limit the effectiveness of certain policies, such as universal gas supply in Istanbul and Sfax. While the question of access to energy is therefore far from settled in terms of network connections and consumer prices for what remains a large swathe of urban populations, very widespread shortages and power cuts affect users already connected to the grid. It is therefore scarcely surprising that, in many cities, supply failures coupled with price hikes generate urban unrest and constitute a "new public issue".

However, there is very little leeway to respond to this at municipal level, as is evidenced equally by secondary cities in Turkey and by a rich metropolis like Cape Town: the technical (choice of primary energy resources) and economic (energy industry structure, heavy infrastructures, spatial planning, etc.) paths of dependency have considerable influence on local capacity for action; the energy sector's institutional architectures still leave little room for the urban scale; the legal and normative frameworks governing the sector remain a national prerogative. While local governments have little control over changes, they suffer the instability linked with national transformations whose timeframes rarely correspond to their own aspirations: if too slow, they inhibit local initiatives; if too fast, they disrupt local economic environments without resources and skills being transferred or transformed.

In addition, these changes raise difficult questions. First, the question of funding: both renewable energies and certain energy efficiency measures are insufficiently profitable in the short term to be financed by the private sector, which may explain the hesitancies and delays in the quest for appropriate frameworks for action, which also affect local players, as in Cape Town. Second, the question of coordinating very heterogeneous measures: saving energy requires efforts in multiple spheres (consumption models, eco-design and retrofitting of buildings and machinery, urbanism and spatial planning, patterns of use, housing equipment). And finally, there is the question of anticipation and supporting urban societies in adapting to changes, notably in a way that protects the most vulnerable households whose consumption patterns are highly constrained (poor insulation, poor ventilation, poor lighting, energy intensive electrical appliances). These changes will not take place without social upheavals and require very early coordination of numerous public policies, for which cities often lack both the competence and the capacity.

Conclusion

While normative rhetoric in favour of energy transition is ubiquitous, this research failed to identify real green urban coalitions that unite economic actors and political elites, let alone organised civil society groups, to drive coherent policies in this sphere. Yet in all the cities studied, actors (elected officials and technicians in local or regional government, industrial groups, environmental lobbies, resident associations) seize on energy issues out of different motives: reduction or even elimination of the subsidies that drain public budgets, territorial marketing and competitiveness, access to available dedicated funding, energy savings, spending reduction, poverty alleviation. The justifications, motives and priorities for local interventions are very closely linked with urban conditions. Energy issues are therefore becoming partially urbanised around factors that have local resonance (security of supply in Buenos Aires, Cape Town, Istanbul), that intersect with other political concerns (green marketing of a service economy in Delhi and Cape Town; review of prices and sociopolitical motives everywhere) and that sometimes coincide with other interests (demand for transparency in local governance in Delhi).

All this confirms that energy issues are increasingly a matter of debate at local scale, that partial public policy responses are being developed (in particular in the sphere of energy efficiency and management); however, it does not demonstrate (at least for the moment) a general surge in municipal interest in the development and implementation of concerted and systematic strategies for energy transition. In this domain, it would seem, they have neither a clear mandate from their populations nor appropriate powers and resources, which remain national or have been conferred on private actors, for which cities are only one field of operation among others.

By showing that the conditions for convergence between local strategies, national policies and a "universal" model of energy transition are apparently not yet in place, the approach favoured in the research illustrates the advantages of complementarity between, on the one hand, sectorial and national analyses of transition and, on the other, analyses of the urban determinants of energy changes. The latter produce a more realistic understanding of the interdependencies between decentralised urban endpoints and the centralised strategic components of energy systems. The research also confirms the growing politicisation of the energy question in and on the part of cities, the emergence of a "public problem" inseparable from the dynamics of resistance, negotiation and opposition manifested by urban societies towards the contemporary forms of energy transition. At the same time, it cast lights on the way in which these crises and controversies can destabilise but also potentially contribute to changes in energy systems as a whole.

Beyond the cities of emerging countries examined in this chapter, this observation should prompt researchers to look again at the implementation of energy transition policies in other contexts, and to explore how, beyond a set of global conditions (resources, climate change, transformation of capitalism), the variety of issues and local trajectories affect the transformation of energy systems.

Notes

1 Energy trajectories in southern metropolitan regions, funded by the French National Agency for Research (scientific director: S. Jaglin, Latts/UPE); and Urban energy governances, North and South, a multiannual project funded by Université Paris-Est Marne-la-Vallée (UPEM) (scientific directors: S. Jaglin and J. Rutherford, Latts/UPE).
2 This chapter is a slightly revised translation of the introductory piece of a thematic issue for the French journal *Flux* (no. 93–94, 2013). We thank John Crisp the translator, as well as the research unit UMR Environnement Ville Société, CNRS-Université de Lyon, which funded this translation.
3 "Energy landscapes" are defined here as follows: "the constellation of activities and socio-technical linkages associated with energy capture, conversion, distribution and consumption" in which the "material landscapes . . . are the product of social processes and the outcomes of conflict and negotiation among different social groups" (Bridge *et al.* 2013, p. 335).
4 Our translation of "S'extraire de l'imaginaire transitionniste n'est pas aisé tant il structure la perception commune de l'histoire des techniques, scandée par les grandes innovations définissant les grands âges techniques. . . . Cette vision n'est pas seulement linéaire, elle est simplement fausse: elle ne rend pas compte de l'histoire matérielle de notre société qui est fondamentalement cumulative" (Fressoz 2013, p.174).

5   This is the case in some projects of urban regularisation of favelas in Rio of Janeiro (Pilo 2015).
6   C40 is a network of cities working to reduce greenhouse gas emissions (see: www.c40.org/).

References

Arik, E., 2013. Chronique d'une transition énergétique en tension: l'universalisation inachevée du réseau de gaz naturel à Istanbul. *Flux*, 93–94(3), pp.56–69.

Barnes, D.F., Krutilla, K. & Hyde, W.F., 2005. *The Urban Household Energy Transition. Social and Environmental Impacts in the Developing World*, Washington: Energy Sector Management Assistance Program, Resources for the Future Press.

Bolzon, H., Rocher, L. & Verdeil, É., 2013. Transitions énergétiques multiples et contradictoires à Sfax (Tunisie). *Flux*, 93–94(3), pp.77–90.

Bridge, G. et al., 2013. Geographies of energy transition: Space, place and the low-carbon economy. *Energy Policy*, 53, pp.331–340.

Bulkeley, H.A. et al. eds., 2010. *Cities and Low Carbon Transitions* 1st ed., London; New York: Routledge.

Droege, P., 2002. Renewable energy and the city: Urban life in an age of fossil fuel depletion and climate change. *Bulletin of Science, Technology & Society*, 22(2), pp.87–99.

Dubresson, A., 2013. À propos d'une initiative municipale verte au Cap (Afrique du Sud): les leçons du Solar Water Heater Advanced Programme. *Flux*, 93–94(3), pp.43–55.

Fressoz, J.-B., 2013. Pour une histoire désorientée de l'énergie. *Entropia. Revue d'étude théorique et politique de la décroissance*, 15, pp.173–187.

Geels, F.W., 2002. Technological transitions as evolutionary reconfiguration processes: A multi-level perspective and a case-study. *Research Policy*, 31(8–9), pp.1257–1274.

Grubler, A., 2012. Energy transitions research: Insights and cautionary tales. *Energy Policy*, 50, pp.8–16.

Hodson, M. & Marvin, S., 2010. *World Cities and Climate Change: Producing Urban Ecological Security*, Maidenhead: Open University Press.

Jaglin, S., 2014. Urban energy policies and the governance of multilevel issues in Cape Town. *Urban Studies*, 51(7), pp.1394–1414.

Jaglin, S., 2017. Métropoles des pays émergents: des acteurs de la transition énergétique? Leçons du Cap (Afrique du Sud). *Géographie, Economie, Société*, 19, pp.5–27.

Jaglin, S. & Subrémon, H., 2015. La transition énergétique à l'épreuve des logiques d'usages: le cas des petites classes moyennes au Cap. In M.-C. Zélem & C. Breslay, eds. *Sociologie de l'énergie. Gouvernance et pratiques sociales*. Paris: CNRS Editions, pp. 293–302.

Leach, G., 1992. The energy transition. *Energy Policy*, 20(2), pp.116–123.

Lee, K.N., 2006. Urban sustainability and the limits of classical environmentalism. *Environment and Urbanization*, 18(1), pp.9–22.

Pérouse, J.-F., 2013. La gouvernance énergétique dans les villes turques: un état des lieux. *Flux*, 93–94(3), pp.70–76.

Pilo, F., 2015. *La régularisation des favelas par l'électricité: un service entre Etat, marché et citoyenneté*. Thèse de doctorat. France: Université Paris-Est.

Piveteau, A. & Rougier, É., 2010. Émergence, l'économie du développement interpellée. *Revue de la régulation. Capitalisme, institutions, pouvoirs*, 7. Available at: http://regulation.revues.org/7734 [Accessed January 8, 2014].

Prévôt-Schapira, M.-F. & Velut, S., 2013. Buenos Aires: l'introuvable transition énergétique d'une métropole fragmentée. *Flux*, 93–94(3), pp.19–30.

Raven, R., Schot, J. & Berkhout, F., 2012. Space and scale in socio-technical transitions. *Environmental Innovation and Societal Transitions*, 4, pp.63–78.

Rojey, A., 2008. *Énergie et climat: réussir la transition énergétique*, Paris, France: Éd. Technip.

Rutherford, J. & Coutard, O., 2014. Urban energy transitions: Places, processes and politics of socio-technical change. *Urban Studies*, 51(7), pp.1353–1377.

Rutherford, J. & Jaglin, S., 2015. Introduction to the special issue – Urban energy governance: Local actions, capacities and politics. *Energy Policy*, 78, pp.173–178.

Sgard, J., 2008. Qu'est-ce qu'un pays émergent? In J.-C. Jaffrelot, ed. *L'enjeu mondial, les pays émergents*. Paris: Presses de SciencesPo-L'Express, pp. 41–54.

Smith, A., Stirling, A. & Berkhout, F., 2005. The governance of sustainable socio-technical transitions. *Research Policy*, 34(10), pp.1491–1510.

Späth, P. & Rohracher, H., 2015. Conflicting strategies towards sustainable heating at an urban junction of heat infrastructure and building standards. *Energy Policy*, 78, pp.273–280.

Thorp, B. & Marvin, S., 1995. Local authorities and energy markets in the 1990s: Getting back into power? *Local Government Studies*, 21(3), pp.461–482.

Verdeil, É., 2014. The energy of revolts in Arab cities: The case of Jordan and Tunisia. *Built Environment*, 40(1), pp.128–139.

Zérah, M.-H. & Kohler, G., 2013. Le déploiement des énergies propres à Delhi aux prises avec la défiance de la société urbaine. *Flux*, 93–94(3), pp.31–42.

8

Geographies of energy intermediation and governance in the context of low carbon transitions

Ralitsa Hiteva

Energy governance in the context of low carbon transitions

The transitions taking place within energy, including those towards low carbon living, are necessitating radically redefining the way we think about energy policy and infrastructure. This chapter makes the argument that in the context of low carbon transitions we cannot think about energy governance without discussing intermediation. In doing so it aids understanding of how geographies of intermediation and geographies of energy governance interact. The chapter puts forward an argument that the strategic role of intermediaries in energy governance challenges established assumptions about how to respond to energy challenges, such as low carbon transitions and uncovers a key role for intermediaries in the creation of spatial patterns through facilitating the flow of energy from its production to its transmission and distribution to a particular location.

Energy governance refers to a wide range of formal and informal activities (including but not limited to regulation, policy and user practices) by different groups of actors (government, industry, civil society, individuals) involved in the everyday use and management of energy. Energy governance is concerned with the inclusion of new actors, and co-ordination between the different stakeholders, activities and levels at which these take place. The process of co-ordination in the context of low carbon transitioning is essentially a process of negotiation and re-ordering of interests, rules and infrastructure at and between different levels and sites (LCS-RNet, 2012). Energy governance in the context of low carbon transitions can be thought of having several key characteristics: being a crowded and complex space; taking place within intersecting and competing sites and levels; and producing blurred interfaces.

A large scope of activities – from user practices like recycling to the introduction of smart grid technologies which involves considerable innovation and coordination between multiple otherwise siloed sectors like ICT, electricity and vehicles – are constitutive of pathways of low carbon transitioning. Often approaches to the latter, especially when discussed in a user context are centered around a continuous lowering of the carbon emissions per unit energy used, improving energy efficiency and curbing energy demand.

'Intermediaries' include individuals, organisations and/or networks, whose strategic location in-between different categories of actors (regulators and regulated, public and private actors) enables them to act across these categories (for example to translate or facilitate between them) (Beveridge and Guy, 2011). Below I outline some of the key characteristics of governing energy infrastructure, before discussing the nature of intermediaries. The geographies of energy intermediation are illustrated through two case studies: one which shows the role of energy intermediaries in natural gas supply in Bulgaria, and one which illustrates the need for energy intermediaries in the transmission of offshore wind in the UK. They illustrate cases of intermediation characteristic of contemporary energy systems in the midst of low carbon transitions. The cases of electricity and natural gas have been selected because of their importance to putting lower carbon energy production and use in practice. The final section reflects on the spatial aspects of intermediary activities in energy governance and how they contribute to understanding the geography of energy transitions.

Governing energy infrastructure

The legacies of privatisation and liberalisation have resulted in a complex but rather fragmented ownership, regulation and processes of energy production. The predominantly centralised nature of energy production (built around large nuclear and fossil fuel power plants) on one hand, and the increasingly interconnected nature of energy infrastructure (Smil, 2010) on the other hand, complicate the complexity of energy governance. Networked energy infrastructures such as natural gas transportation pipelines are capable of unevenly binding spaces together across cities, regions, nations, and international boundaries, creating in the process specific material and social dynamics within and between these spaces (Amin and Graham, 1998). Infrastructure networks lock certain technologies to specific spaces and extend the interests of certain groups through time and space beyond the 'here' and 'now' (Bijker, 1993). Thus they can be thought of as producing geographies of (dis)connection and acting as a 'transmission belt' for national policies and interests (Brenner 1998).

However, the running of infrastructure networks, such as natural gas pipelines and electricity grids, is subject to constant struggles between actors, institutions and companies with varying social, economic, and political power. The infrastructure networks linking producers, traders and distributers of natural gas across space are, in fact, dynamic processes only temporarily stabilised, requiring continuous effort to be maintained. Infrastructure networks can be understood as essentially assemblages of actors, technology and interdependencies, producing specific 'geographies of enablement and constraint' (Law and Bijker, 1992, p. 301), which reflect and reinforce existing relationships of power (such as power concentrated in specific places but missing in others – i.e. off-grid communities). Thus, natural gas pipelines connecting Russian gas with Europe can be thought of as power lines enabling the powerful vested interests of the (Russian) state and the industries that benefit from the access to or lack of natural gas, and serving a particular territorial role (Bouzarovski et al., 2015). Bouzarovski et al. (2015, p. 218) bring to attention the political work of socio-technical assemblages, and their importance in making territory, through 'relational networks' and 'networked socio-technical practices' (Painter, 2010, p. 1090).

So energy infrastructure is simultaneously fragmented and held together by specific sets of relationships at multiple levels and scales, which involve economic and environmental regulation, voluntary and self-imposed organisation, and economic and social activities. Geographies of energy governance are thus defined by these elements and the relationships between them. For example, the energy policy of countries like Bulgaria has transformed from a fairly coherent

Soviet controlled national arena in 1989 to an array of decision-making and power centers simultaneously located at the national, European and Russian spheres by virtue of a transit and import monopoly in gas (Hiteva and Maltby, 2014). Increasingly national energy strategies are influenced or dictated by supranational and sub-national levels, like the Kyoto Protocol, EU climate change strategies and voluntary organisations like the Covenant of Mayors. This means that an important aspect of energy governance is how coordination between these dynamic, intersecting and crowded elements and complexities take place, both formally and informally, to manage the relations between public and private actors, and to ensure interconnectedness and a flow of information, knowledge and energy (i.e. electricity and natural gas) through energy infrastructure networks (Graham and Marvin, 2001).

Furthermore, in the context of low carbon transitions, the large number and complex nature of energy regulations and policy have produced a 'crowded policy space' (Sorrell, 2003) prone to policy overlaps, jurisdictional conflicts and unanticipated consequences. So energy production, distribution and use are thoroughly entwined at multiple levels and sites, and are subject to dynamic interplay, well beyond what we understand as the energy sector, spanning into ICT, transport, waste and water management.

The changing ownership of infrastructure due to the processes of privatisation and unbundling has introduced a range of new actors, creating complex interface spaces between users, providers and regulators. For example, the pursuit of a competitive energy market has led to the introduction of companies in a state of private–public betweenness. Parastatal organisations for example, are usually owned and controlled by the state, and although they are designed to run like commercial companies, they often receive government subsidies to fund capital expenditure, and operate independently of market incentives (Kessides, 1993). Thus, the typical characteristics assigned to the different stakeholders in governing energy are becoming increasingly blurred, as states act as economic agents concerned about competitiveness and acting as entrepreneurs, while firms (especially global firms like Siemens) are important political actors with significant policy influence. The state is also extensively involved in the energy business in most countries through ownership, funding, development and protection of markets (Newell and Paterson, 1998), what Mariana Mazucatto (2013) calls 'the entrepreneurial state'.

Furthermore, the technical implications of the increased interdependencies between different sectors (for example through the increased use of ICT in energy) are also contributing to the blurring of boundaries in energy governance. In practical terms this could mean that one economic regulator (for example Ofgem) could end up regulating the activities of ICT companies in the context of smart grids. Some of the most significant changes to energy are taking place within the electricity sector, a domain that has one of the most significant targets for lowering carbon emissions and energy efficiency improvements. This makes transmission and distribution networks of electricity key sites of low carbon transitions, where connection and disconnection of actors, technologies and policy occurs. Following privatisation and several bouts of structural changes brought upon by EU legislation they have been fragmented into economically distinct activities, carried out by multiple private actors. They have been left in the border in-between the competitive and non-competitive aspects of energy systems. While their individual activities have been siloed by economic regulation, environmental regulation (i.e. on lowering carbon emissions) have attempted to redraw the boundaries between multiple actors and their respective activities. This has also produced blurred spaces in governing energy, where the boundaries between independent units of governance, are being challenged.

The blurriness of energy infrastructure and services can also be traced as result of attempts to move towards lower carbon energy production and use. The ability to produce energy within households and sell excess to the grid, though the use of technologies like solar panels and batteries (in the form of electric vehicles), have produced new categories like prosumers (energy consumers who are also producers). The integration of renewable electricity from wind and solar power into the national grid is central to achieving the objectives of low carbon transitions. It is also a complex technological process involving multiple actors (producers, regulators, transmission operators and distributers) and interfaces between them and between corresponding technologies. For example, connection of offshore wind to the onshore grid in the UK involves a complex interaction between three or four different actors, requiring contractually defining each activity between them beforehand. These along with the fragmentation of the energy systems have generated a demand for new forms of coordination and intermediation activities, and opened up new spaces for intermediation.

Intermediaries in energy governance

The term 'intermediaries' is used to describe individuals, organisations, networks or institutions, strategically located in-between regulators and regulated, public and private actors, or sets of different social interests (Moss et al., 2011), and where existing boundaries between stakeholders are being eroded or redefined (Beveridge and Guy, 2011). Intermediaries can work to facilitate, coordinate, make connections and mediate disputes to enable relationships between different groups of actors. They work by forming a range of (formal and informal) networks and coalitions, following a specific order and/or hierarchy of interests and actors, thus creating new forms of interdependencies and socio-technical assemblages (Medd and Marvin, 2011). By doing so, they are enabling the use of energy infrastructure networks as transmission belts for actors' interests.

However, intermediaries can work against, stall change and contribute to system obduracy (Moss et al., 2011), taking on the role of gatekeepers within a system (Randles and Mander, 2011). This is achieved through maintaining a strategic position to exercising control over access points to products and services, in order to maintain existing market structure in the interest of incumbents and block access for new actors. Furthermore, the role of intermediaries is not neutral (Moss et al., 2011). They are capable of translating and redefining what they convey between sets of actors and interests (Latour, 1993). By translating, intermediaries redefine and reframe, pursuing their own agendas, creating new realities and meanings, and enabling particular socio-technical assemblages.

Intermediaries can operate strategically so that they can support a particular infrastructure and configuration of power. By translating between sets of actors and interests, strategic intermediaries redefine, reframe and transform strategies and interests. In the context of fragmented infrastructure subject to multiple competing agendas and complexities, intermediaries are able to strategically reconfigure relations between different system actors and components in order to advance particular interests, and 'produce an outcome that would not have been possible, or as effective, without their involvement' (Marvin and Medd, 2004, pp. 84–85).

In the context of renewable electricity, points of access and spaces of transmission open up opportunities to re-negotiate, re-order and re-align key interests. Thus they can be considered as intermediary spaces, where boundaries between production and consumption, regulator and regulated 'can be questioned, broken down, reworked and actively reconstructed by intermediaries' (ibid., p. 82). It is within such intermediary spaces that intermediaries possess the power to affect, re-shape and transform infrastructures and policy through formal and informal activities, such as the forming of networks and coalitions.

Blurred interfaces are spaces where the boundaries and responsibilities between the public and the private, the regulator and the regulated, provider and supplier are being challenged and transformed. For example, in the context of integrating renewable energy, these interfaces involve the negotiation of how the uncertainty and costs of renewable power production and transmission will be allocated. By doing so, they affect the material landscapes of energy (i.e. how many substations will be build and what they would look like) and can facilitate changes in the way energy is governed. Thus, blurred interfaces can be seen as having a key role to play in the low carbon transitions as strategic areas for intermediation, where the governance imperative for intermediaries is to translate between interests, agendas (and voltages in the case of renewables) in order to (dis)connect new renewable capacities.

Geographies of energy governance through intermediation: natural gas supply in Bulgaria

The case study of natural gas production in Bulgaria illustrates the role played by intermediaries in energy governance. The case of Bulgaria's natural gas is worth considering because of its location on the border of the European Union and as a gateway to Russian gas in the EU. The production and transportation of natural gas are still controlled by the large integrated utilities that grew out of the former state monopoly, resulting in a complex patchwork of public and private ownership (Bouzarovski, 2009). The natural gas supply system in Bulgaria is an area of contestation between market forces and institutions, and state-centered, power-based geopolitics.

With approximately one third of EU gas imports currently sourced from Russia and total technical capacity for natural gas transit transmission through Bulgaria of 18.7 bcm p.a., the regulation of natural gas in Bulgaria can be seen as of direct relevance to the implementation of EU energy policy. The EU's Energy Roadmap 2050 sets out the objectives of a significant role for natural gas as a transition fuel (toward lower carbon living), diversification of natural gas imports and the full liberalisation of the natural gas market. However, in Bulgaria there have been significant delays in: (1) diversifying natural gas imports; (2) the full liberalisation of the internal market; (3) transposition of EU legislation; and (4) increasing the very low levels of residential gasification. These delays have resulted from problems in the implementation of the Bulgarian national energy policy caused by the presence and operation of two Bulgarian state-owned natural gas intermediaries: Bulgargaz and Bulgartransgaz positioned between upstream (domestic producers like Melrose Resources and the main natural gas importer, Gazprom Export) and downstream companies (gas distribution companies like Overgas).

The natural gas from Russia is supplied exclusively by Gazprom and purchased exclusively by Bulgargaz, via one pipeline. Despite being a small market for imports, over 10 per cent of Russian gas exports to Europe transit through Bulgaria's territory. Once it reaches the Bulgarian border, the natural gas starts flowing through Bulgartransgaz's pipelines until it reaches municipal borders. Bulgartransgaz's pipelines do not enter municipal territory, nor do they reach all municipalities in Bulgaria. Once the natural gas reaches municipal borders, it is sold to the gas distribution companies (like Overgas), which are responsible for constructing the pipeline infrastructure within municipalities in order to deliver the gas to consumers.

The Bulgarian Energy Holding (BEH) 'is very well placed to influence policy and regulatory choices in a way that preserves its dominant position' (Silve and Noël, 2010: 15). Bulgargaz and Bulgartransgaz are strategically positioned intermediaries and their importance for the state have allowed them a privileged position within the natural gas sector and the ability to 'dictate their own terms. Under BEH's umbrella both Bulgargaz and Bulgartransgaz are part of a close circle of institutions under the control of the Bulgarian state, which is able to mobilise Bulgargaz and Bulgartransgaz's resources to achieve its ends. The strategic position of the two companies within BEH and within the natural gas supply and transit in Bulgaria is of high importance to the state. This mode of control is enabled by the regulator SEWRC. For example, despite

Table 8.1 Main actors in natural gas supply in Bulgaria

Actor	Role
Bulgarian Energy Holding (BEH)	A holding company for a group of (previously state-owned) companies engaged in (among other things) natural gas transmission, supply and storage. BEH is wholly owned by the Bulgarian state and is the largest state-owned company in terms of total assets in the country.
Bulgargaz	The state-owned sole public supplier of natural gas for the territory of Bulgaria, responsible for providing customers with an uninterrupted supply of natural gas. Bulgargaz acts as a gas trader, buying natural gas from external suppliers like Gazprom and selling it to natural gas distribution companies.
Bulgartransgaz	The state-owned Transmission System Operator (TSO) performing licensed activities of natural gas transmission and storage. It owns and operates the national gas transmission network to distribution companies and industrial consumers, and the transit transmission network to other countries.
State Energy Regulatory Commission (SEWRC)	Issues and controls licenses for electric/heat power generation; natural gas and electricity transmission, distribution, storage and facilities; as well as for electricity trade, power exchange, natural gas public provision and supply.
Overgas	A private company, jointly owned by the Bulgarian Overgas Holding (50%), Gazprom (0.49%) and GazpromExport (49.51%). The largest private natural gas distribution company in Bulgaria, comprising over 56% of the market. Overgas is involved in the construction, operation and maintenance of gas distribution networks and facilities.
Gazprom	A private company, the largest extractor of natural gas in the world and one of the world's largest companies. The Russian state has a 50% controlling stake in Gazprom.

Source: Hiteva (2013)

numerous requests by the European Commission and distribution companies, including Overgas, for SEWRC to exercise more control over Bulgargaz and Bulgartransg, the regulator has failed to sanction them about the terms of their operation. Contrary to EU regulation (required under the Third Energy package) and national strategy (the Bulgarian state has committed to increase the levels of household gasification in the country as a measure towards moving to lower carbon living both in 2008 and 2011) Bulgartransgaz has been refusing to allow distribution companies like Overgas access to its pipelines for over five years. Bulgartransgaz justifies its decision on technical grounds, claiming that it needs more time in order to allow such access to its pipelines, given the high number of entry and exit points required by distribution companies.

Access to Bulgartransgaz's pipelines for distribution companies like Overgas will enable significantly increasing the gasification rates at household level in the country, currently at 3 per cent. Furthermore, the percentage of Bulgarian municipalities with gas supply is only 15 per cent compared to more than 90 per cent as EU average. The process of gasification involves the construction of the gas pipeline to the point of final consumption, the construction of gas installation and the supply of natural gas to consumers (whether household or industrial). BEH, Bulgargaz and Bulgartransgaz are unable – at present and in the foreseeable future – to update the natural gas infrastructure network and develop the gas transmission and distribution networks. As Bulgargaz's and Bulgartransgaz's power is physically sunk into the ground (i.e. the pipelines that are laid) any changes to the material infrastructure network and how it is used will lead to changes in their intermediary power and ability to fulfil their purpose.

The terms of operation of Bulgargaz and Bulgartransgaz offer the opportunity for more strategic forms of policy-driven intervention through intermediation activities than it is usually possible under neoliberal governance. The state-owned intermediaries Bulgargaz and Bulgartransgaz use an array of formal (such as sole national license) and informal institutional arrangements (such as a *de facto* immunity from SEWRC) to manage the Bulgarian natural gas system in a way that accommodates and excludes specific sets of interests. Although the Bulgarian state has a vested interest in maintaining the position and role of Bulgargaz and Bulgartransgaz, they should not be considered as one and the same as those of the Bulgarian state. Bulgargaz and Bulgartransgaz also possess a level of independence from the state and do not function as passive or neutral means of achieving the state's objectives (in this case higher levels of gasification). State-owned Bulgargaz and Bulgartransgaz are capable of acting in their own self-interest, even if their (in)actions go against the interest of the Bulgarian government.

While the state has a considerable leverage over financial, managerial and organisational aspects of the two intermediaries, it could only maintain it as far as it does not challenge their strategic position as intermediaries. Consequently, it could be argued that in relation to their strategic position as a physical intermediary between the state and Overgas, Bulgargaz and Bulgartransgaz are independent of the state and the latter cannot fully use its infrastructure networks as transmission belts of policy. By refusing access to their pipelines to natural gas distribution companies and to develop the national natural gas infrastructure, Bulgargaz and Bulgartransgaz are blocking the implementation of the national objectives of increased household gasification. Because of the two intermediaries' level of independence from the Bulgarian state to pursue their own interests and their political significance to the state, national policy objectives are unlikely to be successfully implemented if they challenge Bulgargaz's and Bulgartransgaz's positions. Therefore, the two intermediaries represent 'choke points' in the implementation of national energy policy and, given the importance of natural gas as a transition fuel, to lower carbon transitions as well.

Geographies of energy governance through intermediation: offshore wind transmission in the UK

The UK has committed to deliver 15 per cent of its energy consumption from renewable sources by 2020. The UK Renewable Energy Roadmap until 2050 highlights the importance of the electricity sector for lower carbon transitions. The rapid development of offshore wind is one of several preferred pathways for achieving the UK carbon dioxide emission reduction targets and the renewable generation targets. Offshore wind in the UK has significantly increased over the last decade, and since 2010 the UK has the largest offshore wind industry in the world and the fastest growing wind capacity. Thus, with relatively shallow waters and strong winds extending far into the North Sea, offshore wind is expected to make the single biggest contribution towards the UK's 2020 target. In 2009 the UK government announced its intent to facilitate an upscaling of offshore wind capacities from the then levels of 9 GW up to 33 GW of offshore wind by 2020.

To facilitate the required investment of up to £8 billion for the construction of the transmission assets needed to take that electricity onshore, the UK Department of Energy and Climate Change (DECC) introduced a new regulatory framework known as the OFTO regime, within which an Offshore Transmission Network Owner (OFTO) is selected through a competitive tender process, for a period of 20 years, to operate and maintain the transmission infrastructure which takes electricity generated by offshore wind farms to be distributed onshore. The purpose of the OFTO regime is to regulate the construction and operation of offshore transmission assets.

It generally involves offshore platforms with transformers and switch gear equipment where electricity from wind turbines is collected and transformed to high voltage, the subsea cables which transport that electricity to onshore substations, and onshore substations which transform the electricity to the voltage required for transmission through the onshore network.

However, introducing a new actor in the supply chain for offshore wind, strategically positioned between the offshore wind generators and their market onshore, created concerns of increased risk, both technically and financially. While only 10 per cent of OFTO's income depends on whether any electricity is being transmitted or not, 100 per cent of generators' income depends on the electricity transmitted and sold. However, equally important is the technical risk associated with the complexity of the OFTO's interface with generators and the National Grid in relation to operation and maintenance.

The new regime introduced significant changes in the contracts for the design and construction of the offshore wind farms and transmission assets. For example, boundaries were not picked for control equipment, while many developers opted for doubling the control and access equipment in order to prevent a situation where in case of emergency they have to rely on being granted access by the OFTO. Such measures were driven by attempts to mitigate against access uncertainty and led to more complex contractual arrangements between all key stakeholders, the installation of unnecessary transmission equipment and infrastructure, and hence complex interfaces between the parties involved. In practical terms, many generators installed duplicate equipment to ensure access in the future, while ownership of the vital equipment in substations and within wind farms was unclear. Developers have tried to militate against access uncertainty by separating transmission infrastructure unnecessarily and by installing parallel control systems, as in the case of the Briggs Lowe project. Although such systems gather data from the same places, they process it separately, and then send it to two separate customers via two separate routes. Therefore, the introduction of the OFTO regime, at least initially, amplified the blurriness between stakeholders and created blurred interfaces in offshore wind transmission.

For example, whether generators include the offshore substations or not as part of the transmission asset bears significant cost consequence for the generator, and different developers have chosen different limits to what is a transmission asset. Deciding who owns what especially on substations is also a complex issue with considerable access and cost consequences. While some generators include the offshore substation as part of the transmission asset, in other projects (like Robbin Rigg) the OFTO only owns the cables and the onshore connection. Therefore, OFTOs and generators have to negotiate interface agreements, which specify what kind of information and how frequently needs to be passed on from one to the other, as well as what happens if generators require major modification or major repairs outside of the ownership agreement.

Another significant issue of offshore wind production is associated with the reduction of its cost. One intermediary – the Crown Estate – has played a key role in resolving the blurred interfaces associated with offshore wind production and developing a coalition and a strategy for the cost reduction of offshore wind in the UK. The strategic inbetweenness of the Crown Estate within key actors in the offshore wind supply chain stems from its role as an owner of the seabed out to 12 nautical miles, an area referred to as territorial sea. The 2004 Energy Act also gave the Crown Estate the right to the Renewable Energy Zone: a territory of over 200 km^2 spanning to the limit of UK waters, as a way of extending the potential for renewable energy projects such as offshore wind. As a landowner of the seabed, the Crown Estate leases the land necessary for the construction of offshore generation and transmission infrastructure (such as wind turbines, substations and cables).

Although the Crown Estate is mandated to aid the state in achieving policy objectives, it is run as an independent commercial company with an independent mandate to generate income from the estates it manages. Therefore, the objectives of the Crown Estate could be said to be complementary to those of the state, but also on occasions divergent from them. The Crown Estate generates income from offshore wind through renting the seabed to offshore wind generators and OFTOs. The rent consists of a minimum fee independent of the volume of generated renewable electricity, and a larger fee, reflecting the volume of renewable electricity generated. This makes profits for the Crown Estate from offshore wind generation and transmission directly dependent on the volume of renewable electricity generated. The reason why the Crown Estate has developed as a very proactive enabling intermediary is because its income depends on high volumes of offshore wind generated on and transmitted across its land.

As a landlord, the Crown Estate is also strategically positioned in between regulator, the state, the offshore wind supply chain industry, OFTOs and offshore wind generators, which allows it significant scope for enabling the development of the UK offshore wind sector. The Crown Estate invested £100 million into a programme of derisking (removing risks associated with the different stages of offshore wind projects) and accelerating offshore wind (removing barriers and preventing delays to the timely and scaled-up development of the offshore wind industry). As a proactive landlord, the Crown Estate has, among other things, employed a team of experts on the policy regime, planning, the grid, supply chain, skills and project finance that is at the disposal of offshore wind developers to facilitate the development of the industry.

The Crown Estate's intermediary role involves a wide range of tasks and services such as coordination between stakeholders, bridging capacity gaps, mediation and facilitation, and forming and fixing relationships between actors. For example, the Crown Estate facilitates the signing of collaboration agreements for using the seabed and gets involved in negotiating conflicts between generators and OFTOs, often involving issues around blurred interfaces, such as access to equipment. Its intermediary work is aided by several offshore wind industry working groups, some of which include the Crown Estate as a member and some of which were created with its support.

However, by facilitating negotiation between the different stakeholders, the Crown Estate is also 'translating' the agendas and interests between those involved. Translation is a powerful process of re-representation and reordering of actors through the construction of new networks, which has the capacity to alter policy objectives (Callon, 1986; Beveridge and Guy, 2011). One example of this process of translation can be found in the network mobilised to define the cost reduction strategy for the offshore wind industry in 2011 and 2012. When the Renewable Obligation Certificate (ROC) band for offshore wind was reduced in 2011, the Crown Estate facilitated the production of two reports. The first, the Crown Estate Offshore Wind Cost Reduction Pathways Study, identified areas for reducing the cost of offshore wind, mainly focused on interventions in the supply chain to create economies of scale and vertical and horizontal collaboration; and renegotiating existing contracting arrangements. Building on these findings, the Offshore Wind Cost Reduction Task Force report ordered by DECC and led by industry, including the Crown Estate came out in June 2012. Both reports were used by DECC to formulate the future national cost reduction policy, while their production established a legitimate coalition for cost reduction in offshore wind in the UK, consisting of (apart from the Crown Estate, Ofgem and DECC) the industry lobbyist RenewableUK, the Carbon Trust, investors and supply chain companies. The findings of both reports align considerably with the interests of the supply chain companies and, since their introduction cost reduction has become almost synonymous with developing a national supply chain. While the Crown Estate was instrumental in building up the national coalitions for cost reduction, the two reports provided legitimation for the continued public and private investment and support behind offshore wind. Therefore, through its intermediary role the Crown Estate can be seen as enabling low carbon transitions in the UK.

Conclusion

This chapter has illustrated the important role of intermediaries in the governance of the crowded, blurred and complex space of energy, in the context of low carbon transitions. Intermediaries in energy systems can affect both energy policy, as in the case of Bulgargaz and Bulgartransgaz in Bulgaria, and the material infrastructure, as in the case of creating blurred interfaces through the governing of offshore wind in the UK. Intermediaries in energy governance can take a variety of forms. What the intermediaries discussed here have in common is their strategic embeddedness within the materiality of energy governance. Through their strategic position between actors and interests, intermediaries are able to shape energy infrastructure, facilitate, as well as block change associated with the objectives of low carbon transitions.

Intermediary activities represent a 'new reality' in the governance of energy. The case studies in Bulgaria and the UK demonstrate that intermediary work is key to achieving low carbon transitions objectives like gasification and integrating renewable energy into the grid, as well as governing the 'blurriness' of socio-technical change, and with it the growing number of intermediary spaces it creates. Therefore, the chapter also argues for the introduction of the concept of energy intermediaries and recognition for their importance in understanding energy governance. Intermediaries are instrumental in providing a framework for energy governance through which co-ordination can take place through organising a dialogue among policy stakeholders, working towards acceptance and/or coherence of interests and policy, and shaping the understanding of strategic capacities and interests of actors (as evident in the case of offshore wind in the UK).

The growing use of intermediaries in energy governance and the profound effect that they have on the implementation of policy poses the question: To what extent is existing understanding of energy governance sufficient in explaining the range of processes under way in low carbon transitions? Although energy infrastructure is supposed to serve as transmission belts for state policies, intermediaries embedded within energy infrastructure act to protect their interests, thus contributing to the realisation of certain outcomes, but in so doing sometimes precisely precluding the realisation of others. Thus, it is important to understand more clearly the limits and potential of intermediaries to redefine and follow a particular trajectory, which might challenge the rationale of the energy systems and with it, their strategic position.

Analysis of the two Bulgarian intermediaries shows how the process of intermediation can redefine the objectives of the state and thus, misinterpret and oppose a particular strategy for low-carbon transitions. Bulgargaz and Bulgartransgaz maintain a strategic position within the gas transmission network, controlling aspects of the energy infrastructure and limiting the significance of natural gas as a transition fuel within the territory of Bulgaria. While the Crown Estate was pivotal in enabling the production and transmission of offshore wind in the UK, thus facilitating the development of a specific low carbon trajectory around offshore wind. Thus, by strategically configuring relations between different system actors and components, intermediaries play a pivotal role in the making of assemblages across space and the spatial distribution of energy flows.

Parastatal intermediaries Bulgargaz and Bulgartransgaz did not seem to be influenced by regulation alone, while the Crown Estate seemed to respond well to financial incentives, pointing to the fact that intermediaries need to be explicitly considered in policy design in the context of low carbon energy transitions, and targeted with a mixture of market and regulatory mechanisms. Their level of independence from government policy and relative power due to strategic inbetwenness, makes them a complex actor, rather than just a benevolent accelerator of policy and technology.

Acknowledgement

The Bulgarian case study on natural gas was discussed in Hiteva, R. and Maltby, T. 'Standing in the way by standing in the middle: The case of state-owned natural gas intermediaries in Bulgaria'. *Geoforum* 54, (2014) 120–131.

Bibliography

Amin, A. and S. Graham. 'The ordinary city', *Transactions of the Institute of British Geographers*, 22, (1998): 411–429.

Beveridge, R. and S. Guy. 'Innovation to intermediaries: translating the EU Urban Wastewater Directive'. In T. Moss, S. Guy, S. Marvin, and W. Medd (eds.) *Shaping Urban Infrastructures: Intermediaries and the Governance of Socio-technical Networks*, London: Earthscan, 2011, 92–107.

Bijker, W. 'Do not despair: there is life after constructivism'. *Science, Technology and Human Values* 18, no. 1 (1993): 113–138.

Bouzarovski, S. 'East-Central Europe's changing energy landscapes: a place for geography'. *Area* 41, no. 4 (2009): 452–463.

Bouzarovski, S., M. Bradshaw, and A. Wochnik. 'Making territory through infrastructure: the governance of natural gas transit in Europe'. *Geoforum* 64, (2015): 217–228.

Brenner, N. 'Between fixity and motion: accumulation, territorial organisation and the historical geography of spatial scales'. *Environment and Planning D: Society and Space* 16, (1998): 459–481.

Callon, M. Some elements of a sociology of translation: domestication of the scallops and the fishermen of St Brieuc Bay. In J. Law (ed) *Power, Action and Belief: A New Sociology of Knowledge?* London: Routledge and Kegan Paul, 1986, 196–233.

Graham, S. and S. Marvin. *Splintering Urbanism. Networked infrastructures, Technological Mobilities and the Urban Condition*. London: Routledge, 2001.

Grübler, A. and N. Nakicenovic. 'Decarbonizing the global energy system'. *Technological Forecasting and Social Change* 53, no.1 (1996): 97–110.

Hiteva, R. 2013, 'Geographies of energy governance – negotiating low-carbon and energy secure futures in the European Union', Doctoral Thesis submitted at the School of Environment and Development, University of Manchester.

Hiteva, R. and T. Maltby. 'Standing in the way by standing in the middle: the case of state-owned natural gas intermediaries in Bulgaria'. *Geoforum* 54 (2014): 120–131.

Kessides, C. 'Institutional options for the provision of infrastructure'. *World Bank Discussion Paper* 212. Washington, DC: World Bank, 1993.

Latour, B. *We Have Never Been Modern*. Cambridge, MA: Harvard University Press, 1993.

Law, J. and W. Bijker. 'Postscript: technology, stability and social theory'. In W. Bijker and J. Law, *Shaping Technology, Building Society: Studies in Sociotechnical Change*, London: MIT Press, 1992.

LCS-RNet (International Research Network for Low Carbon Society). Achieving a low carbon society. Sharing knowledge to meet a common challenge. Synthesis Report of LCS-RNet Fourth Annual Meeting, 2012.

Marvin, S. and W. Medd. 'Sustainable infrastructures by proxy? Intermediation beyond the production-consumption nexus'. In D. Southerton, H. Chappels and B. van Vliet (eds) *Sustainable Consumption. The Implications of Changing Infrastructures of Provision*. Cheltenham, UK: Edward Elgar, 2004, 81–94.

Mazucatto, M. *The Entrepreneurial State: Debunking Public vs. Private Sector Myths. Anthem Other Canon Economics*, Anthem Press: London, 2013.

Medd, W. and S. Marvin.' Strategic intermediation: between regional strategy and local practice'. In T. Moss, S. Guy, S. Marvin, and W. Medd (eds.) *Shaping Urban Infrastructures. Intermediaries and the Governance of Sociotechnical Networks*. Earthscan: London, 2011, 143–159.

Moss, T., S. Guy, S. Marvin, and W. Medd. 'Intermediaries and the reconfiguration of urban infrastructures: an introduction'. In T. Moss, S. Guy, S. Marvin, and W. Medd (eds.) *Shaping Urban Infrastructures. Intermediaries and the Governance of Socio-technical Networks*, London: Earthscan, 2011, 1–13.

Newell, P. and M. Paterson. 'A climate for business: Global warming, the state and capital.' *Review of International Political Economy* 5 (1998): 679–704.

Painter, J. 'Rethinking territory'. *Antipode* 42 (2010): 1090–1118.

Randles, S. and S. Mander. 'Mobility, markets and 'hidden' intermediation: aviation and frequent flying'. In T. Moss, S. Guy, S. Marvin, and W. Medd (eds.) *Shaping Urban Infrastructures. Intermediaries and the Governance of Socio-technical Networks*. London: Earthscan, 2011, 124–137.

Rohracher, H. 'Constructing markets for green electricity: the "soft power" of intermediaries in transforming energy systems'. In T. Moss, S. Guy, S. Marvin, and W. Medd (eds.) *Shaping Urban Infrastructures. Intermediaries and the Governance of Socio-technical Networks*. Earthscan: London, 2011, 75–91.

Silve, F. and P. Noël. 'Cost curves for gas supply security: the case of Bulgaria' 2010, Cambridge Working Paper in Economics (CWPE), at: www.repository.cam.ac.uk/handle/1810/242081 (accessed 7 January 2016).

Smil, V. *Energy Transitions. History, Requirements, Prospects*, Oxford: Praeger, 2010

Sorrell, S. 'WHO OWNS THE CARBON? Interactions between the EU Emissions Trading Scheme and the UK Renewables Obligation and Energy Efficiency Commitment'. *Energy and Environment* 14, no. 5 (2003): 677–703.

Part 2

Energy landscapes and the public

Introduction

Martin J. Pasqualetti

Energy is a public matter. Publics provide supply. Publics create demand. Efforts to match demand with supply can strongly influence global politics, international trade, spatial patterns of energy ethics, and the competing forces of globalization. In short, energy is not just a public matter. It is also a matter of geography. As such, energy involves issues of space and distance. Given the growing population of the world, the buffering space once more common between publics and the effects of energy development continue to shrink. For this reason, all energy projects – present and proposed – tend to influence health, safety, lifestyle and public well-being. Such influence is manifest most commonly in the form of 'energy landscapes'.

Often such landscapes are visual affronts to personal aesthetics – such as might occur from the installation of wind farms and transmission pylons. Other times they produce risks to public safety, such as often accompany mine subsidence and acid mine drainage. Still other times they can also include gross environmental injustices – such as from coal mining northeastern China and eastern Kentucky, and oil development in the Niger River delta.

The people living in these areas have long 'suffered in silence' with little attention from outsiders. For example, in Western countries landscape transformations attracted little attention up to the 1960s when concerns raised by environmental movements started to be recognized by larger publics. This happened in the context of abundant reserves of fossil fuels, low fuel prices, and nuclear power that was thought of as 'too cheap to meter', and a global communication network as yet too feeble to have global influence. Local protests rarely attracted outside attention. Thus, for many publics, energy landscapes in all their many forms either have been accepted as 'normal' or have been literally out of sight and out of mind. They were simply isolated consequences of the energy resources that various publics needed to use in order to survive and prosper.

Today, however, energy landscapes are now almost ubiquitous, more visible than ever before. In the West, public awareness of energy landscapes started to change with a series of well-publicized events that included the grounding of the Torrey Canyon in 1967, a blowout in the Santa Barbara Channel in 1969, the grounding of the Amoco Cadiz in 1978, an alarming oil embargo in 1979, accumulating acid-rain damage to forests in the early 1980s, the reverberations of the devastating nuclear accident at Chernobyl, Ukraine, in 1986, and the growing unease with the construction of large hydro-electric dams in the US and elsewhere. With the realization

135

that fossil fuel combustion was increasing the levels of heat-trapping gases in the atmosphere, the question of landscape transformation for energy production has become more and more contested.

Owing to the growing capabilities of worldwide communication, new publics have increasingly been seeing energy landscapes that had until then escaped widespread notice. Photographs of the wrecked Torrey Canyon and its aftermath on the coasts of England and France appeared on magazine covers and newspapers from Iceland to Argentina. Images of blackened beaches and choking birds from the Santa Barbara mishap shocked the world. Oil from the Amoco Cadiz coated the Brittany coast only a few years after it had been scrubbed clean of oil from the Torrey Canyon. These visualizations, plus oil embargoes imposed by Organization of the Petroleum Exporting Countries (OPEC), raised public awareness of the environmental costs of energy demand while simultaneously illustrating the globalization of its supply.

Recognition of the growing proliferation of energy landscapes grew from there. Soon, there were reports that forests were dying in New York State, Norway, Sweden, southern Germany, and elsewhere. The cause was acidification, but the question was where was the acid coming from?

Eventually, geographers and climate scientists solved the mystery by zooming out on their maps. The problem, they found, originated with the sulfur dioxide emissions spewing from coal-burning power plants hundreds of miles upwind.

While the symptom was the visible decline in the forests, the origin of the pollution was so far away that it had escaped easy detection. The living example of acid rain's effect on the landscape taught the valuable lesson that distance alone was an insufficient protection from the costs of power. Attractive, untrammeled, and often bucolic landscapes were being transformed into energy landscapes with little expectation or active responsibility.

The forests being decimated by coal burning were just one further representation of energy's new spatial order. It was soon to also include nuclear energy. In 1986, radioactivity from the explosion and meltdown of Chernobyl unit 4 changed the lives of publics over much of Europe and the former Soviet Union. It resulted in the creation of energy landscapes of abandoned buildings, heaps of rusting emergency equipment, and forestlands scraped clean, fenced and placed off limits to human use. For the first time, we were talking about energy landscapes as 'dead zones', an area expected to remain uninhabitable into the foreseeable future. Similarly, the search intensified for permanent energy landscapes where we could dispose of spent nuclear fuel from routine commercial operations.

In contrast to the creation of energy landscapes from the invisible hazards of nuclear power, large hydroelectric projects offer up something different. They produce landscapes that are both noticeable and direct. They create lakes by submerging lands, and they result in new settlements as the displaced publics relocate to drier ground. As hydro projects create energy landscapes they often generate strong and sustained public resistance and international condemnation. Such has been the case in places including southern Turkey, western India, eastern China, and most recently Ethiopia.

Once commonplace, it is no longer possible to hide the impacts that energy demand has on the landscape. This is especially true considering the trajectory of their scale. In less than half a century, they have evolved from localized examples – such the headgear from deep coalmines in West Virginia, all the way to global concerns – such as global warming – that should worry everyone on the planet.

The quest for energy, wherever it is manifest, has become a visible and public matter in the form of energy landscapes which provide unprecedented visual evidence of the costs we pay for the energy demand we create. It has also produced a conundrum we have yet to resolve:

How can we meet energy demands that the public generates without creating energy landscapes the public would rather avoid?

The trend we are observing is pressing all of us in several ways. First, profiting from little self-restraint and insufficient official oversight, energy developers continue seeking resources wherever reserves are suspected, frequently with full disregard for the associated environmental and human costs. This has witnessed a move from locations that are convenient but in decline to areas that are more abundant but within environments that are more sensitive to damage. The Arctic Ocean, permafrost areas of northern Siberia, Alberta's boreal forests, tropical rain forests in Ecuador, and aboriginal lands in Australia serve as several examples.

Second, we continue to struggle with the future role of nuclear power. Should we favor building the next generation of power plants because they produce electricity carbon free, or should we avoid nuclear power because of the near-term risks of another major accident and the long-term risks of nuclear waste storage?

Third, technical adjustments, inventions, and refinements are expanding our access to greater proportions of existing resources through improvements in hydraulic fracturing, deep-water recovery, and in-situ mining, all of which are raising the prospect of creating more intense and more frequent energy landscapes.

Fourth, we continue developing hydropower despite the fact that such projects can transform landscapes, increase seismic activity, force massive human relocations, and result in greatly increased evaporation rates even as such projects promise the advantages from economic development.

These, among other questions, are addressed in Part 2. The seven chapters here address the most salient issues of the energy landscapes that have become part of our modern existence. Typically, they arise from attempting to meet our energy demand in a world that is increasingly overtaxed. They encompass issues of public opinion, democratic process, participatory practice, environmental justice, as well as central geographic themes of siting, scale, distribution, and land use.

Karen Bickerstaff and Phil Johnstone start us off with the perennial dilemma of nuclear facility siting. They highlight key tensions in how participation and democracy are being enacted and scaled at the front end (power generation) and back end (waste management) of the nuclear fuel cycle. They question the means and ends of participatory practice in the nuclear sector at a time when many countries are looking towards new nuclear build as a solution to the twin policy challenges of climate change and energy security.

Susan Christopherson examines many of the concerns of shale gas development, a newly refined approach that coaxes natural gas out of formations previously considered uneconomic. Its implementation has produced substantial public resistance, especially where fossil fuel developments have not previously been common. She asks how geographers can help guide the choices we make about our energy futures.

Richard Cowell addresses issues of place attachment and landscape change in the UK's 'dash for gas', as natural gas is being substituted for coal. One of the questions being encountered in crowded places such as the UK is where to put such facilities. Cowell suggests reusing existing energy sites, a form of site 'recycling', as a means to avoid public resistance.

Patrick Devine-Wright and Matthew Cotton evaluate a citizen deliberation workshop that was used to engage residents about a local proposal to site a high-voltage transmission line in the UK. These transmission lines, much like the introduction of any new energy facility, commonly stimulate strong objections from affected communities. Transmissions line routing is a particularly knotty problem because it produces such a noticeable impact on the lands of many different uses. They disrupt viewsheds, divide land use and suggest health risks. Generally, the public do not like them nearby. The authors conclude that, while public participation of this kind does enable learning about technology proposals, it can also increase feelings of threat about those same proposals, in contexts of mistrust in power line companies.

Using the example of Most, Czech Republic, Bohumil Frantál highlights issues of environmental injustice that dominate the development of the huge lignite mines underlying villages a few dozen miles northwest of Prague. These shallow and extensive deposits have a long history of development and continued expansion. Already in the past, accessing coal has required the relocation of populations. Frantál asks whether the coal reserves most coveted by developers are those that just happen to underlay relatively powerless Roma people.

Marina Frolova's location is Andalusia, Spain, where she analyzes the varying discourses and practices of local stakeholders in the context of hydroelectric developments. She raises some of the critical questions that arise in all areas of proposed hydropower development, especially issues of place attachment and alteration. Her objective is to show how both landscapes and hydropower are institutionalized and constructed as shared, collective entities.

Returning to nuclear waste disposal, Michael Greenberg, Molly Coon, Matthew Campo, Jennifer Whytlaw and Edward Bloustein propose the acronym CLAMP (Choosing Locations at Major Plants) to abbreviate the option of storing commercial high-level nuclear waste on the sites where nuclear defense waste already exists. Such treatment reinforces the fact that nuclear power is an energy technology that produces waste no one wants and everyone should fear, and one that holds hazards for the public for thousands of years. CLAMP would seem one possible way to reduce the problem we have created.

In sum, Part 2 presents seven explorations into the relationships between our energy demand and the landscapes that we have all had a hand in creating. They highlight many of the interactions that have become a common source of worry and conflict in today's crowded world. However, they also serve to give us hope. Now that such landscapes are more publically recognized, it has made us more alert than ever to the costs of energy. Cataloging, understanding, and mitigating these costs are services geographers are particularly suited to provide.

9

The re-scaling of energy politics

UK nuclear facility siting in historical context

Karen Bickerstaff and Phil Johnstone

UK nuclear policy: a history of antagonistic politics

In the UK, a defining moment that tied together UK policy on nuclear waste and (new) power was the 1976 report on *Nuclear Power and the Environment*, published by the Royal Commission on Environmental Pollution (RCEP), under the Chairmanship of Lord Flowers. In this report, the RCEP expressed serious concerns about the state of nuclear waste[1] management – the accumulation of wastes at nuclear sites and the absence of any clearly formulated policy for the disposal of the waste. The conclusion of the report was that it was 'morally wrong' to keep generating nuclear waste without a demonstrably safe way of storing it. The RCEP called for a national disposal facility, and stated that 'there should be no commitment to a large programme of nuclear fission power until it has been demonstrated beyond reasonable doubt that a method exists to ensure the safe containment of long-lived highly radioactive waste for the indefinite future' (Royal Commission on Environmental Pollution, 1976: 131). From this point on, dealing with the waste legacy has been viewed as the Achilles Heel of the nuclear industry, an apparently insoluble problem continuing into the far future, and halting the onward progression of nuclear energy (Blowers, 2010).

The RCEP report was also pivotal in highlighting the need for greater levels of public participation in relation to the issue of nuclear power. The post-1947 planning system gave the power of planning decision-making explicitly to politicians, and largely to politicians at the local authority level. Successive Planning Acts have retained this critical procedural assumption (TCPA, 2012). In tandem, there has been a traditional, and to a degree unchanging, way in which major infrastructure projects (major roads, power transmission lines, waste water facilities, airports and power stations) have been planned and given permissions. Normally some sort of wider strategic thinking existed, which then led to the identification of projects. But this had to move through a process of local planning consents and public inquiry – the latter with cross-examination of witnesses as well as written evidence (Marshall, 2011).[2] Anyone could observe a public inquiry, but only active participation had the potential to affect the inquiry's outcome – a daunting form of engagement that required considerable preparation. A planning inspector would send the Secretary of State a summary of the evidence, along with a recommendation. The Secretary of State took the final decision, and did not have to accept the recommendation, provided that he gave good reasons. Wynne (2010) makes clear that during the 1970s and 1980s a debate emerged around what the function of the inquiry should be in terms of its role in wider policy formation: for government and industry the purpose of the inquiry was often seen in modest terms, as a space to carry out technological assessments to inform pre-ordained decisions, whereas increasingly various public interest groups saw the inquiry as a political space for changing the direction of policy itself. The 1980s saw the public inquiry being utilised by a variety of non-governmental groups, particularly around road building plans where they were used as a space to challenge policy. In the late 1970s and 1980s nuclear power became synonymous with the 'big public inquiry' (Kemp et al., 1984). Before these inquiries, nuclear decision making was essentially a private affair for uniquely protected governmental institutions. But, due to the formal separation of civilian and military nuclear matters in the early 1970s and the rise of environmental NGOs, nuclear was increasingly scrutinised within the available institutional spaces of the British Planning system including hearings and inquiries (Patterson, 1979). The 'big' public inquiries included The THORP inquiry (1977–78) into the construction of a controversial reprocessing facility (Wynne, 2010), the Sizewell B inquiry taking place between 1982–85 (O'Riordan, 1988) and the Hinkley C inquiry which occurred from 1988–89 (Aubrey, 1991). The latter two concerned the construction of new nuclear reactors. The complex scalar politics and particularly the tension between *national* policy and *local* particularities, and the degree to which each scale of political decision-making should be engaged within the inquiry setting became a key theme. Purdue et al. (1984) point out, in relation to the Sizewell B inquiry, that the bulk of evidence was coming from national environmental pressure groups as opposed to the local community. The Sizewell inquiry running from 1982–85 was a milestone in terms of the 'opening up' of the nuclear black box to non-governmental scrutiny, with debate around a plethora of issues that had rarely been engaged with by the public (Kemp et al., 1984). In particular, notions of 'tolerable risk' were deliberated on extensively as many of the complexities involved in running, maintaining, and ensuring the safety of a nuclear power plant were publicised for the first time through the cross-examination process (O'Riordan, 1988). The 'political opportunities' made available by the inquiry – opportunities that were not available through the normal channels of representative democracy, at least in terms of nuclear issues – thus represented the politicisation of the civilian nuclear power (Wynne, 2010).

Of course, this only tells one side of the public inquiry story: many saw the inquiry as being a tool of government to allow activist groups to 'blow off steam' without influencing the direction of policy rather than a space of democratic potential, (Rough, 2011); and inquiries remained highly technical and have been described as being 'tortuous' experiences due to their length and complexity (Owens, 1985). For others, such as the Confederation of British Industry (CBI), concerns were raised that the public inquiry was causing inefficiencies in planning due to the fact that they were increasingly being used as sites to challenge the basis of national policy which industry believed should be fixed prior to any engagement with the public inquiry, a fora that should be limited to 'local' issues. During the Hinkley C inquiry, in the late 1980s, it was expressed by the Central Electricity Generating Board (the nationalised electricity supply industry) that the inquiry should only focus on 'local' issues and not discuss underlying arguments related to alternative energy policies, economic evaluations and so on (Johnstone, 2013). Hinkley C obtained planning permission in 1990 following a lengthy public inquiry, but was subsequently abandoned as uneconomic when the electric power industry was privatised.

Over the same period the UK saw successive unsuccessful attempts at finding a disposal route for higher level nuclear wastes – a strategy led by the Nuclear Industry Radioactive Waste Executive (Nirex), a body funded by organisations that produced nuclear waste. This culminated in efforts in the 1990s to develop a geological disposal (GD) option at Sellafield in West Cumbria.

In 1992 Nirex identified a need for a Rock Characterisation Facility (RCF), an underground laboratory to investigate the detailed properties of the potential host rock. A borehole investigation programme was undertaken, but increasingly Cumbria County Council,[3] as Planning Authority, was concerned about the suitability of the proposed site and refused consent for the RCF in 1995. The County's case was essentially that this was 'a poor site – chosen for the wrong reasons [i.e. social acceptability]'. Local (authority) opposition was, at the time, borne of a lack of trust in the transparency and openness of those leading early processes in the national site selection – notably the Nirex Board decision to *decide* on Sellafield and then *defend* it, even as doubts arose (Simmons et al., 2006). Nirex appealed, resulting in a public inquiry which upheld the Council's decision. The then Secretary of State, John Gummer, in 1997, confirmed the inquiry's conclusions, stating that he remained: 'concerned about the scientific uncertainties and technical deficiencies in the proposals presented by Nirex [and] about the process of site selection and the broader issue of the scope and adequacy of the environmental statement' (Styche, 1997).

Taken together, the abandonment of Hinkley C power station and the failure of Nirex's proposal for an RCF in West Cumbria appeared to signal the beginning of the end of the nuclear industry in the UK. Our purpose in bringing these two arenas of UK nuclear policy and planning together is to address the differential modes of participatory engagement, and the rescaling of political opportunities, that followed the siting failures at Hinkley and Sellafield. We begin with a brief overview of key geographical themes in existing work on energy infrastructure siting in general and nuclear infrastructure siting in particular.

Geographical engagements with nuclear siting

Much of the existing geographical literature on energy infrastructure siting policy in general and nuclear sites in particular has centred on the notion of NIMBY 'not in my back yard' – defined in terms of locally organised, often emotional, campaigns opposing a locally unwanted land use. Specifically, NIMBY describes opponents of new developments who recognise that a facility is needed but are opposed to its siting within their locality (Burningham, 2000). Such action can then be dismissed as selfish, irrational, and costly to society – making essential projects impossible to site. In the substantial literature on the siting of nuclear and hazardous facilities in the UK during the 1980s and 1990s, the concept went largely unchallenged and gained wide usage to describe the public's response to the search for suitable nuclear waste disposal sites (Kemp, 1990). The application and validity of NIMBY has come under widespread scrutiny and critique – for its limited conceptualisation of community and the manifold (political, cultural, and geographical) contexts that influence opposition to a siting decision (Wolsink, 2006) as well as for a lack of empirical support (Devine-Wright, 2011). For Devine-Wright (ibid.), NIMBY is a pejorative label attributing hostility to the ignorance, irrationality, or prejudice of members of the public involved in technology controversy – homogenizing and simplifying their arguments. Indeed, the language of NIMBY retains a simplistic scalar framing of the problem, centring on competing scales (notably national versus local) of social interests. The critique of NIMBY arguments has been focused, in part at least, on accounts of siting decision-making processes connected to breaches of social and spatial justice. On this point, there has been a considerable engagement with justice concepts and, in distributional terms, the direct or indirect targeting of vulnerable groups such as indigenous and economically marginal communities for the siting of risky energy infrastructures and industries (Butler & Simmons, 2013; Cotton et al., 2014; Blowers and Leroy, 1994; Cowell et al., 2012; Bell and Rowe, 2012; Ottinger, 2013; Wolsink, 2007; Conde and Kallis, 2012).

Concerns have also been raised about the democratic adequacy of procedures for the siting of new energy infrastructure – perceptions of the planning process as opaque and unfair, and a critique of the instrumentalism driving public participation activities – occurring too late in decision-making and primarily oriented to the delivery of 'socially acceptable' outcomes (e.g. McLaren et al., 2013; Wolsink, 2007). Cotton et al. (2014: 434), in the context of shale gas exploitation in the UK, refer to the prevalence of 'deliberative speak': 'the rhetoric of engagement not matched by mechanisms to ensure community involvement in decisions'. Recent attention has also rested on procedural responses to unfair outcomes in the form of policies of volunteerism, in which a community hosting a facility must be a willing partner, attracted by the economic gains on offer (e.g. Bickerstaff, 2012 on nuclear waste). It has been suggested that community volunteerism expands opportunities for local publics and stakeholders to meaningfully participate in siting decisions and that compensation measures offer a corrective to unfair distributional outcomes. Some have, however, contested this view – for instance, on the grounds that only already disadvantaged communities are likely to volunteer to host a nuclear facility (Bickerstaff, 2012; Cowell et al., 2012). Others have developed a post-political critique of the siting process for 'nationally significant' infrastructure planning – highlighting systematic structural and democratic inadequacies (Johnstone, 2014; Swyngedouw, 2010). The 'postpolitical' condition is conceived as that in which politics proper, identified as 'dissensus' involving competing ideologies of socioeconomic trajectories, is foreclosed, establishing a 'consensual' policy framework driven by the dominant ideological convictions of neoliberalism in the guise of a 'value-free', technocratic governance regime. An emerging feature of the post-political condition is the way in which agreement can be negotiated through the technique of appealing to universal themes, such as 'sustainable development', and 'climate change' which encourage, even necessitate, agreement and support in principle (Allmendinger and Haughton, 2012). Critics have stated that this will further marginalise the opportunities for opponents of major applications to gain a voice. Furthermore post-political arguments emphasise how the management of dissent is characterised by carefully allocating the handling of controversial issues to alternative methods and scales of planning (ibid.; Johnstone, 2014).

In what follows, and taking our lead from post-political arguments, we critically compare these two facets of nuclear (siting) policy in order to shed light on the relationship between the scaling of decision-making and the nature of nuclear politics. In doing so, we distinguish a period marked by considerable divergence in participatory processes (broadly up until 2013) and a more recent period of convergence around the rescaling of participatory opportunities.

Divergence and conflicting participatory tendencies in UK nuclear policy

The various reviews and inquires that came at the end of Nirex's search for a GD site make clear that the 'Decide, Announce, Defend' approach, by what were seen to be remote bodies (as Nirex was perceived to be), had failed. Blowers (2014: 546) refers to a 'palpable shift from confrontation to consensus' in the search for a solution to the problem of nuclear waste. Furthermore, 'decentralisation', 'participation', and 'community empowerment' had become touchstones of the post-1997 Blair Government, and there are 'striking parallels' between the language of New Labour and the aims of collaborative planning. Collaborative planning inspired from insights from Habermasian 'communicative rationality' had become increasingly popular as a means of promoting more participatory forms of planning as opposed to the top down approaches which had often dominated British Planning in the past. In response, 'dialogue' techniques were widely embraced, to clarify and promote engagement and reflection on the issues at stake. The Managing Radioactive Waste Safely (MRWS) process initiated in 2001 (DEFRA, 2001), leading up to and beyond a decision regarding how to manage higher activity wastes, can be read precisely as an embracing of a deliberative and participatory style of decision-making. MRWS established in 2003 the independent Committee on Radioactive Waste Management (CoRWM) – with a brief to develop and advise government on a UK long-term radioactive waste management programme which inspires public support. Between 2003 and 2006 a wide range of options for how to deal with the UK's higher activity radioactive waste were considered, involving extensive consultation with the public and expert groups, in efforts to clarify issues and achieve some level of agreement (or consensus). The work conducted by CoRWM included eight specially recruited discussion groups, 568 self-selecting discussion groups, four citizens' panels, a schools project, an open access web-based discussion guide, as well as stakeholder round tables and public meetings at 14 nuclear sites (Chilvers and Burgess, 2008). The Committee's recommendations, identifying geological disposal as the most appropriate end point and proposing an approach to finding a site based on cooperative arrangements (CoRWM, 2006), were largely accepted by Government. In June 2008 a white paper was published setting out a framework for implementing geological disposal through principles of volunteerism, in which communities, effectively local councils, express an interest in taking part in the process leading up to the siting of a facility (DEFRA, 2008).

Devine-Wright (2011) has remarked on the ways in which policy makers and developers involved in the siting of renewable technologies view the contexts of large-scale projects in simplistic, territorially, and temporally bounded terms. It is of note, then, that while a GDF would be a repository for the UK's higher activity radioactive waste, participatory engagement has become increasingly delimited to the comparatively small geographical area surrounding a possible facility – 'a specific community in a specific area' (DECC, 2013a: 25). In this regard, the UK's formal strategy for identifying a site has placed considerable emphasis on local participation, intended to contribute to the development and maintenance of community confidence – a process that rested on two key local decision points: (i) an Expression of Interest (EOI) – at which *local communities* would register their interest in discussions with government; and (ii) a Decision to Participate – at which local councils (as the decision-making bodies) would make a formal commitment to participate. From the Decision to Participate, local representatives would work with the Nuclear Decommissioning Authority (NDA)[4] to construct a mutually acceptable proposal. A key element of the process was the Right of Withdrawal (or veto) – in place up until a *late stage*, prior to construction. In effect, these local decision-making powers presented a new vehicle for collective agency on the part of democratic institutions that were effectively frozen out of previous decision-making processes. Community benefits packages -measures put in place to enhance the social and economic well-being of a host community – were introduced in efforts to spatially bound and compensate those (most) affected by a GDF (in terms of their exposure to risk – broadly defined).

The definition of volunteer communities and their associated decision-making powers reflected a simple three-way characterisation of 'host community' (the community in which any facility will be built); 'decision-making body' (the local government decision-making authority); and 'wider local interests' (other communities that have an interest). So, while local government would have decision-making authority for their community, and therefore be the body that could exercise the veto, the 2008 white paper did not specify which tier (or tiers) of local government should have this decision-making authority. In 2008–9, three formal Expressions of Interest were received by Government – from Allerdale Borough Council, Copeland Borough Council and also Cumbria County Council as the local planning authority (in respect of the areas of Allerdale and Copeland[5] – i.e. West Cumbria). As such the three councils constituted the decision-making bodies, and they entered the West Cumbria Managing Radioactive Waste (MRWS) Partnership with a range of other local organisations. For three years (2009–12) the Partnership sought to deepen its understanding on matters such as inventory, geology, planning and ethics, producing a final report in 2012 as a basis for the three councils to decide whether or not to proceed (Blowers, 2014). On 30 January 2013, the three councils took their decisions: Allerdale and Copeland Borough Councils both voted in favour of proceeding; Cumbria County Council voted against, thus bringing the process to a halt (DECC, 2014).

For Cumbria County Council, the process lacked 'credible public support . . . [reflecting a] lack of support from Parish Council, environmental organisations, concerned citizens and sectors of the business community' (Cumbria County Council, 2013). As the Government review of the siting process makes clear, the experience in Cumbria exposed the difficulties of the localisation agenda – and inevitable scalar tensions in establishing decision-making over national infrastructure as essentially a local matter (DECC, 2013a; Bickerstaff, 2012).

Turning to nuclear power, during the 1990s no significant proposals were put forward to build new power plants: the industry was in turmoil, and the contradictions caused by the dual aims of privatisation and preserving a new build agenda proved incompatible. The nuclear option was not reconsidered until the Energy Review of 2002 leading to an Energy White Paper the following year (DTI, 2003) which concluded that due to unresolved issues around waste and the poor economics of nuclear it was not being considered at that time, and a move towards renewables, energy efficiency, and a decentralised energy model was the best pathway to pursue. By 2007 a full U-turn in energy policy had occurred: after two controversial energy reviews (Johnstone, 2013) new nuclear was seen as an increasingly attractive option in the context of climate change mitigation and energy security.

With the 1980s' public inquiry experience of Sizewell as a key reference point, the UK government made clear the view that the planning system had 'frustrated' rather than 'facilitated' new nuclear (Hutton, 2008). This mirrored concerns being raised by industry: EDF, as a leading nuclear utility, had identified planning controls as a 'common barrier in the development of many energy related projects', that more consistent decisions were needed to 'balance the national and local interests where there is conflict' and that planning inquiries should focus on 'local issues' (EDF, 2005). Elsewhere, notable reviews of planning such as the Barker review (Barker, 2006) of land use planning and the Eddington transport study (Eddington, 2006), both published in 2006, focused on the need to 'streamline' and modernize planning, developing stronger national policy directives in order to enable the development of important infrastructure through the market in the increasingly competitive and challenging context of globalisation. The recommendations of these reports were taken up by government in the 2007 white paper 'Planning for a Sustainable Future' (HM Government, 2007).

In 2008, new nuclear construction was made official policy. However now, given the 'liberalised' nature of the UK energy market, it was up to private investors to come forward and for new nuclear to be built 'without subsidy', and there were significant concerns that the industry could be curtailed by 'barriers' in the planning system. A radical overhaul of the planning system, involving the abandonment of the public inquiry system, was widely seen as the solution to the delays that had previously befallen the siting of new nuclear (HM Government, 2007). The *Planning Act 2008* (HM Government, 2008) was a key piece of legislation designed to speed up planning for large scale, nationally significant, infrastructures. The *Planning Act* put in place a new decision-making process based on National Policy Statements (NPSs) and a series of hearings. The NPSs are produced by Government. They give reasons for the policy set out in the statement, and must include an explanation of how the planned infrastructure takes account of Government policy relating to the mitigation of, and adaptation to, climate change. They include the Government's objectives for the development of nationally significant infrastructure in a particular sector. Crucial to our analysis are changes to the planning system, as amended by the *Localism Act 2011*, which passed responsibility for dealing with development consent orders for nationally significant infrastructure projects (NSIP) from local authorities to an Infrastructure Planning Commission (IPC) in 2009 – a non-departmental public body responsible for examining and deciding on proposed nationally significant infrastructure projects in England and Wales. The IPC closed in April 2012, with its functions transferred to the Major Infrastructure Unit of the Planning Inspectorate, which examines applications and makes recommendations to the Secretary of State for decisions. As such, decision-making effectively bypasses any form of local planning authority control. NPSs do undergo a process of public consultation and parliamentary scrutiny, before being designated (i.e. published). For nuclear power, a Strategic Siting Assessment (SSA) would also locate suitable sites for new nuclear power which would not be open to further deliberation once approved. For new build, the SSA criteria linked to the identification of (potential) new nuclear sites were devised by government, and advised by consultants with minimal input from local environmental groups, councils or communities (Blowers, 2010). Responses to the consultation, many of them heavily critical, were not openly considered by government and very few changes were made to the revised criteria (ibid.). The project developers are also required to carry out extensive pre-application public consultation prior to submitting their application direct to the Planning Inspectorate. 'Local' consultations would deal only with the specificities of the particular development in question. These measures were designed to militate against developmental risks, and encourage investment through reducing the substantial upfront construction costs of large-scale infrastructure. There was also a belief that 'the adversarial nature of the inquiry system for major infrastructure projects can be intimidating and make it difficult for local government, non-governmental organisations [NGOs] and members of the public to participate effectively' and claims were made that 'the time and costs involved means it often favours the well-resourced and well-organised over less well-off communities and citizens' (HM Government, 2007: 14–15).

The discussion thus far reveals how the history of nuclear policy in the UK has been marked not only by intense opposition but also, crucially, contradiction in terms of planning processes – with an increasing divergence in opportunity structures available to (local and national) publics to engage with and oppose projects. Moves to greater deliberation, partnership working and volunteerism in relation to GD planning sit uncomfortably with the centralist, top-down tendencies of decision-making regarding new nuclear power (cf. Blowers, 2010). There is a temporality here of course, with no sustained efforts to site nuclear power stations in the 1990s and early 2000s. We would also point to the decoupling of the two issues – so tightly bound in the 1976 RCEP report – and the necessity of establishing 'a demonstrably safe way of storing' nuclear waste to sustain a new build programme. That said, what connects the MRWS process (for waste) and reformulated planning arrangements for nationally significant infrastructure (for new build) is a desire to centralise decision-making over matters of 'need' and localise (and bind) participation to matters of implementation. In the next section we develop a comparison of the siting process and participatory opportunity structures associated with more recent nuclear infrastructure planning, specifically following the 2008 Planning Act – and consider the significance of an emerging *convergence* in the formulation and scaling of participation.

A post-political convergence? (Re)scaling decision-making

Hinkley Point C nuclear power station is the first of the eight UK sites announced by government to progress through the revised planning system. The rhetoric surrounding the new framework – separating national policy decisions from local consultation – was that the process would enhance public participation while increasing efficiency (DECC, 2013b). For Hinkley C, national policy discussions in Bristol, London and Manchester, had addressed, and apparently solved, the justificatory basis of new nuclear in terms of the substantive issues surrounding the technology including safety, economics, proliferation, and nuclear wastes produced from new build (discussed at length in Johnstone, 2013). It must be noted, however, that many NGOs and activists pointed out that the conclusions made during the National Policy Statement consultations, which were designed to settle matters of policy, relied on further consultations that were yet to take place or hypothetical 'solutions'. For example, it was argued that issues pertaining to the future production of nuclear wastes will be solved because a final repository *would* be built. The framing that repeatedly justified what some participants in the NPS consultations referred to as 'putting the cart before the horse' was the 'urgency' in dealing with climate change and securing energy supply as the overriding priority against which all other issues had to be assessed (Johnstone, 2013).

When it came to the local consultations, these were solely concerned with the 'site specific' issues and not matters related to nuclear policy. So to emphasise, at the point of the local consultations, the national 'need' for Hinkley C had already been established and fixed. Between 2009 and 2012 EDF hosted a range of exhibitions, meetings, and events where the emphasis was on a duty to 'demonstrate [the] business case' for the construction of Hinkley C by setting out detailed plans for construction, community benefits, and employment opportunities. Of course, the flip side to the increased fairness of the onus being on the developer to demonstrate their business case is that many people from communities surrounding Hinkley C saw these events being dominated by emphasis on job creation and substantial investment in community centres, and colleges that gave a one-sided picture of the impacts the development would have to the surrounding area (Johnstone, 2013). Over the course of the consultations controversies emerged around a range of issues including road congestion, noise disruption, the housing of workers for Hinkley C, to which EDF were forced to provide more information.

The sheer quantity of events put on around the Hinkley C development demonstrates there was a high level of visible public 'participation'. However, the question that remains is what kind of participation this constituted, and how the new scalar separation between the 'national' and the 'local' changed the qualities of participation with new nuclear development. The case of 'national need' concluded during the NPS consultations influenced proceedings related to the various decision-making stages of the Hinkley C development in more or less direct ways. For one, this relates to the rhetorical framing of the overriding priority of climate change mitigation. This was clear with a particular decision that was to be taken by the relevant local authority, to grant EDF permission for 'preliminary works' on the proposed site prior to planning permission being granted. It was argued that when it came to the recommendations made by the IPC, a recommendation in favour of the Hinkley C development was far more likely for a site where substantial works had already begun, rather than a greenfield site. However, the rationale put forward by EDF for this exceptional application was that it was necessary, 'to expedite the construction programme so that the new power station can be operational as soon as possible. [. . .] This is in accordance with the Government's policy to encourage the early deployment of new nuclear power stations' (EDF Energy, 2011). It was added that 'accelerating the completion date of Hinkley Point C by twelve months would result in a saving of approximately twelve million tonnes of carbon dioxide, which would otherwise be emitted from power stations burning fossil fuels'. While the decision was under the *Town and Country Planning Act 1990*, and thus the Local Planning Authority, the national policy and 'need' for early deployment in relation to CO_2 mitigation, impinged on the decision meetings of Sedgemoor District Council in important ways. Notably, the planning meeting that would decide this application opened with a senior Civil Servant from DECC, stating that the decision was that of Sedgemoor District Council alone, however the reason the application was being given was due to an exceptional case of 'national need' (Johnstone, 2013). The permission was granted, and heightened the growing controversy around the Hinkley development, and underlined how the rationale around CO_2 emissions, the topic of the 'national' justification, was being utilised as a means of 'speeding up' processes related to the specifics of a local development application.

A more direct example of the complex scalar politics surrounding Hinkley C relates to decision making with regard to the final application by EDF for consent for the two new reactors and the role of the IPC (now MIPU). The guidelines for the IPC arguably provide a crucial insight into how the scalar separation of nuclear issues serves to depoliticise nuclear policy. It was clearly outlined that the MIPU in assessing the case of EDF's application for Hinkley Point were not, during the planning hearings, considering 'matters of policy', which had been decided at the national level. Nor was it their remit to consider 'external' events that were occurring in the tumultuous world of energy policy (Johnstone, 2013). However, it was stated by DECC (2011: 8) that:

> when considering an application for a new nuclear power station that is capable of deployment by a date significantly earlier than the end of 2025, the IPC [now MIPU] should give substantial weight to the benefit (including the benefit of displacing CO_2 emissions) that would result from the application receiving development consent.

So, on the one hand, the MIPU must 'give weight' to a certain aspect of the nuclear debate, but ignore other potentially relevant factors in deciding whether the project should go ahead, or not. This was at a time when utility companies were pulling out of nuclear power in the UK, evidence was mounting of spiralling costs of the technology, the Fukushima disaster had raised issues regarding the intricacies of nuclear safety, and new information regarding the length of time that high level nuclear waste would remain at the Hinkley site was emerging. At MIPU hearings, as these substantive issues were raised, it was continually pointed out that it was not in the remit of these fora to discuss matters of policy (Johnstone, 2013). Policy was thus 'insulated' from the growing realities of nuclear elsewhere, in a manner that could be seen as an exemplary case of the 'foreclosure' of the political (Žižek, 1999). Increasingly, the MIPU proceedings, as well as the earlier NPS consultations, were not seen by NGOs and activist groups, by now including 'local' activist groups, as a place where any kind of worthwhile participation could take place (Johnstone, 2013) and thus through the scalar separation of nuclear issues, the 'national' discourse related to rapid deployment and CO_2 reduction was the issue under which all others were effectively 'trumped' (Johnstone, 2014), where 'certainty' and 'consensus' is produced through processes of rescaling.

Turning to the siting policy around GD, the UK context has changed somewhat since the MRWS process was initiated in 2001 and CoRWM began its deliberations. The governmental demands for acceleration of the national programme of site decommissioning and a policy of economic austerity that has resulted in widespread cuts in public expenditure, all created pressure to 'speed up the process' (Bergmans et al., 2015). Reformulated government proposals for implementation of a GDF were set out for consultation in 2013 (DECC, 2013a) following the withdrawal of West Cumbria from the voluntary siting process. These new proposals sharply reveal efforts to rescale decision-making, to co-opt or neuter a more confrontational mode of politics and to achieve specific planning outcomes. In the consultation document much is made of the principle of subsidiarity – to guarantee a degree of independence for a lower authority in relation to a higher body or for a local authority in relation to central government (DECC, 2013a). In the UK context, this principle has been reinforced by the *Localism Act 2011* which contends that '*power should be exercised at the lowest possible practical level* – close to the people who are affected by decisions, rather than distant from them' (Communities and Local Government, 2011: 4, italics added). In the context of the siting process for GD in the UK, and determining the lowest *practical* level, not only must the proximity of the local government body to the people affected be considered, but also the General Power of Competence, referred to in the *Localism Act*. On this basis, Government proposals stated that there should be one representative level of local government that holds the Right of Withdrawal and has the final decision on proceeding, subject to demonstration of community support, and that that should be the relevant District Council in England (DECC, 2013a). A similar, but more cautious, point is made in the 2014 white paper on Implementing Geological Disposal, which recognises that local representative bodies – including all levels of local government – will need to have a voice in this process, but that 'no one tier of local government should be able to *prevent* the participation of other members of that community' (DECC, 2014: 37). The response to policy failure in West Cumbria thus appears to be one of re-scaling decision-making responsibilities in order to facilitate a 'democratic' consensus. Yet any such site would have spatially extensive ramifications, linked to increased traffic volume, waste delivery and noise pollution. The shift in the scaling of local democracy, could be perceived as a simple switch of political scales of decision making in order to get a positive decision for the GDF, which was not reached last time (Johnstone et al., 2013). As Stewart Young, the leader of the County Council, responded to the proposals: 'It looks as though the government didn't like that decision and so they are inventing a new process that will exclude that [County] level of council' (BBC News, 2013).

The 2014 white paper retains a community right of withdrawal from discussions with the developer at any stage in the siting process. However, historically and geographically complex issues surrounding community participation – the roles and responsibilities of community representatives, how and when a test of public support will be made, and what options there will be for disbursement of community investment – are consigned to the deliberations of a new technical expert panel – the Community Representation Working Group (CRWG). The CRWG will help develop practical processes for community representation, the test of public support, and community investment. Alongside this deferral of complex (and controversial) issues to expert committees, the principle of volunteering is now firmly positioned within an amended *Planning Act 2008* which brings a GDF in England within the definition of nationally significant infrastructure. As such the decision-making process is pre-defined – 'involving objective examination by the Planning Inspectorate, which recommends to the Secretary of State who makes the final decision whether or not to grant development consent' (DECC, 2014: 37). If the community withdraws from discussions with the developer prior to the test of public support, the siting process in that community will stop. If the community's response to the test of public support is positive, the development proceeds, with the developer applying for planning consent for a GDF. The definition of a GDF as a nationally significant infrastructure, and the lack of public debate about such a 'big' change, raises concerns that 'in principle' decisions would foreclose debates at an early stage (Blowers, 2014: 552). It also sits uncomfortably with the localism principles enshrined in volunteerism – effectively it strips local authorities of the ability to stop a facility being sited within their boundaries (Johnstone et al., 2013).

Conclusions

In this chapter we have highlighted points of divergence and convergence in the trajectories of participation with nuclear waste governance and new nuclear build, with issues of scale playing a crucial role in redefining the ways in which participation is carried out. We have noted an 'opening up' of nuclear waste policy and a move towards greater participation, and 'horizontal' forms of decision-making – responding in large part to a history of costly siting failures. While these moves would not be enough to satisfy many theorists of 'post-politics' who would maintain that they were simply cynical exercises in the articulation of already foreclosed decisions (Wilson and Swyngedouw, 2014) this moment nevertheless enabled views to be aired and to influence (albeit in a short-lived manner) the direction of policy and dissensus to be enunciated. These can be considered as a brief, and partial, rupture point, where political issues are aired in the notoriously technocratic realm of nuclear siting policy (Mackerron, 2011). As the need to find 'solutions' to policy challenges around energy security and climate change were thought to become more urgent, the 'rescaling' of policy through the fixed separation of 'national' policy and the particularities of 'local' decisions, became established through the *Planning Act 2008*. The consequence of this rescaling, in the context of new build, and as witnessed in the NPS consultations and subsequent local consultations around the Hinkley C site, has been to curtail debate, even when new 'external' information regarding the state of nuclear in general came to light. Thus rescaling processes through the *Planning Act* could be considered as a case of the 'foreclosure' of the political with arguably serious 'post democratic' consequences for energy policy in the UK more broadly. The rescaling processes in relation to nuclear waste disposal are more recent, and could be characterised as a situation where in the face of not getting the desired siting decision in 2013 (Wainright, 2013) it is possible to observe the rescaling of the decision-making framework – to greater centralisation in site selection alongside a simultaneous localisation in the classification of a (volunteer) community. With new nuclear however, the foreclosure of the political, has seen the displacement of various 'returns' of the political to other sites and democratic tools within the broad landscape of energy policy, including direct action and networked forms of protest at the European level.

The other lesson learned from the new rescaled framework for new nuclear is that the entire rationale for the removal of dissensus from planning – speeding up the development of nuclear – has not been delivered. The Hinkley C development will take longer to come online than the last nuclear development in the UK, Sizewell B. So, there are serious considerations in terms of whether the rationale for the foreclosure of political dissensus, in the name of averting apocalyptic climate change (Swyngedouw, 2010) necessarily leads to the 'speeding up' of policy. This is yet to be seen with regards to nuclear waste policy in the UK, though certainly efforts to position GD as a 'local' issue have failed to deliver desired policy outcomes. Beyond curtailing proper political engagement, rescaling (upwards) decision making for major energy infrastructure, and tightly delimiting the scale and scope of public participation, is very unlikely to serve as a panacea for the intensely complex range of issues that confront nuclear power and nuclear waste disposal policy.

Notes

1   In the UK, radioactive waste arises from a variety of sources – predominantly military and civil nuclear programmes – and are classified according to the radioactive content of the waste and its half-life as high-level (HLW), intermediate-level (ILW), or low-level wastes (LLW).
2   All other infrastructure projects were determined by the relevant local *planning authorities* – the local authority that is empowered by law to exercise statutory planning functions for a particular area of the UK.

3   Local government in England operates under either a one-tier system – unitary authorities, or a two-tier system – county or district councils. In the two-tier system county councils' responsibilities include waste disposal and strategic planning. Some parts of England have a third tier of local government – town and parish councils – responsible for smaller local services such as parks.

4   With responsibility, on behalf of the Government and taxpayers, for overseeing the clean-up and decommissioning of 17 of the UK's civil public sector nuclear sites spread across the UK. The NDA is specifically responsible for implementing Government policy on the long-term management of nuclear waste.

5   The Sellafield nuclear complex sits within the Copeland boundaries and neighbours Allerdale.

References

Allmendinger, P. and Haughton, G., 2012. Post-political spatial planning in England: a crisis of consensus? *Transactions of the Institute of British Geographers*, 37(1), pp.89–103.

Aubrey, C., 1991. *Meltdown: The Collapse of the Nuclear Dream*, London: Colllins & Brown.

Barker, K., 2006. *Barker Review of Land Use Planning*, Norwich: HMSO.

BBC News, 2013. Nuclear waste repository: Cumbria attacks government move, 25/11/2013, www.bbc.co.uk/news/uk-england-cumbria-25041302

Bell, D. and Rowe, F. 2012 *Are Climate Policies Fairly Made?* York, UK: Joseph Rowntree Foundation.

Bergmans, A., Drago. K., Simmons, P., & Sundqvist, G., 2015. The participatory turn in radioactive waste management: deliberation and the social–technical divide. *Journal of Risk Research*, 18(3), pp.347–363.

Bickerstaff, K., 2012. "Because we've got history here": nuclear waste, cooperative siting, and the relational geography of a complex issue. *Environment and Planning A*, (11), pp.2611–2628.

Blowers, A., 2010. Why dump on us? Power, pragmatism and the periphery in the siting of new nuclear reactors in the UK. *Journal of Integrative Environmental Sciences*, 7(3), pp.157–173.

Blowers, A., 2014. A geological disposal facility for nuclear waste – if not Sellafield, then where. *Town & Country Planning*, 83(12), pp.545–553.

Blowers, A. and Leroy, P., 1994. Power, politics and environmental inequality: a theoretical and empirical analysis of the process of "peripheralisation", *Environmental Politics*, 3, pp.197–228.

Burningham, K., 2000. Using the language of NIMBY: a topic for research, not an activity for researchers. *Local Environment*, 5, pp.55–67.

Butler, C. and Simmons, P., 2013. Nuclear power and climate change: just energy or conflicting justice. In K. Bickerstaff, G. Walker, & H. Bulkeley, eds. *Energy Justice in a changing climate*, London: Zed Books, pp.139–158.

Chilvers, J. and Burgess, J., 2008. Power relations: the politics of risk and procedure in nuclear waste governance. *Environment and Planning A*, 40(8), pp.1881–1900.

Communities and Local Government, 2011. *A Plain English Guide to the Localism Act*, London: The Stationery Office.

Conde, M. and Kallis, G., 2012. The global uranium rush and its Africa frontier. Effects, reactions and social movements in Namibia. *Global Environmental Change*, 22(3), pp.596–610.

CoRWM, 2006. *Managing our Radioactie Waste Safely: CoRWM's Recommendations to Government*, London: Committee of Radioactive Waste Management.

Cotton, M., Rattle, I. and Van Alstine, J., 2014. Shale gas policy in the United Kingdom: an argumentative discourse analysis, *Energy Policy*, 73, pp.427–438.

Cowell, R., Bristow, G. and Munday, M., 2012. *Wind Energy and Justice for Disadvantaged Communities*, York: Joseph Rowntree Foundation.

Cumbria County Council, 2013. Letter from Leader and Deputy Leader of Cumbria County Council to Secretary of State for Energy and Climate Change, 7/2/2013, www.cumbria.gov.uk/news/2013/February/07_02_2013-135301.asp.

DECC, 2011. *National Policy Statement for Nuclear Power Generation (EN-6)* Volume I of II, London: The Stationery Office.

DECC, 2013a. *Consultation Review of the Siting Process for a Geological Disposal Facility*, London: The Stationery Office.

DECC, 2013b. *Leading the Way: The Uk's New Nuclear Renaissance*, London: The Stationery Office. www.gov.uk/government/speeches/leading-the-way-the-uks-new-nuclear-renaissance.

DECC, 2014. *Implementing Geological Disposal: A Framework for the Long-Term Management of Higher Activity Radioactive Waste*, London: The Stationery Office.

DEFRA, 2001. *Managing Radioactive Waste Safely: Proposals for Developing a Policy for Managing Solid Radioactive Waste in the UK*, London: The Stationery Office.

DEFRA, 2008. *Managing Radioactive Waste Safely: A Framework for Implementing Geological Disposal*, London: The Stationery Office.

Devine-Wright, P., 2011. Public engagement with large-scale renewable energy technologies: breaking the cycle of NIMBYism. *Wiley Interdisciplinary Reviews: Climate Change*, 2(1), pp.19–26.

DTI, 2003. *Our Energy Future: Creating a Low Carbon Economy*, London.

Eddington, R. 2006. *The Eddington Transport Study – Transport's role in Sustaining the UK's Productivity and Competitiveness*, main report from the Eddington Transport study, London: HMSO.

EDF, 2005. *Response to the 2005 DTI energy review*. Formerly Online (no longer available): www.edfenergy. com/report/report2005/downloads/edfenergy_energyreviewsub mission.pdf [Accessed December 1, 2010].

EDF Energy, 2011. *Hinkley C: Development Consent Order application: application summary document*, www.edfenergy.com/file/1663/download.

HM Government, 2007. *Planning for a Sustainable Future*, London: The Stationery Office.

HM Government, 2008. *Planning Act 2008*, London: The Stationery Office. Available at: www.legislation. gov.uk/ukpga/2008/29/pdfs/ukpga_20080029_en.pdf\nhttp://www.legislation.gov.uk/ukpga/2008/29/contents.

Hutton, J., 2008. Nuclear new build: government action. *BERR nuclear investor conference*. Available at: http://webarchive.nationalarchives.gov.uk/+/http://www.berr.gov.uk/pressroom/Speeches/page46589. html [Accessed March 5, 2012].

Johnstone, P., 2013. *From Inquiry to Consultation: Contested Spaces of Public Engagement with Nuclear Power.* Unpublished Thesis. University of Exeter.

Johnstone, P., 2014. Planning reform, rescaling, and the construction of the postpolitical: the case of the *Planning Act 2008* and nuclear power consultation in the UK. *Environment and Planning C: Government and Policy*, 32, pp.697–713.

Johnstone, P., Gross, M., MacKerron, G., Kern, F. and Stirling, A., 2013. *Response to the DECC Consultation of the Siting Process for a Geological Disposal Facility*, Sussex University: SPRU.

Kemp, R., 1990. Why not in my backyard? A radical interpretation of public opposition to the deep disposal of radioactive waste in the United Kingdom? *Environment and Planning A*, 22, pp.1239–1258.

Kemp, R., O'Riordan, T. and Purdue, M., 1984. Investigation as legitimacy: the maturing of the big public inquiry. *Geoforum*, 15(3), pp.477–488.

Mackerron, G., 2011. Foreword. In B. Wynne, ed. *Rationality and Ritual*. London: Earthscan, pp.x–xi.

Marshall, T., 2011. 'Reforming the process for infrastructure planning in the UK/England 1990–2010', *Town Planning Review*, 82(4), pp.441–467.

McLaren, D., Krieger, K. and Bickerstaff, K. 2013. Procedural justice in energy system transitions: the case of CCS in Bickerstaff, K., Walker, G., Bulkeley, H. eds. *Energy Justice in a Changing Climate*, London: Zed, pp.158–181.

O'Riordan, T., 1988. *Sizewell B: An Anatomy of Inquiry*, Basingstoke: Palgrave Macmillan.

Owens, S., 1985. Potential energy planning conflicts in the UK. *Energy Policy*, 13(6), pp.546–558.

Ottinger, G., 2013. The winds of change: environmental justice in energy transitions. *Science as Culture*, 22: 222–229.

Patterson, W., 1979. Environmental involvement in British civil nuclear policy. *Journal of the Institute of Nuclear Engineering*, 20(5), pp.1–6.

Purdue, M., Kemp, R. and O'Riordan, T., 1984. The context and conduct of the Sizewell B Inquiry. *Energy Policy*, 12(3), pp.276–282.

Rough, E., 2011. Policy learning through public inquiries? The case of UK nuclear energy policy 1955–61. *Environment and Planning C: Government and Policy*, 29(1), pp.24–45.

Royal Commission on Environmental Pollution, 1976. *Sixth Report. Nuclear Power and the Environment*, London: The Stationery Office.

Simmons, P., Bickerstaff, K. and Walls, J., 2006. *Country report – United Kingdom report to CARL project on stakeholder involvement in radioactive waste management*, Available from the authors on request.

Styche, P. C., 1997. Secretary of State's decision letter on the Nirex Appeal, 17 March 1997 (Authorised by the Secretary of State for the Environment to sign in that behalf), Government Office for the North West, 17 March 1997, para 9.

Swyngedouw, E., 2010. Apocalypse forever? Post-political populism and the spectre of climate change. *Theory, Culture & Society*, 27(2–3), pp.213–232.

TCPA, 2012. *TCPA Briefing: The Growth and Infrastructure* Bill, London: TCPA.

Wainright, M., 2013. Cumbria rejects underground nuclear dump, 30/01/2013. *Guardian* online. Available at: www.theguardian.com/environment/2013/jan/30/cumbria-rejects-underground-nuclear-storage [Accessed January 30, 2013].

Wilson, J. and Swyngedouw, E., 2014. Seeds of dystopia: post-politics and the return of the political. In J. Wilson and E. Swyngedouw, eds. *The Post-Political and its Discontents: Spaces of Depoliticisation, Spectres of Radical Politics*. Edinburgh: Edinburgh University Press, pp.1–24.

Wolsink, M., 2006. Invalid theory impedes our understanding: a critique on the persistence of the language of NIMBY. *Transactions of the Institute of British Geographers*, 31(1), pp.85–91.

Wolsink, M., 2007. Wind power implementation: the nature of public attitudes: equity and fairness instead of "backyard motives'. *Renewable and Sustainable Energy Review*, 11, pp.1188–1207.

Wynne, B., 2010. *Rationality and Ritual: Participation and Exclusion in Nuclear Decision-Making* 2nd ed., London: Earthscan.

Žižek, S., 1999. *The Ticklish Subject*, London: Verso.

10

Re-framing the shale decision

How do we evaluate regional costs and benefits?

Susan Christopherson

The shale gas and oil revolution: what is happening?

Since 2010, the US has experienced a dramatic industrialization process connected with the utilization of new technologies to commoditize a fossil fuel source that had hitherto been too difficult and expensive to extract (Curtis, 2002). The importance of the shale gas and oil transformation is evidenced by its rate of growth, which rose from 4.1 percent of total US production in 2005 to 23.1 percent in 2010 (Wang and Krupnick, 2013). By contrast with other resource extraction processes, shale development will impact a significant portion of the US, including possible extraction regions in 30 states (Christopherson, 2015; Finkel, 2015).

The shale revolution in the US is portrayed as an economic success and a model for other countries to emulate (Crooks, 2015). Despite decreasing prices for shale gas and oil, US production has continued, particularly in high productivity shale plays where extraction costs are lower. There is ample evidence that shale gas drilling has reduced the cost of natural gas, though this reduction has predominantly benefited industries such as petro-chemicals for which natural gas by-products are a feedstock, not residential customers. Lower oil prices have benefited refineries and lowered fuel prices for car owners. In addition, the US federal government has benefited from the tax revenues produced by shale development, an estimated $14 billion US in 2013.

But because of the fragmentation of policymaking in the US, and a growing awareness that there are environmental costs wherever the extraction industry develops its infrastructure, the industry has had to demonstrate benefits at the regional scale as well as at the "societal" or national scale (CBO, 2014; Mason et al., 2015). The need for a regional demonstration of benefit is magnified by the fact that the US has no national energy policy that could provide social license to oil and gas developers or provide cross-jurisdictional research to inform cost-benefit analysis (CBO, 2014).

While the environmental quality and public health consequences have been widely debated, other local costs and risks are frequently ignored in discussion of the shale decision because of an assumption: that local economies *always* benefit from the development of extraction industries on balance, despite negative externalities.

Recent research findings, however, raise questions about that assumption, and about how the costs and benefits to places affected by extraction industries have been weighed. New approaches have emerged for assessing costs and benefits of fossil fuel development, especially for extraction regions, and these new approaches are changing the questions that researchers ask about energy and energy development. Their thinking includes broader definitions of the "environment" to include community character, a more sophisticated conception of the geographic spaces involved in shale development, and a call for more transparency about methods for predicting, and then documenting, its social and economic consequences.

This chapter draws on this work to examine how the interpretation of energy development costs and benefits is being affected by what researchers have learned from looking carefully at the local and regional impacts in shale extraction regions. I probe the dichotomy that has been used to frame the shale decision: economic benefits versus environmental costs. I examine the contested results from analyses of the local and regional impacts of shale development using the narrowly defined perspective of job creation, tax revenues, and in some instances, negative externalities. I demonstrate how reframing taken-for-granted concepts about environment, space and time can change how researchers assess the costs and benefits of shale development.

How is the shale decision being framed?

There is now an extensive body of research into the potential impacts and risks attendant to the development of shale gas and oil in the United States (Finkel, 2015; Burnett, 2015; Small et al., 2014), and there is also broad agreement about *some* of what needs to go into the shale decision.

For example, a survey of experts from government agencies, the oil and gas industry, academia, and nongovernmental organizations sought to catalogue the environmental risks related to shale development, and the results indicate a consensus that the risks of greatest concern are contamination of surface water, methane leaks, failure of drilling well casings and cementing, and damage to habitats (Krupnick and Gordon, 2015). While the survey results indicate general agreement about the nature and extent of environmental risks associated with the development of shale gas and oil, there is disagreement about who should be responsible and accountable for ameliorating those risks. Non-industry survey participants identify industry as primarily responsible for solving the problems, while industry participants place responsibility on government and government regulators (ibid.).

But, evidence of potential environmental damage and economic, social and public health risks indicates the importance of examining the argument that shale gas and oil drilling has significant economic benefits for the nation and for regions in which the extractive activities take place. The degree, type, distribution, and sustainability of economic benefits are critical to the debate over shale gas and oil development. The debate is framed as one in which economic benefits – to the nation and to extraction regions as well – are weighed against evidence of social and environmental costs. As Burnett (2015: viii) suggests, the parsing of economic benefits is crucial to this debate because of the potential consequences, for example in polluted or depleted water supplies.

Hence, much of the literature on the economics of shale gas and oil development implicitly poses the question of whether the economic benefits of shale gas and oil development justify the acknowledged costs, the documented risks, and the potential damage (Mason et al., 2015).

Of course, as with any complex calculation of costs and benefits, a wise answer would be: It depends.

It depends on assumptions about how economic costs and benefits are to be counted (what is included and what significance it is accorded); about what is encompassed in the concept of "environment"; about the geographic scale at which costs and benefits are to be assessed; and about the relevant time frame for analyzing economic impacts.

What have we learned from regional analyses of the costs and benefits of shale development?

Regional studies form the majority of the US literature on the economic impact of shale development (this kind of analysis is much less prevalent in the UK because the national government owns the resource and has jurisdiction over the extraction process throughout the supply chain), and that literature in turn demonstrates an evolution in understanding what should be encompassed in an assessment of the costs and benefits of any kind of energy development, and what questions need to be answered in composing a meaningful cost–benefit analysis (Sunstein, 2015).

Many of the early analyses of regional economic impacts of shale gas and oil development were industry-supported efforts intended to foster a positive policy environment at the state and federal level. The analysts used models that would demonstrate the benefits of shale development, so as to garner support for extraction activities and development of the national infrastructure required to supply the extraction sites, dispose of waste materials, and transport the extracted product to market.

These analyses took the form of input/output (I/O) models that made projections of employment gains and tax revenues, using data packages designed for projection exercises such as IMPLAN[1] (IHS Global Insight, 2011; Considine et al., 2010, 2011). While very useful as a rough guide to the relative impact of new economic investments or events, I/O models have limitations as a tool to evaluate the overall impact of an industrial activity. Input/output models are used to estimate the *size* of the positive impact on jobs and taxes from an investment in a region or nation, but because they do not account for costs, the results are *always* positive. What is not widely understood among policymakers, the press or the public is that I/O models are *not* a cost-benefit analysis, and that they are particularly problematic if they are not modified by realistic assumptions based on local knowledge.

In the absence of actual employment statistics, these models provided a very powerful argument for moving ahead with shale development. Neither policy makers nor the media had any real understanding of how the models worked, and so they frequently used the *estimated* job and revenue figures from the models as realistic accounts of what would occur.

In Pennsylvania, for example, the initial job numbers used by the media to describe the economic impact of shale development were estimates from models developed by oil and gas industry affiliates. A press release in 2010 by an industry group, the Marcellus Shale Coalition, claimed:

> The safe and steady development of clean-burning natural gas in Pennsylvania's portion of the Marcellus Shale has the potential to create an additional 212,000 new jobs over the next 10 years on top of the thousands already being generated all across the Commonwealth.

These job projections spurred enthusiasm for shale development in Pennsylvania, and gave many people the impression that oil and gas industry employment would lead Pennsylvania quickly out of the recession. That did not happen. Pennsylvania's unemployment remained higher than the US average throughout the state's gas boom, and job creation has slowed significantly with decreasing gas prices.

In part, inaccurate estimates are related to the small size of direct employment in the oil and gas industry. By contrast with industries such as health care or education, which employ millions of Americans, the oil and gas industry creates relatively few jobs – just over 283,000 between 2005 and 2012, according to a multi-state coalition of economic analysts who closely examined the jobs claims (Mauro et al, 2013). And, these jobs are highly concentrated in a few US states – Texas, Oklahoma and Louisiana. An examination of data from the Quarterly Census of Employment and Work from the US Bureau of Labor Statistics indicates that in states hosting shale extraction, direct job growth is rarely over 30,000 jobs during the boom period when drilling is taking place, and that jobs decline dramatically after drilling ceases (author's calculations, 2015; Jacquet, 2011).

So, the big variable in job projections is how many additional jobs are credited to oil and gas development beyond the relatively small number of people directly employed in oil and gas extraction or direct support services.

In 2013, Pennsylvania's Department of Labor and Industry reported that just over 28,000 people were employed in oil and gas industry jobs. That figure was higher than the US federal data indicated, but not unreasonable. Then the Department also attributed another 200,000 jobs to shale development by adding 30 "related" industries. Because these related industries included such major employers as construction and trucking, this additional employment total included every construction worker and truck driver in the state. A driver delivering for FedEx and a housing construction worker were thus "claimed" as jobs produced by the shale industry. The Commonwealth of Pennsylvania attributed seven additional jobs to each one created in the oil and gas industries. A generous standard multiplier is 2 (Beattie and Leones, 1993).

In these initial studies, the primary source of projected indirect and induced job creation was not the extraction process itself. According to Considine et al. (2010), the most significant source of capital flowing into the extraction regions and creating new jobs and expenditures would be the royalties paid to land owners. Assumptions about who owns the mineral rights, where they reside, and when and where they will spend their royalties, all tended to inflate predictions of economic expansion in the extraction regions.

These very positive results also spurred a set of critical analyses that raised questions about the assumptions underlying the findings, and about what such models could and could not say about the economic impact of shale gas and oil development (Barth, 2012, 2013; Kinnaman, 2011; Kay, 2011; Kelsey et al., 2011).

Kelsey et al. (2011) constructed an alternative input/output model based on research into the residence of mineral rights owners and drilling workers in the extraction counties of Pennsylvania, and produced very different findings. While Considine et al. (2011) estimated the annual impact of shale gas extraction on the state of Pennsylvania economy at approximately $14.5 billion in 2011, Kelsey et al. (2011) estimated the annual economic impact at approximately $3 billion. The difference lay in assumptions about where people earning money from extraction could be expected to spend that money. An alternative I/O model for Ohio, conducted by Weinstein and Partridge (2011), also came to modest conclusions regarding the employment impacts on that state. Another note of caution about the results from the original industry studies was raised by Weber (2012), who constructed I/O models with more realistic assumptions for Arkansas, Wyoming and Texas. And Kinnaman (2011) noted that if an economy is at full employment when shale development is introduced into an area, there would not be a significant economic impact at all. He argued that if such a situation occurs,

The economy has simply shifted resources from the production of other goods and services towards the extraction of natural gas. Economic resources necessary to fuel a growing industry would either relocate from other regions of the country or shift from local industries within the region.

(Kinnaman, 2011: 1247)

So, while jobs are created in drilling regions during the boom period, they may divert workers from established local employers, whose hiring and other costs increase – a process called "crowding out". In most drilling regions, the two industries most likely to be affected by "crowding out" are agriculture and tourism (Rumbach, 2011).

In general, those reviewing the results of different input/output models used to predict economic impacts from shale development note that industry-sponsored models tend to produce much higher estimates of economic impact than those models constructed by academics using the same models (Barth, 2013; Kinnaman, 2011; Mauro et al., 2013). The difference lies in the assumptions underlying the models.

In addition to critical analysis of input/output models and the development of I/O models based on more realistic assumptions, researchers began to ask questions about the *costs* of shale oil and gas extraction to local and regional economies – the "negative externalities". A series of policy briefs and local studies, many originating at Pennsylvania State University, identified and documented local costs including increased public safety needs, traffic accidents, road congestion and damage from trucks, and demand on local social service and health facilities (Christopherson et al., 2011; Brasier et al., 2014). These studies were complemented by others examining the capacity of local and state governments to respond to the demand for services (Davis et al., 2014). The general consensus from this work was that while drilling for shale oil and gas brought a surge of economic activity and workers into the extraction regions, many localities did not have the governance capacity to address these new demands. In states such as Pennsylvania where extraction is not taxed, the state government also lacked the capacity to respond to boomtown needs. Although increased property tax receipts brought new revenue to these communities, many were always one step behind in keeping roads in good repair or hiring enough public sector workers to respond to the rapid rise in demand for services.

The bust phase of the cycle

The shale gas and oil revolution entered a new phase in 2012. With declining oil and gas prices, activity in many US shale boomtowns began to level off or to decline over time. This recent phase has been one in which researchers have turned to actual employment data, now available for up to ten years, to assess the economic impact of shale gas and oil development (Cosgrove et al., 2015; Munasib and Rickman, 2015; Wrenn et al., 2015). Although these studies come at the question of regional economic impact from different perspectives and using different models, they arrive at surprisingly similar conclusions. They find that the first wave of enthusiastic industry-supported models overstated the positive economic impacts of shale gas and oil development. They caution that economic impacts can differ from one locality to another, depending on the fiscal capacity of the locality or the policy framework within which extraction activities take place. They also agree that local and regional economic impacts from shale gas development are likely to be modest. Ultimately, these findings recalibrate what has been revealed to be an inflated earlier depiction of the economic benefits of shale development.

If policymaker and public expectations of economic benefits are critical to their willingness to grant social license to unconventional gas and oil development, then these more modest and contextualized results may affect thinking about the value of shale development to the regions in which extraction takes place.

However, research on extraction regions and the broader processes associated with the development of shale resources has reached beyond conventional analysis of economic impacts. It has raised a set of broader questions about how we frame analysis of energy development, its costs and benefits, at the regional and national scale. Cost–benefit analyses of shale development in the US have been used to simplifying complex decisions about the economy and environment by monetizing the trade-offs (Kinnamon, 2011; Mason et al., 2015). As the authors of cost–benefit analyses note, this approach has serious limitations because many of the externalities associated with hydraulic fracturing are difficult to monetize. In addition, as will be discussed in the next sections, cost–benefit approaches neglect the temporal and spatial dimensions of the development process.

In the following sections, we look at some of the research that is informing that analytical context, raising issues about the definition of "environment" and the spatial and temporal frames to be used in examining shale development's impact.

How do we define the "environment"?

Impacts on the physical environment at drilling sites (spills, noise, flaring), in extraction regions (water and air pollution), and beyond (water extraction, sand mining, toxic waste disposal) are critical to assessing the costs and benefits of shale development. In some required environmental impact evaluations, however, the concept of environment is expanded beyond the geophysical attributes of a region to include its aesthetic and cultural characteristics. This broader concept follows from established environmental impact policy in which the "environment" to be assessed encompasses "community character" and the cultural landscape.

So, for example, in an environmental assessment under the State Environmental Quality Review Act (SEQRA), the New York State Department of Environmental Conservation (NYSDEC) asks the question: Will the proposed action impair the character or quality of the existing community?

In describing what should be included in answering this question, they define community character as follows:

> Community character is defined by all the man-made and natural features of the area. It includes the visual character of a town, village, or city, and its visual landscape; but also includes the buildings and structures and their uses, natural environment, activities, town services, and local policies that are in place. Development can cause changes in several community characteristics including intensity of land use, housing, public services, aesthetic quality, and the balance between residential and commercial uses.
>
> (New York State Department of Environmental Conservation, 2015)

To some observers, the determination of impact on community character is impossible because it is based on subjective judgments. There has been an effort, however, to find ways to measure the significance of impacts on community character based on (1) the magnitude of an impact – its severity or extent, and (2) its importance – how many people are going to be affected by an action, over what geographic area, and over what period of time, including cumulative effects

(Ghilain, 2009). The legal basis for community character assessment is rooted in methods for understanding and interpreting cultural landscapes as practiced in environmental and cultural geography (Flad, 2015; Groth, 1997). Flad (2015: 1) defines the significance in an analysis of the impact of a shale gas development facility on community character:

> Significant adverse impact on community character can be ascertained through examination of potential impact on the visual environment, aesthetic and historic resources, and potential socio-economic resources, including heritage . . .

In making a case for assessing the significance of impacts on community character in environmental impact analysis, Ghilain (2009) points to the balance that must be struck between accepting the inevitability of change and recognizing the value of a sense of place to the quality of life for people who live in that place or region, and to those who visit as tourists. In the US, the protection of local quality of life via control over the physical environment has historically been the role of local zoning and land use regulation. State or national regulations allowing oil and gas development to supersede local land use controls threaten even this limited source of local governance over land use to protect social and cultural resources based in community landscape. Thus, one of the most notable rulings on shale development – the 2015 judgment in favor of local jurisdiction over land use in New York State – laid the basis for the moratorium on shale development that was announced by New York State's governor in the summer of 2015 (*Wallach vs. Town of Dryden*, 2015).

That judgment relied upon research conducted in regions where shale development was occurring that documented significant impacts on local and regional quality of life and community character. These studies included evidence of increases in crime, road congestion and noise, and increased costs for policing, health care and public services (Christopherson and Rightor, 2012; Brasier et al., 2014; Headwaters, 2012; Gold and McGinty, 2013). Documentation of these impacts was uneven and largely anecdotal because of the absence of state policies to collect data and measure local impacts, yet the consistency of community character impacts across sites supported a broader view of environmental impact.

Other research on recent shale gas and oil development has altered and expanded the way in which the shale development *space* is perceived. The next section examines the spatial boundaries that are used to view the risks associated with shale development in the US, and the governance questions they raise.

The spatial frame

One way in which contemporary shale gas and oil extraction differs from other fossil fuel resource extraction activities lies in the spatial scale of the industry. Research on shale development has raised questions about the spatial scope of activities, and the efficacy of the jurisdictional boundaries by which they are governed (Bradshaw, 2013; Bridge et al., 2013). For example, by contrast with the regionalized exploitation of coal, there are potential shale extraction sites in 30 US states that raise policy issues for those state governments, thousands of local governments, and for the federal government as well (Finkel, 2015). When the entire industrial value chain is accounted for, almost all states in the US are affected in some way – as sources of inputs such as fracking sand or water, as storage sites, as locations for well waste disposal, or as transport routes.

Thus, the shale development industry has a nationwide footprint (Christopherson, 2015). The environmental, health, and seismic risks of shale development extend outward from the extraction sites to communities which supply them with drilling material and water, or in which toxic waste from wells is deposited in injection wells, or along the truck routes to and from the drilling sites. The risks are present all along the networks of pipelines and pumping stations, or railroads and transfer ports, that carry the oil and gas products to refineries and coastal liquid natural gas terminals. Shale gas and oil extraction requires copious quantities of water, sand and chemicals that have to be trucked to the drilling site. In the US, with the exception of water, these inputs generally come from regions far beyond the drilling region, transforming many non-drilling communities for better or worse. In short, the US shale oil and gas value chain is a new national industrial activity.

For example, mining for the particularly desirable silica (sand) in Western Wisconsin and Eastern Minnesota has transformed what were once small rural Midwestern towns whose economies centered on agriculture and tourism, into mining centers supplying distant drilling sites in the Western and Eastern shale plays (Deller and Schreiber, 2012).

At the other end of the industrial process, drilling tailings, fracking fluid, and other liquid waste – the byproducts of shale gas and oil extraction – must be disposed of safely. In the US, the primary method for disposal of contaminated water from shale development is the injection well, a bored shaft through which the fluids are forced deep underground into porous rock formations.

Like so many of the environmental, health and safety issues associated with shale development, the role of injection wells needs to be understood in the context of the extraction process as a whole. The US has approximately 680,000 waste and injection wells for disposal of hazardous waste. While this method of disposal has been used for decades, the 21 billion barrels of contaminated water produced in a year by high volume hydraulic fracturing clearly exceeds the available capacity of disposal sites.

Much of the contaminated water from Pennsylvania's Marcellus shale drilling sites has been trucked to injection wells in neighboring Ohio. In 2012, Ohio injection wells handled 588 million gallons of wastewater, most of it from Pennsylvania. Now, Ohio is beyond its capacity to handle wastewater from Pennsylvania, and the toxic waste must be shipped to other sites. So, the disposal of toxic waste from shale development in Pennsylvania and Ohio now must travel to a wider range of far-flung sites, including injection wells in Gulf Coast states. The waste material will be transported to these injection wells via barges on the Mississippi River. Thus, because of the search for new locations for unwanted drilling outputs, one of the most widely distributed products of shale development may be toxic waste.

As the preceding example demonstrates, risk-bearing activities associated with the shale development process may occur far from the drilling sites, including regions where drilling is highly regulated or prohibited. And, as in the case of waste disposal in Ohio, distinctive risks are being concentrated in particular regions, creating a set of externalities different from those created in extraction regions. Thus, the geography of hydraulic fracturing is complex, with the populations of some states and regions bearing more or different risks than others.

The uneven spatial distribution of waste disposal activities from shale gas and oil development is in part geologic: some areas offer better conditions for containing the waste produced by this type of fossil fuel development. Another factor, more common in explaining the expanded "footprint" of industrial activities connected with this kind of fossil fuel development, is the fragmentation of government responsibility and inconsistency of regulations governing the interrelated activities in the gas and oil development process. This fragmentation, characterized in the US by appreciably different state-level policies, allows for "venue shopping", whereby shale well operators find less-regulated locations for environmentally damaging aspects of the process.

This emerging picture of the geographic scope of shale development activities and regional differences in impacts and outcomes implicitly undermines the conception of shale gas development as taking place in a few locations while providing widely distributed societal benefits. For example, in the introduction to a special issue devoted to the question of assessing the costs and benefits of fracking, Burnett (2015) notes the tremendous heterogeneity in local impacts from unconventional oil and gas development in the US. This heterogeneity, however, is a consequence: it reflects the ways in which the extraction process and its supply and distribution chain draw on different places to construct the overall national landscape of shale development.

The other important feature of this new industrial landscape and its regional consequences is its particular temporal characteristics, reflected in a new version of the boom-bust cycle typical of extractive industries.

The temporal frame

The question of *time* is also regularly neglected in the narrative of shale development, and in assessments of costs and benefits (Mason et al., 2015). Time is significant in two respects:

1   the volatile development cycle for this extraction industry, given its speculative nature, tighter margins, a variable limit to the *economically viable* resources available at any given time, and a finite limit to the *extractable* resources available in any given shale play; and
2   the likely effects of shale extraction on the longer term economic conditions in an extraction region (Feser and Sweeney, 1999; Jacquet, 2009).

The majority of extraction industries are characterized by a volatile development cycle. Volatility, in the form of a "boom–bust" cycle, occurs because: (a) resources are eventually depleted or are too expensive to extract, given the existing technology and market; and (b) prices for natural resource commodities fluctuate, and price drops lead the industry to pull back on extraction activities. The "boom" is characterized by a rapid increase in economic activity, the "bust" by a rapid decrease.

The rapid increase occurs when drilling crews and other gas-related businesses move into a region to extract the resource. During this period, the local population grows and jobs in construction, retail and services increase. Costs to communities also rise significantly, for everything from road maintenance and public safety to schools. When drilling ceases because the commercially recoverable resource is depleted or extraction becomes economically inefficient, there is an economic "bust" – population and jobs depart the region, and fewer people are left to support the boomtown infrastructure.

Predictions of *continuous* development were particularly important in pitching shale development to investors, who bought into what they perceived as a long-term, low risk proposition (Crooks, 2015; Urbina, 2011; Zeller, 2014). Long-term development was predicated on a national extraction campaign, in which some regional shale plays might produce for only a few years, but in which rigs could move easily from areas with short-lived production to other, presumably more fruitful, regions.

By 2014, the Barnett Shale play in Texas was acknowledged to be in decline and the Fayetteville Shale play in Arkansas not living up to expectations. Rather than a pattern of slow decline and a long "tail", in which wells produce over a long period, many wells were characterized by a pattern of hyperbolic decline, with a rapid drop-off in production after the initial drilling phase (Inman, 2014). A study on the rates of decline by Hughes (2014) found that while the average rate of decline for conventional oil wells is approximately 5 percent per year, the average rate of decline for shale oil wells in the Bakken shale play is 44 percent per year, with some wells declining at rates as high as 70 percent.

Thus, despite assertions early in the shale revolution that the US could expect 100 years of continuous shale-based gas and oil, independent industry experts have predicted a best-case scenario of shale extraction lasting until 2040 (Gülen et al., 2013).

In addition, the projections of longer regional shale booms and a century-long national shale boom did not account for intermediate withdrawal of drilling rigs from one region to develop new wells in another (taking the economic boom with them) in response to a shortage of investment capital, as occurred in 2008, or a glut of natural gas on the global market, as has been the case since 2012. In 2015, all the US extraction regions are in the throes of a "bust". Drilling rigs are being moved from "dry gas" plays to those where more profitable shale oil is present, or pulled from the oil and gas fields altogether.

The question of scale has always been important; national impacts do not directly translate into regional impacts – what will happen to particular places during shale development. Even the original industry predictions were based on overall productivity in US shale plays, and did not account for what might be differences among regional shale plays, their productivity, and their economic prospects over time. Now, production has declined, albeit unevenly. The western slope of Colorado and the northern tier of Pennsylvania began to shut down production in 2012 as the price for natural gas fell below the level necessary to support the costs of hydraulic fracturing for dry gas. Other regions, such as the Bakken Shale play in North Dakota, continue to produce because of the higher productivity of the play and because it can produce higher value shale oil.

Until declining oil and gas prices began to alter the economies of shale boomtowns, the narrative of shale development in the US underplayed the potential for volatility that is inherent to resource extraction industries. To some extent, a concept of "sustainable shale development" influenced research on its economic benefits. Beyond the input/output models that made projections for a timeline that stretched into an undefined future, assessments of economic benefits neglected costs, and the prospect of a "bust". Because the boom–bust cycle is unpredictable, an assumption that present-day economic patterns would continue was perhaps understandable. Researchers focused on extraction regions only, and only on the boom period and its economic impacts, as though growth would continue indefinitely (Weber, 2012). Moreover, problems associated with the boom, such as the lack of human resource capacity or sharply increased demands on public services, were dismissed as temporary because it was assumed that communities would have time as the growth continued to adapt. This depiction persists in regional studies that depict shale gas and oil development as in an early phase of continuous development (Environmental Law Institute, 2014).

Although the temporal cycle was ignored in much of what was written about shale gas and oil development, some of the boomtown communities were more sophisticated about what could (and eventually did) happen because they had experienced a "bust" before, most recently in the 1980s.

In addition, historical research on resource extraction communities has documented the long-term development trends and the volatility of rural extraction economies (Cortese and Jones, 1977; Freudenburg and Wilson, 2002; Haefele and Morton, 2009; Kassover and McKeown, 1981). Studies have assessed the long-term consequences for regions that have experienced the ups and downs of resource extraction and found that, on average, extraction regions experience only very modest permanent employment gains (2 percent or less) including those during the boom periods (Wrenn et al., 2015).

The *ex post facto* analysis of similar resource extraction communities through the boom and bust cycle, in combination with historical studies on the impact of resource extraction on regions worldwide, has tempered expectations of long-term economic development in shale gas and oil extraction regions. Although these regions can expect some job growth in conjunction with drilling activities, this growth will drop off with the end of the drilling phase, however soon that may occur. A longer-term perspective on how local and regional economies are affected by any resource development provides a different, and more modest, picture of what can be expected by the many regions involved in shale development than the inflated expectations produced by a focus on the "boom" period only, in extraction regions only, that we have seen heretofore.

Conclusion

This chapter has addressed the question of whether the framing of the shale decision around regional environmental costs and economic benefits sufficiently captures important dimensions of the decision. Three aspects of this "frame" were discussed. First, what can we anticipate with respect to economic benefits at the regional scale? Second, what are the implications of confining the environmental and economic externalities we associate with shale gas development to those at the extraction site? How do we address externalities produced throughout the shale development supply chain? Finally, what time frame should be used to examine the costs and benefits of shale development? The spatial frame that is typically used measures benefits nationally and globally – for "consumers" – and rarely ties those benefits to costs in the extraction region or to the spaces affected by gas extraction far from the well site.

There is evidence that a broader spatial scale and a longer time horizon change the calculation between costs and benefits, at least at the regional scale. As researchers move toward a more sophisticated "life cycle" approach to the economic as well as the environmental impacts of shale development, we are likely to see models developed that can more fully inform the shale decision, as well as a range of other energy transition issues.

Acknowledgements

I would like to thank New Rightor and the editors for thoughtful comments that contributed to this paper.

Note

1   IMPLAN is software developed by MIG, Inc. that uses economic input-output analysis along with social accounting matrices and multiplier models to generate economic assessments of local or regional economies (e.g. a municipality, city, county, or state). The IMPLAN database consists of economic statistics at county, zip code, state, and federal levels. Data selection can be customized to provide regional data sets, and the model can be used to measure effects (including multiplier effects) on a regional or local economy from a change or event in the economy's activity.

References

Barth, J. M. *The Economic Impact of Shale Gas Development: Can New York Learn from Texas?* Catskill Citizens for Safe Energy (5 May, 2012), www.catskillcitizens.org/barth/JMBTEXAS.pdf. 18 January, 2016.
Barth, J. M. "The economic impact of shale gas development on state and local economies: benefits, costs and uncertainties." *New Solutions* 23/1 (2013): 85–101.

Beattie, B.R. and J. Leones. "Uses and abuses of economic multipliers." *Community Development Issues* 1/2 (1993): 1–5.

Bradshaw, M. 2013. *Global Energy Dilemmas*. New York: Wiley (Polity).

Brasier, K., L. Davis, L. Glenna, T. Kelsey, D. McLaughlin, K. Schafft, K. Babbie, C. Biddle, A. Delessio-Parson, and D. Rhubart. *The Marcellus Shale Impacts Study: Chronicling Social and Economic Change in North Central Pennsylvania*. Final Report to the Center for Rural Pennsylvania, 2014.

Bridge, G., S. Bouzarovski, M. Bradshaw and N. Eyre. "Geographies of energy transition: space, place and the low carbon economy." *Energy Policy* 53 (2013): 331–340.

Burnett, J. Wesley. "Unconventional oil and gas development: economic, environmental, and policy analysis." *Agricultural and Resource Economics Review* 44/2 (August 2015): iii–xvii.

CBO. *The Economic and Budgetary Effects of Producing Oil and Gas from Shale*. Washington, DC: CBO, 2014, www.cbo.gov/publication/49815. 3 October 2015.

Christopherson, S. "Risks beyond the well pad: the economic footprint of shale gas development in the US." In *The Human and Environmental Impact of Fracking: How Fracturing Shale for Gas Affects Us and Our World*, Madelon L. Finkel, (ed.) Santa Barbara CA, Denver CO, and Oxford UK: Praeger, 2015.

Christopherson, S. and Rightor, N. "How shale gas extraction affects drilling localities: lessons for regional and city policy makers." *Journal of Town & City Management* 2 (2012): 250–368.

Christopherson, S. et al. *The Economic Consequences of Marcellus Shale Gas Extraction: Key Issues*. CaRDI Reports 14. Ithaca, NY: Cornell University Community and Regional Development Institute (14 September 2011), www.greenchoices.cornell.edu/resources/publications/regions/Economic_Consequences.pdf.

Considine, T.J., R. Watson, and S. Blumsack. *The Economic Impacts of the Pennsylvania Marcellus Shale Natural Gas Play: An Update*. College of Earth and Mineral Sciences, Department of Energy and Mineral Engineering, Pennsylvania State University, 24 May 2010, http://marcelluscoalition.org/wp-content/uploads/2010/05/PA-Marcellus-Updated-Economic-Impacts-5.24.10.3.pdf. 19 January 2016.

Considine, T.J., R. Watson, and S. Blumsack. *The Pennsylvania Marcellus Natural Gas Industry: Status, Economic Impacts and Future Potential*. College of Earth and Mineral Sciences, Department of Energy and Mineral Engineering, Pennsylvania State University, 20 July 2011, http://marcelluscoalition.org/wp-content/uploads/2011/07/Final-2011-PA-Marcellus-Economic-Impacts.pdf. 19 January 2016.

Cortese, C.F., and B. Jones. "The sociological analysis of boomtowns." *Western Sociological Review* 8/1 (1977): 75–90.

Cosgrove, B.M., D.R. LaFave, S.T.M. Dissanayake, and M.R. Donihue. "The economic impact of shale gas development: a natural experiment along the New York/Pennsylvania Border." *Agricultural and Resource Economics Review* 44/2 (2015): 20–39.

Crooks, E. "Shale looks more like Dotcom boom than Lehman debt bubble." *Financial Times* (6 May 2015), www.ft.com/intl/cms/s/0/b8345a94-f408–11e4-bd16–00144feab7de.html#axzz3pstIM4Vi. 10 October 2015.

Curtis, J. "Fractured shale gas systems." *AAPG Bulletin* 86/11 (November 2002): 1921–1938.

Davis, Kylie R., Timothy W. Kelsey, Leland L. Glenna, and Kristin Babbie. *Local Governments and Marcellus Shale Development*. The Marcellus Impacts Project Report #7, Center for Rural Pennsylvania, 2014.

Deller, Steven C. and Andrew Schreiber. *Frac Sand Mining and Community Economic Development*. Staff Paper Series, no. 565. Madison, WI: University of Wisconsin, Department of Agricultural and Applied Economics, May 2012, http://buffalo.uwex.edu/files/2011/12/Revised-Economic-growth-mining-version-1–4.pdf. 19 January 2016.

Environmental Law Institute. *Getting the Boom Without the Bust: Guiding Southwestern Pennsylvania Through Shale Gas Development*. Washington & Jefferson College Center for Energy Policy and Management, 2014.

Feser, E. and S. Sweeney. "Out-migration, population decline, and regional economic distress." Report prepared for the US Economic Development Administration. Washington, DC: US Department of Commerce, 1999.

Finkel, Madelon L. (ed.). *The Human and Environmental Impact of Fracking: How Fracturing Shale for Gas Affects Us and Our World*. Santa Barbara, CA, Denver, CO, and Oxford, UK: Praeger, 2015.

Flad, H. *Community Character Analysis, Finger Lakes LPG Storage, LLC Town of Reading, Schuyler County, N.Y.* (2015), http://gasfreeseneca.com/wp-content/uploads/2015/01/Ex.-5-Flad-Report-Exhibit-PUBLIC.pdf. 24 October 2015.

Freudenburg, W.R. and L.J. Wilson. "Mining the data: analyzing the economic implications of mining for nonmetropolitan regions." *Sociological Inquiry* 72 (2002): 549–557.

Ghilain, K. "Improving community character analysis in the SEQRA environmental impact review process: a cultural landscape approach to defining the elusive 'community character'." *New York University Environmental Law Journal* 17 (2009): 1194–1242.

Gold, R. and T. McGinty. "Energy boom puts wells in America's backyards hydraulic fracturing largely driving transformation of the nation's landscape." *Wall Street Journal* (25 October 2013), www.wsj.com/news/articles/SB10001424052702303672404579149432365326304. 20 January 2016.

Groth, P. "Frameworks for cultural landscape study." In *Understanding Ordinary Landscapes*, Paul Groth and Todd W. Bressi (eds.). New Haven, CT: Yale University Press, 1997.

Gülen, G., J. Browning, S. Ikonnikova, and S. W. Tinker. "Well economics across ten tiers in low and high Btu (British thermal unit) areas, Barnett Shale, Texas." *Energy* 60 (October 2013): 302–315, www.sciencedirect.com/science/article/pii/S0360544213006464. 20 January 2016.

Haefele, Michelle and Pete Morton. "The influence of the pace and scale of energy development on communities: lessons from the natural gas drilling boom in the Rocky Mountains." *Western Economics Forum* (Fall 2009), http://ageconsearch.umn.edu/bitstream/92810/2/0802001.pdf. 10 January 2010.

Headwaters Economics. *Benefiting from Unconventional Oil: State Fiscal Policy is Unprepared for the Heightened Community Impacts of Unconventional Oil* (2012), www.ag.ndsu.edu/ccv/documents/benefiting-from-unconventional-oil.Plays.* Stanford University, The Bill Lane Center for the American West (April 2012), http://headwaterseconomics.org/wphw/wp-content/uploads/ND_Unconventional_Oil_Communities.pdf. 20 October 2013.

Hughes, J. D. *Drilling Deeper.* Post Carbon Institute (2014), www.postcarbon.org/publications/drilling deeper/. 20 October 2015.

IHS Global Insight. *The Economic and Employment Contributions of Shale Gas In the United States.* Report prepared for America's Natural Gas Alliance, Englewood, CO (2011), http://anga.us/media/235626/shale-gas-economic-impact-dec-2011.pdf. 30 January, 2012.

Inman, M. "Natural gas: the fracking fallacy." *Nature* 516 (3 December 2014): 28–30, www.nature.com/news/natural-gas-the-fracking-fallacy-1.16430. 20 October 2015.

Jacquet, J.B. *Energy Boomtowns & Natural Gas: Implications for Marcellus Shale Local Governments & Rural Communities.* NERCRD Rural Development Paper No. 43 (2009). University Park, PA: Northeast Regional Center for Rural Development, Pennsylvania State University, http://nercrd.psu.edu/Publications/rdppapers/rdp43.pdf. 30 January, 2010.

Jacquet, J.B. *Workforce Development Challenges in the Natural Gas Industry.* Working Paper Series: A Comprehensive Economic Impact Analysis of Natural Gas Extraction in the Marcellus Shale (February 2011). Ithaca, NY: Department of City and Regional Planning, Cornell University, www.greenchoices.cornell.edu/resources/publications/jobs/Workforce_Development_Challenges.pdf.

Kassover, J. and R.L. McKeown. "Resource development, rural communities and rapid growth: managing social change in the modern boomtown." *Minerals and the Environment* 3/1 (1981): 47–57.

Kay, D. *The Economic Impact of Marcellus Shale Gas Drilling What Have We Learned? What are the Limitations?* Working Paper Series: A Comprehensive Economic Impact Analysis of Natural Gas Extraction in the Marcellus Shale (April 2011). Ithaca, NY: Department of City and Regional Planning, Cornell University, www.greenchoices.cornell.edu/resources/publications/jobs/Economic_Impact.pdf.

Kelsey, T. W., M. Shields, J.R. Ladlee, M. Ward, T.L. Bundage, J.F. Lorson, L.L. Michael, and T.B. Murphy. *Economic impacts of Marcellus shale in Pennsylvania: Employment and income in 2009.* Marcellus Shale Education & Training Center (August 2011), www.shaletec.org/docs/EconomicImpactFINAL August28.pdf. 20 January 2016.

Kinnaman, Thomas C. "The economic impact of shale gas extraction: a review of existing studies." *Ecological Economics* 70/7 (2011): 1243–1249, http://works.bepress.com/thomas_kinnaman/9/. 30 January 2012.

Krupnick, A.J. and H.G. Gordon. "What the experts say about the environmental risks of shale gas development." *Agricultural and Resource Economics Review* 44/2 (2015): 106–119.

Mason, C., L. Muehlenbachs and S. Olmstead. *The Economics of Shale Gas Development.* Discussion Paper, Resources for the Future (February 2015), www.rff.org/files/sharepoint/WorkImages/Download/RFF-DP-14–42.pdf. 20 October 2015.

Mauro, Frank, M. Wood, M. Mattingly, M. Price, S. Herzenberg and S. Ward. *Exaggerating the Employment Impacts of Shale Drilling: How and Why.* Multi-State Shale Research Collaborative (2013), www.multistateshale.org/shale-employment-report. 20 January 2016.

Munasib, A. and D.S. Rickman. "Regional economic impacts of the shale gas and tight oil boom: a synthetic control analysis." *Regional Science and Urban Economics* 50 (2015): 1–17.

New York State Department of Environmental Conservation. *Community Character*. Short Environmental Assessment Form Workbook (Part 2), Q.3 (2015), www.dec.ny.gov/permits/91399.html. 12 October 2015.

Rumbach, Andrew. *Natural Gas Drilling in the Marcellus Shale: Potential Impacts on the Tourism Economy of the Southern Tier*. Report prepared for the Southern Tier Central Regional Planning and Development Board (July 2011), www.greenchoices.cornell.edu/resources/publications/regions/Impacts_on_Tourism _Economy.pdf.

Small, M., P.C. Stern, Elizabeth Bomberg, Susan M. Christopherson, Bernard D. Goldstein; Andrei L. Israel; Robert B. Jackson; Alan Krupnick; Meagan S. Mauter; Jennifer Nash; D. Warner North; Sheila M. Olmstead; Aseem Prakash; Barry Rabe; Nathan Richardson; Susan Tierney; Thomas Webler; Gabrielle Wong-Parodi; Barbara Zielinska. "Risks and risk governance in unconventional shale gas development." *Environmental Science and Technology* 48/15 (2014): 8289–8297.

Sunstein, C. "Cost-benefit analysis and the knowledge problem," Regulatory Policy Program Working Paper RPP-2015-03. Mossavar-Rahmani Center for Business and Government, Harvard Kennedy School, Cambridge, MA, 2015.

Urbina, Ian. "Insiders Sound an Alarm Amid a Natural Gas Rush." *New York Times* (25 June 2011), www.nytimes.com/2011/06/26/us/26gas.html?pagewanted=all&_r=2. 20 October 2015.

Wallach versus Town of Dryden, 16 N.E.3d 1188 (N.Y. 2014). *Harvard Law Review Case Notes, Environmental Law* (15 March 2015), http://harvardlawreview.org/2015/03/wallach-v-town-of-dryden/. 13 October 2015.

Wang, Z. and A. Krupnick. *US Shale Gas Development: What Led to the Boom?* Resources for the Future Issue Brief 13–04 (May 2013), www.rff.org/files/sharepoint/WorkImages/Download/RFF-IB-13-04.pdf. 20 October 2015.

Weber, J.G. "The effects of a natural gas boom on employment and income in Colorado, Texas, and Wyoming." *Energy Economics* 34/5 (2012): 1580–1588.

Weinstein, Amanda and Mark Partridge. *The Economic Value of Shale Natural Gas in Ohio*. Summary and Report, the Swank Program in Rural-Urban Policy (December 2011). Columbus, OH: Ohio State University, Department of Agricultural, Environmental, and Development Economics, http://aede.osu.edu/sites/aede/files/publication_files/Economic%20Value%20of%20Shale%20FINAL%20Dec%202011.pdf. 30 December 2013.

Wrenn, D., T. Kelsey and E. Jaenicke. "Resident vs. nonresident employment associated with Marcellus shale development." *Agricultural and Resource Economics Review* 44/2 (August 2015): 1–19.

Zeller, T. "Is the US shale boom going bust?" *Bloomberg View* (22 April 2014), www.bloombergview.com/articles/2014-04-22/is-the-u-s-shale-boom-going-bust. 20 October 2015.

11

Siting dynamics in energy transitions

How generating electricity from natural gas saves cherished landscapes

Richard Cowell

Introduction

Around the world, the 'high politics' of energy is dominated by a trilemma of geopolitical and technological concerns: about security of supply, long-term sustainability (especially decarbonisation), affordability and access, which together dominate much national and international strategising. When our attention turns from strategic goals to building the infrastructures that might be required to address these concerns, it becomes clear that delivering future energy pathways encounters messy and complex social and economic worlds. It is in infrastructure siting and development that the priorities of developers and governments runs up against myriad social attachments to places and landscapes that these projects might affect. Given this, concerns about siting and energy landscapes should be seen as intimately connected to wider questions about the energy pathways that we do, or should, pursue (Nadaï and van der Horst 2010), and it is such connections that set the agenda for this chapter.

Policy-makers, energy businesses and researchers around the world have become increasingly interested in the connections between siting and transitions to sustainability but, to date, their interpretations of what 'the problem' is has been driven by the most visible conflicts (Aitken 2010). Thus significant attention has been given to the often protracted public conflicts arising from on expanding highly visible, spatially extensive forms of renewable energy, especially on-shore wind (Szarka et al. 2012; Nadaï and van der Horst 2010), for which key problems have been the effects on landscape and the risks that public opposition could lead to projects being delayed or rejected. Driven by wider concerns to foster renewable energy, significant attention has thus been given to better mechanisms for public engagement, tools and strategies for improved siting, improved project designs, and to modes of development that better compensate local publics for receiving development in their midst (Cowell et al. 2011). It is difficult to disagree with the dominant logics underlying this work: that renewable energy must be expanded, that fostering social acceptability is important, and that finding a better 'fit' between technologies and particular local conditions is a valuable means to achieve these ends.

What has been less considered in relation to energy siting conflicts and landscape is the flip-side of sustainability transitions, neatly encapsulated by Shove and Walker (2007: 764), that if we wish to mainstream new pathways we must also 'figure out how currently dominant sociotechnical regimes might be dislodged and replaced'. In the energy context, such dominant regimes involve the bulk provision of fossil fuel and nuclear power, supplying energy to consumers at cross-national scale through massive grid infrastructures (Lovins 1977; Szarka 2007). Compared to new renewables, relatively little attention has been given to how issues around infrastructure siting and energy landscapes shape the persistence of presently dominant energy technologies. Siting conflicts have sometimes provided arenas in which the sustainability of such pathways has been questioned, as concerns for the impacts of specific projects – such as coal-fired power stations – elicit objections that embrace local impacts alongside wider judgements about environmental risk, cost, technology choice and the merits of meeting rather than managing energy demand. However, if local and landscape concerns can sometimes trigger more destabilising protests, they do not always do so, and there are other ways in which concerns about energy infrastructure, siting and landscape can be combined to make wider, strategic arguments. Thus, analysts have projected the notional space requirements of renewables like wind to quantify and question their potential contribution to future supply (Mackay 2009). The land use nexus is also mobilised by advocates of sustainability pathways based on high growth and technological intensification; such as the so-called 'eco-modernists', advocating 'power dense' energy sources, with minimal land take, because they 'make more room for nature' (Asafu-Adjaye 2015: 6), for which the apogee is nuclear power. Such arguments raise challenging questions, then, about the direction that concerns for landscape fit and ease of siting might take us.

In this chapter, these questions are illustrated by a somewhat cautionary tale – the expanding use of gas for electricity generation, with particular reference to the experience of the UK from the late 1980s to 2015. Over this period, investment in gas-fired power capacity exceeded investment in all renewables, with significant implications for pathways to decarbonised energy futures, yet has received remarkably little academic attention (though see Mounfield 1990; Garrone and Groppi 2012). At the time of writing, further investment in gas-powered electricity generation remains a key part of the UK government's energy policy. It is argued here that the siting dynamics of gas-powered electricity are important, being characterised by a remarkable absence of disruptive siting conflicts (and, indeed, for many projects, little opposition at all). The main aim of the chapter is to explain why this is so, by identifying why the physical form of gas-fired power stations, the environmental claims that might be made for them, when linked to the kinds of sites selected, reduced the likelihood of effective public opposition. This analysis also considers how the numerous potentially critical claims that could be made of gas-fired power stations – to specific projects, their impacts and as a wider energy pathway – were insulated from legitimate challenge by the institutional structures of consenting procedures.

The structure of the chapter is as follows. First, the rise of gas as a source of energy is outlined, flagging up the environmental issues this brings with it, especially with respect to climate change. Following this is a closer review of the potential intersections between energy infrastructure siting and wider political conflict over energy policy. From this follows the main part of the chapter, a review of the expansion of gas-fired power stations in the UK, the siting choices made, and the conflicts they created. The main data sources are the consent letters issued for some 90 power station projects, along with associated planning documentation. Gas-fired power stations are shown to have emerged mainly by re-inhabiting the spaces vacated or created by previous fossil fuel-based industry, and the conclusions thereby review the wider implications.

The rise of natural gas

The use of gas has become an important component of the global dominance of fossil fuel energy systems, but also at times a somewhat liminal presence.

Gas first appeared as a major fuel in industrialising nations in the nineteenth century, as a product manufactured from coal by various heating and gasification processes; all of them highly polluting. Natural gas – extracted straight from the ground – was initially often 'flared off' as an unwanted by-product of oil extraction, but began to be exploited in quantity as a fuel in its own right from the 1960s. Expanding supplies and falling prices have facilitated significant expansion in natural gas consumption, aided by the enhanced scope for transportation, as discussed below. World production tripled between 1973 and 2014 from about 1200 billion cubic metres to 3500 billion cubic metres (IEA 2015). By 2012, natural gas had risen to 21 per cent of World primary energy supply (IEA 2014). This expansion reflects the rapid growth of different uses in different countries: as a replacement for coal or oil in space heating; in industrial processes, as a feedstock for petrochemical processes and fertiliser production; and in electricity generation. Globally, gas use for electricity generation doubled from 1992 to 2012, from 750 billion cubic metres to over 1400 billion cubic metres. In the OECD, the consumption of gas for electricity production rose from 250 billion cubic metres to a peak of 600 billion cubic metres in 2012, before growth in renewable energy technologies began to become significant.

The expansion of gas has co-evolved with a series of wider political and environmental concerns. With growing use has come dependency, such that maintaining gas availability has become important to debates about energy security and prompted intensifying industry and political attention to trade links and to the capacity, reach and reliability of supply infrastructure (Bradshaw 2014). In many regions of the world these concerns about security of supply have been intermeshed with agendas of privatisation and market integration, which in infrastructural terms often pushes in the same direction i.e. towards the extension, interconnection and standardisation of supply systems. 'European gas grid' concepts are one example (see Van der Vleuten and Högselius 2012).

If gas is an ethereal substance, its provision has a wide array of material effects. It requires specialized technologies and facilities for extraction and to process it for standards suitable to its end use. To distribute it over any distance requires costly equipment, such as high pressure transmission pipelines, or facilities to cool it into a liquid (i.e. liquefied natural gas [LNG]) so that it can be shipped between centres of supply and demand, stored, and converted back into a vapour. Moreover, like coal gas before it, natural gas is a hazardous material, being both toxic and highly combustible, and so its handling creates risks of leakages and explosions. As a result, even if the siting of gas facilities generally has not created the kind of controversy associated with, say, nuclear power, individual facilities have generated conflicts, centred on processing and storage facilities, and pipelines (see Van der Vleuten and Högselius 2012; Barry 2013; Marsden and Markusson 2011; Groves et al. 2013; Sovacool and Cooper 2013).

As a fossil fuel, addressing climate change has become the dominant environmental concern, but here gas has come to occupy a contested position between different pathways. Natural gas has lower carbon intensity than other fossil fuels, embodying only 56 per cent and 71 per cent of the carbon of coal and oil respectively. In addition, gas combustion generates negligible acidifying sulphur emissions, lower levels of nitrogen oxides than coal and oil, and no ash waste to dispose of. Consequently, compared to other fossil fuels, gas can be represented as *relatively green*. We might say that gas has greater 'eco-efficiency' (Jacobs 1991), in that for each unit of energy delivered it creates less waste and pollution (and thereby lower environmental impacts).

The expanded use of gas has been represented as delivering more sustainable energy systems: notably in the UK in the 1990s, where the rapid expansion of gas-fired power stations, replacing coal and oil generation, enabled successive UK governments to claim success in attaining carbon emission reductions. Comparable dynamics can be seen elsewhere, as in the US. At the time of writing (January 2015), gas is frequently presented as a 'bridging fuel' by which greenhouse gas emissions can be reduced while giving time for renewables and other lower carbon alternatives to develop. Indeed, gas sector interests have been keen to promote gas as a solution to environmental problems: as a power generation fuel that can be made more sustainable through the use of carbon capture and storage; as an alternative to petrol and diesel in transport, as well as pushing the expansion of various biomass-based gases as substitutes for natural gas (Taanman 2012). A key feature of these technological pathways is that they easily 'fit the current infrastructure and institutional setup' (Van der Vleuten and Högselius 2012: 89), offering the prospect that sustainability problems can be solved with little need to adjust present patterns of energy consumption. Such sanguine positions have also been challenged. The scale of reduction in greenhouse gas emissions now deemed to be necessary to avoid dangerous climate change is enormous (IPCC 2014), and decarbonising electricity production is often seen as more achievable in the short term than progress in other sectors (see for example HM Government 2009). Given this, a critical question is whether continually 'bridging' to medium-term energy futures with gas is actually maintaining a dangerous level of 'lock-in' to fossil-fuel based, industrial-scale models of energy provision that ought to be subjected, much sooner, to more radical change (Chignell and Gross 2013).

This brief account shows that there is much about the expanded exploitation and consumption of gas that is potentially contestable. Indeed, the sector has not been without conflict, some of it focused on infrastructure projects, but in most instances these conflicts have not prevented the projects being realised (Marsden and Markusson 2011; Arapostathis et al. 2014). Important questions arise from this. How might the experience of gas-fired electricity generation shape our understanding of how conflicts over infrastructure siting fit into the wider politics of change in energy pathways? What is the role of energy landscapes here – might the persistence of fossil fuel landscapes shape the outcomes of infrastructure siting?

Sites, places and politics

Interpreting 'siting problems'

Geographers and other social scientists have sought to develop explanations of public responses to infrastructure projects (Calvert 2015), and concepts of landscape and place feature prominently. Through a 'place attachment' perspective, Devine-Wright and Howes (2010) identify how some infrastructures – like wind turbines – may threaten attachments in some locations, encouraging the public to object, but encounter less opposition in places where attachment to the existing environment is lower, or where the technology resonates with existing feelings about place. As Pasqualetti (2012) points out, beliefs about place and landscape are interwoven with ways of life, rooted in social and economic activities that may be more or less compatible with the arrival of new energy facilities. A recurring fault line surrounds conceptions of the rural and the urban or industrial. Woods (2003) noted how particular conceptions of 'the rural', as a natural idyll free of certain 'industrial' intrusions, often informed opposition to infrastructure like wind turbines that threatens this spatial ordering.

From an instrumental perspective then, one important factor affecting the smooth realisation of energy infrastructure is whether project proponents prove capable of navigating the multiple social, economic and environmental relations bound up with candidate sites and the landscapes in which they are enmeshed. Viewed simply, 'success' for developers (in the sense of attaining a swift and positive consent decision) may reflect their scope to identify and exploit locations where conditions are more conducive to acceptance of the proposed facilities. Research suggests that the chances of successful outcomes can also be raised where processes of public engagement are seen to be open and inclusive, fostering trust. Public opinion can turn against projects where the actions of developers or decision-makers are seen to lack transparency, or to give the public insufficient opportunity to express their views (Wolsink 2007). Public opposition can also be fomented where it is felt that infrastructure siting has unfair distributive effects, in that local publics bear the costs while other constituencies – company shareholders and consumers – gain the benefits (Gross 2007).

However, as was noted in the introduction, much of the siting literature has adopted a framing of the problem in which public resistance is a problem to the extent that it risks the delivery of technologies assumed to be desirable. This gives only a partial perspective, as it underplays the diverse ways in which infrastructure siting intersects with wider energy transitions.

Attention needs to be given to those circumstances in which the siting of major new infrastructure, despite potential risks, encounters little reaction from local publics. Researchers have identified the social, economic and political processes by which places that have become dependent on environmentally risky industrial activities tend to be more accepting of further such investments (Blowers and Leroy 1994). Such acquiescence can create 'pollution havens' in which such industries can more easily persist, with any questions about the very need for such activities remaining off local political agendas (Crenson 1971). The tendency of environ-mental risky activities to concentrate over time in less affluent areas, occupied by more socially marginalised groups that often have less capacity to assess or challenge projects successfully, is a central theme of the environmental justice literature (Walker 2011; Bickerstaff 2012). It shows how notionally 'successful' siting decisions, from a developer's perspective, can contribute to distributive unfairness.

Viewing the issue simply as a 'siting problem' is also rather simplistic in that the problem is reduced to finding a 'local' site for facilities that fulfil some wider need that is taken to be self-evident. Indeed, siting conflicts are full of spatially questionable terminology which rein-forces this perspective. The accusation that opponents are NIMBYs ('not in my back yard') implies that they are selfish, seeking only to ensure that a facility is not built near them. That the facility should be built somewhere is unquestioned. Such concepts have been heavily criticised for their pejorative and reductionist treatment of public responses to infrastructure projects (see Burningham 2000; Devine-Wright 2011; Wolsink 1994). In practice, the arguments that circulate around proposed new energy installations can combine numerous lines of dispute, connecting the immediate effects on places and landscapes to wider concerns about technological safety, whether alternative technologies have been considered, or whether the underlying demands (in this context, for energy) could not be met in other ways. Indeed, concerns about the landscape impacts are often bound up with a wider set of issues.

Herein lie more fundamental, more politicised links between siting and energy develop-ment pathways. The history of energy development is dotted with instances where siting conflicts over new facilities become fulcrums for the wider contestation of the direction of development, prising open the key assumptions that shape energy policy (e.g. Owens 1985; Sovacool and Cooper 2013). However, not all energy infrastructure projects become political in this way, and there is an important task for geographers in understanding the con-textual conditions in which siting disputes open up such questions, with disruptive effects on wider policy agendas, and where they do not. Landscape issues can be factors, and the extent to which potential impacts foment opposition, but so too are the structures of decision-making processes.

Infrastructure decision-making and the containment of dissent

Whatever publics may feel about the prospect of new energy infrastructure in their midst, the interplay of arguments for and against do not play out across a blank sheet of paper, but within arrangements for decision-making that are institutionally structured in myriad ways, seeking to define what level of impact is acceptable, and which issues legitimately are up for discussion (Aitken 2010; Cowell and Owens 2006). Here then we can see how consenting procedures occupy an important position in the persistence or transformation of energy systems. The rules that consenting processes set and the standards they apply may represent opportunity structures for asking challenging questions, or for delimiting them.

One can see this pattern in the consents that proponents of new energy infrastructure may need to acquire. The adjudication of whether a facility is 'safe' or has impacts that are 'acceptable', draws heavily on the judgements embodied in existing standards and policies, be that for pollution abatement (are emissions levels acceptable, has the required abatement technology been adopted?), noise, wildlife (are designated habitats or species affected?), or landscape, where whether an area has warranted protective designation structures judgements about the seriousness of any effects. Rules may delimit developer responsibilities in other ways too. Thus proponents of a power station may be legitimately responsible for emissions from their own plant, but not necessarily for cumulative effects with emissions from other sources, or prospective future projects.

Another important aspect is whether the need for a facility can be questioned and, if at all permissible, how thorough such questioning can be. An individual piece of energy infrastructure may draw its justification from wider policy statements supporting particular energy technologies, or targets for reducing carbon emissions, or strategies to improve energy security, but it does not automatically follow that the extent to which infrastructure proposed project delivers on that policy, or even the wisdom of the policy, is legitimately up for discussion when project consents are determined. National governments may regard issues of need as their prerogative to determine, or decide that such decisions should be left to market actors, reinforcing divisions of decision-making labour that seek to confine infrastructure consenting to local issues.

To acknowledge that arguments about energy infrastructure take place in an institutionally structured context is not to suggest institutional stability. Indeed, it is a dynamic front, as the rules governing project siting, land use planning and the assessment of impacts can themselves be a focus for contestation and change. As Barry (2001, 17) suggests: '[T]hrough regulation, or political activism, it is possible that the borders of the plant will be reconfigured, and connections drawn between what goes on inside the physical site of the plant and what goes on outside its perimeter fence'. Concerned parties may argue that standards are too weak; that standard assessment procedures are blind to the specific qualities of place, or that new risks have been omitted or under-played. Consenting processes can be a particular focus for such pressure, because publics often get more engaged in energy questions when faced with specific projects, which are more tangible in their consequences, than in response to abstract energy policy statements. Consents procedures typically require that proposals are made available for public consultation (Cowell and Owens 2006; Marsden and Markusson 2011). In such conflicts one can see how antagonisms over infrastructure siting can be manifestations of deeper societal and ideological tensions around future energy pathways (Calvert 2015).

However, governments have not always responded to such pressures by extending environmental scrutiny of (or tightening the environmental standards on) energy infrastructure. One could characterise the period since the late 1960s as one where, in many democratic countries, public concerns about the impacts of projects did have such positive and progressive environmental effects, exemplified by requirements that major developments should be subject to environmental impact assessment. From the start of the 21st century, however, governments across Europe, North America and beyond have increasingly been animated by concerns about 'delay' in infrastructure 'delivery'. Industry, concerned about finance risk, has exerted pressure to make consenting processes smoother, swifter and more predictable (Lee et al. 2012). Many states have introduced 'speed up' legislation to streamline procedures, pass decisions from local government to central government, introduce time limits and, often, remove issues of 'need' from project-specific determinations (Garrone and Groppi 2012). In the UK, it has long been the case that the need for large power stations (50MW and over) and grid infrastructure is determined by central government, but nevertheless steps have been taken to accelerate the time taken to make decisions (Marshall and Cowell 2016).

From this brief review, our conception of how energy siting conflicts play into the wider politics of energy transition becomes more multi-dimensional. It is clear that infrastructure siting and consenting offers potential arenas for questioning the direction of energy development, but they do not automatically do so. Resistance is not ubiquitous. It varies from location to location in its extent or effectiveness, and the relationship between new projects and existing energy landscapes may be critical here. There is also a need to consider how far any resulting conflicts lead to institutional changes that shape the wider governance of the technology or sector. In certain contexts governments and industry may be able to 'hold the line' of decision-making procedures, delimiting the effects of dissent, and ensure that infrastructure is delivered, but not with others. To understand in more detail the contextual conditions that shape how far energy infrastructure becomes contentious, we turn now to the expansion of gas-fired electricity generation in the UK.

Where did the UK 'dash for gas'?

Technologies, sites and environmental claims

To energy companies in the UK, from the end of the 1980s, gas fired power stations rapidly became the technology of choice for new generation investment. By the end of 2014 the installed capacity of gas-powered electricity generation was almost 34,000MW,[1] approximately 40 per cent of total UK plant capacity (DECC 2015),[2] not counting the use of gas in electricity generation operated by other industries (oil refining, iron and steel, paper mills).

How then did so much capacity come forward in less than thirty years? Technology and markets are often given a central role in this story (see Watson 1997; Patterson and Grubb 1996), particularly the availability of both cheaper natural gas and the combined-cycle gas turbine (CCGT) technology. From the developer's perspective, key features of the CCGT are the lower capital cost per unit of output compared to coal, oil or nuclear plant, which has made them attractive investments in the liberal, short-termist economic environment of a privatised electricity system. CCGT technologies are also amenable to modular construction, available in a variety of sizes and combinations, meaning that they can be 'plugged in' to a variety of settings, require less water for cooling, and be quicker to build. Compared to alternative plant, they are also more straightforward to switch off and start up. The combination of these qualities gives

CCGTs more flexibility, which in turn makes them less risky and therefore easier to finance (Patterson and Grubb 1996).

It was noted above that generating electricity from natural gas is relatively more eco-efficient (Jacobs 1991) than coal and oil, and this relative greenness is further enhanced by the extra efficiencies of CCGT technologies, which use gas turbines to generate electricity but then re-use the hot exhaust gases from the gas turbines to drive an additional, electricity-generating steam turbine. So, whereas typical coal-fired power stations achieve 35 per cent efficiency (comparing the energy content of the fuel with the electricity produced), CCGTs raised this towards 50 per cent and, in some projects, towards 60 per cent. As a result, electricity from CCGTs emits 400g of CO_2 emissions per kilowatt-hour (kWh) of electricity output compared to 1000g/kWh for electricity generated from coal (Patterson and Grubb 1996).

Importantly, these logics of eco-efficiency extend also to the direct use of land. The mode of generation and the fact that gas is piped straight to the facility, needing few on-site handling facilities, with no ash to dispose of, means that that CCGT power stations demand less land than equivalent thermal power sources (Watson 1997; Patterson and Grubb 1996). This, in addition to other factors, is presumed to ease siting decisions. So, while coal projects and, especially, new nuclear capacity appeared mired in inexorably extending decision making processes as the 1980s wore on (Owens 1985), gas power offered the prospect of quicker, easier and cheaper construction.

A key question, however, is how and to what degree these abstract claims about the relative eco-efficiency of CCGTs would actually enable developers to navigate the complex, place-specific and more strategic concerns that could arise as actual projects get proposed for particular sites (Shove 1998). Data on the consenting processes suggests that gas-fired stations encountered relatively few difficulties in gaining permission. Figure 11.1 shows the significant volumes of consented capacity (not all of which was developed) that has come forward. Other research has shown an average time from application to consent of about two years, with very few applications having to undergo a public inquiry (Marshall and Cowell 2016). Under consenting legislation that prevailed until 2008, insurmountable objections from local councils would trigger a public inquiry, yet between 1988 and 2008 only three CCGTs went to public inquiry.[3] All three were subsequently permitted.

What has been particularly advantageous to gas-powered generation, however, are the sites available to be exploited for new facilities. Table 11.1 summarises the land use conditions, and clearly shows the very high association with former energy generating sites and previous or existing heavy industry. Approximately 45 per cent of capacity has been consented on sites either previously occupied by coal- or oil-fired power stations, or which has previously received planning consent for major fossil-fuel electricity generation. The connections to the carbon economy increase further if one considers CCGTs developed on former coal gas works, or on former or current oil refineries. This conferred practical advantages, as elaborated below, but also symbolic and representational advantages. In their environmental statements, developers extolled the relative eco-efficiency of CCGTs compared to oil or coal plant, but these arguments could achieve particular resonance for local councils where oil or coal plant had in fact been the previous site use. The material history of these sites helped claims about relative eco-efficiency to 'stick' (Nadaï and van der Horst 2010). On greenfield sites, locally, relatively greener technologies might more likely be seen as causing net environmental deterioration.

In institutional terms, power generation or similar industrial activities were recognised as 'an established use' of these sites in local planning processes. Developers could thus claim that they were re-using existing power generation sites, with technologies of lessor impacts than their predecessors, and also making effective use of brownfield land rather than consuming greenfield

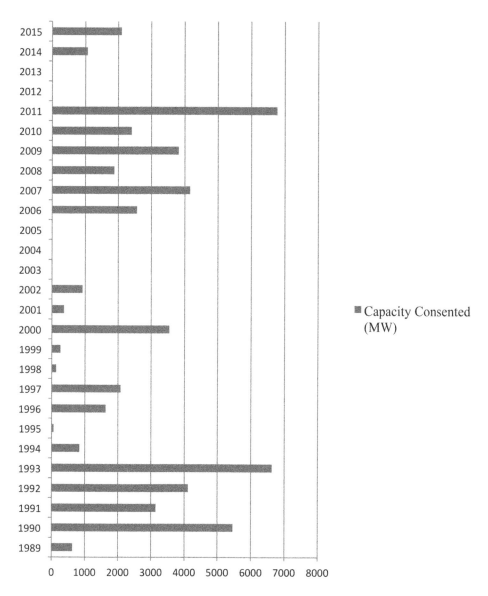

Figure 11.1 Annual capacity additions to gas-fired electricity generation in the UK, 1989–2015
Source: Richard Cowell.

sites. Even where projects did encompass patches of previously undeveloped land, the previous patterns of use meant that this land often had limited public access, reducing the basis on which social attachments and claims about conservation might form (van der Horst and Vermeylen 2011). By and large, because of previous land uses the sites did not fall within areas attracting large numbers of visitors. Many such sites also enjoyed the massive technical and cost advantages of proximity to existing high capacity grid connections (Garrone and Groppi 2012), and gas pipelines too in many cases. This is important because new high voltage grid connections can be as much if not more controversial than the power plant itself.

Table 11.1 Sites of gas-fired electricity generation, UK, 1988–2015

	Site category (total capacity consented, MW)	Number of projects (% of total capacity consented)
Electricity generation (former coal/oil power station site; consented for such a use; or associated use e.g. ash disposal area)	25 (24,188)	45%
Former industrial use (ex iron production, chemical plant, gas works, sugar processing, etc)	17 (12,264)	23%
Adjacent, within and/or connected to existing industrial complex (chemical plant, sugar processing, oil refinery, paper mill)	41 (15,596)	29%
Greenfield site (no previous industrial use of site, or of adjacent land)	3 (1,159)	2%

Source: derived inter alia from power station consent decision letters and environmental impact statements.

The next major category of site selected was within or adjacent to existing major industrial complexes: steelworks, paper mills, chemical plants. CCGTs and other gas-fired technologies offered companies ways of meeting their electricity needs themselves, often also replacing older on-site power plant and offering a surplus for export to the grid, as well as supplying steam for industrial processes. Again, in siting terms, new gas generation capacity has been developed within existing, 'private' industrialised spaces.

The siting dynamics of CCGTs thus contrast with renewable energy technologies such as onshore wind, in which the spatially extensive nature of the energy source has long been recognised to recast traditional geographies of energy provision towards the exploitation of new, predominantly rural locations (Walker 1995). With gas, we see a concentrated re-inhabiting or 'recycling' of old energy landscapes – former fossil energy and heavy industry spaces – presenting limited challenges to landscape norms and their wider societal associations. Landscapes cherished for their rurality or visual qualities were rarely threatened. Data from decision letters show that many CCGT applications attracted very few written objections from the public: the 750MW scheme at Coryton, within Mobil's refinery site attracted no objections at all; just one resident wrote in with an objection to the 1270MW project for West Burton B, adjacent to the existing coal-fired power station.

This siting geography may be a key explanation as to why gas-powered electricity has encountered relatively few siting problems. Nevertheless, siting processes around CCGTs certainly warrant closer attention. As was noted above, achieving 'successful fit' between new infrastructures and the landscape is not simply a functionalist exercise in matching past and future land uses, but also reflects the way in which consenting procedures may delimit discussion of 'mis-fitting' wider issues.

Delimiting debate?

Environmental justice scholars have alerted us to the potentially adverse social implications of facilities location. With CCGTs we may well be seeing another instance where rounds of energy infrastructure investment tend to end up in places socially and economically less likely to exercise voice (Garrone and Groppi 2012; van der Horst and Toke 2010). Figure 11.2 also shows how the aggregate effect of individual project decisions created marked spatial concentrations of CCGT development (see Figure 11.2), around the Thames Estuary east of London, as well as north and south of the Humber, and around the Dee and Mersey in north-west England and north-east Wales. Such large-scale mapping cannot fully capture the scale of development channeled into particular small areas. One example is the area around the villages of North and South Killingholme on the southern side of the Humber, which has seen two CCGTs developed, much of it on an area previously consented for an oil-fired power station, with another at South Humber Bank only 5km away, and others consented in 1996 and 2015, augmenting an industrial landscape already dominated by petro-chemical complexes.

If there is scant evidence that public objections or environmental concerns slowed the expansion of gas generation, publics and pressure groups were not always acquiescent either. Objections were raised, to individual projects and to the expansion of gas-fired generation in general, but the way in which they were handled is revealing of the scope for such concerns to exercise influence through consenting processes.

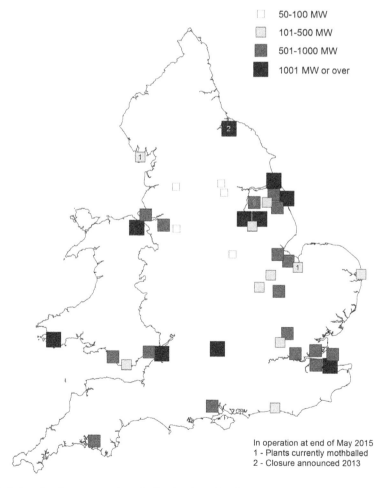

Figure 11.2 Gas-fired power stations in England and Wales

Source: Richard Cowell.

As one might expect, developments proposed for greenfield sites often attracted high levels of local opposition. Projects at Rugby (for a site allocated in the local plan for light industry) and at Ardleigh (a greenfield site in open countryside, not far from landscapes painted by John Constable), were both dropped after significant local objections. Such outcomes reinforce the wider point: where CCGTs did not re-inhabit landscapes of heavy industry or older fossil fuel power stations then claims about their relative eco-efficiency cut little ice. However, even when former coal-fired power station sites were being exploited, the relative eco-efficiency claims for CCGT technology could not match the way in which local environmental expectations had changed. At Plymouth, the 1990 application to develop a CCGT on a former power station site foundered on local authority concerns with emissions to air and landscape, intensified by the fact that wider regeneration had upgraded environmental quality in the area. A proposal for Greenwich in London, to replace existing electricity generators with a (higher capacity) CCGT was also withdrawn, as any net increase in pollution still caused local concerns in a densely populated urban area.

Viewing CCGTs as eco-efficient in terms of their use of land – the direct land take of the power station site – also underplays the various ways in which such projects have effects that spill beyond site boundaries. The control of construction traffic raised public concerns in numerous sites; so too did the handling of cooling water. Many coastal energy sites are juxtaposed with wildlife sites protected under European Union Directives; the highest tier of protection in the UK. Constructing cooling water pipes and the abstraction and discharge of water within protected wetland habitats often required careful attention (Cowell 1997). To give an example, protracted disputes about the ecological consequences of discharging warm water from a 2000MW CCGT at Pembroke into the sensitive waters of Milford Haven explain why that project took more than four years to receive consent, despite re-using the site of a former oil-fired power station.

Such impact management concerns could delay decision-making while mitigation measures were negotiated, but they rarely proved an existential threat to projects or to the expansion of gas-fired power plants generally. In some instances, this is because of the fragmented structure of consenting processes, which insulated consenting of the power station from its wider infrastructural consequences. A remarkable illustration is the 1700MW Wilton CCGT on Teesside, consented in just four months in 1990, where the 400kV grid capacity enhancements associated with it infringed on a National Park, were immensely controversial, and took seven years to consent. The fact that power stations and grid connections are developed by different companies has tended to thwart efforts to promote comprehensive assessment (Glasson et al. 1998).

This fragmentation reflects the dominant UK planning doctrine of 'treating each application on its merits', and it has delimited developer responsibilities in other respects too. Where large numbers of CCGTs have been consented, concerns have arisen about the cumulative effects of air pollution. After all, the relative eco-efficiency of gas-powered generation in terms of NOx (nitrogen oxide) emissions does not imply that the effects, locally and in aggregate with other projects, might not breach air quality limits. This became the concern of planning authorities around the Thames Estuary, east of London, who in the 1990s found themselves faced with a multiplicity of combustion projects including CCGTs and waste incinerators (see Figure 11.2). To understand the interactive effects, local planning authorities themselves took on the task of modelling the cumulative effects on air quality (Street 1997).

Challenging need?

Concerns emanating from power station siting procedures have had relatively little effect on the overall speed and scale of the 'dash for gas' in the UK in general, even if certain projects have faced delay and cancellation. Nevertheless, the strategic economic and environmental implications of such rapid investment in the exploitation of gas has led to more fundamental objections which have, in turn, been mobilised within the consenting processes for individual projects.

The effect of the dash for gas on the UK's domestic coal industry prompted challenges at local and national levels. The coal industry was declining rapidly in the late 1980s/early 1990s, but fuelling electricity generation remained by far the major market. The Coalfield Communities Campaign submitted objections to a number of CCGT applications, and managed to organise more concerted letter-writing campaigns and petitions against particular projects. The UK government has occasionally been sensitised to the interactions between exploiting gas and domestic coal, such as when it was taking steps in the 1990s to privatise the coal industry. As Kuzemko (2014) has noted, the energy security implications of becoming more reliant on gas is an issue that often brings avowedly 'market decisions' or 'technical choices' into the political realm.

On the environmental front, Friends of the Earth ran national campaigns through the 1990s that CCGTs should not be developed without more assiduous efforts to capture and use the waste heat, to significantly further improve the eco-efficiency of the technology. Often it was a lone voice on this issue, though objectors to specific power station applications sometimes picked up the heat wastage issue. What we do see, however, is that when central government became anxious about the speed at which gas-fired power stations were being developed, the main policy tool by which it sought to apply the brakes have been environmental. Steps were taken at the end of the 1990s to restrict the consenting of new gas-fired capacity to schemes meeting stringent requirements for the use of waste heat.[4] One such period, ending in November 2000, did lead to some CCGT project cancellations and delays, but other developers just rode out the restrictions, with a spike in new consented capacity of almost 3500MW when the restrictions ended (see Figure 11.1). This experience affirms the views of other siting conflict analysts that shifts in government policy are often a bigger source of delay than planning and consenting procedures (Owens 1985).

A common thread, however, is that challenges to the need for CCGT investments – targeted at particular applications, but questioning the wider merits of further fossil fuel investment – have made little headway. Where objectors challenged the 'need' for gas-fired electricity, government decision letters invariably rebuffed them with words to the effect that 'as a general rule, the need for a generating station remains a commercial matter for the applicant'.[5] This position has persisted and, if anything, tightened over the period. Steps taken in the Planning Act 2008 to further smooth and streamline the issuing of major energy infrastructure consents included the issuing of National Policy Statements, designed precisely to issue a firm government-backed statement on the need for various categories of energy infrastructure, including new fossil fuel capacity, coupled with the advice that this need case should not normally be challenged when individual project consents are determined (DECC 2011, para 2.2.4).

From 2009 one can see the introduction of measures that begin to place the development of gas-fired power stations within absolute, environmental constraints. The environmental agenda driving this is not about sites or landscape, but efforts to decarbonise the electricity sector. In fact, the various measures have an ambiguous relationship to the dynamics of project siting. Stringent national climate change legislation – the Climate Change Act 2008 – includes provision for carbon budgets, setting a cap on the amount of greenhouse gases emitted by the UK over a five-year period. The UK government has constantly resisted calls that new power station applications should be assessed directly against these budgets, placing faith instead in the capacity of developers to form their own investment judgements in the light of market-based instruments like the EU Emissions Trading Scheme, which seek to attach a price to carbon emissions. If this stance disconnects greenhouse gas emissions from infrastructure siting, others connect them. It has become government policy that all new thermal power plants should either include CHP, or – if not – provide evidence that the possibilities for CHP have been fully explored.[6] However, if a careful investigation reveals little realistic demand for heat users in the vicinity, this is not a bar to the electricity generating station being consented.[7]

The Government has also come to require that all commercial scale combustion power stations (300MW capacity or over) would need to demonstrate that they are 'carbon capture ready' if they are to obtain consent. Dimensions of 'readiness' include that the development site has space available for the technology, and can access suitable geological storage. As it is not yet clear that CCS is wholly technically feasible, the implications of this measure are unclear, but that developers see these measures as a potential risk to future gas-fired generation is evidenced by the fact that a high proportion of the most recent consents (at Hirwaun in south Wales, Eye in East Anglia) offer a capacity of 299MW, just below the threshold at which this policy would apply.

Conclusions

This chapter has focused on an under-examined dimension of modern energy systems – the siting of gas-fired power stations in the UK. Although it is difficult to compare the 'levels of controversy' that different energy technologies have encountered, it is reasonable to claim that siting processes for gas-fired stations have been much less problematic than nuclear or coal fired power stations (of which there have been few applications in the last three decades) or on-shore wind, especially when you consider the large volume of new CCGT capacity that has been built. Whereas siting controversies have been presented as a constraint to the expansion of renewable energy, it is hard to sustain this conclusion for gas. A major explanation is the remarkable extent to which CCGT technologies have been able to re-inhabit the landscapes vacated by previous rounds of coal- and oil-fired electricity provision, or be developed on sites enclosed within the domains of heavy industry, thus avoiding the disruption of 'pristine' 'un-industrialised' landscapes that has been a feature of much public opposition to renewable energy.

One could view the story of gas-fired power in the UK as a product of technical siting success; in which relatively eco-efficient technologies have been channelled towards well-chosen brownfield sites, reinhabiting previous fossil fuel landscapes, making good use of existing grid capacity (for the most part) and consequently raising few public concerns. Certainly, the reliance here primarily on documentary evidence from the consenting process means that there is much scope to refine our understanding of public concerns, based on wider data. Nevertheless, there is sufficient reason to be dissatisfied with an over-sanguine view of 'siting success'.

First, an absence of public response is ambiguous in its meaning and ethical valency (Marsden and Markusson 2011). The siting choices of gas-fired power stations reinforce energy geographies in which environmentally exploited and industrialised spaces get reproduced over time. This raises issues of environmental justice, even if – or perhaps especially because – local publics felt little scope to raise objections. Moreover, as the analysis has shown, the UK's dash for gas has raised a whole series of challenges over the past thirty years, about the effects of CCGTs on environments and resources at a variety of scales, but the structuring of decision-making processes has kept many such challenges *ultra vires* to project-specific decisions. Importantly, institutional and landscape effects have been mutually reinforcing. The institutional arrangements of consents decision-making sought to confining legitimate public debate to 'local issues' – in effect, to siting questions – yet on the narrow issue of siting, on landscape 'fit', CCGTs were often difficult to challenge. This outcome is important in terms of environmental politics, as it is often where locality-specific concerns find common cause with opposition to the wider direction of development– as with roads policy, minerals extraction or aviation – and where local, national and international pressure groups are all consistently involved, that siting conflicts form part of an effective pressure politics of wider change (Owens and Cowell 2010). Indeed, such multi-scalar pressure politics has undermined the growth of onshore wind in the UK. Where opposition to individual gas-fired power stations has arisen, there has been little sign of protest 'jumping scale' (Cox 1998), to raise existential questions about the merits of continually reinvesting in such fossil fuels.

Second, the tale of gas-fired power stations throws down some challenges to the way in which landscape is considered in analyses of energy transitions. On a *prima facie* basis, it seems desirable that one should pursue energy pathways that cause the least disruption to existing societal attachments to places and landscapes, yet the pursuit of land use 'fit' may undermine agendas of change and innovation. In the UK at least – an 'old country' (Wright 1985), in which preserving the past and rural Arcadian vistas bears heavily in contemporary politics – gas-fired power continues to offer the achievable prospect of maintaining electricity provision without any need to confront difficult choices about landscape change or place attachment, or indeed about electricity consumption. Indeed, continuing to generate electricity in bulk, in safely confined in 'industrial spaces', sustains Arcadia as we know it. The considerable scope to re-use old energy landscapes needs to be viewed as a significant advantage to the incumbent fossil fuel sector. Even if it is not causal, landscape here is at least complicit in reproducing the status quo, fostering the power of fossil fuel path dependencies. For analysts of energy and landscape, wishing to connect their work to transitions towards more sustainable energy pathways, the dynamics of siting quietude and stability warrant as much sustained critical attention as overt conflict.

Notes

1   Setting aside use of gas in non-CCGT power stations.
2   2014 is not even necessarily the peak year for gas-fired electricity generation in the UK. From 2010 plants constructed in the 1990s began to be mothballed and closed.
3   With the Planning Act 2008 all major power station applications had to go through an examination which might entail an element of public hearings, whereas prior to this, public inquiries only arose when the local planning authority sustained an objection. However, the 2008 Act places a statutory time limit of six months on those examinations.
4   By the policy on power station consents set out in the Conclusions of the Review of Energy Sources for Power Generation and Government response to the 4th and 5th Reports of the Trade and Industry Committee (Cm 4071) (HM Government 1998).
5   Decision letter for Killingholme, issued 31 October 1996.
6   See paragraph 4.6 in DECC 2011, noting that such measures are based on a regulation issued in 2006, by the then DTI, applying to Section 36 consents.
7   A 1020MW CCGT was consented at Kings Lynn in February 2009 without, at that point, any requirement being imposed to supply otherwise wasted heat to prospective users.

References

Aitken, Mhairi. 'Why we still don't understand the social aspects of wind power: a critique of key assumptions within the literature'. *Energy Policy* 38, no.4 (2010): 1834–1841.
Arapostathis, Stathis. 'UK natural gas integration in the making, 1960–2010: complexity, transitional uncertainties and uncertain transitions'. *Environmental Innovation and Societal Transitions* 11, (2014): 87–102.
Asafu-Adjaye, John et al. 'An eco-modernist manifesto'. 2015, at: www.ecomodernism.org (accessed 1 December 2015).

Barry, Andrew. *Political Machines. Governing a Technological Society.* London: Athlone Press, 2001.

Barry, Andrew. *Material Politics. Disputes Along the Pipeline.* Oxford: Wiley Blackwell, 2013.

Bickerstaff, Karen. '"Because we've got history here": nuclear waste, cooperative siting, and the relational geography of a complex issue'. *Environment and Planning A* 44, (2012): 2611–2628.

Blowers, Andrew. Leroy, Pieter. 'Power politics and environmental inequality: a theoretical analysis of the process of peripheralization'. *Environmental Politics* 3, (1994): 197–228.

Bradshaw, Mike. *Global Energy Dilemmas.* Cambridge, UK: Polity Press, 2014.

Burningham, Kate. 'Using the language of NIMBY: a topic for research, not an activity for researchers'. *Local Environment* 5, (2000): 55–67.

Calvert, Kirby. 'From "energy geography" to "energy geographies": perspectives on a fertile academic borderland'. *Progress in Human Geography*, online first, (2015): 1–21.

Chignell, Simon, Gross, Robert J K. 'Not locked-in? The overlooked impact of new gas-fired generation investment on long-term decarbonisation in the UK'. *Energy Policy* 52, (2013): 699–705.

Cowell, Richard. 'Stretching the limits: environmental compensation, habitat creation and sustainable development'. *Transactions of the Institute of British Geographers* 22, (1997): 292–306.

Cowell, Richard, Owens, Susan. 'Governing space: planning reform and the politics of sustainability'. *Environment and Planning 'C',* Government and Policy 24, No. 3 (2006): 403–421.

Cowell, Richard, Bristow, Gill, Munday, Max. 'Acceptance, acceptability and environmental justice – the role of community benefits in wind farm development'. *Journal of Environmental Planning and Management* 54, no. 44 (2011): 539–557.

Cox, Kevin R. 'Spaces of dependence, spaces of engagement and the politics of scale, or: looking for local politics'. *Political Geography* 17, (1998): 1–23.

Crenson, Mathew A. *The Un-Politics of Air Pollution: A Study of Non-Decision Making in the Cities.* Baltimore, MD: Johns Hopkins Press, 1971.

DECC (Department of Energy and Climate Change). Overarching National Policy Statement for Energy, EN-1. London: DECC, 2011.

DECC (Department of Energy and Climate Change). Electricity: Chapter 5, Digest of United Kingdom Energy Statistics (DUKES). 30 July 2015, London: DECC, 2015.

Devine-Wright, Patrick. 'Public engagement with large-scale renewable energy technologies: breaking the cycle of NIMBYism'. *Wiley Interdisciplinary Reviews: Climate Change* 2, (2011): 19–26.

Devine-Wright, Patrick, Howes, Yuko. 'Disruption to place attachment and the protection of restorative environments: a wind energy case study'. *Journal of Environmental Psychology* 30, (2010): 271–280.

Garrone, Paola, Groppi, Angelamaria. 'Siting locally-unwanted facilities: what can be learnt from the location of Italian power plants'. *Energy Policy* 45, (2012): 176–186.

Glasson, John, Therivel, Riki, Chadwick, Andrew. *Introduction to Environmental Impact Assessment.* London: UCL Press, 1998.

Gross, Catherine. 'Community perspectives of wind energy in Australia. The application of a justice and community fairness framework to increase social acceptance'. *Energy Policy* 35, (2007): 2727–2736.

Groves, Chris, Munday, Max, Yakovleva, Natalia. 'Fighting the pipe: neo-liberal governance and barriers to effective community participation in energy infrastructure planning'. *Environment and Planning 'C',* Government and Policy 31, no. 1, (2013): 340 – 356.

HM Government. *The UK Renewable Energy Strategy,* Cm7686. London: The Stationery Office, 2009.

IEA (International Energy Agency). *Key World Energy Statistics.* Paris: IEA, 2014.

IEA (International Energy Agency). 'Key natural gas trends' (Excerpt from Natural Gas Information). Paris: IEA, 2015.

IPCC (Intergovernmental Panel on Climate Change). *Climate Change 2014.* Synthesis Report. Contribution of Working Groups I, II and III to the Fifth Assessment Report of the Intergovernmental Panel on Climate Change (Core Writing Team, R.K. Pachauri and L.A. Meyer [eds.]). Geneva, Switzerland: IPCC, 2014.

Jacobs, M. *The Green Economy.* London: Pluto Press, 1991.

Kuzemko, Caroline. 'Politicising UK energy: what "speaking energy security" can do'. *Policy and Politics* 42, (2014): 259–274.

Lee, Maria et al. 'Public participation and climate change infrastructure'. *Journal of Environmental Law* 25, (2012): 33–62.

Lovins, Amory. *Soft Energy Paths: Toward a Durable Peace.* Cambridge, MA: Ballinger, 1977.

Mackay, David. 'Sustainable energy – without the hot air'. 2009, at: www.withouthotair.com/ (accessed 10 February 2016).

Marsden, Wendy, Markusson, Nils. *Public Acceptance of Natural Gas Infrastructure Development in the UK (2000–2011)*, Final Case Study Report. November, Edinburgh: UK Energy Research Centre, 2011.

Marshall, Tim, Cowell, R. 'Planning, infrastructure and the command of time'. *Environment and Planning C: Government and Policy* 34, (2016): 1843–1866.

Mounfield, Peter R. 'Electricity production after privatization'. *Geography* 75, (1990): 374–378.

Nadaï, Alain, van der Horst, D. 'Introduction: landscapes of energies', *Landscape Research* 35, (2010): 143–155.

Owens, Susan. 'Energy, participation and planning: the case of electricity generation in the United Kingdom'. In Calzonetti, F. and Soloman, B. (eds), *Geographical Dimensions of Energy*. Dordrecht: Reidel, 225–253, 1985.

Owens, Susan, Cowell, Richard. *Land and Limits: Interpreting Sustainability in the Planning Process*. 2nd Edition. London: Routledge, 2010.

Pasqualetti, Martin J. 'Opposing wind energy landscapes: a search for common cause'. *Annals of the Association of American Geographers* 101, (2012): 907–917.

Patterson, Walt, Grubb, Michael. *Liberalizing European Electricity: Impacts on Generation and Environment*. Briefing Paper No. 34, November, London: RIIA, 1996.

Selman, Paul. 'Learning to love the landscapes of carbon-neutrality'. *Landscape Research* 35, (2010): 157–171.

Shove, Elizabeth. 'Gaps, barriers and conceptual chasms: theories of technology transfer and energy in buildings'. *Energy Policy* 26, (1998): 1105–1112.

Shove, Elizabeth, Walker, G. 'Commentary. CAUTION! Transitions ahead: politics, practice and sustainable transitions management'. *Environment and Planning A* 39, (2007): 763–770.

Sovacool, Benjamin, Cooper, Christopher. *The Governance of Energy Mega-Projects*. Cheltenham: Edward Elgar, 2013.

Street, Elizabeth, 'EIA and pollution control', in: Weston, Joe (ed.), *Planning and Environmental Impact Assessment in Practice*. Harlow: Addison Wesley Longman, 165–179, 1997.

Szarka, Joseph, *Wind Power in Europe. Politics, Business and Society*. Basingstoke: Palgrave Macmillan, 2007.

Szarka, Joseph, Cowell, Richard, Ellis, Geraint, Strachan, Peter A., Warren, Charles, eds. *Learning from Wind Power. Governance, Societal and Policy Perspectives on Sustainable Energy*. Basingstoke: Palgrave, 2012.

Taanman, Mattijs, 'Working in the science-policy interface: transition monitoring in the Dutch energy transition program', In: Verbong, Geert, Loorbach, Derk (eds), *Governing the Energy Transition. Reality, Illusion or Necessity?* New York: Routledge, 251–276, 2012.

Van der Horst, Dan, Toke, David. 'Exploring the landscape of wind farm developments: local area characteristics and planning process outcomes in rural England'. *Land Use Policy* 27, (2010): 214–221.

Van der Horst, D., Vermeylen, S. 'Local rights to landscape in the global moral economy of carbon'. *Landscape Research* 36, (2011): 455–470.

Van der Vleuten, Erik, Högselius, Per. 'Resisting change? The transnational dynamics of European energy regimes', in: Verbong, Geert, Loorbach, Derk (eds), *Governing the Energy Transition. Reality, Illusion or Necessity?* New York: Routledge, 75–100, 2012.

Verbong, Geert, Loorbach, Derk (eds). *Governing the Energy Transition. Reality, Illusion or Necessity?* New York: Routledge.

Walker, Gordon. 'Energy, land use and renewables: a changing agenda'. *Land Use Policy* 12, (1995): 3–6.

Walker, Gordon. *Environmental Justice: Concepts, Evidence and Politics*. London: Routledge, 2011.

Watson, Jim. 'The technology that drove the "dash for gas"'. *Power Engineering Journal* February, (1997): 11–19.

Wolsink, Maarten. 'Entanglement of interests and motives: assumptions behind the "nimby-theory" on facility siting'. *Urban Studies* 31, (1994): 851–866.

Wolsink, Maarten. 'Planning of renewable schemes. Deliberative and fair decision-making on landscape issues instead of reproachful accusations of non-cooperation'. *Energy Policy* 35, (2007): 2692–2704.

Woods, Michael. 'Conflicting environmental visions of the rural: windfarm development in mid Wales'. *Sociologia Ruralis* 43, (2003): 271–288.

Wright, Patrick, *On Living in an Old Country. The National Past in Contemporary Britain*. London: Verso, 1985.

12

Experiencing citizen deliberation over energy infrastructure siting

A mixed method evaluative study

Patrick Devine-Wright and Matthew Cotton

Introduction

Partly in response to the threat of climate change, energy infrastructures are in flux. In developed countries, policies to reduce carbon emissions are leading to widespread changes to centralised systems of energy provision, which in turn impact upon particular landscapes and communities. This relates to a central theme of this book, which applies geographical frameworks to study multiple facets of security, sustainability, space and place as they relate to energy. Our focus is upon the high voltage overhead transmission lines (hereafter HVOTLs) that connect new nuclear and large-scale renewable energy projects to the grid and that frequently stimulate strong objections from affected communities. Organisations that manage grid infrastructure – transmission system operators (hereafter TSOs) – for example National Grid plc in the UK, have sought to lessen objections and increase community acceptance through a range of measures, including earlier consultation with local communities and proposals to distribute community funds.

This chapter addresses the ways that communities engage with local proposals to site HVOTLs. In particular, we evaluate the experiences of citizens that took part in a two-day deliberative engagement workshop that was instigated by the authors, as part of a research project, in a town in South West England directly affected by a power-line siting proposal. Our aim was to trial a method that up to now has rarely been used by grid companies to engage with communities, with the aim of better understanding how participants felt about taking part in energy infrastructure deliberations, and in doing so, to make recommendations both for future research and for grid companies about the potential value of and role for this method of engagement in future siting contexts. To evaluate participant experiences, we used a combination of qualitative and quantitative methods including in-workshop surveys and interviews conducted six months after the workshop concluded. In terms of outcomes, the research aimed to capture whether participation fostered social learning, as evidenced by changes in opinions about the proposals, or wider beliefs about planning and energy technologies, and led to commitments to adopt new practices. In terms of process, the research aimed to capture participants' views on the workshop overall and specific elements within it.

We adopt Rowe and Frewer's (2000) typology of engagement types, identified by the flow of information occurring between the parties and by its significance in decision-making processes. *Communication* involves one-way information flow from the sponsor (typically a government organisation or private company) to publics, who are simply informed of decisions with little or no influence upon these. *Consultation* involves two-way information flow between sponsor and publics, but the information flows back without any dialogue. *Participation* involves a two-way exchange of information between sponsor and public with the possibility for transformed opinions in both parties.

Here, we focus upon one form of participation: *deliberative engagement*. This method provides citizens with opportunities to learn about the issues at stake, to question experts and to discuss potential outcomes, enabling informed opinions and the voicing of those opinions by citizens within a particular decision-making context. To date, the opportunity to participate in deliberative engagement has rarely been provided to local citizens affected by proposals to construct HVOTLs. Research into the engagement practices of TSOs has revealed that information provision and some degree of consultation is the norm (Cotton and Devine-Wright 2011). This has led to grievances by communities of tokenist attempts to involve them in decision-making, evidenced by the importance of procedural justice in research investigating the factors underlying public objections to HVOTL siting (Keir et al. 2014). Research in the UK and Norway has revealed public concerns over limited information being made available; a short amount of time to respond, and doubts concerning whether public responses have a genuine influence upon decision-making (Devine-Wright 2013; Knudsen et al. 2015; Cotton and Devine-Wright 2013).

As well as the need to trial more participatory forms of citizen engagement in contexts of HVOTL siting, there is also a need to assess the experiences of those taking part (Halvorsen 2001; Webler et al. 2001). This can provide evidence for the formulation of effective and context sensitive community involvement practices that can potentially benefit both developers (in terms of reduced costs and planning delays following opposition actions from locally affected citizens), policy-makers (in terms of streamlining applications for nationally significant infrastructure and thus achieving sustainable development goals) and community actors (in terms of ensuring fairer decision outcomes in planning processes). However, despite some recent exceptions (Cotton and Devine-Wright 2011; Cotton 2011; Groves et al. 2013), few studies have assessed the ways in which developers involve locally affected community actors from the perspective of those actors – including the methods and engagement 'tools' that are used. Our research addresses this gap.

When it comes to evaluating deliberation, Rowe and Frewer (2000) suggest that evaluation of deliberative processes is difficult because of the complexity of the concept, and the lack of widely held criteria, agreed-upon evaluation methods and measurement tools for gauging success. Their evaluative framework for the structure and implementation of deliberative processes focuses upon impact or outcomes (however these are defined) on decision-making as criteria from which to measure success. Indeed a broader feature of the literature on evaluating public participation is that it is dominated by a focus upon either outcome-based criteria or process criteria. However, there is still a lack of quantitative or wholly objective terms to measure the process aspects of deliberative quality (Wilson and Schooler 1991; Bohman 2000; Chambers 2004), primarily because of the relatively vague conceptual precepts that underpin it (Burkhalter et al. 2002).

As far as outcome and process evaluation is concerned, decision-outcomes based evaluation criteria are inappropriate in this study, since the deliberative experiment was run as part of an academic research project, rather than play a formal role in a decision or policy making process. Instead, our emphasis is upon the (potentially) transformative effects of deliberation on the participants themselves. These effects include the development of 'civic expertise' (Bäckstrand

2004) or motivating pro-environmental behaviour (see for example Barr 2003). Process-based outcomes such as participant reflections upon what makes good deliberation (Webler et al. 2001), include indicators of fairness and competence (Webler 1995), and also attention to issues such as evaluating deliberative quality (Graham and Witschge 2003); the efficacy of methods (Rauschmayer and Wittmer 2006); experiences of comfort and satisfaction (Halvorsen 2001) and reflections upon longer term participant learning. Given that community involvement is often cited as providing opportunities to promote active citizenship through a transformation of values and preferences in response to encounters with other deliberators (D'Entrèves 2002: 25), it is important to empirically investigate the potentially transformative power of engagement to produce long-term 'beyond process' learning (Bull et al. 2008); in particular in this context, for the fostering of pro-environmental behaviours (Kollmuss and Agyeman 2002), and transforming public responses to project proposals under consideration in the workshop.

In terms of learning and emotional responses, deliberative processes are frequently posited as promoting learning about public problems and possibilities. In this study we evaluated aspects of learning using qualitative and quantitative methods, including both technical elements (such as learning about electricity transmission systems, electricity production and consumption and knowledge of planning processes and developer activities), and social learning – a process by which changes in individuals and social systems are obtained through mutual observation and social interaction (Bandura 1977). Related to this learning process are issues of emotion, whereby negative emotions about developer organisations, project plans and their impacts on particular places become influential factors in community opposition (Cass and Walker 2009; Devine-Wright 2009). Research has shown that emotional attachments to places (positive emotional relationships that individuals feel about valued environments, see Altman and Low 1992) can serve to both motivate support and opposition to infrastructure developments in specific places, depending upon whether (in this case HVOTL) developments are characterised as threatening and thus disruptive of place attachments, or as opportunities for local development and thus strengthening localised place attachments (Devine-Wright 2009; Cotton and Devine-Wright 2013; Devine-Wright 2013).

Moving on to HVOTL proposals, in September 2008, EDF announced proposals to construct a new 3.6GW nuclear power plant at Hinkley Point in Somerset, South West England. In response to this, National Grid plc (hereafter NG), the TSO for England and Wales, proposed a 37-mile 400kV connector transmission-line upgrade (the Hinkley Point C Connector, hereafter HPCC) to reinforce the local grid between the towns of Bridgwater and Avonmouth. From October 2009 to July 2010, NG implemented a public consultation programme on the corridor options. They contacted 37,000 households, advertised in the local press, held 17 public exhibitions attended by over 4000 local people, and briefing meetings for MPs, county, ward and parish councillors. Throughout the consultation phase, NG provided website and phone-line contacts for community input. Feedback revealed strong public preference for an under-sea option, and NG responded in a second round of 24 public consultation events in May 2010, by providing further information about the chosen route corridors and their justification for not taking the under-sea route forward (National Grid 2010). NG did not utilise deliberative methods to engage with communities during this period.

Context

Nailsea is a town with approximately 18,000 residents, situated eight miles to the South East of the city of Bristol in the West of England. Nailsea was encircled by both route corridor options initially identified by NG when it began public consultation, and strong public opposition emerged. A local protest organisation 'Save Our Valley' (SOV) was formed, which ran its own campaign, distributing leaflets, setting up a 'graffiti wall' for community responses to proposals, alongside meetings, sponsored walks along the line route and written responses to local politicians and NG. In light of this, Nailsea was selected as a case study site by the researchers on the grounds that the strength of public opposition to proposals was indicative of high issue salience among local residents. The workshop was held as part of a wider academic research project (Supergen FlexNet) that aimed to investigate the role of public participation in contexts of HVOTL siting. Other findings of the project (for example the outcome of in-depth interviews with grid company representatives) are reported elsewhere (Cotton and Devine-Wright 2011, 2012, 2013; Devine-Wright 2013).

The workshop was held across two Sundays in November and December 2010. Day 1 of the workshop aimed to inform participants about issues of line-siting, planning legislation and the proposed route corridors, with presentations on these topics delivered by the authors. To provide information about the local proposals, information was provided through media clips from National Grid outlining the HPCC proposal, the technical and cost implications, and NG's position on undersea routing, and from a BBC documentary about electricity grids, alongside a presentation and question and answer sessions by a senior Electronic and Electrical Engineering academic from the University of Bath (we refer to these collectively as the 'technical presentations' below). The workshop leaders also ran presentations on aspects of planning and politics from their previous research. NG was invited, however they refused to participate, expressing concerns that the event would be confused with their official consultation programme. Following the researcher presentations, participants were broken up into a series of sequential discussion groups (7–8 per group, participants alternating in each session). Each was independently facilitated using participatory methods in the following format:

Day 1

a   Identification of actors, technologies and discussion of resultant socio-economic, governance, health and environmental issues.
b   Roleplay of the stakeholder actor perspectives identified in the first session.

Day 2

a   Examination of local impacts of siting in relation to the town of Nailsea, and other communities affected by the power line.
b   Identification and discussion of challenges and problems to be addressed in the planning process.
c   Identification of potential solutions to issues raised and points of further research and discussion.

Discussions within each session were prefaced with participant-led identification of issues recorded on sticky-notes that were then clustered into groups around commonly identified themes. These themes then framed focus-group-type discussion activities.

In terms of the evaluation methodology, we adopted a mixed method approach to evaluate participants' experiences of the deliberative workshop. Questionnaires were distributed at the beginning and end of day 1 and at the end of day 2, and consisted of closed and open-ended questions. Follow-up interviews were conducted with a sample of participants ($n = 15$) six months after the workshop concluded (March–April 2011). Questions in the surveys were intended to capture both process and outcome dimensions of participants' experiences of the workshop, as well as more general beliefs and feelings about the power line proposal and levels of trust in the TSO. Qualitative analysis involved coding the transcribed utterances *in vivo* using MaxQDA data analysis software and then performing a thematic analysis of coded sections. Utterances are anonymised and participants are denoted as either male or female speakers (M/F) and numbered for differentiation.

For the sample, 39 local residents took part in the workshop (dropping to 38 by the end of day 1 and 37 due by the end of day 2 due to illness). These were recruited by a professional marketing company who was commissioned by the research team to deliver a varied sample of participants representative of the town of Nailsea where the workshop took place. The sample shows diversity in gender, age, educational attainment and tenure; though the larger number of females to males, and fewer younger participants, means this is not a fully representative sample. Participant demographic details are shown in Table 12.1.

Findings

The results of our evaluation are presented in two sections, first referring to process aspects; second to outcome aspects. In each case, both quantitative and qualitative data is drawn upon to provide an overall picture of participants' experiences and whether these may have changed over time.

Table 12.1 Characteristics of the workshop participants

Sex	n
M	17
F	22
Age	
18–29	6 (15.4%)
30–44	10 (25.6%)
45–60	10 (25.6%)
> 60	13 (33.3%)
Education qualifications	
None	7
GCSE	16
A-level	8
University degree	8
Tenure	
Owner/occupier	27
Tenant	5
Private tenant	4
Other	3
Length of residence in Nailsea	
Less than one year	1 (2.6)
1–5 years	5 (12.8)
6–10 years	4 (10.3)
Over 10 years	29 (74.8)

Section 1. Process evaluation

Findings from the surveys suggest that participants had overwhelmingly positive opinions about the deliberations, in terms of feeling able to understand the purpose of the workshop, to discuss the issues, and to understand the information provided (see Table 12.2 below). Statistical analyses showed that response to only one of the questions changed significantly over time: the item

Table 12.2 Opinions about the process of deliberation

	Day 1					Day 2				
	Strongly Agree	Agree	Neither	Disagree	Strongly Disagree	Strongly Agree	Agree	Neither	Disagree	Strongly Disagree
The event was well organised	13 / 34.2	23 / 60.5	1 / 2.6	1 / 2.6	0 / 0	17 / 45.9	18 / 48.6	2 / 5.4	0 / 0	0 / 0
I understood the purpose of the event	13 / 35.1	23 / 62.2	1 / 2.7	0 / 0	0 / 0	14 / 37.8	21 / 56.8	2 / 5.4	0 / 0	0 / 0
I understood how the results of the event will be used	13 / 34.2	20 / 52.6	5 / 13.2	0 / 0	0 / 0	7 / 18.9	27 / 73	3 / 8.1	0 / 0	0 / 0
There was enough time to fully discuss the issues	7 / 18.4	29 / 76.3	2 / 5.3	0 / 0	0 / 0	10 / 27	23 / 62.2	1 / 2.7	3 / 8.1	0 / 0
I was able to discuss the issues that concern me	9 / 23.7	27 / 71.1	2 / 5.3	0 / 0	0 / 0	13 / 35.1	22 / 59.5	2 / 5.4	0 / 0	0 / 0
The information provided was relevant	9 / 23.1	29 / 76.3	0 / 0	0 / 0	0 / 0	8 / 21.6	27 / 73	1 / 2.7	1 / 2.7	0 / 0
There was enough information provided	3 / 7.9	24 / 63.2	5 / 13.2	5 / 13.2	1 / 2.6	Question not asked				
I understood the information provided	5 / 13.2	30 / 78.9	3 / 7.9	0 / 0	0 / 0	9 / 24.3	26 / 70.3	2 / 5.4	0 / 0	0 / 0
I enjoyed taking part	15 / 39.5	19 / 50	4 / 10.5	0 / 0	0 / 0	23 / 62.2	14 / 37.8	0 / 0	0 / 0	0 / 0

'*I enjoyed taking part*' was rated significantly more positively at the end of the second day in comparison to the end of the first day (($t(35) = 2.94$, $p<.006$, with a mean of 1.67 for day 1 (s.d. 63, $n=36$) and a mean of 1.39 for day 2 (s.d. 49, $n=36$)). Sixty-two per cent of respondents to the second day survey 'strongly agreed' that they had enjoyed taking part, suggesting that the workshop was an enjoyable experience.

This finding was supported by post-event interviews. All of the interviewed respondents ($n=15$) reported positive satisfaction with taking part in the workshop overall. The most common descriptors (used by all respondents) were that they found the workshops either 'interesting' or 'enjoyable', or both. Some were highly positive about expressing their enjoyment of the day, for example:

> *Is there anything in particular that you enjoyed about them?*

> iF7: Well just everything. Informative-wise and because obviously there were people there I knew, and it was just good fun to be doing things like that with people I wouldn't normally do things like that with. Yeah, they were very good.

When asked about specific aspects of the workshop which did not work well, the most commonly expressed problem was an exercise on day 1 where participants were asked to take the roles of key industry, environmental and local community stakeholders identified in the first session of stakeholder mapping. Some suggested that they were deeply uncomfortable with the idea of taking part in roleplaying, going so far as to state that they would not have agreed to be involved in the workshop at all had they known that roleplaying was involved. The reasons for the relative failure of the roleplaying exercise were mostly described as being down to the personalities of the participants, the level of personal comfort from 'performing' to an unfamiliar audience and the lack of preparation that they had. Given that some of the stakeholder roles they were asked to perform had a technical component (such as NG), this was problematic for those who felt their technical knowledge was insufficient to present the informational adequately.

> *You know that we ran those two workshops in December and November last year, when you think about them is there anything that immediately comes to mind? Anything that you remember, anything that stands out?*

> iF5: The role play, I didn't like that, that stands out big time.

> *Well what about the role play in particular? Why was that not so good?*

> iF5: Because most people didn't want to join in and it was left to a few of us that weren't that keen but who did it anyway. It would've been different if everybody had participated and they didn't so it was quite embarrassing really when you had people who wouldn't even consider doing it. And for the people that did, it was quite embarrassing for them. That's the only reason, really.

Others felt that the role play exercise gave them an opportunity to consider the different sides of the issue and learn from those with a non-technical perspective on power lines issues.

> iF4: I could understand more clearly how the systems work. I think I've got a pretty good idea now of how the whole business of deciding these things works. I don't necessarily

agree with the processes. I still think that they are not right. That they are not as open as they could be but I can see how it works. Some of the role plays that we did when we had to be the other side as it were. That taught me a lot I can see how they work. I think I just understand it better.

Though the workshops were enjoyed overall, a number of specific process-based criticisms arose. In particular with regards to consistent facilitation, allowing a fair and balanced discussion where all participants can contribute freely and have adequate opportunities to speak:

So why do you think it was the risk of going the wrong way. Was it just the facilitator or are there any other factors?

iM5: No I think primarily it was the facilitator didn't facilitate she allowed things to drift and I'm not having in that sort of environment having looked at you and as I was looking at your colleagues in a critical view having done all that you think oh for goodness sake what you should be doing is this.

Other criticisms were based on the overall length of the workshop, how the depth of discussion leads to repetition of the same issues:

Just thinking back to the workshops in November and December last year, is there anything off the top of the head that you can remember or that really stands out for you?

iF6: My brain's gone since then. They went on probably too long because all the things we were discussing we were going over and over again.

So you think they should be shorter overall?

iF6: Yes.

Overall, the findings suggest that the workshop was a pleasant, interesting and informative experience for the participants. However, it also revealed that the role-play exercise, in particular, was controversial for some, as well as some variability in small-group facilitation, with the perceptions of facilitator *competence* – through discussion control and impartiality – paramount in satisfaction with the experience.

Section 2. Outcome evaluation

LEARNING

Analysis of quantitative data from the surveys gives a broad indication that participants learnt new information and changed their views as a result of taking part. Levels of reported familiarity with the power line proposals increased significantly over time. Statistical analysis showed significantly higher levels of perceived familiarity at the end of day 2 by comparison to the start of day 1 (($t(33)$ = 4.461; $p<.000$) (mean levels: Day 1: 2.44 ($n=34$) s.d. =1.02; Day 2: 1.65 ($n=34$) s.d. =0.49). Moreover, participants reported increases in learning (see Table 12.3). For example, 44.8 per cent agreed on day 1 that attending the event had changed their views. By day 2, this had increased to 67.5 per cent. On day 1, 91.9 per cent agreed that they had learnt something they didn't know before, increasing to 97.3 per cent by the end of day 2. Paired sample *t*-tests showed that these scores had become significantly more positive over time

Table 12.3 Change in opinions about the outcomes of deliberation

Attending this event changed by views	Strongly Agree	Agree	Neutral	Disagree	Strongly Disagree
End of day 1	5	12	16	4	1
	13.2	31.6	42.1	10.5	2.6
End of day 2	8	17	7	5	0
	21.6	45.9	18.9	13.5	0
I learnt something I didn't know before					
End of day 1	14	20	1	2	0
	37.8	54.1	2.7	5.4	0
End of day 2	22	14	1	0	0
	59.5	37.8	2.7	0	0

(Attending the event changed views: $t(35) = 2.41$, $p<.021$; Learnt something didn't know before: $t(34) = 2.78$, $p<.008$).

Analysis of qualitative data gives depth to these findings. Technical learning is an issue that emerges in relation to the presentation of information on a range of aspects of how transmission networks operate. The aforementioned technical presentations were understood very differently by different participants. Though the majority of interview respondents (n=8) reported that they found the presentations to be helpful in understanding key issues around power generation, a smaller number (*n*=5) found them either to be confusing due to the level of technical understanding that the was required, or else not balanced enough in their coverage of the issues. Responses to the information provision aspect included critical appraisal of the science communication abilities of the presenter. Positive comments highlighted the academic's capacity to communicate to a lay audience, condensing complex information and inspiring interest in the workshop processes itself:

> iM2: Yeah, I did. Well, I mean I just learnt all about the system, I couldn't imagine the complexity of it without being, you know, first of all, the professor or the third chap who spoke to us –
>
> [*interviewer gives name of presenter*]
>
> Yeah, [name anonymised]. He was very good, I thought he gave us a right good insight to start with which to me got me interested. He was the one that started me off giving me the interest for the rest of the two days. You know, the AC/DC current and where it was coming from, . . . and what they were looking at and all this business and it was put over in layman's terms rather than just too technical.

However, the capacity for technical learning was not uniformly welcomed:

> iF5: When you had the computer running and the slides. Some of it was interesting but a lot of it – especially when you had another professor in, an engineer or something – was quite difficult for a lot of people to understand because obviously we're not that way inclined, we're just Joe Public and he was going on about things that I couldn't under-

stand and my friends, the people I knew there, they didn't understand really what he was saying. So that was more your side of things, you could understand but maybe Joe Public couldn't.

So do you think there are any ways in which we could –

iF5: You could simplify it more. . . . Just put it in simpler terms because I think a lot of the time, especially with the engineer, he was using words that we wouldn't even know and you'd think 'well what does that mean?' because we didn't know what it meant and he didn't actually explain what it meant.

Given that the presentations were the primary form of technical information provision upon which the participants grounded their deliberations, the quality of those presentations was key to their subsequent experience within the group. The differences in background knowledge and personal comfort in tackling complex engineering topics was highly variable across participant responses. In addition to the presented information, two interview respondents stated that they had gone on to research power line issues before and after coming to the first workshop and in between the two workshop sessions, and the use of smart phones enabled participant-led in-workshop research to take place. The majority however did not do any extra research.

Impact upon feelings and beliefs about the power line proposals

EMOTIONS ASSOCIATED WITH THE POWER LINE PROPOSALS

Analysis of survey data showed differences between positive and negative emotions associated with the power line. Responses to positively worded items (e.g. excited, happy, proud, curious) were consistently low throughout. However, participants reported feeling more frustrated, shocked, threatened and angry about the proposals after day 2 in comparison to the beginning of day 1 (see Table 12.4). This was an unexpected finding, and suggests that learning more about the proposals increased negative emotions associated with them. This contrasts with the emotional associations felt by the participants with the workshop itself (see above) and indicates the value of future studies ensuring separate ways of capturing these responses.

TRUST

Analysis of survey data suggested that the workshop led to an increase in trust in the local action group ($t(35) – 3.58$, $p<.001$) with a mean level of 3.75 at the end of day 1 ($n=36$, s.d. $=1.01$) and a mean level of 2.78 at the end of day 2 ($n =36$, s.d. $=1.15$). By contrast, there was no significant difference in trust in National Grid: levels were consistently low at both time points. This finding was also reflected in the follow-up interviews, as it was clear that none of the interviewees reported that levels of trust in National Grid had increased in the intervening period

Table 12.4 Negative effects associated with the proposed power line over time

	Beginning day 1			End day 2			T (p value)
	mean	n	s.d.	mean	n	s.d.	
Frustrated	3.38	34	1.44	2.76	34	1.42	2.41 (.022)
Shocked	3.91	34	1.31	3.32	34	1.41	1.97 (.058)
Threatened	3.57	35	1.19	3.00	34	.60	2.72 (.010)
Angry	3.37	35	1.26	2.97	35	1.15	2.17 (.037)

between the workshop and the follow-up interviews (6–7 months). One unexpected, and indeed the most common response, was that participating in the workshop had reinforced their perception that NG was an untrustworthy organisation, partly due to its size, or due to the fact that decision making processes were considered a *fait accompli* – made in advance of their mandated public consultation process. One interview participant suggested that it was EDF, the builder of the proposed Hinkley Point C power station that was to blame, not NG.

> *Having taken part in the workshops themselves, has that changed your opinion on any of those organisations that were mentioned, like the government or National Grid or EDF?*

> iF5: No, it's not changed my mind about them at all. I still think what I thought at the beginning, that whatever they've decided then they're still going to do in the future. Whatever they say, whatever we've done, I don't think it's going to make a blind bit of difference.

Longer term impacts

In the follow-up interviews, the majority of respondents (n=9) pointed out that the power line issue had 'gone quiet' in recent months. The local campaign group (Save Our Valley) had stopped producing materials and canvassing local people. NG had sent a pamphlet explaining their stage of the decision-making process, but this had not moved further, and the local press had not run any stories recently. This led to a number of participants stating that the issue had become less salient for the local community. Those that had done research or been actively contacting NG had been dissuaded by the lack of response and the issue 'going quiet'. A number of them spoke about how other aspects of their life had 'gotten in the way'. Examples included dealing with chronic illness and other life stresses that changed their perspective on the importance of things such as power lines.

> *What about since then? Have you come back to doing research afterwards?*

> iM4: To be honest, no I haven't. I've been so busy with other things that it's kind of dropped off the front burner into the back burner at the moment. But as far as I'm aware it's gone quite quiet on any sort of development side of things. I haven't seen anything local posted about further discussion groups or protest walks or anything of that nature that we've had previously. As soon as I become aware of it I'll definitely become active in it again.

> *Since taking part have you talked with any other people about pylon issues or anything that came out of the workshop?*

> iM1: Only very briefly. I've talked to our son-in-law about it because he came to the meetings as well but not to a great extent. The family have been rather overwhelmed by my wife's illness and the fact that on Saturday this week our oldest grandchild is getting married. All of which seems to have rather pushed other matters into the background.

Pro-environmental attitudes and behaviours

A small number of respondents (n=5) reported that their involvement in the workshops had a significant effect on their reported environmental attitudes or associated pro-environmental behaviours, particularly to local environmental protection and energy use. Some reported that taking part had given them a greater awareness of electricity issues, and that this had encouraged them to conserve electricity (and opt for a smart meter), others reported a greater awareness of conservation issues in the local area.

> iF3: Yes, it has. In fact we've got a little thing from our supplier so now I know exactly how much electricity we're using because we've got one of those little gizmo things. It was free but that's what spurred me on to get one so I know now. So yes, I'm more inclined to say to the boys 'turn the lights out, don't leave this on'.

> If4: I think it has made me more aware of local groups that deal with things. My son has volunteered to help with the they need to do compassing and things like that around here and trimming hedge ways and clearing dykes and stuff like that to keep it all viable. To keep the wildlife viable and he has actually volunteered to do that so it has pushed us a bit that way to preserving . . .

The evidence suggests that taking part in the workshops increased the relative *visibility* of electricity system connection and energy use (Devine-Wright et al. 2010) to local participants, prompting reflection upon domestic energy consumption patterns and social practices. Moreover, taking part in the workshop stimulated awareness of local environmental changes; the perceived threat posed by the HVOTL proposals manifests as a salient influence upon the way participants structure their relationship with their surroundings and take part in local community environmental conservation activities (see for example Rogan et al. 2005).

Discussion

Against a backdrop of one-way information provision by UK grid companies towards local communities (Cotton and Devine-Wright 2012), this chapter reports an attempt to trial a deliberative engagement workshop in the context of real-world HVOTL siting, and to evaluate how this was experienced by the participants. Here we reflect upon potential deliberative engagement practices for TSO and other energy infrastructure developers that go beyond one-way information provision activities (e.g. leaflet campaigns). Given that the workshop arose from an academic project, it is inappropriate to use conventional outcome measures for deliberation (such as impact on decision-making). Rather, we evaluate participant experiences and associated personal impacts upon perceptions, behaviours and learning through on-the-day questionnaire and six-month follow-up interviews with a selection of participants.

In terms of the research context, timing the workshop to coincide with a real planning process for a power line revealed the local salience of the issues discussed, and general overall satisfaction with their collective participation. However, our findings reveal a number of significant design and facilitation factors that both positively and negatively influence overall satisfaction with the process of engagement. These were multifaceted, encompassing participant perceptions of key organisations (such as TSOs), of the communicative competence of expert speakers, the experience of taking part in deliberative activities (such as role play) and the clarity of facilitation. In our evaluation, we found that the types of information presented, and the perceived communicative competency of the speakers was viewed variably by the participants. From this we can conclude that it behoves the organisers of engagement processes around the technical aspects of energy infrastructure developments to get immediate feedback to gauge citizen uptake of scientific and technical information from the presented material and then relay these responses back to the target audience in order to gauge the efficacy of the communication strategy.

The types of activities (notably the use of role-play, whereby participants pretended to be different stakeholders in order to empathetically engage with ideas expressed by competing interests) was controversial because it demands a particular degree of deliberative competence and technical expertise which many within the group felt they did not have. This suggests the need for long-term deliberation as a form of technical and social learning. If participants can engage over a longer period of facilitated dialogue than was possible in this workshop, evidence from other processes suggests that their competencies and confidence rise, encouraging lay citizen members to challenge technical experts more easily (Burgess and Clark 2006). Moreover, we see some evidence of post-workshop transformations in environmental perceptions and involvement in pro-environmental social practice (such as involvement in energy saving behaviours or conservation activities), yet there is also evidence that concern with line siting diminishes. What these findings suggest is that the longer term impact of citizen experiences is moderated by the salience of the issue in that community context. In this case, the period following the workshop contrasted markedly with the controversy and activity that had preceded it, reducing the perceived importance of line siting issues. The finding underscores the challenges involved in engaging publics in infrastructure siting within decision-processes and/or construction that can run over years rather than weeks or months. It suggests the need for research into ways of maintaining the salience of infrastructure siting proposals across relatively long time spans.

The decision by the UK TSO not to engage with the workshop was a significant issue to consider. In some respects, this engagement would have improved the deliberative capacity of citizens engaging in the process, by providing participants with the opportunity to question TSO staff directly. It would also have afforded opportunity for us to invite other stakeholders to speak, including the local action group (in order to give representative balance of pro and anti-development perspectives). Involvement of experts and decision-makers in the deliberative process is frequently seen as best practice – producing an integrative or analytic-deliberative (Renn 2004) approach. The direct involvement of experts has the advantage of bringing citizens into the loop of governance, providing opportunities for mutual learning to occur. Representatives can tap into the experiences and expertise of the public and citizens can come to understand the complexities and dilemmas of planning, policy and technical development (Coleman and Gøtze 2001).

However, this too has potential pitfalls. Pre-existing distrust of NG was a feature of participant perspectives on the process – a problem that got worse as participants learnt more about the proposals through their engagement within the workshop. This is likely an effect of Slovic's (1993) asymmetry principle: that trust is much easier to destroy than to create. As Poortinga and Pidgeon (2004) have shown in deliberative engagement over genetically modified organisms, there are a number of biases through which information is integrated into participants' understanding of technological proposals which, in turn, influences their trust in technical/scientific authorities. For example, influences include the effect whereby negative information is perceived as 'more informative' than positive information (reflecting a negativity bias) or else the lack of trust in NG reflects the influence of people's prior attitudes toward the issue and the organisations involved (a confirmatory bias). Given that our results show increased trust in the local action group (Save Our Valley) following the workshop, but no increased trust in NG, it suggests likely evidence of these biases at work within the deliberative process. Non-involvement of NG likely exacerbated this problem and so underlines the need for two-way direct dialogue between developer and community in building such trust relationships.

The workshop failed to build trust relationships with the TSO, but also failed to stimulate positive emotional engagement with the issue. It is notable that our results concerning emotions show clear expression of satisfaction with the *process* of deliberation, but also feelings of threat, anger and shock when learning more about the subject under study. As Cass and Walker (2009) find, expressions of emotion are frequently perceived by development organisations as an unwanted intrusion into an otherwise rational process; yet emotional arousal is a near inevitable consequence of threats to place attachment and place identity, perceived procedural injustice around decision-making which lead to social opposition (see also Devine-Wright 2009, 2013); which our broader study of the participants' engagement with the HVOTL planning process confirms (Cotton and Devine-Wright 2013). We thus reiterate the need for greater developer understanding of and engagement with emotional elements of siting processes for energy infra-structures, with processes designed to tap into and explore citizen perspectives beyond the 'bounded' rationality of technical expertise. We also call for further research into the development of specific methods for capturing, expressing and sharing emotional responses to siting or policy proposals within deliberative engagement procedures.

In conclusion, our findings suggest that well-designed deliberative engagement is positively experienced by participants and leads to technical and social learning, both immediate and over the longer term. However, this requires careful pre-workshop planning, notably of how technical information is to be communicated, along with frequent feedback from participants. Given this, we call for TSOs and other energy infrastructure developers to reconsider conven-tional consultation practices. Adopting deliberative engagement techniques to foster community engagement with power line proposals may help to address trust deficits and negative emotions over power line proposals stemming from place impacts and concerns over procedural justice. One suggestion that might link research and practice might be to set up a long-standing com-munity panel throughout a multi-year planning process that could provide informed feedback to a TSO, enable more challenging engagement techniques such as role-play, and overcome difficulties caused by periods of lack of activity in consultation processes. This could be tracked by research similar to that presented here, drawing on both quantitative and qualitative methods to capture process and outcome aspects of participant experience. This proposal resonates with broader themes captured in this book, in particular the potential role that geographers can play both as a forceful critic of current practices *and* to explore means of opening up new pathways for sustainable futures. Moving beyond information provision towards greater citizen deliberation in the siting of energy infrastructures has the potential to increase energy justice. Therefore, future research of the kind presented here can contribute to cutting edge policy thinking and provide a robust case for changing current practices.

Acknowledgements

This project was funded by the Engineering and Physical Sciences Research Council, Supergen FlexNet: Thinking Networks project (EP/EO4011X/1). The authors wish to thank the partici-pants and facilitators in the Nailsea workshops.

References

Altman, I. and Low, S., *Place Attachment*. New York: Plenum Press, 1992.
Bäckstrand, K., 'Scientisation vs. civic expertise in environmental governance: eco-feminist, eco-modern and post-modern responses'. *Environmental Politics* 13(4): 695–714, 2004.
Bandura, A., *Social Learning Theory*. Englewood Cliffs, NJ: Prentice-Hall, 1977.
Barr, S., 'Strategies for sustainability: citizens and responsible environmental Behaviour'. *Area* 35(3): 227–240, 2003.
Bohman, J., *Public Deliberation: Pluralism, Complexity and Democracy*. Cambridge, MA: MIT Press, 2000.

Bull, R., Petts, J. and Evans, J., 'Social learning from public engagement: dreaming the impossible?' *Journal of Environmental Planning and Management* 51(5):701–716, 2008.

Burgess, J. and Clark, J., 'Evaluating public and stakeholder engagement strategies in environmental governance.' In *Interfaces Between Science and Society*, edited by A.G. Peirez, Vas, S.G., Tognetti, S. London: Greenleaf Press, 2006.

Burkhalter, S., Gastil, J. and Kelshaw, T., 'A conceptual definition and theoretical model of public deliberation in small face-to-face groups.' *Communication Theory* 12(4):398–422, 2002.

Cass, N. and Walker, G., 'Emotion and rationality: characterising and understanding opposition to renewable energy projects'. *Emotion, Space and Society* 2(1):62–69, 2009.

Chambers, S., 'Behind closed doors: publicity, secrecy, and the quality of deliberation'. *The Journal of Political Philosophy* 12(4):389–410, 2004.

Coleman, S. and Gøtze, J., 2001. *Bowling Together: Online Public Engagement in Policy Deliberation*. London: Hansard Society, 2001.

Cotton, M., 'Public engagement and community opposition to wind energy in the UK.' In *Sustainable Systems and Energy Management at the Regional Level: Comparative Approaches*, edited by M. Tortoro. Hershay, PA: IGI Publishing, 2011.

Cotton, M. and Devine-Wright, P., 'Discourses of energy infrastructure development: a Q-method study of electricity line siting in the UK'. *Environment and Planning A* 43(4):942–960, 2011.

Cotton, M. and Devine-Wright, P., 'Making electricity networks "visible": industry actor representations of "publics" and public engagement in infrastructure planning'. *Public Understanding of Science* 21(1): 17–35, 2012.

Cotton, M., and Devine-Wright, P., 'Putting pylons into place: a UK case study of public beliefs about the impacts of high voltage overhead transmission lines'. *Journal of Environmental Planning and Management* 56(8):1225–1245, 2013.

D'Entrèves, M.P., 'Political legitimacy and democratic deliberation.' In *Democracy As Public Deliberation: New Perspectives*, edited by M.P. D'Entrèves. Manchester: Manchester University Press, 2002.

Devine-Wright, P., 'Rethinking NIMBYism: the role of place attachment and place identity in explaining place-protective action'. *Journal of Community and Applied Social Psychology* 19(6):426–441, 2009.

Devine-Wright, P., 'Explaining "NIMBY" objections to a power line: the role of personal, place attachment and project-related factors'. *Environment and Behavior* 45(6):761–781, 2013.

Devine-Wright, P., H. Devine-Wright, and F. Sherry-Brennan, 'Visible technologies, invisible organisations: an empirical study of public beliefs about electricity supply networks'. *Energy Policy* 38 (8):4127–4134, 2010.

Graham, T. and T. Witschge, 'In search of online deliberation: towards a new method for examining the quality of online discussions'. *Communications* 28:173–204, 2003.

Groves, C., M. Munday and N. Yakovleva, 'Fighting the pipe: neo-liberal governance and barriers to effective community participation in energy infrastructure planning.' *Environment and Planning C: Government and Policy* 31(2):340–356, 2013.

Halvorsen, K.E., 'Assessing public partcipation techniques for comfort, convenience, satisfaction, and deliberation.' *Environmental Management* 28(2):179–186, 2001.

Keir, Laura, Richard Watts, and Shoshanah Inwood. 'Environmental justice and citizen perceptions of a proposed electric transmission line.' *Community Development* 45(2):107–120, 2014.

Knudsen, J.K., L.C. Wold, Ø. Aas, J.J.K. Haug, S. Batel, P. Devine-Wright, M. Qvenild, and G.B. Jacobsen, 'Local perceptions of opportunities for engagement and procedural justice in electricity transmission grid projects in Norway and the UK'. *Land Use Policy* 48:299–308, 2015.

Kollmuss, Anja, and Julian Agyeman, 'Mind the gap: why do people act environmentally and what are the barriers to pro-environmental behavior?' *Environmental Education Research* 8(3):239–260, 2002.

National Grid. 2010, 'The Hinkley Point C connection project.' National Grid Accessed 02/03/2011. www.nationalgrid.com/uk/Electricity/MajorProjects/HinkleyConnection/.

Poortinga, W. and N.F. Pidgeon, 'Trust, the asymmetry principle, and the role of prior beliefs.' *Risk Analysis* 24(6):1475–1486, 2004.

Rauschmayer, F., and H. Wittmer, 'Evaluating deliberative and analytical methods for the resolution of environmental conflicts'. *Land Use Policy* 23(1):108–122, 2006.

Renn, O., *Analytic-Deliberative Processes of Decision-Making: Linking Expertise, Stakeholder Experience and Public Values*. Stuttgart: University of Stuttgart, 2004.

Rogan, R., M. O'Connor and P. Horwitz, 'Nowhere to hide: awareness and perceptions of environmental change, and their influence on relationships with place.' *Journal of Environmental Psychology* 25(2):147–158, 2005.

Rowe, G. and L.J. Frewer, 'Public participation methods: a framework for evaluation.' *Science, Technology & Human Values* 25(1):3–29, 2000.

Slovic, P., 'Perceived risk, trust and democracy.' *Risk Analysis* 13:675–682, 1993.

Webler, T., '"Right" discourse in citizen participation: an evaluative yardstick.' In *Fairness and Competence in Citizen Participation*, edited by O. Renn, Webler, T., Wiedemann, P., 35–86. Dordrecht: Kluwer, 1995.

Webler, T., S. Tuler and R. Krueger, 'What is a good public participation process? Five perspectives from the public.' *Environmental Management* 27(3):435–450, 2001.

Wilson, T.D. and J.W. Schooler, 'Thinking too much: introspection can reduce the quality of preferences and decisions.' *Journal of Personality and Social Psychology* 60(2):181–192, 1991.

13

Under the curse of coal

Mined-out identity, environmental injustice and alternative futures for coal energy landscapes

(The case of the Most region, Czech Republic)

Bohumil Frantál

Introduction and theoretical departures

> "Can we be called a cultural nation if we convert in few years the historical heritage of centuries into energy to be exported abroad?"
>
> Václav Havel, former President of the Czech Republic, 2005

During the last two decades, environmental and security concerns have led to a rapid and widespread development of renewable energies. Coal still plays a vital role in electricity generation worldwide, however. Coal-fired power plants currently provide about 40 percent of global electricity, but in some countries coal fuels an absolute majority of electricity production, e.g., in South Africa, China, Australia, Kazakhstan, India, Poland, Serbia or the Czech Republic (International Energy Agency, 2012). It has been even assumed that coal's share of the global energy mix will continue to rise, and by 2017 it will come close to surpassing oil as the world's primary energy source. The World Resources Institute identified some 1200 new plants in the planning process, with about three-quarters of those projects in China and India, and 130 projects in Europe (Yang & Cui, 2012). Nevertheless, the public resistance against coal mining, new power plants and carbon capture and storage infrastructure projects increases across nations around the world (Pape, 2013).

This chapter investigates and discusses long-term negative consequences of coal energy production on the regional level in terms of the resource curse and environmental injustice theories using an example of the Most region in the Czech Republic (see Figure 13.1). The Czech Republic is a country with significant coal mining tradition dating back to the Middle

Ages. The industrial development of coal mining is associated with the construction of the railway network in the country in the mid-nineteenth century, which connected major industrial regions with the locations of coal deposits. The biggest mining boom, however, came in the second half of the 20th century. During the era of communism (1948–1989) coal was regarded as the national "black gold" and the "life blood" of metallurgical, heavy and energy industries, which have been centrally supported and developed as dominant sectors of national economy.[1] This planning orientation resulted in environmental devastation of several regions, particularly in Northern Bohemia, including the regions of Most and Sokolov, which have been extensively exploited on the basis of brown coal mining[2] and linked industries at the expense of other economic activities, natural environment, existing built environment, social structures and public health (Říha et al., 2005).

More than twenty years after the break of communism, in spite of a general economic restructuring, decline of coal mining and heavy industries, and investments in environmental restoration, the coal mining regions still suffer from the "curse of coal" being characterized by a long-term exploitation and commodification of landscape through mining and coal combustion with draining profits from energy sales out of the region (Barton, 2013), negative environmental and health impacts, persistently high levels of unemployment and poverty, social deprivation, and the absence of realistic alternatives for diversified development (Frantál & Nováková, 2014). Nevertheless, the wind energy development, which proved to be significantly more accepted by both regional policy-makers and local communities living in environmentally stricken and

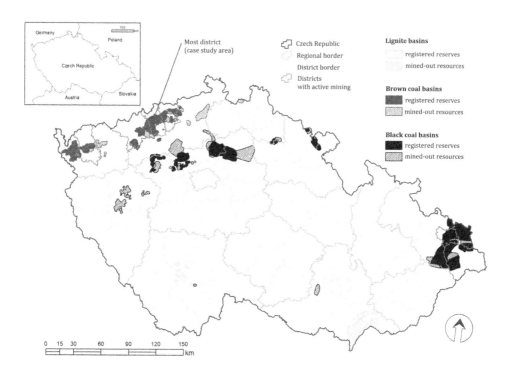

Figure 13.1 Area under study: map of the Czech Republic with the current state of coal mining

Source: data from Czech Geological Survey (2013); map adapted from Frantál & Nováková (2014).

structurally depressed regions (Frantál & Kunc, 2010), may represent an alternative way of sustainable development for coal mining landscapes.

The historical role of coal for industrialization, urbanization, creation of new jobs, and regional economic development is indisputable (see e.g., Domenech, 2008; Latzko, 2011). The economic benefits of coal for host regions have been in the long-term view outweighed by negative environmental, health and social externalities (Lockie et al., 2009; Riva et al., 2011), however, they have been typically subject to "boom and bust" cycles (Black et al., 2005; Shandro et al., 2011). In this sense, the coal industry has been frequently associated with the resource curse theory (see, among others: Freudenburg 1992; Perdue & Pavela, 2012), stressing that resource-dependent regions whose development has been strongly dependent on the extraction of natural resources (specifically non-renewable resources such as minerals and fossil fuels) are characterized by economic vulnerability, demographic instability, negative health and socio-economic impacts, increasing geographic isolation, imbalances of scale and power with respect to extractive industries, and the absence of realistic alternatives for diversified development. In the context of economic and social impacts of coal mining on local communities, Petrova and Marinova (2013) speak about the culture of transiency and dependency.

Coal mining traditionally took place underground. Since the late 1960s, surface mining methods have become more common, and today they account for more than half of total worldwide coal extraction (Maxwell, 2004). The methods of surface mining have made the coal industry more efficient by increasing production while reducing workforce, but they also drastically increased negative impacts on the topography, vegetation and water resources of the affected landscapes. Although coal mining has always had a negative effect on the surrounding environment and people, surface mining has shown a notable increase in these ecologically damaging effects (Sipes, 2010). Subsequently, the massive coal combustion in thermal power plants has become the most polluting and extensive manner of electricity production (Epstein et al., 2011).

In this respect, environmental economists speak about the apparent (explicit or internalized) costs and the hidden (secondary or externalized) costs, which together compose the social cost of energy (Epstein et al., 2011). Social costs arise when any costs of production or consumption are passed on to third parties, like future generations or society at large (Hohmeyer, 1988). Most of the impacts of coal energy production (whether environmental or socioeconomic) are cumulative (Franks et al., 2010), they extend well beyond the geographic locations of operating mines and power plants and they bring about other direct, indirect and unintended consequences at higher spatial levels, from regional to global. Most of the impacts – whether positive or negative – are also spatially and socially unevenly distributed, which raise a question of the environmental injustice of coal energy (Maxwell, 2004).

The glory and the curse of coal in the Czech Republic

Alexander von Humboldt, the famous German geographer, climbed the Milešovka mountain in 1819. He described the view over the Central Bohemian Uplands and the range of Ore Mountains following the Czech north-west border with Germany as the third most beautiful in the world (Ehrlich, 1929). We find today, almost two centuries later, that the region and its landscape have changed a lot. Nowadays, this area evokes surface coal mining, smoking power plants, chemical factories, and social and health deprivation. Glassheim (2007: 83) described this region as a "surreal panoply of ecological and social destruction", comprising half-eaten mountains, vast pits inhabited by massive earth-devouring machines, agglomerations of belching smokestacks, row upon row of decaying prefabricated apartment buildings.

Figure 13.2 Coal mine of the Czechoslovak Army, city of Most, Czech Republic
Source: photo by Věra Kailová, reprinted with permission.

The region of North-West Bohemia has always attracted the attention of the extractive industries for their rich mineral deposits. The successive regimes of national socialism and communism have contributed to the transformation of the region from what was once the protective gateway from Saxony into the Bohemian kingdom via the old royal city of Brüx (Most) with its surrounding expanse of fruit orchards, to today's panorama that is one part opencast "lunarscape", one part recultivated slag heap, and one part uniform socialist-realist cityscape (Barton, 2013: 2).

During the era of communism (1948–1989), Czechoslovakia, as a member of the former East European COMECON group of countries, was designated the "forge of the socialist camp" with a centrally controlled dominance of metallurgical and energy-intensive heavy industries. In that period, the production of brown coal as the main source of energy increased about five times (see Figure 13.3) and electrical power generation about twenty times as documented by the Czech Statistical Office (2012). The focus on heavy industries and energy production affected overall national economy and resulted in environmental devastation of several regions, most extensively in the Most brown coal basin, which covers four districts: Most, Chomutov, Teplice and Louny. From 1945 to 2005, the share of the Most basin in the total amount of mined coal in the country has always been above 70 percent. In 1984 (the most productive period of mining), just in the district of Most, 38.6 million tons of coal were produced and 11 mines were in operation (Chytka & Valášek, 2007).

As communist planners continually increased norms for energy and coal production in the 1950s through 1970s, the sprawling surface mines, constructions of large thermal power plants and related infrastructural projects forced the abandonment of many human settlements. Over a hundred villages and parts of several larger cities have been destroyed and over 90,000 people

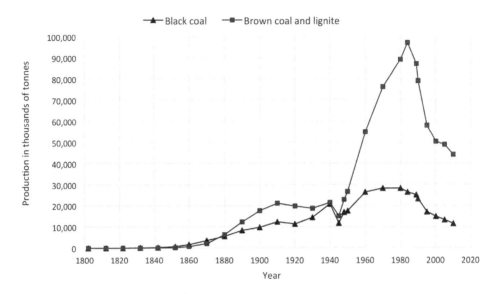

Figure 13.3 The rise and fall of coal mining in the Czech Republic

Source: data from Czech Statistical Office (2012), author's elaboration.

were relocated due to mining and related activities from 1949 to 1980 (Říha et al., 2010). Infamously, the entire historic center of the medieval city of Most was obliterated as a "decaying capitalist relic" in order to expose over 85 million tons of coal (Glassheim, 2007). The combination of a regime of unrestrained mining of brown coal at any cost and a militantly atheistic philosophy proved to be an especially poisonous cocktail for the region's religious structures: more than 500 churches, chapels, monasteries, synagogues and Jewish cemeteries were destroyed in the region during the communist era (cf. Barton, 2013).

The ecological stability of landscapes and their agricultural and forestry potentials were disrupted. While land regeneration has been successfully carried out in many cases (e.g. the regeneration projects of a motor-racing circuit, racecourse or the lake in the new city of Most; see Figure 13.4 and Figure 13.5), the scope of devastation in the entire region is much greater. It has been estimated that the surface mining within the Most brown coal basin has affected so far about 250 sq.km, while only about 95 sq.km was reclaimed (Zahálka et al., 2008).

After the fall of communism, the newly established Federal Ministry of Environment prepared programs to restore the environment of the Most basin as the most environmentally affected area in the country. As a result, so-called "territorial ecological limits for mining" were established by Government Decree No. 444/1991 (Říha et al., 2005). By restricting exploration, mining and other coal mining-related activities beyond certain spatial limits, the government established a balance between economic and ecological interests, but it also ignited a fierce political debate that has been smouldering ever since (Kotouš & Jurošková, 2013).

The mining of brown coal in the Most basin, the dominant supply for Czech thermal power plants, will reach the current legal territorial limits with reserves estimated to be available by 2018. Breaking of these limits, which have been recently contested by coal mining companies and the central government, would cause a demolition of several municipalities, forced relocation of thousands of people, and ecological devastation of a valuable foothill landscape. A potential

Figure 13.4 The racecourse with trail for in-line skating which was built on a recultivated coal dump, city of Most

Source: photo by Bohumil Frantál.

Figure 13.5 Anthropogenic lake made during the recultivation of the Ležáky mine, the location of historical city of Most; the surrounding area is designated for construction of new family houses

Source: photo by Bohumil Frantál.

territorial expansion of coal mining in the Most region has become a point of contention among different ways of seeing, various interests, value judgments, ideologies, myths, and discourses (see Table 13.1).

The current Czech energy policy remains highly dependent on conventional resources. Overall electricity generation is based predominantly on thermal or combined-cycle power plants burning brown coal (41 percent), black coal (6 percent), gas and other fuels (6 percent), nuclear power plants (35 percent), and renewable energy sources (12 percent) (Energostat, 2014). The Czech Republic has also been regularly among the biggest world net exporters of electricity, with the annual net export about 17 TWh. This annual export represents approximately 15 million tonnes of coal burnt in thermal power plants with low energy efficiency (some of the still operational power plants which were constructed already during the 1960s operate with only 30 percent efficiency).

The burning of coal and exportation of electricity abroad has been considered a form of landscape commodification and exportation, which raised questions of environmental injustice or the uneven spatial and social distribution of benefits (economic profits for mining companies, energy producers and stakeholders) and costs (environmental, economic and social impacts on host regions and local communities) of coal energy. Today the coal mining companies pay only 1.5 percent of the profits from extracted brown coal (and only 0.5 percent in the case of black coal) into the state budget, and only three quarters of this profit go back to the region and municipalities affected by mining.

Table 13.1 Competing arguments related to potential coal mining expansion in the Most region

Advocates	Opponents
Coal mining and coal energy contribute to national energy security and economic competitiveness (coal as a brown gold)	Coal mining and coal combustion contribute to global climate change and environmental degradation of host regions (coal as a curse)
Electricity and heating generated from domestic coal is cheap (in contrast with the subsidized renewable energy or imported natural gas)	Negative externalities are not included in market prices of coal energy (a higher carbon tax should be applied for coal energy)
Tax revenues from coal mining and coal energy contribute to state and regional budgets and to local development	Economic benefits from coal energy are unevenly distributed with minor tax revenues guaranteed for the state, regions and municipalities
Coal mining and related activities prevent an increase of regional and local unemployment rates	Breaking the limits will have negligible effects on a long-term employment (the loss of jobs in demolished municipalities is more significant)
Economic profits from the sale of coal reserves will allow greater investments in landscape reclamation and regeneration projects	Investments in regeneration will never rehabilitate the extent of landscape devastation and impacts on people's health and well-being
Coal will not play a role of key strategic raw materials in the future such as today and it is going to lose its price (we should exploit it until it has a price)	Current coal combustion technology is little energy efficient and the use of coal is ineffective (we should keep reserves for future generation)
	Coal mining contributes to the continued stigmatization of host regions, a decline of real estate market, a decline of tourism, etc.

Sources: Author's conceptualization based on the review of Glassheim (2007), Houda (2013), KŽP AV ČR (2013), Říha et al. (2005), Wikipedia (2016).

Environmental injustice of coal energy

Active coal mining currently affects the area of six Czech districts (see Figure 13.1). Brown coal mining takes place at eight quarries within the Most and Sokolov basins, with the annual production in 2013 accounted for 40.6 million tonnes. Underground mining of black coal takes place in four mines in the Ostrava-Karvina basin only, whose total production in 2013 amounted to 8.8 million tons (Energostat, 2014).

To provide empirical evidence of the "curse of coal" hypothesis, a comparative analysis of selected indicators (including environmental indicators, population vital and health statistics, quality of life indicators, labour market data, and social cohesion indicators) was made for groups of districts with active coal mining and districts without mining. The statistical testing has been provided using the analysis of variance (ANOVA) to analyze differences between group mean values. Out of 30 tested indicators (representing mostly data from the 2011 Census), we have found a statistically significant correlation with coal mining for 15 indicators, suggesting that coal-mining districts are characterized by poorer living conditions, worse population health, inadequate labor market and weak social capital. The significant differences between districts according to mean values are summarized in Table 13.2.

The most significant differences between coal mining and non-mining districts are revealed in differences in air quality, which was measured by the concentration of basic pollutants

Table 13.2 Mean values of selected indicators as related to mining and non-mining districts

Indicators	Category			
	Non-mining districts	Mining districts	Most district	Eta
Air pollution (emissions of SO_2 + NO_x + CO tonnes/km^2)	7.4	42.4	48.2	0.39[b]
Male life expectancy (years)	74.1	72.1	72.2	0.54[a]
Abortion rate (abortions per 100 births)	34	43	49	0.44[b]
Infant mortality (‰)	2.7	4.3	1.8	0.22[c]
Property value (average price of flats in mil. CZK)	1.296	0.660	0.435	0.46[a]
Average monthly pension (CZK)	10.148	10.316	10.508	0.30[c]
Average monthly wage (CZK)	16.475	17.309	18.299	0.21[c]
Flats with central heating (%)	70	82	86	0.52[a]
Unemployment rate (%)	9.1	12.4	15.6	0.40[b]
Business activity (business units per 1000 population)	236	194	199	0.37[b]
Persons with basic or no formal education (%)	19	23	24	0.44[a]
Roma ethnic people per 1000 population	0.4	1.6	5.0	0.76[a]
Crime rate (ascertained offences per 1000 population)	24	34	35	0.35[b]
Regional elections turnout (%)	37	32	33	0.44[a]
Population change (2005–2011 per 1000 population)	+21	–23	–16	0.22[c]

Notes: Mining districts category includes six districts where coal mining is still active (including Most district), the non-mining category includes all other districts of the Czech Republic. Measures of association (Eta) are significant at: [a] $p < 0.001$; [b] $p < 0.01$; [c] $p < 0.05$.
Sources: Czech Statistical Office (2011), Institute of Regional Information (2008); author's calculation.

(SO_2, NO_x, CO). The air quality is six times worse in coal mining districts (i.e., 7.4 vs. 42.4 tonnes/km^2). There is also a significant association between coal mining, air quality and some population vitality and health indicators, such as a lower life expectancy, higher rates of abortions and infant mortality. However, specifically the Most district reported below the average numbers of infant mortality.

Significant differences among coal mining and non-mining districts are also related to key labor market characteristics, i.e. unemployment rate and business activity, the former being higher and the latter lower in districts whose economy has been dependent on coal industry. The Most district has the third largest unemployment rate in the whole country. The results also indicate the unemployment rate is connected with other negative social phenomena such as a higher rate of crime and out-of-district migration. The higher-than-average incomes and pensions indicated the coal industries have brought about positive economic effects to local employees. However, we can assume the mentioned negative social phenomena indicate that the economic benefits have been socially unevenly distributed. Moreover, the differences in average incomes and pensions are negligible in terms of practical life (the differences are in the order of 100 Czech crowns which is only about 5 percent of the total salary).

The analysis revealed that districts affected by coal mining are characterized by higher concentration of ethnic minorities and lower levels of education. For example, the district of Most has the largest concentration of Roma people among all Czech districts, and the third largest proportion of people with basic education or uneducated people. The coal industry has contributed district heating. However, there is a negative association between mining activity and the average price of flats, with the lowest prices in the district of Most.

Figure 13.6 Housing estate Chanov – built in the mid-1970s for people from the destroyed old Most, currently being inhabited mostly by Roma people

Source: photo by Bohumil Frantál.

The rise of anti-coal attitudes and searching for an alternative future

While in the period of communism, the public's opportunity to participate in decision-making processes was minimal, there are significantly more legal options now. In 2012, the so-called *Mining Act* (No. 44/1988 Coll.) was amended, by which this legislation has deleted paragraphs allowing the mining companies to expropriate the property and land located at so-called reserved mineral deposits. Therefore, the *Mining Act* gives new priority to the concerns of local communities before the interests of mining companies or public interest, and it provides greater opportunities for negotiations on financial compensation. The companies and political authorities should now take more account of the views of local communities, and the factors that shape public opinion.

In 2012, the author distributed a questionnaire survey to inhabitants of Horní Jiřetín and Litvínov – the towns which would be directly affected by mining if the current territorial limits were to be broken. Horní Jiřetín is a small historical town, which would be completely destroyed by such a mining expansion. Litvínov is a district city with a long-term tradition of coal mining and chemical industries. Its suburb, part Janov, is located approximately 3.5 kilometres west from the city centre. A dominant part of Janov is represented by the housing estate of prefabricated houses built during the 1970s and 1980s, partially for inhabitants from the neighboring villages, which were destroyed due to coal mining. The inhabitants of Janov would find themselves only 500 metres from the edge of the future open-pit mine if the current mining limits were to be broken (see Figure 13.7).

More than half of the respondents responded that the coal production should be significantly restricted, the mining areas should be cancelled and the coal reserves written off, and we should

Figure 13.7 Mine of the Czechoslovak Army approaching the border of territorial limits of mining

Source: photo by Bohumil Frantál.

develop more renewable energy. About a third of the population are inclined to the view that current available coal reserves are sufficient for our national consumption, so that limits can be maintained and there is no need to expand the mining. Only 14 percent of respondents consider coal mining as a traditional branch of the Czech economy, contributing to the energy independence of the state, and therefore existing mining limits should be broken. The restriction of coal mining is more likely to be supported by people with university education and by the younger generation. The level of place attachment proved also to be a significant predictor of attitudes.

More than half of the respondents have been involved in public protests against the breaking of mining limits and the expansion of mining in the region. The most common protest activity was to sign a petition against breaking the mining limits. Respondents also participate in other forms of protest such as meetings and membership in protest organizations. However, just about two thirds of actively opposing people believe that protest activities such as petitions, blockades or other civic campaigns against the expansion of mining have the importance and power to affect decisions of political authorities.

Relations and negotiations between local communities and the management of mining companies and political authorities are obviously characterized by asymmetries of power and control (Morrice & Colagiuri, 2013). Coal mining industry benefits from greater political capital, due to the fact that they have control over specific knowledge and access to resources. The greater political capital can even be used to push for changes in the laws governing disposal of natural resources and land-use planning, including proposals for the exclusion of local communities from the decision-making process.

The problem of procedural injustice is often related with the problem of social deprivation, low socio-political efficacy, and low level of civic engagement of communities living in environmentally stricken and economically depressed areas. Then whatever hazardous or environmentally significant facilities (such as power plants, chemical factories or waste incinerators) are often planned for these "sacrificed for public good" regions where projects face less public resistance either due to a greater effect of financial compensation or because of the passivity of stigmatized and socially deprived populations. This problem is particularly evident in the Most region, where some people have been forced to move two or more times during their lives due to the expanding surface mining. Many of them have lost a positive tie (or place attachment) to their place of living apart from benefiting from the landscape's productive functions.

On the other hand, some studies from the United Kingdom (as cited in Van der Horst, 2007: 2709) reported a relationship between the industrial character of a location and the local people's perception of environmental risks and their higher acceptance of alternative technologies, such as renewable energy production systems. According to our survey, two-thirds of people accept the construction of wind farms in their backyards. A significant number of opponents would still agree to the construction if the project brought some economic benefits for their home or community. This is a significantly higher rate of acceptance than the studies demonstrated from other, non-mining regions in the country (Frantál & Kučera, 2009; Frantál & Kunc, 2010). We can hypothesize that people living in environmentally stricken areas consider wind energy a better alternative to the devastating coal mining. It is also likely that in the economically disadvantaged and poor areas the economic motivation (money that communities receive from developers) has a greater effect on the social acceptance of projects than in wealthy regions with unspoiled landscapes.

A validity of the hypothesis of higher social acceptance of renewable energy in environmentally stricken areas is supported even by the actual regional differences in the implementation

Figure 13.8 Wind farm in Nová Ves v Horách municipality which is located in the hills just about 5 km above the border of surface mine

Source: photo by Bohumil Frantál.

Table 13.3 Differences in the implementation of wind energy potential in mining and non-mining districts

Indicators	District category		
	Non-mining	Mining	Eta
Average realizable potential of wind energy [MW]	32.7	41.5	0.063
Average installed capacity [MW]	2.6	17.2	0.447*
Total realizable potential of wind energy [MW]	2.285	249	–
Total installed capacity [MW]	180	103	–
% of realization	8	41	–

Notes: Mining districts category includes six districts where coal mining is still active, the non-mining category includes all other districts of the Czech Republic. Measures of association (Eta) are significant at *p <0.001.

of wind energy projects in the Czech Republic (see Table 13.3). While there is not a statistically significant difference between the realizable wind energy potential in coal mining and non-mining districts, there is a significant difference between the implemented potential (i.e., installed capacity). While mining districts have already implemented more than 40 percent of overall realizable potential, non-mining districts have implemented only 8 percent of their realizable potential. These numbers indicate there is higher political and social acceptance of renewable energy development in coal mining regions.

Discussion and conclusions

The case of the Most region supports the validity of the theories of resource curse and environmental injustice of coal energy. Although the coal mining and linked industries contributed to slightly above average regional incomes and pensions (which are actually significant statistically but not of practical relevance), and provided households with some technical services such as district heating, these positives have come at high environmental and health costs (worse air quality, lower life expectancy, higher infant mortality, etc.) paid by the local population. Above average rates of unemployment and crime in coal mining districts also indicate that the economic benefits have been unevenly distributed socially.

We found a significant negative association between mining activity and the average price of flats, with the lowest prices in the district of Most. This probably indicates a generally lower standard of apartments (large proportion of housing estates with prefabricated houses built during the 1970s and 1980s) and their worse marketability in the context of low demand for living in a region characterized by inferior quality of life – evidence of which is also a significant loss of population during the last decade.

Our analysis revealed that coal-mining districts have a higher proportion of people with only basic education and early school leavers and also a higher proportion of ethnic minorities, particularly Roma people. Such results suggest that coal energy is environmentally and socially unjust. This finding, however, does not confirm the theory of "disproportionate siting" which suggests that polluting industries and toxic facilities are deliberately planned and localized in areas with higher concentrations of poor and minority populations (see e.g., Pastor et al., 2001). The coal basins in the Czech Republic are located mostly in borderland areas. These border areas were characterized on the one hand by distinct depopulation (caused by the expulsion of the German population after World War II), on the other hand, by the growing demand for labor for the massively expanding mining and metallurgical industries from the 1950s to the late 1980s. As a result, less educated, and minority populations migrated to extensively industrialized and urbanized areas with relatively affordable housing in new prefabricated housing estates – a so-called uneven migration of minorities (or "disproportionate minority move-in"). Moreover, the migration of the Roma population from rural areas to large cities in Northern Bohemia and Northern Moravia during the 1970s and 1980s has usually had a form of forced resettlement (see e.g., Glassheim, 2006).

As compared to the few studies on the issue from other countries, our findings are partially in accordance and partially in conflict with results reported by Hajkowicz and colleagues (2011), which affirmed positive impacts of mining activities on incomes, housing affordability, communication access, education and employment across regions in Australia, but negative impacts on life expectancy. They did, however, highlight the fact that while their data were valid at an aggregate level, there is often an uneven income distribution within mining regions and that certain sub-groups in regional and remote communities are more vulnerable to mining activities. Another Australian study by Taylor and Scambary (as cited by Hajkowicz et al., 2011) reported that indigenous communities, resident in mining regions, in particular were excluded from the socio-economic benefits of adjacent mining operations.

In the context of the on-going public debates about possible changes to the current territorial limits of mining in the Czech Republic and about the potential adoption of a carbon tax for electricity produced from fossil fuels, our findings suggest that the actual long-term environmental and socioeconomic cumulative effects of coal mining and coal combustion should be taken into account more responsibly, and that market prices should reflect the real social price of coal energy to a greater extent. In terms of the environmental justice, the economic profits from coal should be more fairly redistributed to compensate for the negative impacts in affected areas.

A recent survey provided by the author revealed that only about a tenth of residents living in areas threatened by potential expansion of surface coal mining consider coal mining as a traditional sector of Czech economy, which contributes to national energy security and should be further developed over the current ecological limits. About a third of the local population are inclined to the view that current available coal reserves (within the stated legal limits) are sufficient for the national consumption, so that limits can be maintained and there is no need to expand the mining. More than half of the population are of the opinion that the coal production should be significantly restricted, the mining areas should be cancelled and the coal reserves written off, and more renewable energy projects should be implemented. The restriction of coal mining and support for renewable sources increase with higher levels of education and in particular with a stronger local place attachment. In other words, people showing higher rates of place attachment are more likely to be against the expansion of coal mines. Conversely, the support of mining is more common by people whose profession is connected to the coal industry.

The preservation of local jobs in regions that are characterized by higher rates of long-term and structural unemployment remains the main argument for breaking of the limits and expansion of mining. According to recent studies of non-governmental organizations (Greenpeace, 2012; Piňos, 2013), however, breaking the mining limits within the Czechoslovak Army mine would not bring any new jobs but only maintain the existing approximately 900 jobs. On the other hand, almost 800 jobs in the town of Horní Jiřetín would perish in case of the mining expansion. Moreover, the problem of long-term reductions in employment in the coal industry would be just postponed for several years.

Conversely, in case of keeping the limits and successive moving from extensive mining and energy industry the region can focus on developing energy efficiency, renewable energy development and investments in the implementation of "green technologies" (i.e., energy renovation and home insulation, replacing old boilers, installation of small on-roof solar systems, etc.). According to some expert estimates (Beranovský et al., 2012), this alternative way could bring about 2300 new jobs. The significantly higher implementation of wind energy projects in coal mining districts can be regarded as demonstrating that local communities and decision makers living in environmentally affected areas are more likely to support alternative technologies.

The key word will therefore be the Czech government – especially its decision about the future energy strategy, environmental policy and structure of the "energy mix". According to the Commission for the Environment of the Czech Academy of Sciences (2013), the further focus of Czech energy policy on coal mining and coal combustion is considered as wrong way not only with respect to the global context of climate change and the efforts to carbon emissions, but also due to the economic profile and long-term and structural unemployment of the Most region. Conversely, it is highly desirable to maintain coal reserves behind the borders of current ecological limits for next generations and for more considerate methods of coal extraction protecting values at the surface. Experts of the Commission assume that in the coming decades coal will become a valuable chemical feedstock for smarter uses than the extensive burning in power plants with low energy efficiency and export of electricity (and therefore the Czech landscape) abroad.

The main focus of this chapter, and presented case study, was at the regional level; however, the impacts of coal energy exceed regional and national levels. The emphasis paid to coal by McKibben (as cited by Freese, 2003), given the particular chemistry of global warming, is instructive: the decisions we make about coal in the next two decades may prove to be more important than any decisions we have ever made as a species.

Notes

1 The communist poet Vlastimil Školaudy wrote in the daily Rudé Právo (as cited in Glassheim, 2007: 455) that "Coal is today a crown jewel of our land . . . the generating wind of our factories, the rhythm of labour, the warmth of our homes . . . the blood pouring into the arteries of our industry".

2 There are four basic types of coal according to the content of carbon: (i) lignite (with 30–50 percent of carbon), (ii) brown coal (50–80 percent), (iii) black coal (80–90 percent), and (iv) anthracite (over 90 percent). In the Czech Republic, there are deposits of lignite (mining in South Moravia was terminated in 2009), brown coal (still active mining in Most and Sokolov brown coal basins), and black coal (still active mining in Ostrava-Karvina coal basin); see Figure 13.1.

References

Barton, A. (2013). Environmental mining limits in the North Bohemian Lignite Region. *Envigogika*, 8(4), 1–7.

Beranovský, J., Truxa, J., & Srdečný, K. (2012). *Možnosti využití energetických úspor a obnovitelných zdrojů energie v Ústeckém kraji* [online]. [cit. 30.01.2017] Retrieved from: www.alies.cz/wp-content/uploads/Moznosti_vyuziti_energetickych_uspor_a_ OZE_v_Usteckem_kraji.pdf.

Black, D., McKinnish, T., & Sanders, S. (2005). The economic impact of the coal boom and bust. *The Economic Journal* 115(503), 449–476.

Chytka, L., & Valášek, V. (2007, October 9). The past and present of mining brown coal in Northern Bohemia. And the Future? *All For Power*. [online]. [cit. 30.01.2017] Retrieved from: www.allforpower.com/clanek/378-the-past-and-present-of-mining-brown-coal-in-northern-bohemia-and-the-future/.

Commission for the Environment of the Czech Academy of Sciences. (2013). *Updated opinion of the Commission for the Environment of the Czech Academy of Sciences about the issue of 'territorial ecological limits of mining' in the North Bohemian brown coal basin (in Czech)* [online]. [cit. 30.01.2017] Retrieved from: www.koreny.cz/news/aktualizovane-stanovisko-komise-pro-zivotni-prostredi-akademie-ved-cr-k-problematice-tzv-uzemnich-ekologickych-limitu-tezby-v-severoceske-hnedouhelne-panvi-shp-/.

Czech Geological Survey. (2013). Map applications. [online]. [cit. 30.01.2017] Retrieved from: www.geology.cz/extranet-eng/maps/online/map-applications.

Czech Statistical Office. (2012). *Historická ročenka statistiky energetiky. Vývoj těžby uhlí od roku 1782 na území České republiky* [online]. [cit. 30.01.2017] Retrieved from: www.czso.cz/csu/czso/8113-12-n_2012-01.

Czech Statistical Office. (2013). *Districts of the Czech Republic* [online]. [cit. 30.01.2017] Retrieved from: www.czso.cz/csu/2013edicniplan.nsf/engpubl/1303-13-eng_r_2013.

Domenech, J. (2008). Mineral resource abundance and regional growth in Spain, 1860–2000. *Journal of International Development*, 20(8), 1122–1135.

Ehrlich, R. (1929). Der Donnersberg in geschichtlicher, kartographischer und touristischer Hinsicht. *Erzgebirge Zeitung, Donnersberg-Sonderheft*, 52, 55–60.

Energostat. (2014). *Energetics of the Czech Republic – basic data* [online]. [cit. 30.01.2017] Retrieved from: http://energostat.cz/elektrina.html.

Epstein, P. R., Buonocore, J. J., Eckerle, K., Hendry, M., Stout, B.M., Heinberg, R., Clapp, R.W., May, B., Reinhart, N.L., Ahern, M.M., Doshi, S.K., & Glustrom, L. (2011). Full cost accounting for the life cycle of coal. *Annals of the New York Academy of Sciences*, 1219, 73–98.

Franks, D., Brereton, D., & Moran, C. (2010). Managing the cumulative impacts of coal mining on regional communities and environments in Australia. *Impact Assessment and Project Appraisal*, 28(4), 299–312.

Frantál, B., & Kučera, P. (2009). Impacts of the operation of wind turbines as perceived by residents in concerned areas. *Moravian Geographical Reports*, 17(2), 34-45.

Frantál, B., & Kunc, J. (2010). Factors of the uneven regional development of wind energy projects (a case of the Czech Republic). *Geographical Journal* (Slovak), 62(3), 183–201.

Frantál, B., & Nováková, E. (2014). A curse of coal? Exploring unintended regional consequences of coal energy in the Czech Republic. *Moravian Geographical Report*, 22(2), 55–65.

Freese, B. (2003). *Coal: A Human History*. Cambridge: Basic Books.

Freudenburg, W. R. (1992). Addictive economies: extractive industries and vulnerable localities in a changing world economy. *Rural Sociology*, 57(3), 305–332.

Glassheim, E. (2006). Ethnic cleansing, communism, and environmental devastation in Czechoslovakia's Borderlands, 1945–1989. *The Journal of Modern History*, 78(1), 65–92.

Glassheim, E. (2007). Most, the town that moved: coal, communists and the 'gypsy question' in post-war Czechoslovakia. *Environment and History*, 13(4), 447-476.

Greenpeace. (2012, October 31). Několik souvislostí dnešní demonstrace zaměstnanců firmy Czech Coal za zachování vyvlastňovacích paragrafů v horním zákoně a za těžbu uhlí za zbourání Horního Jiřetína. *Britské Listy*. [online]. [cit. 30.01.2017] Retrieved from: http://blisty.cz/art/65822.html.

Hajkowicz, S. A., Heyenga, S., & Moffat, K. (2011). The relationship between mining and socio-economic well-being in Australia's regions. *Resources Policy*, 36(1), 30–38.

Havel, V. (2005). *An open letter to the Prime Minister Jiří Paroubek* [online]. [cit. 30.01.2017] Retrieved from: www.vaclavhavel.cz/showtrans.php?cat=pr&val=61_pr.html&typ=HTML.

Hilson, G. (2002): An overview of land use conflicts in mining communities. *Land Use Policy*, 19(1), 65–73.

Hohmeyer, O. (1988). *Social Costs of Energy Consumption*. Berlin: Springer-Verlag.

Houda, P. (2013, February 26). Raději ať se hůř dýchá. *Česká pozice*. [online]. [cit. 30.01.2017] Retrieved from: http://ceskapozice.lidovky.cz/radeji-at-se-hure-dycha-0bd-/tema.aspx?c=A130226_093147_pozice_98965.

International Energy Agency. (2012). *Medium-Term Coal Market Report 2012: Market Trends and Projections to 2017*. Paris: OECD/IEA.

Institute for Regional Information. (2008). *Price maps* [online]. [cit. 30.01.2017] Retrieved from: www.iri.name/Article.asp?nDepartmentID=196&nArticleID=116&nLanguageID=2.

Kotouš, J., & Jurošková, L. (2013, May 17). Beyond the limits. *Mining Journal*. [online]. [cit. 30.01.2017] Retrieved from: www.mining-journal.com/reports/beyond-the-limits.

Latzko, D.A. (2011). Coal mining and regional economic development in Pennsylvania, 1810–1980. *Economies et Sociétés, Serie 'Histoire Economique Quantitative'*, 44, 1627–1649.

Lockie, S., Franettovich, M., Petkova-Timmer, V., Rolfe, J., & Ivanova, G. (2009). Coal mining and the resource community cycle: a longitudinal assessment of the social impacts of the Coppabella coal mine. *Environmental Impact Assessment Review*, 29(5), 330–339.

Maxwell, N. I. (2004). Environmental injustice of energy facilities. In C. J. Cleveland & R. U. Ayres (Eds.), *Encyclopedia of Energy*, pp. 503–515. Amsterdam: Elsevier.

Morrice, E., & Colagiuri, R. (2013). Coal mining, social injustice and health: a universal conflict of power and priorities. *Health & Place*, 19, 74–79.

Pape, R. (2013, December). Anti-coal movement in Europe is growing. *Acid News*. [online]. [cit. 30.01.2017] Retrieved from: www.airclim.org/acidnews/anti-coal-movement-europe-growing.

Pastor, M., Sadd, J., & Hipp, J. (2001). Which came first? Toxic facilities, minority move-in, and environmental justice. *Journal of Urban Affairs*, 23(1), 1–21.

Perdue, R. T., & Pavela, G. (2012). Addictive economies and coal dependency: methods of extraction and socioeconomic outcomes in West Virginia, 1997–2009. *Organization & Environment*, 25(4), 368–384.

Petkova-Timmer, V., Lockie, S., Rolfe, J., & Ivanova, G. (2009). Mining developments and social impacts on communities: Bowen basin case studies. *Rural Society*, 19(3), 211–228.

Petrova, S., & Marinova, D. (2013). Social impacts of mining: changes within the local social landscape. *Rural Society*, 22(2), 153–165.

Piňos, J. (2013). *8000: Tolik pracovních míst chrání limity těžby uhlí* [online]. [cit. 30.01.2017] Retrieved from: www.hnutiduha.cz/aktualne/cislo-tydne-8000-tolik-pracovnich-mist-chrani-limity-tezby-uhli.

Říha, M., Stoklasa, J., Lafarová, M., Dejmal, I., Marek, J., & Pakosta, P. (2005). *Environmental Mining Limits in North Bohemian Lignite Region*. Praha: Společnost pro krajinu.

Riva, M., Terashima, M., Curtis, S., Shucksmith, J., & Carlebach, S. (2011). Coalfield health effects: variation in health across former coalfield areas in England. *Health and Place*, 17(2), 588–597.

Saha, S., Pattanayak, S. K., Sills, E. O., & Singha, A. K. (2011). Under-mining health: environmental justice and mining in India. *Health and Place*, 17(1), 140–148.

Shandro, J. A., Veiga, M. M., Shoveller, J., Scoble, M., & Koehoorn, M. (2011). Perspectives on community health issues and the mining boom–bust cycle. *Resources Policy*, 36(2), 178–186.

Sipes, T. (2010). *The Polluting of a Nation: Surface Coal Mining in America* [online]. [cit. 30.01.2017]. Retrieved from: www.academia.edu/285527/The_Polluting_of_a_Nation_Surface_Coal_Mining_In_America.

Taylor, J., & Scambary, B. (2005): *Indigenous people and the Pilbara mining boom: a baseline for regional participation. Centre for Aboriginal Economic Policy Research Monograph, No. 25*. Canberra: ANUE Press.

Van der Horst, D. (2007). NIMBY or not? Exploring the relevance of location and the politics of voiced opinions in renewable energy sitting controversies. *Energy Policy*, 35(5), 2705–2714.

Wikipedia (2016). *Brown coal mining limits in North Bohemia* [online]. Retrieved from: https://en.wikipedia.org/wiki/Brown_coal_mining_limits_in_North_Bohemia.

Yang, A., & Cui, Y. (2012). *Global Coal Risk Assessment: Data Analysis and Market Research. WRI Working Paper*. Washington, DC: World Resources Institute.

Zahálka, J., Farský, M, & Měsíček, L. (2008). Severočeská hnědouhelná pánev: determinace a disparity vývoje krajiny. *Studia Oecologica*, 2008(1), 61–67.

14

Construction of hydropower landscapes through local discourses

A case study from Andalusia (southern Spain)

Marina Frolova

Introduction

Since 1880–1890 when the first plants were built, hydropower has become an important source of electricity in many different countries all over the world (Koch 2002; Oud 2002). Hydropower was the first renewable energy to be developed on a mass scale and is now considered to be one of the most cost-effective means of producing clean renewable electricity. It is generally more efficient, more reliable and has a higher capacity factor than solar, wind, and marine energy (wave and tidal) technologies (Fraenkel et al. 1991; Mishra et al. 2011) and often offers a better energy payback ratio than other power generation technologies (British Hydropower Association and IT Power 2010).

Hydropower was developed and applied in quite a specific manner that evolved over time (Oud 2002). It appeared in different contexts from other kinds of renewables, making it today one of the most widespread but also the most 'traditional', well-established renewable energy technologies (Briffaud et al. 2015; Frolova et al. 2015a).

In the early stages, hydropower plants were built as part of the electrification process, so establishing close links between hydropower production and the access to and consumption of electricity. It is the only existing renewable energy technology that emerged and expanded in a completely decentralized context, in an era when energy production and consumption took place at a local territorial scale (Frolova et al. 2015b). In more recent hydropower development however (in Western countries at least) these close links between electricity production and consumption no longer apply, in that even remote rural areas of Europe are already connected to the grid. Hydroelectricity and landscape however remain interlinked. In many areas the development of hydropower infrastructures with all their associated elements has altered the natural landscape to create a specific form of energy landscape (Briffaud 2014; Briffaud et al. 2015; Frolova et al. 2015b). The process by which these hydropower landscapes are formed provides useful lessons for understanding the relations underlying the co-construction of renewable energy capacity and culturally shared landscapes (Frolova et al. 2015b).

How has hydropower development affected the landscapes of the mountain areas of southern Spain? How have hydropower, water and landscape resources been 'co-constructed' in this context?

The objective of this chapter is to show how both landscape and hydropower are institutionalized and constructed as shared, collective entities. To illustrate our points we focus on the nexus between hydropower, water and landscape resources and their role in local territorial practices through case studies of the Sierra Nevada mountain area (Andalusia, Spain). These allow us to examine the driving forces behind the evolution of local landscapes and local perceptions of the hydropower landscape. This chapter explores these forces through analysis of the different discourses and practices of local stakeholders in the Sierra Nevada mountain range.

Our study is based, first of all, on textual analyses of a range of sources at different geographical scales, including energy and environmental policy documents for the whole of Spain, for the Autonomous Region of Andalusia, and documentary information on hydropower in Sierra Nevada and second, on fieldwork and semi-structured qualitative interviews (conducted between 2012 and 2014) with the different stakeholders and residents in our study area. During the survey we interviewed 44 people, including town councillors and mayors, hydropower plant managers, tourism entrepreneurs and other local stakeholders. In the interviews we explored their perception of landscape and local landscape practices, local production of hydroelectricity and its relationship with other local activities, such as tourism, nature protection, agriculture, water use etc. From the original 44 interviews, we then selected 15 of particular interest for our research purposes, which we analysed in great detail.

The evolving perception of hydropower and its impacts

Hydroelectricity was the first renewable energy to experience the effects of our continually evolving value systems (Koch 2002). For a long time, it was considered to be one of the cleanest sources of energy and many governments nationalized the hydropower facilities in their countries on the basis of their important public benefits (Frolova et al. 2015a). Increased public concerns about significant negative environmental and social consequences of large dams however led to the adoption in 2001 of the first EU Directive on Renewable Electricity, in which large scale hydroelectric systems were removed from the official list of 'clean energy' technologies (Abbasi & Abbasi 2011; Ansar et al. 2014).

Since then many different countries have labelled 'large hydro' as non-renewable and non-sustainable (Koutsoyiannis 2011) and only small[1] hydropower systems continue to be classified as green technologies. Although as some researchers point out the definition of small-scale hydropower is somewhat arbitrary (Egré & Milewski 2002; Bakken et al. 2014) and the threshold separating small ('renewable') and large ('non-renewable') hydropower plants varies from one European country to the next (e.g. 50 MW in Spain, 15 MW in Greece, 5–10 MW in the UK, 5 MW in Germany, 1–10 MW in Norway) (Koutsoyiannis 2011; Espejo Marín & García Marín 2010; Bakken et al. 2014; Bracken et al. 2014). Even though small-scale hydropower systems (SHS) have until recently been portrayed as a relatively benign technology in accordance with the view that 'small is beautiful and large is bad' (Koutsoyiannis 2011; Bakken et al. 2014), the latest research is revealing that they too have adverse effects on wildlife, the transfer of sediments, flood risk and recreational activities (for example, Abbasi & Abbasi 2011; Bakken et al. 2014; Bracken et al. 2014). Even their visual and landscape impact, which for a long time was seen as less significant than that of large-scale hydropower systems, is now the subject of controversy (Slee et al. 2011; Bakken et al. 2014). First, suitable sites for small hydro schemes are often located in environmentally sensitive areas perceived as pristine and natural (Malesios & Arabatzis 2010), and second, in contrast to large hydropower where much of the necessary infrastructure (such as water transportation tunnels, power houses and penstocks) is installed underground, in small-scale hydropower plants most of the infrastructure is located above surface, and even the penstock/pipeline is often visible from the surface despite being buried (Bakken et al. 2014). As a result these plants are often perceived as a threat to 'landscape attractiveness' (Slee et al. 2011).

Finally, if the environmental problems caused by small hydro are analysed on the scale of impact per kilowatt of power generated it becomes evident that if SHS were more widespread it would cause just as many, equally serious problems as those caused by large hydropower projects (Abbasi & Abbasi 2011), they would just be more dispersed. Various policy documents have also recognized the negative impact of massive development of SHS. For example, in the Spanish Autonomous Region of Galicia, one of the areas in Spain where small hydro has been most extensively developed, a special River Law passed in 2006 put many of these projects on hold, citing negative landscape and ecological impact as an important argument against the construction or restoration of small hydropower plants (Frolova 2010).

There is therefore an important scientific debate as to whether the most environmentally friendly strategy is to develop a lot of small-scale hydropower projects or a few large ones (Bakken et al. 2014), a debate that we will be exploring here with particular reference to mountain areas.

Mountain areas are a very good example of how hydropower can bring about dramatic changes in the landscape (Ferrario & Castiglioni 2015), something that has happened in most European mountain ranges (Briffaud et al. 2015). This resulted in the appearance of a set of specific, interrelated elements such as hydropower stations, water reservoirs, dams, pipelines, water diversion channels etc. that have together formed what today can be perceived as authentic energy landscapes (Briffaud 2014; Briffaud et al. 2015; Frolova et al. 2015b). Hydropower infrastructures and the landscapes they help produce have also participated in the construction of specific identities or sense of place, and have become part of our collective memory and history (Varaschin & Bouvier 2009), and of our cultural heritage (Frolova et al. 2015b). Finally, hydropower has developed over a long period during which it has at times been welcomed and accepted by local people, while at others it has been a source of conflict. The hydropower landscapes that have emerged from this process can therefore be viewed as culturally constructed objects (Briffaud et al. 2015; Ferrario & Castiglioni 2015; Frolova et al. 2015a). As Serge Briffaud emphasized (Briffaud 2014; Briffaud et al. 2015), hydroelectric energy can be considered as the 'resource of the resource', exploitation of which allows us to activate other resources such as water, landscapes, etc., or changes the ways they are exploited. Hydropower therefore offers us a useful guide for understanding the complex set of relations on which the co-construction of renewable energy capacity and culturally shared landscapes is based (Frolova et al. 2015b).

The nexus between hydropower and landscapes

The nexus between hydropower and landscape is an issue that goes far beyond the impact of infrastructures (dams, water reservoirs, power plants, pipelines, penstocks, etc.) and helps us understand the values and practices of local stakeholders in relation to their territory and landscape.

If we analyse the *energy–landscape nexus* in a similar way to the *water–energy nexus* (Scott et al. 2011), our first conclusion is that energy and landscape are primarily interlinked in terms of resource use. On the one hand, landscape is a useful tool for understanding local place perspective and the socio-cultural dimension of hydropower resources and projects, while if landscape is viewed through the prism of energy what we see is a socio-territorial system that allows us to produce and develop hydropower (Briffaud 2014). In addition, landscape is often an important factor in discourses on the environmental and social impacts of hydropower, either as an object to be protected, or as an element mobilized by certain stakeholders, in particular by the 'protectionists' (Slee et al. 2011; Ferrario & Castiglioni 2015).

In this chapter we will be focusing specifically on the effects of hydropower development on mountain areas. In the following sections we analyse hydropower development in Spain at a national level and its relationship with landscape and water policies. Then through case-studies of different towns and villages in the Sierra Nevada mountain area we examine how hydropower landscapes have been constructed through local discourses and practices.

National context

Spain like many other European countries (including Germany, France, Italy, Switzerland, Sweden) has already developed almost all its economically viable large-scale hydro-potential (80–100 percent) (Koutsoyiannis 2011). Hydropower once played a leading role in electricity generation such that in 1940 more than 90 percent of all Spanish electricity (3617 GWh) came from hydropower (Frolova 2010). Since then the relative contribution of hydropower to Spain's overall electricity supply has declined, although in absolute terms hydropower generation continued to increase until the early 2000s. In 2013 the installed capacity of large-scale hydropower was 17,766 MW, which represented 16.4 percent of Spain's total installed capacity, while small-scale hydropower with its installed capacity of 2,102.5 MW represented 1.9 percent (Red Eléctrica de España 2014).

Most of Spain's large dams were built before the concept of Environmental Impact Assessment was introduced into Spanish law in 1986. Thus, hydropower projects were carried out with no regard whatsoever for river landscapes, which were adapted according to the needs of each project.

Although landscape became an important issue in Spain's land use regulations and environmental policy in 2000 when it signed the European Landscape Convention (ELC), landscape policies are still out of step with renewable energy policies. Prior to the ELC, the real impact of hydropower infrastructures on river landscapes had rarely been assessed or even considered (Frolova 2010). In 2000, the European Union (EU) also issued the Water Framework Directive (WFD), in which it announced a new water policy that placed strong emphasis on the integral water management of river basins, a landscape which had hitherto earned only a marginal position in government policy and academic research on hydropower in Spain (Frolova 2010).

Public perception of the hydropower landscape in Spain is intimately related to the level of acceptance of water policy and water management. For many years, the perception of hydropower in Spain was strongly influenced by the 'water management paradigm' (*paradigmahidraulico*). This paradigm consisted of a centralized state-based water regulation system with the ultimate objective of ensuring the availability of cheap water and electricity to permit economic growth (Saurí & del Moral 2001; Frolova 2010; Castán Broto 2015). Prior to Franco's dictatorship (before 1936) it was part of the rhetoric of agricultural development but in reality played little part in water management. But during the 1940s and 50s a system of large, modern infrastructures based on water reservoirs, dams for generating hydroelectricity and networks of irrigation channels was promoted by the State, and the hydraulic paradigm became the basis for Spain's hydropower development policy (Frolova 2010; Castán Broto 2015). Although the influence of this paradigm can still be seen in some aspects of water policy in Spain (del Moral 2006; Oñate & Peco 2005), landscape concerns, including the ecological changes caused by dams, are becoming increasingly important in current approaches to water management and hydropower development (Frolova 2010).

Hydropower, landscape and water in the Sierra Nevada mountain range

Our study area encompasses nine municipalities (Güejar Sierra, Pinos Genil, Monachil, Dílar, Nigüelas, Lecrín, Capileira, Bubión and Pampaneira) in the Province of Granada in the foothills of Western and Southern Sierra Nevada, located in the valleys of the River Genil (one of the principal tributaries of the most important river in Southern Spain – the Guadalquivir) and its offshoots, the Monachil, the Maitena, the Poqueira, the Dilar and the Torrente (Figure 14.1). The relatively abundant supply of water in Sierra Nevada made it a key factor in the process of settlement and occupation over the centuries, and also in local identity and sense of place. During the Nasrid era (13th – 15th century) a complex system of water channels (*acequias*), transfers and artificial recharge was created in each river valley in order to control the local hydrological cycle and so guarantee sufficient flow levels in the spring and summer. Water management benefited from the natural regulation produced by snowfalls and also by the infiltration of streams created by melted snow, which fed natural springs and other water sources lower down (Castillo Martín 2010). Water management was therefore a driving force behind the evolution of local landscapes (Frolova et al. 2015a) and 'water culture' has become an important element of local people's sense of place.

Until the 1940s the dominant model for hydropower development in Sierra Nevada was small plants, as happened in other Spanish mountain areas. 'Small-scale electrification' (Núñez 1998: 268) predominated even prior to concerns about the environmental impacts of large dams. This was due to various factors that limited the development of large hydropower systems: low

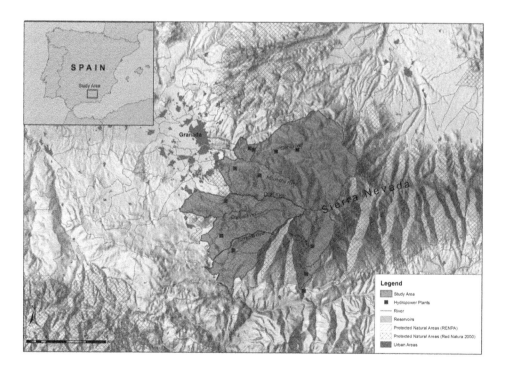

Figure 14.1 Map of study area with hydropower plants and Protected Natural Areas
Source: Belén Pérez-Pérez.

flow rate due to annual summer droughts, technical problems for the construction of large dams and historically low demand for electricity in an area of small towns and villages with little or no industry (Frolova et al. 2015a).

As in other regions of Spain these plants were owned by numerous small electric companies with private capital, which developed small continuous current circuits and generated electricity near the place of consumption. Some of these plants were managed as communal property by neighbours' associations and known as *fábricas de luz* ('electricity factories') (Castán Broto 2015). These 'electricity factories' proliferated in Sierra Nevada in the early 20th century supplying on the one hand all the street-lighting for the small, neighbouring villages and indeed for the city of Granada, and on the other the electricity required for small-scale industrial activities that were often housed within the plant itself (Núñez 1994).

In Sierra Nevada there has been substantial development of small hydropower facilities with more than 30 plants installed since the 1890s (Requena Galipienso 2012, 2013), more than half of which were built between the 1890s and the 1920s (Table 14.1). With the setting up of a centralized national distribution network all over Spain in the 1940s, production became increasingly concentrated in the hands of a few large companies driven by public investment (Castán Broto 2015). This also affected hydropower development in Sierra Nevada through the centralization and construction of bigger plants (up to 10–12.8 MW), some of which have remained in operation, while others have been abandoned or refurbished.

Today there are twelve hydropower plants in service in Sierra Nevada (Figure 14.1) and their installed capacity varies from about 0.019 MW (Eléctrica del Blanqueo plant) to 12.8 MW (Duque plant and Pampaneira plant) (Table 14.1). Most of these plants are served by free-flowing rivers (Frolova et al. 2015a). Hydropower remains the most important renewable energy source in our study area with a total installed capacity of almost 63 MW, while wind power produces 16 MW and solar photovoltaic power a mere 115.7 KW (ibid.).

Hydropower was developed most intensely in the River Monachil Valley, the Lecrín Valley and the Poqueira Valley, where the infrastructures (water-tanks, headraces, buildings, etc.) were adapted to the topography, and in the valleys of the Rivers Genil and Maitena where the only large reservoir (Canales) is located.

Since the 1980s, however, when the emblematic landscapes of Sierra Nevada were protected as a National Park, the process of hydropower development has come to an almost complete standstill, as national and regional legislation declared industrial uses of the Sierra's resources incompatible with its conservation. The creation of the Sierra Nevada National Park made increasing its hydropower capacity very difficult, and in some cases caused it to fall. The only real possibilities for development nowadays lie in restoring abandoned plants or renovating existing ones.

Many projects have been rejected for breaching the regulations protecting the Sierra Nevada National Park, although there have been some exceptions such as the council-owned hydroelectric plant set up in Lancha de Cenes built in 1995, or the plant built in Nigüelas in 1996 on the Torrente River (Frolova et al. 2015a).

Construction of hydropower landscapes through local discourses

The first important hydroelectric site in our study area was the Genil Valley, where hydropower was developed hand in hand with tourism. The valley has become an archetype of a tourist landscape based on energy production. Hydropower plants were built along the River Maitena – a tributary of the River Genil – in the 1920s to provide electricity for the Sierra Nevada Tram Railway (1925–1974) (Frolova et al. 2015a), which opened up the Sierra Nevada to tourism. Later the construction of the Canales reservoir with its hydroelectric plant (1975–1988) created another popular tourist attraction for the Sierra.

Table 14.1 Hydropower plants in Sierra Nevada

Name	Municipality	River	Installed capacity (MW)	Period in service	State/Conservation
Genil-Maitena Valley					
Pinos Genil	Pinos Genil	Genil	No data	1897–1978	Abandoned-In ruins
Eléctrica del Blanqueo	Pinos Genil	Aguas Blancas	0.019	1907–1931	Abandoned-In ruins
El Castillo	Güéjar-Sierra	Genil	6.52	1923–1989	Abandoned-Disappeared
Maitena	Güéjar-Sierra	Maitena	1.17/1.92	1923–1967/ restored 1996	In service
Rosario	Güéjar-Sierra	Genil	0.095	no data-1935	Abandoned-Disappeared
Canales	Pinos Genil	Genil	8.8	1989	In service
Nuevo Castillo	Güéjar-Sierra	Genil	3.48/4.36	1989	In service
Lancha de Cenes	Granada	Genil	0.96	1995	In service
Monachil Valley					
Diéchar	Monachil	Monachil	0.66/0.8	1919–1966/ restored in 1989	In service
La Vega	Monachil	Monachil	1.7/2.4	1904–1967/ restored 1990	In service
Trola I	Monachil	Monachil	0.22	1909-no data	Abandoned-Disappeared
Trola II	Monachil	Monachil	0.06		
Tranvías	Monachil	Monachil	4.32/1.90	1907–1974/ restored in 1991	In service
Sierra Nevada	Monachil	Monachil	no data	no data	Disappeared
La Estrella	Monachil	Monachil	no data	no data	No data

continued . . .

Table 14.1 Continued

Name	Municipality	River	Installed capacity (MW)	Period in service	State/Conservation
Dílar- Lecrín Valley					
La Espartera-Dílar	Dílar	Dílar	1.43/3.36	1925	In service
Nigüelas	Nigüelas	Torrente	2.98	1996	In service
Molino de Doña Juana/ Fábrica de Luz	Dúrcal	Dúrcal	no data	1905	Disappeared
1° Fábrica de Luz Cozvíjar-Eléctrica de San Antonio	Cozvíjar (Vllamena)	Dúrcal	no data	1908	Disappeared
2° Fábrica de Luz Cozvíjar-Eléctrica de San Antonio	Cozvíjar (Villamena)	Dúrcal	no data	1908	Disappeared
Molino De Melegís/ Fábrica de Luz	Melegís (El Valle)	Dúrcal	no data	1908/1930	Disappeared
Dúrcal (Estación Eléctrica San José)	Dúrcal	Barranco de la Rambla	2.72	1923	In service
Fábrica de Luz del Padul	El Padul	Barranco del Anciano	no data	1909	Abandoned-Disappeared
Poqueira Valley					
Poqueira	Capileira	Poqueira	10.4	1957	In service
Pampaneira	Pampaneira	Poqueira	12.8	1956	In service
Duque	Pampaneira	Poqueira	12.8	1981	Abandoned
Molino del Granadino	Órgiva	Guadalfeo	no data	no data	Abandoned-Disappeared
Eléctrica de Órgiva	Órgiva	Guadalfeo	no data	1902–1960	Abandoned-Disappeared
Eléctrica Nuestra Señora de Las Nieves	Trevélez	Trevélez	10	1950-no data	Abandoned-Disappeared

Sources: Agencia Andaluza de Energía (2012); Nuñez (1993, 1998); Requena Galipienso (2012); in-depth interviews.

Hydropower development reached its peak in the Poqueira river valley between the 1940s and 1980s, during Franco's dictatorship. Three large hydropower plants (Poqueira, Pampaneira and Duque) with the highest installed capacity in the study area were constructed during this period, creating their own hydropower landscapes (Frolova et al. 2015a) (Figure 14.2). Even after the Declaration in 1982 of three villages in the Poqueira Valley (Capileira, Pampaneira and Poqueira) as a Group or Area of Regional Historic/Artistic Importance (BIC) and the Declaration of Sierra Nevada as a Biosphere Reserve (1986), a Natural Park (1989) and finally a National Park (1999), this cultural landscape of the Alpujarra has peacefully coexisted with its hydropower infrastructures, in spite of their significant impact on the landscape.

In Monachil, an important regional centre for mountain tourism and downhill skiing (Pradollano), there are three hydroelectric plants (La Vega, Tranvías and Diéchar) which were constructed prior to the protective measures that prevented new hydroelectric facilities from being installed in this area. The most emblematic of these is the Tranvías plant which was constructed in 1907 and reconstructed in 1991. This is an exceptional case for our study area in that the plant belongs to the local community.

Another hydropower plant which belongs to the community is the one in Nigüelas (on the Torrente River) (Figure 14.3). It is located in a traditional agricultural area, which is currently suffering a severe crisis. Today this area is characterized by the development of scattered low-intensity rural tourism and the installation of other forms of renewable energy (solar and wind). In these landscapes, which are considered of high cultural value and visited by large numbers

Figure 14.2 The Poqueira hydropower plant and view of Capileira village, 2013
Source: Marina Frolova.

of tourists, renewable energy infrastructures sometimes create a climate of conflict. There were some protests by ecologists about the damage caused during the construction of the hydroelectric plant in Nigüelas, in spite of it being widely supported by local residents, who saw it as a source of income for the community. The Nigüelas area of the Lecrin Valley has also seen the contested development of other kinds of renewables (wind and solar farms), which are not always welcomed by local residents.

In the next sections, we will be contrasting the results of our preliminary analysis with those of the semi-structured interviews we conducted in the study area in order to illustrate the processes through which hydropower, water and landscape resources are co-constructed through different discourses of local stakeholders. This analysis focuses particularly on fifteen interviews with local stakeholders: six mayors and local councillors, three tourism entrepreneurs, two Council technicians, one hydroelectric plant manager, one employee of the Granada Provincial Energy Agency, one farmer and one warden from the Sierra Nevada National Park.

Perception of hydropower as a local resource

Most local stakeholders consider hydroelectricity production a real privilege given Sierra Nevada's geographical location in the far south of Spain, and in view of the differences in rainfall between the valleys situated on the Sierra's windward side and its semi-arid eastern end. It is generally viewed as one of the area's most important energy resources. The hydroelectricity

Figure 14.3 The Torrente hydropower plant
Source: Marina Frolova.

produced by the relatively small plants in Sierra Nevada (the most common plants in our study area) is perceived as green energy, apparently producing a lot less impact than wind energy and also than the energy produced by photovoltaic installations (above all due to their proximity to the Sierra Nevada National Park and because of their visual impact). As an owner of a rural hotel said:

> Hydroelectricity is the most natural form of energy [. . .] Hydroelectric plants are the best. They make good use of the water, and they are the most ecological. They can also be well integrated, as happens in Nigüelas where the pipeline is camouflaged and they have made an attractive building that sits well in the landscape.

Many of them see hydroelectric power as very efficient and 'economic': 'it is a very efficient form of energy, because a very high percentage can be used' (Councillor responsible for Tourism). Certain stakeholders mentioned that prior to the 1950s the plants were small and decentralized and that if this form of energy production were taken up again today, it could become a real driving force for the development of local villages. Local stakeholders emphasize that the hydroelectricity produced in the private plants feeds a grid outside their area and is not used to meet local needs. Many of them also underline that this energy only serves to cover external needs and creates almost no benefits for the local population.

Hydropower as an economic resource

Hydropower has a special place in local discourses because of the economic benefits it produces and is positively viewed as a way of helping maintain or develop the valleys. Those questioned about the possibilities of developing new forms of energy production often suggest the construction of small and micro hydropower plants. The Tourism Councillor of Monachil boasts:

> The restoration of the Tranvías hydroelectric plant [. . .] was something that brought huge benefits for the village first of all, because there are people who come here specifically to see it, and second because it helped us conserve that special relationship with water by socializing its use . . . There are a lot of villages that have lost it [. . .] we haven't [. . .].
> Hydroelectricity is a very economical form of energy and provides a lot of benefits. The Council is receiving an average of 300,000 to 400,000 Euros in net annual income from energy production. We are among the fortunate few that can say that we have electricity produced by water from the river [. . .]

Different local stakeholders told us that for them 'hydropower' was a 'synonym' for Nigüelas, because the plant 'provided jobs for almost half the village population' and income to fund its economic development (Frolova et al. 2015a). 'The hydropower plant in the village provides a framework for the economic system of the village, investment, jobs [. . .]' Mayor 1.

Perception of the water resource

It seems that the perception of hydroelectric energy is closely associated with the vision of water as an important resource and an essential element of landscape and of local identity in this area of southern Spain. When we asked the mayors of most of the villages we studied in Sierra Nevada what the main energy resource exploited in their villages was, almost all replied 'water'.

Water is highly valued by local stakeholders. The relatively abundant water resources of Sierra Nevada from both rivers and springs have become a feature of its identity and a key factor in the historic process of settlement of the area and in the life of the population (agriculture, livestock farming, energy production, tourism production). Almost all the stakeholders we asked mentioned water as one of the most important local resources (together with tourism).

However the use of water for hydroelectricity production is not the number one priority in Spain, coming in third behind human consumption and agriculture (crop irrigation). Since the application of the Water Act of Andalusia, approved by the Regional Government in 2010, which established minimum water levels or 'ecological flow' in Andalusian rivers, the already fierce competition for water in the study area has intensified. This is due to the fact that a lot of crops require irrigation and that the farmers' irrigation associations play a very important role in water management. The amount of water available for energy production depends above all on irrigation. If there is not enough water to irrigate the crops, no hydroelectricity is produced.

The Sierra Nevada ski resort also puts considerable pressure on the water levels in the Monachil and Dilar rivers by storing large quantities of water for producing artificial snow and supplying tourist needs. This makes it difficult to maintain the ecological flow levels and the ecosystems of these two rivers, both of which belong to the Sierra Nevada natural space, and goes against the grain of the conservationist policy applied in this area since the 1980s (Frolova et al. 2015a).

Perception of the landscape resource

Hydropower has an important role in this vision of landscape. The importance for tourism of some hydropower landscape features should also be noted. These include for example, the Canales reservoir with its picturesque walk for tourists, a true landscape milestone in the Genil Valley (Figure 14.4). There is a similar case in the Monachil Valley, where the Tranvías hydropower plant has become a tourist attraction in itself.

From these conversations, it would seem that hydropower plants are generally highly regarded and appreciated as a part of mountain landscapes. For many of the people we interviewed, the dams and to be more precise the lakes are positively viewed landscape features in the sense that they are seen as old and established parts of the landscape that local people have come to regard as part of the landscape they know, whether natural or man-made.

Hydroelectric developments seem to be part of the landscape for the people we met, none of whom had anything really negative to say about them. There are many reasons for this, related with the advantages offered directly by hydropower or indirectly by the economy it generates. Local stakeholders' perception of landscape is often linked to their view of it as a tourist resource. Many of the people we spoke to have noted that over the course of the last thirty years their villages have moved from a situation of subsistence in which their landscapes were purely an agricultural resource to a new situation in which they are an important means of attracting tourism.

Local residents have a positive opinion of their local hydropower facilities, especially in villages where the hydroelectric plants belong to the community. This seems to instil a feeling of proximity, a sense of rapport with these plants, which can develop into a feeling of ownership or at the very least an assimilation of hydroelectric facilities into local landscape discourses.

We have the Diechar, La Vega and Tranvía plants, and in the past there were two others called la Trola and la Fabriquilla. But almost all of them disappeared, except for Diechar and La Vega, which were bought up by private interests and put back into service. And the Tranvía Plant was bought by Monachil Town Council and put into service.

(Owner of a rural hotel)

Figure 14.4 The Canales reservoir, 2012
Source: Marina Frolova.

For the local stakeholders we interviewed, the impacts of hydroelectricity are more acceptable than those of other renewables. Hydropower in Sierra Nevada is also viewed as 'more natural', probably due to the fact that it is a more 'traditional' form of energy. It is as if it had already been assimilated into the landscapes and other territorial assets of this area, in that it is accepted by the majority of the people we interviewed, and this in protected areas in which the hydropower plants had been built before the areas were granted protected status (Figure 14.5).

> [The Tranvías Hydropower plant] is an old-fashioned kind of construction made of stone, which fits perfectly into its surrounding environment.
>
> (Tourism councillor)

> These power stations appeared at the beginning of the last century [. . .] Later they were abandoned. These buildings were restored and in fact it is important to restore them because [. . .] the canals were already made and the lines were marked out [. . .] and the works were [. . .] integrated into the landscape.
>
> (Owner of rural hotel)

In fact some of the interviewees went so far as to say that 'all renewable energies are beneficial for the landscape if the project is conceived in an intelligent way' (Mayor 1). 'In any work on renewable energies, the aim is to benefit the landscape [. . .]'.

Figure 14.5 The Tranvía hydropower plant restored in 1991 and opened for guided tours, 2012
Source: Marina Frolova.

By way of conclusion

Our analysis has shown that attitudes towards hydropower in Sierra Nevada are associated with local perceptions of landscape and water. Hydropower, landscape, water and tourism in Sierra Nevada have been closely connected throughout the 20th and 21st centuries.

Establishing the *hydroelectric power–water–landscape nexus* is therefore a useful means of approaching these resources, which are interlinked in terms of their use by different stakeholders. In addition, exploring how hydropower, water and landscape resources are co-constructed through the different discourses and practices of local stakeholders is a useful tool for assessing energy projects in all their territorial, environmental and cultural dimensions. By integrating landscape, water and energy in the same overview and showing the territorial, environmental and social effects of energy development, this analysis of landscape and energy perceptions provides a way of integrating energy projects into local territorial planning.

The long historical process of development of this kind of renewable has contributed to the establishment of hydropower landscapes as culturally constructed objects, in both material and symbolic terms. In a similar way, these landscapes have participated in the construction of local identities, collective memory and history, all of which has helped make hydropower landscapes part of our cultural heritage. Hydropower infrastructures now play a significant part in the local environment and in water use in Sierra Nevada. They have also become an important feature of the landscape.

As our case study shows, although preservation of landscape character is often used as an argument in the conflicts that develop around renewable energy projects, its relationship with these projects is not always conflictive. As our analysis for hydropower shows, this depends on the way in which both landscape and renewable energy developments in general are institutionalized and constructed as shared, collective entities. In a similar way, as recent studies show (Frolova et al. 2015b) other renewable energy infrastructures could participate in the construction of local sense of place and become an important feature of local scenery. The relationships of the local population with renewable energy are not produced by its purely visual perception or aesthetic qualities and their attractiveness to tourists (Briffaud et al. 2015), but require a multidimensional view of the renewable energy landscape that brings together practices, experiences and perceptions as well as physical properties.

Acknowledgements

The research presented in this chapter was conducted within the framework of the Project 'Ressourcespaysagères et ressourcesénergétiquesdans le montagnessud-européennes' (Research program IMRIgnismutat Res), chaired by Serge Briffaud and funded by the French government. I am also grateful to the Spanish Ministry of the Economy and Competitiveness (CSO2011–23670) for supporting this research. My special thanks go to Alfredo Requena-Galipienso for collecting quantitative data, to Belén Pérez-Pérez for her help with cartography, students of Environmental Sciences from the University of Granada, who did some of the interviews, to Serge Briffaud, Yolanda Jiménez-Olivencia and Miguel-Ángel Sánchez-del-Árbol for their valuable comments. I am also grateful to Martin Pasqualetti, whose comments have helped me improve the presentation and arguments.

Note

1   In academic literature the word 'small' is normally used to refer to renewable hydropower plants and is related to their installed capacity. The installed capacity below which a plant is considered 'small', or renewable, is usually specified by law and varies from one country to the next.

References

Abbasi, Tasneem, & Abbasi, S.A. 'Small hydro and the environmental implications of its extensive utilization'. *Renewable and Sustainable Energy Reviews* 15, no. 4 (2011): 2134–2143.

Ansar, Atif, Flyvbjerg, Bent, Budzier, Alexander, & Lunn, Daniel. 'Should we build more large dams? The actual costs of hydropower megaproject development'. *Energy Policy* 69, (2014): 43–56.

Bakken, T.H., Aase, A.G., Hagen, D., & Sundt, H. 'Demonstrating a new framework for the comparison of environmental impacts from small- and large-scale hydropower and wind power projects'. *Journal of Environmental Management* 140, (2014): 93–101.

Bracken, L.J., Bulkeley, H.A., Maynard, C.M. 'Micro-hydro power in the UK: The role of communities in an emerging energy source'. *Energy Policy* 68, (2014): 92–101.

Briffaud, S. (ed.). Ressources paysagères et ressources énergétiques dans les montagnes sud-européennes. Histoire, comparaison, expérimentation, Rapport final de recherche, Programme 'Ignis mutatres. Penser l'architecture, la ville et les paysages au prisme de l'énergie', Paris: Atelier international du Grand Paris, 2014.

Briffaud, Serge, Heaulmé, Emmanuel, André-Lamat, Véronique, Davasse, Bernard, & Sacareau, Isabelle. 'The nature of resources. Conflicts of landscape in the Pyrenees during the rise of hydroelectric power' in M. Frolova, M.-J. Prados & A. Nadaï (eds), *Renewable Energies and European Landscapes: Lessons from Southern European Cases*, New York, London: Springer, 2015, pp. 135–153.

British Hydropower Association and IT Power. England and Wales Hydropower Resource Assessment, Department of Energy & Climate Change (DECC) & Welsh Assembly Government (WAG), 2010.

Castán Broto, Vanesa. 'Innovation territories and energy transitions: Energy, water and modernity in Spain, 1939–1975'. *Journal of Environmental Policy and Planning* 18, no. 5 (2015): 712–729.

Castillo Martín, Antonio. 'El papel de las surgencias en los regadíos de Sierra Nevada'. In J.R. Guzmán Álvarez & R. Navarro Cerrillo (eds), *El agua domesticada. El paisaje de los regadíos de montaña en Andalucía*, Agencia Andaluza del Agua, 2010, pp. 80–84.

del Moral, Leandro. 'Territorio mediterráneo y planificación hidrológica'. In *Actas del I Congreso Andaluz de Desarrollo Sostenible: El Agua, Granada*, Universidad de Granada, 2006, pp. 61–67.

Egré, Dominique, & Milewski, Joseph C. 'The diversity of hydropower projects'. *Energy Policy* 30, (2002): 1225–1230.

Espejo Marín, Cayetano, & García Marín, Ramón. 'Agua y energía: Producción hidroeléctrica en España'. *Investigaciones Geográficas* 51, (2010): 107–129.

Ferrario, Viviana, & Castiglioni, Benedetta.'Hydropower exploitation in the Piave River Basin (Italian Eastern Alps): A critical reading through landscape'. In M. Frolova, M.-J. Prados & A. Nadaï (eds), *Renewable Energies and European Landscapes: Lessons from Southern European Cases*, New York London: Springer, 2015, pp. 155–172.

Fraenkel, Peter, Paish, Oliver, Bokalders, Varis, Harvey, A., Brown, A., & Edwards, R. *Micro-Hydro Power: A Guide for Development Workers*, London: IT Publications Ltd, 1991.

Frolova, Marina. 'Landscapes, water policy and the evolution of discourses on hydropower in Spain'. *Landscape Research* 35, no. 2 (2010): 235–257.

Frolova, Marina, Jiménez-Olivencia, Yolanda, Sánchez-del Árbol, Miguel-Ángel, Requena-Galipienso, Alfredo, & Pérez-Pérez, Belén. 'Hydropower and landscape in Spain: Emergence of the energetic space in Sierra Nevada (Southern Spain)'. In M. Frolova, M.-J. Prados & A. Nadaï (eds), *Renewable Energies and European Landscapes: Lessons from Southern European Cases*, New York London: Springer, 2015a, pp.117–134.

Frolova, Marina, Prados, María-José, & Nadaï, Alain, 'Emerging renewable energy landscapes in Southern European countries' in M. Frolova, M.-J. Prados & A. Nadaï (eds), *Renewable Energies and European Landscapes: Lessons from Southern European Cases*, New York London: Springer, 2015b, pp. 3–24.

Koch, Frans H. 'Hydropower – the politics of water and energy: Introduction and overview'. *Energy Policy* 30 (2002): 1207–1213.

Koutsoyiannis, Demetris. 'Scale of water resources development and sustainability: Small is beautiful, large is great'. *Hydrological Sciences Journal* 55, no. 4 (2011): 553–575.

Malesios, Chrisovalentis, & Arabatzis, Garyfallos. 'Small hydropower stations in Greece: The local people's attitudes in a mountainous prefecture'. *Renewable and Sustainable Energy Reviews* 14, (2010): 2492–2510.

Mishra, Sachin, Singal, S.K., & Khatod, D.K. 'Optimal installation of small hydropower plant-A review'. *Renewable and Sustainable Energy Reviews* 15, (2011): 3862–3869.

Núñez, Gregorio. 'Origen e integración de la industria eléctrica en Andalucía y Badajoz' in J. Alcaide, A.M. Bernal, E. García de Enterría, J. M. Martínez Val, G. Núñez, & J. Tusell (eds), *Compañía Sevillana de Electricidad. Cien Años de Historia*. Seville: Fundación Sevillana de Electricidad, 1994, pp. 126–159.

Núñez, Gregorio. 'La hidroelectricidad en pequeña escala' in M. Titos (ed.), *Historia económica de Granada*. Granada: Cámara de Comercio, Industria y Navegación, 1998, pp. 267–282.

Oñate, Juan J., & Peco, Begoña. 'Policy impact on desertification: Stakeholders' perception in southeast Spain'. *Land Use Policy* 22, (2005): 103–114.

Oud, Engelbertus. 'The evolving context for hydropower development'. *Energy Policy* 30, (2002): 1215–1223.

Red Eléctrica de España, 2014. *El sistema eléctrico español 2013*. Madrid.

Requena Galipienso, Alfredo. Paisaje, energía hidroeléctrica y turismo en un ámbito de Sierra Nevada. Valles del Alto Genil y Maitena. Memoria de Master, Granada: Universidad de Granada, 2012.

Requena Galipienso, Alfredo, 'Implicaciones sociales y paisajísticas de la implantación de energías renovables en el Valle de Lecrín, Sierra Nevada, Granada' in D.A. Fabre Platas and C. Egea Jiménez (eds) *Socializando saberes en un primer encuentro internacional de posgrados*. Granada: Universidad de Granada, 2013, pp. 83–98.

Saurí, David, & del Moral, Leandro. 'Recent development in Spanish water policy. Alternatives and conflicts at the end of the hydraulic age'. *Geoforum* 32, no. 3 (2001): 351–362.

Scott, Christofer A., Pierce, Suzanne A., Pasqualetti, Martin J., Jones, Alice L., Montz, Burell E., & Hoover, Joseph H. 'Policy and institutional dimensions of the water-energy nexus'. *Energy Policy* 39, (2011): 6622–6630.

Slee, Bill, Whitfield, Rachel, & Whitfield, Stephen. 'Discourses of power. The development of small-scale hydropower in North East Scotland'. *Rural Society* 21, no. 1 (2011): 54–64.

Varaschin, Denis, & Bouvier, Yves. 'L'électricité, diva à Divonne' in D. Varaschin & Y. Bouvier (eds), *Le patrimoine industriel de l'électricité et de l'hydroélectricité, Actes du colloque international de Divonne-les-Bains et de Genève*. Chambéry: Université de Savoie, 2009, pp. 7–12.

15

Finding locations for endurably objectionable energy-related facilities

The CLAMP policy

Michael R. Greenberg, Molly Coon, Matthew Campo and Jennifer Whytlaw

Introduction

This chapter considers the challenge of finding locations for energy production and waste management facilities that society needs, but that are objectionable to many people. These facilities are among the most enduringly objectionable locally unwanted land uses (Greenberg et al. 2012) that have led to the acronyms NIMBY ('not in my back yard'), NIABY ('not in anyone's back yard'), NIMTOO ('not in my term of office'), BANANA ('build absolutely nothing anywhere near anyone'), and LULU ('locally unwanted land use') (O'Hare et al. 1983; Portney 1991; Macfarlane and Ewing 2006).

An earlier paper focusing on nuclear power plant permit applications and coal plant sites demonstrated that companies and governments had focused on trying to locate these facilities in existing plant locations, that is, concentrating locations at major plants (CLAMP) (Greenberg 2009). That paper also showed more public acceptance of LULUs at locations that already had nearby plants than those that did not. This chapter expands that research by addressing two questions:

1   Is the CLAMP policy really a product of the last few decades, or has it existed for a far longer period of time? What kinds of electrical energy production facilities does it apply to?
2   What is the reaction of people living near sites where nuclear defense waste is already stored to the idea of storing nuclear waste from commercial nuclear power plants in area, that is, applying CLAMP to nuclear waste management?

Research question 1

It is conceivable that companies might have used the CLAMP policy more often in the early and mid-20th century when the distribution network was more limited, which forced companies to site electrical generating units nearer to users. Also, natural gas has become a far more prominent source of electrical energy than in the past, and it may be that natural gas plants are able to fit into spaces that coal and nuclear plants could not. In other words, new sites could be found within existing locations. Accordingly, we studied the changing geography of electrical generating sites in five states with different energy histories and with which the authors were familiar:

- Arizona, New Mexico, and Texas: large western states with a mix of electrical energy fuel sources, and rapid population and economic growth; and
- New Jersey and Indiana: an eastern and a midwestern state, respectively, with the first heavily dependent on nuclear energy and the second with a history of heavy coal dependence.

The authors obtained data from the U.S. Energy Information Administration (EIA) about the number of energy sites and when they opened, as well as the number of units at those sites, closed sites and units, and total generation capacity. Ratios of numbers of units and sites were compared pre-1990 (older sites) and 1991–2013 (newer sites) (Table 15.1).

The first units were built in the 1920s and many more were built immediately after the Second World War and during the 1950s. The average site has more than 3.5 electrical generating units, with a range of 4.15 in New Jersey to 2.86 in New Mexico. In 2013, there is evidence of more clamping in Arizona, Indiana, and Texas and slightly less in New Jersey and New Mexico. In short, the idea of clustering multiple units at a single site is not new and clamping new units at existing sites has been occurring at many sites.

The results in Table 15.1, however, are slightly misleading. Table 15.2 shows the aggregate results for the five states divided into natural gas and all other non–renewable sources. Companies

Table 15.1 Number of units and sites for electrical power generation in selected states, 1990 and 2013

State	1990 Sites	1990 Units	1990 Units/sites	2013 Sites	2013 Units	2013 Units/sites
Arizona	19	76	4.00	34	160	4.71
Indiana	33	123	3.72	46	178	3.87
New Jersey	33	137	4.15	51	202	3.96
New Mexico	14	40	2.86	22	57	2.59
Texas	95	316	3.33	134	534	3.99
Total, 5 states	194	692	3.57	287	1131	3.94

Source: U.S. EIA-860 detailed data (www.eia.gov/electricity/data/eia860/)

Table 15.2 Comparison of natural gas and other sources, new units added 1991–2013

State	Natural gas Added to Total	pre-1990	1991–2013	% pre-1990	Other Sources (coal, oil, nuclear) Added to Total	pre-1990	1991–2013	% pre-1990
Five States	502	88	414	18	90	71	19	79
New Jersey	59	10	49	17	36	29	7	81
Texas	271	54	217	20	30	20	10	67

Source: U.S. EIA-860 detailed data (www.eia.gov/electricity/data/eia860/); calculations by the authors

added relatively few coal, nuclear, and oil units, and 79 percent of those were added at sites opened prior to 1991. In contrast, over 500 new natural gas units were added, and over 80 percent were located at new sites. Clustering is occurring at these new natural gas sites, but most of them are new sites.

New Jersey is a good example of the changeover currently occurring in some fossil fuel poor areas of the United States. Until about 1970, utilities generated more than two-thirds of the New Jersey electrical energy from coal. This proportion has steadily dropped. New Jersey imports electrical energy, and the split between nuclear and natural gas is about 50:50. O'Neill (2014) states that this makes New Jersey a leader in reducing greenhouse gas emissions. Along with four nuclear power plants, it has added gas-fired or combined gas-coal facilities, along with hydroelectric, wind farms, photovoltaic, solar, and trash to steam plants.

Table 15.2 also shows, however, that a similar location pattern has been followed in Texas where oil and gas supplies are plentiful and yet like New Jersey new sites with multiple units have been the pattern for the gas sites, whereas clamping has been the choices for other sources.

Research question 2

There is no denying that finding a solution to storing commercial nuclear fuel rods and other high-level hazardous nuclear waste is a dilemma, indeed an embarrassment. In 1982, the United States government identified Yucca Mountain (see Figure 15.1), located about 90 miles north of Las Vegas, Nevada, as the site for a permanent repository for used nuclear fuel, which is fuel that has been removed from a nuclear reactor because it can no longer sustain power production for economic or other reasons (Macfarlane and Ewing 2006). It is not clear whether that site will become politically viable (U.S. GAO 2011).

The default policy has been to leave used nuclear fuel where it was used, i.e., at more than 60 locations where 100 operating and 16 closed nuclear power plants are located. To transport the used fuel from the origin plant, it must be converted from liquid to a solid form so as to be placed in transport/storage casks.

On March 11, 2011 multiple nuclear plant failures at the Fukushima-Daiichi facilities in Japan alerted people to the reality that so-called "spent" nuclear fuel can be extremely hazardous. For example, prior to Fukushima, a survey found that over 90 percent of U.S. residents did not know what happens to used commercial nuclear fuel. Indeed, most did not even provide an answer and 20 percent—one out of five—incorrectly believed it was buried at Yucca Mountain (Greenberg and Truelove 2010).

While Yucca remains a plausible option for storing used nuclear fuel, and the default is to continue to store nuclear waste at the power plant that generated the waste, energy policy managers must consider other plausible options. Two viable options are (1) finding new locations and constructing new facilities; and (2) using one of the U.S. Department of Energy's (DOE) existing defense waste sites. While both options were surveyed for public input, there is a set of reasons compelling the use of existing nuclear defense waste sites: (a) there is existing on-site technical expertise in managing nuclear waste; (b) the land is already owned by the DOE; (c) the sites are large and relatively remote from large urban centers; and (d) some of the local population works at sites, and/or has friends or relatives that do, and consequently may at least be willing to consider expanded nuclear activity in their region. This would be a waste management illustration of CLAMP.

While the idea of CLAMP requires consideration of public health, ecological, political, social, ethical, economic, and other implications, public willingness to accept or at least tolerate such a facility is critical. Accordingly, to explore the public's preferences and perceptions on this

Figure 15.1 Location of selected US DOE nuclear waste management sites

issue we conducted a random-digit-dial survey and sampled the general, non-institutionalized, adult population living within a 50-mile radius of four major DOE defense sites, which are all CLAMP sites. They were asked two questions:

1 What proportion of the public living in a CLAMP region would favor hosting the storage and management of used nuclear fuel at the DOE defense site in their region?
2 What attributes are associated with either support, or at least a willingness to consider, a policy that would move used nuclear fuel in casks to the DOE defense site in their local region?

Factors associated with considering hosting the storage of used nuclear fuel

We expected widespread opposition to the notion of even "temporarily" hosting and managing used nuclear fuel in the respondents' regions (Greenberg et al. 2012; Macfarlane and Ewing 2006; Portney 1991; O'Hare et al. 1983; Kemp 1990; Gerrard 1994). Yet, characterizing a nuclear waste management facility as impossible to site is simply incorrect; after all, sites have, indeed, been found for all of these types of LULUs, including nuclear waste management facilities and suggestions for government programs to test these options have been offered (Ewing 2013; Macfarlane and Ewing 2006; Inhaber 1998; National Research Council 2003).

We need to know how many people residing in selected DOE CLAMP regions would support the idea of hosting such a storage facility or at least be neutral about it. For example, the Waste Isolation Pilot Plant (WIPP) is a nuclear waste management facility sited near Carlsbad, New Mexico, which opened in 1999 to manage transuranic wastes (Figure 15.1). By building a relationship with state officials and regulators, WIPP leadership has built a cadre of government officials and public supporters (Jenkins-Smith et al. 2011). In addition, we looked at the Savannah River, Oak Ridge, and Hanford sites – all large sites with major nuclear defense waste missions at the current time and for the foreseeable future.

The environmental psychology, risk analysis, and social science literatures offer strong clues about the attributes of those who would favor moving the waste to caskets as soon as feasible and be willing to at least consider hosting the site. Briefly, the authors divided the factors into three categories: (1) location and demographic characteristics; (2) trust of DOE and its contractors; and (3) preferences, perceptions and interest in the site.

Regarding location, previous studies show that those who reside in a host county are more likely to be familiar with the site, may work at the site, or have friends, family and others who do, and directly or indirectly benefit from it economically. Because, on average, they should be less concerned about the risk and more motivated by jobs and other economic benefits (Greenberg 2009) residents proximate to an existing facility should be among those willing to consider a new role for their DOE site. With regard to demographics, the attributes of likely supporters are those associated with the so-called "white male effect" (Finucane et al. 2000; Flynn et al. 1994). As such, those who are affluent, have more formal education, are male, and self-identify as White have been shown to be less concerned about risk and more interested in economic benefits. In turn, they should be willing to consider the expansion of the role of the DOE site in their region.

We were less certain about age and political party identification. The broader context question is what political affiliations and age groups would be likely to be more open to new ideas and emerging technologies such as nanotechnology, genetically modified organisms, use of embryonic stem cells, as well as new technologies to manage and store used commercial nuclear fuel. The literature suggests that a cadre of educated, mobile, and politically liberal or independent people is settling into small to medium-sized urban areas. The profile of that group suggests that they would disproportionately be part of the group supporting or at least willing to consider the two policy options (Fischer et al. 2013; Brooks, 2000; Florida 2002; Robb et al. 1997; Scheufele and Lewenstein 2005; Rudzitis and Streatfeild 1993). Hence, while not certain, our expectation was that supporters of the policies would be independents, Democrats, and the relatively young.

Strengthening this expectation is the fact that three of the four sites in this study are in the West: Boise (ID), Albuquerque (NM) and Seattle (WA) are in Richard Florida's (2002) list of top 10 small, medium-sized, and large cities in the United States that are attracting many well educated and creative young people. These cities are in states that host the Idaho, WIPP, and Hanford DOE defense waste sites, respectively, although they are 220, 300, and 225 miles from those three highly rated cities. Would young and educated members of Richard Florida's creative class of highly educated and mobile people with the ability to change an economy be willing to consider locating a facility to manage used commercial nuclear fuel in their states? We thought that they disproportionately would at least consider the idea.

Trust was the basis for a second set of explanations. We expected that anyone willing to support or consider moving used commercial fuel to the DOE site in their region would exhibit competence-related trust, i.e., that they would trust DOE and its contractors to safely manage new technologies at the sites. Moreover, we anticipate they would also exhibit communications-related trust, by reporting trust in DOE and its contractors to communicate honestly with the nearby communities (Love et al. 2013; Poortinga and Pidgeon 2003).

The third set of considerations focus on respondents' attitudes toward the future of the site, as well as their motivations to be involved in site-related activities. Those who have a vision of the site as a place to explore new science, including nuclear, and who are motivated by local economic development and less worried about contamination, should be more willing to consider the idea of a new type of facility (Greenberg 2012). While, to be sure, preferences are a signal of interest, an even stronger signal is the active desire for information. If a respondent is interested enough to want information from the DOE about new site activities through newspapers, television, radio, as well as electronic media and social network sources (Williams and Gulati 2007), they are more likely to be interested to the point of involvement in discussions of the management, safety, and security of the DOE site. This last expectation ties back to the group of creative, highly educated, and mobile individuals who are motivated to secure information from multiple sources.

Data and methods

Survey interviews were gathered from respondents who live within 50 miles of the four sites, a compromise between relying too heavily on areas immediately adjacent to the site (e.g., within 15–20 miles) or too far (100+ miles), both can bias the results (Greenberg 2012). The survey recorded county of residence, which researchers used to represent a host county effect, and it directly asked respondents about their familiarity with the local site.

Survey design

The survey used a dual-frame sample design, with landline (75 percent) and cell phones (25 percent) proportions, and industry-standard random digit dialing (RDD) protocols. The sample size target was 920 (i.e., 230 completed survey interviews at each of the four sites), drawn from the population of non-institutionalized persons 18 years and over living within 50 miles of the above four DOE sites. The margin of sampling error for the full sample of 920 is plus-or-minus 3.2 percent (with 95 percent confidence at 50 percent/50 percent answer proportions); and it was 6.4 percent for the four sites, where the sample size was 230.

The survey followed an eight-call design, meaning that the company attempted to call each number up to eight times in order to increase the likelihood of response. To reduce bias, interviewers made phone calls at different times of the day and days of the week, including weekends and administered the survey in English with a Spanish-language option. Interviewers recorded and catalogued refusal responses. In order to prevent question order bias, the questions were randomized. Samples never perfectly mirror the population from which they are drawn because some sub-populations are less likely to respond. Accordingly, the researchers examined the preliminary sample and then, using post-stratification weighting for age and race, was adjusted for differences between the sample and population of the study regions as described below under results.

Questions

In two separate questions, respondents were asked if they favored moving the waste to concrete casks as soon as feasible and then they were asked to choose one of four preferences for location. The first two options involved transferring "all used fuel in casks to three or four locations in the United States that. . . ." (1) ". . . already manage and have U.S. defense-related nuclear waste," or, (2) ". . . would be developed as used fuel storage facilities." The second two options involved moving "all used fuel in concrete casks to . . ." (3) ". . . a single deep underground repository at Yucca Mountain, Nevada that was partly built but not completed," or (4) ". . . a single deep underground repository location that has not yet been identified."

We then we asked how respondents "feel about transferring used nuclear fuel in casks from other sites to [the site in their region (identified by name)] for temporary storage and management?" The choices were: (1) strongly favor that idea; (2) somewhat favor that idea; (3) neither favor nor oppose the idea; (4) somewhat oppose that idea; and (5) strongly oppose that idea; and "don't know" and refusal responses were noted.

Multiple questions were asked about demographic attributes, location, trust, perceptions, preferences, and personal interest in the site, familiarity with their local site, and concern about environmental issues (see Table 15.3). The final set of eight questions focused on perceptions, preferences, and interest in the site. Here we asked respondents about their preferred information sources, i.e., to indicate how they would like to learn about DOE's efforts to improve sustainability at its local site and reduce its on-site resource use. We deliberately chose sustainability as the exemplar for the information sources probes because it is a relatively new program that we believe would appeal to those concerned about the future of the site. Seven options to receive information were provided: local newspaper, local radio, local television, DOE web site, local county fair, community meetings, and an "other" option, to be specified by the respondent. We operated under the assumption that anyone who chose multiple options was more interested than a respondent who selected only one or no information sources. On that basis, we assumed that a multiple-information-source respondent would be extremely interested. Along the same lines, any respondent providing additional information sources above and beyond the main list, would be interested and motivated to explore new uses for the site in their region.

Results

The survey was pilot-tested on June 24, 2013. After a few adjustments, data were collected during the period June 26, 2013 to August 30, 2013. A total of 922 responses were collected: 231 each at INL and Hanford and 230 each at SRS and WIPP. The response rate was 26.1 percent for the full sample; the co-operation rate was 44.3 percent.

Table 15.3 shows the distribution of respondents on the basis of willingness to host the waste in their site region. Strong opposition to hosting the facility in their site-region was (21 percent) and somewhat opposed to hosting the facility was (16 percent). Adding in the almost 8 percent of the respondents who did not answer the question as opponents to hosting the facility, meant that 82 percent of the sample (including those who did not want the waste in casks as soon as practically achievable) would not consider hosting used fuel storage in their site-region at this time, while 18 percent (i.e., the aggregate of "strongly favor," "somewhat favor," and "neither favor nor oppose") would at least consider "getting to maybe." By site, the most positive aggregate response was from WIPP respondents (24 percent) and the least was from Idaho (14 percent).

Correlates of the willingness to consider the option of storing the waste

Table 15.4 presents three models that test the associations described earlier. The table displays the B-values of the binary logistic regression and the odds ratios. In this presentation, we emphasize the odds ratios, which are relatively easy to interpret. Generally speaking, coefficients greater than 1 indicate more support for the policy and those less than 1 indicates less support. An odds ratio of 2.0, for instance, means that the odds of someone with that characteristic are twice the

Table 15.3 Willingness to host spent fuel storage in their site-region, 2013

Option (n=922)	%
Strongly favor	4.1
Somewhat favor	10.4
Neither favor nor oppose	3.9
Somewhat opposed	16.1
Strongly opposed	21.4
Maintain current policy not to move waste	36.4
Don't know or refused	7.7
Total	100.0
Strongly favor, somewhat favor, or neutral to hosting in their region	
All sites	18.4
Hanford	19.1
Idaho	14.3
Savannah River	15.2
WIPP	23.9*

*WIPP proportion significantly higher than Idaho and Savannah River at p<.05.

odds as someone without that characteristic to report a particular outcome.

Assuming we only knew basic demographic and location information about the respondents, Model 1 shows that 6 of the 10 personal attributes were statistically significant predictors (p<.05) of the willingness to consider storing and managing used nuclear fuel from other sites at the DOE site in their region. The strongest were being a resident of the WIPP region (OR=2.12), self-identifying as a Democrat (OR=2.03), and being 18 to 49 years old (OR=1.84). These were followed by self-identification as a political independent (OR=1.63), having an annual family income of $75,000+ (OR=1.49), and being very familiar with the DOE site (OR=1.36). All of these findings were anticipated.

Model 2 added the six trust metrics, expecting that supporters would have to trust the DOE and contractors to manage any new facilities. The results agree in part with those expectations. Only two of the six trust question results were statistically significant correlates at p<.05. One was that respondents trust the DOE to manage new nuclear-related activities (OR=1.39). The other two trust-related DOE relationships, while not statistically significant, were in the expected direction. In comparison, all the contractor relationships were negative, that is, respondents tended not to trust the contractors. This, indeed, was a surprising finding that merits follow-up.

Model 3 added eight indicators of perception, preference, and site-related interest. Supporters of hosting used nuclear fuel at the DOE site in their region did not typically declare as strong environmental protection supporters, nor were they very concerned about environmental threats posed by their local DOE site, nor did they place a high priority on continuously monitoring the air and water at the site. They did, however, support the idea of working with local government to improve the local economy. We had assumed these relationships would have been stronger, but none of these four turned up as significant correlates.

The four significant predictor variables were, first, lack of support for excluding new nuclear-related activity on the site (OR=0.87), and making sure that the site is owned by the federal government until all hazards have been eliminated (OR=1.18). The two most interesting results in model 3, however, are that those willing to consider the hosting option wanted to

Table 15.4 Attributes of respondents who support or are neutral to hosting used nuclear fuel in their region

Indicator	Model 1		Model 2		Model 3	
	B-value	OR	B-value	OR	B-value	OR
Exogenous						
4-year college	.186	1.21	.227	1.26	.261	1.30
Annual family income, $75,000+	.400*	1.49	.360	1.43	.106	1.11
Male	.126	1.13	.139	1.15	-.001	1.00
Self-identified White	-.149	0.86	-.139	0.87	-.246	0.78
Age 18–49	.610**	1.84	.569**	1.77	.542*	1.72
Live host county	.072	1.08	.121	1.13	.152	1.16
Very familiar with the site	.308*	1.36	.270*	1.31	.105	1.11
Identify with Democrat Party	.705**	2.03	.692**	2.00	.776**	2.17
Identify as Independent	.489*	1.63	.572**	1.77	.602**	1.83
WIPP region resident	.752**	2.12	.833**	2.30	.935**	2.55
Trust						
DOE will make sure underground radioactive and chemical materials will not pollute outside the site boundaries (1=strongly disagree, . . . 5=strongly agree)			.016	1.02	-.065	0.94
Site contractors will make sure underground radioactive and chemical materials will not pollute outside the site boundaries (1=strongly disagree, . . . 5=strongly agree)			-.044	0.96	-.027	0.97
DOE communicates honestly with people in the area(1=strongly disagree, . . . 5=strongly agree)			.096	1.10	.052	1.05
Site contractors communicate honestly with people in the area (1=strongly disagree, . . . 5=strongly agree)			-.181*	0.83	-.138	0.87

	B	Exp(B)	B	Exp(B)	B	Exp(B)
DOE will effectively manage any new nuclear-related activity (1=strongly disagree, . . . 5=strongly agree)			.327**	1.39	.351**	1.42
Site contractors will effectively manage any new nuclear-related activity (1=strongly disagree, . . . 5=strongly agree)			−.003	.099	−.084	0.92
Site-related perceptions, preferences, interest						
Self-identify as active supporter of efforts to preserve and protect the environment					−.042	0.96
Number of ways they want information about DOE's sustainability programs (0–7)					.114*	1.12
Offered suggestion about way of receiving information from DOE about sustainability programs#					2.230**	9.30
Concern about local DOE site (1=not at all worried, . . . 10=extremely worried)					−.038	0.96
High priority to continuously sample the quality of the air and water at the site (1=very low priority, . . . 10=very high priority)					−.031	0.97
High priority to working with local officials on economic development of the area by improving transportation and securing jobs (1=very low priority, . . . 10=very high priority)					.007	1.01
High priority to not allowing any new nuclear-related activity on the site (1=very low priority, . . . 10=very high priority)					−.140**	0.87
High priority to make sure that the site remains owned by the federal government until all the hazards are removed (1=very low priority, . . . 10=very high priority)					.164**	1.18
Constant	−3.08		−3.90		−3.79	
Model n	878		878		805	
Nagelkerke-pseudo r2	.082		.111		.226	

*Significant at p<.05
**Significant at p<.01
#B is the estimated logit coefficient; Exp(B) is the odds ratio of the coefficient.

be to be informed in multiple different ways about DOE's on-site efforts to reduce resource use (OR=1.12), and offered suggestions on how to provide information as one of their first three responses (OR=9.30). That 9.3 odds ratio means that the odds that a respondent offering at least one information source suggestion will support the hosting option are 9.3 times the odds that a respondent not offering at least one suggestion will support the hosting option. Suggestions offered by these respondents at least twice were that DOE send e-bulletins, post data on the web, use Facebook, Twitter and other media, provide data at local libraries, lecture at schools and offer site tours. It is a very telling fact that 71 percent of these respondents were 18–40 years old compared to 39 percent of the sample population.

Discussion

Before discussing the key findings from these two studies, several caveats are in order. Question 1 results include only five states, and the next step is to repeat the analysis for all 50, and add renewable sources to the data. The major limitation of the waste management study is that the data were gathered prior to two events at the WIPP site that caused WIPP's shutdown and may have undermined the credibility of the DOE not only to consider the WIPP site for expansions to non-defense nuclear waste but also for defense waste. Clearly follow-up surveys are required, even while repairs occur at the site.

With these limits noted, the CLAMP policy has existed for many decades, and even now seems stronger with regard to coal and nuclear sources. Companies have added many natural gas units at new sites and have also added or replaced units fueled by other energy sources at existing sites.

With regard to relocating commercial high-level nuclear waste at one of the existing major DOE sites, realistically these data, even before the WIPP event, suggest that developers have steep political and public perception mountains to climb. 18 percent were supportive or at least neutral with regard to the notion of hosting a used nuclear fuel facility at the existing site. We can interpret these results as a signal that even in regions that have and benefit from nuclear waste, the strongest correlates of being supportive or at least neutral to the hosting option are clearly in the minority.

The 18 percent who are least willing to consider the idea are disproportionately young, college-educated, seem to trust DOE (but not its contractors) to manage new projects, and are interested enough to ask for much more information. Notably, 43 percent said that they would be interested enough in DOE's environmental management efforts to "serve on a community committee," or "like the idea but need more information," compared to 35 percent of all other respondents. Could members of this group be persuaded to work together with DOE to evaluate the factors that would make such a site more acceptable for the hosting option? Could members of this group "get to maybe" on this critical nuclear policy issue? We think such an effort is worth pursuing.

Toward that end, we suggest pursuing the following three sets of issues. While, on the evidence we currently have, the group does not seem to be unusually motivated by economic use of the site, it is important to determine what the larger public thinks about the economic benefits of such a program at their sites. Along with economic benefits, other research has shown residents of these areas to possess a higher degree of patriotism than others (Greenberg 2012), and it would be good to know if patriotic feelings are part of their interest in the proposed policy.

Transportation and approaches to on-site storage constitute a second pair of issues to pursue with this group. Research has shown that many members of the public are more worried and interested in the off-site transportation-related risks than on-site waste management (Greenberg

2009, 2012). How important are these issues for the public? Also, how important is below-ground versus above-ground storage and stored-waste-irretrievability (Aparicio 2010; Nuclear Energy Agency 2012; National Research Council, 2003)?

Finally, a key policy consideration to which this group could have meaningful input is the development of an organizational structure to inform and involve the public and manage facility operations. The Blue Ribbon Commission (2012) suggested several alternatives to current management structure. The interested group is likely to be motivated to help DOE determine the importance and priority of organizational issues as well as exploring viable alternatives.

Our primary goal was not to determine if the United States is better off with more nuclear power plants. This paper has focused on one key element of the nuclear power issue: long-term storage and maintenance of spent nuclear fuel, which is increasingly seen as a major constraint. Thanks to the rapid development of natural gas facilities and the ability to cluster them, as well as the use of clamping for other energy facilities, the United States has more time to ponder a politically viable solution to the nuclear waste management issue.

Acknowledgments

This research was prepared with the support of the U.S. Department of Energy, under Cooperative Agreement Number DE-FC01–06EW07053 entitled The Consortium for Risk Evaluation with Stakeholder Participation III awarded to Vanderbilt University. The authors thank David Kosson and Charles Powers, who have been Co-PIs of the CRESP project for many years for their support of this ongoing work. The opinions, findings, conclusions, or recommendations expressed herein are those of the author and do not necessarily represent the views of the Department of Energy or Vanderbilt University, or any of the people acknowledged. This report was prepared as an account of work sponsored by an Agency of the United States Government. Neither the United States Government nor any agency thereof, nor any of their employees, makes any warranty, express or implied, or assumes any legal liability or responsibility for the accuracy, completeness, or usefulness of any information, apparatus, product, or process disclosed, or represents that its use would not infringe privately owned rights.

References

Aparicio, Luis, ed. *Making Nuclear Waste Governable: Deep Underground Disposal and the Challenge of Irreversibility*. New York: Springer, 2010.

Blue Ribbon Commission on America's Nuclear Future. *Final Report to the Secretary of Energy*. http://cybercemetery.unt.edu/archive/brc/20120620121/http:/broc.gov/. Accessed February 3, 2012.

Brooks, David. *Bobos in Paradise: The New Upper Class and How They Got There*. New York: Simon and Schuster, 2000.

Ewing, Rodney. "Letter to the Honorable Rodney F. Frelinghuysen, Chairman, Subcommittee of Energy and Water Development." Committee on Appropriations, U.S. House of Representatives. September 16, 2013.

Finucane, Melissa, Paul Slovic, CK Mertz, James Flynn, and Theresa Satterfield. "Gender, Race, and Perceived Risk: the 'White Male' Effect." *Health, Risk & Society*. 2(2000), 159–172.

Fischer, Arnout, Heleen Van Dijk, Janneke de Jonge, Gene Rowe, and Lynn Frewer. 2013. "Attitudes and Attitudinal Ambivalence Change Toward Nanotechnology Applied to Food Production." *Public Understanding of Science*. 22(2013), 817–831.

Florida, Richard. "The Rise of the Creative Class." *Washington Monthly*, May 2002. www.washington monthly.com/features/2001/0205.florida.html. Accessed October 18, 2013.

Flynn, James, Paul Slovic, and CK, Mertz. "Gender, Race, and Perception of Environmental Health Risks." *Risk Analysis*. 14 (1994), 1101–1108.

Gerrard, Michael. *Whose Backyard, Whose Risk*. Cambridge, MA: MIT Press, 1994.

Greenberg, Michael. "NIMBY, CLAMP and the Location of New Nuclear-Related Facilities: U.S. National and Eleven Site-Specific Surveys." *Risk Analysis*. 29(2009), 1242–1254.

Greenberg, Michael. *Nuclear Waste Management, Nuclear Power and Energy Choices: Public Preferences, Perceptions, and Trust*. New York: Springer, 2012.

Greenberg, Michael and Heather Truelove. "Right Answers and Right-Wrong Answers: Sources of Information Influencing Knowledge of Nuclear-Related Information," *Socioeconomic Planning Sciences*. 44(2010), 130–140.

Greenberg, Michael, Frank Popper, and Heather Truelove. "Are LULUs Still Enduringly Objectionable?" *Journal of Environmental Planning and Management*. 55(2012), 713–731.

Inhaber, Herbert. *Slaying the NIMBY Dragon*. New Brunswick, NJ: Transaction, 1998.

Jenkins-Smith, Hank, Carol Silva, Matthew Nowlin, and Grant deLozier. "Reversing Nuclear Opposition: Evolving Public Acceptance of a Permanent Nuclear Waste Disposal Facility." *Risk Analysis*. 31(2011), 629–644.

Kemp, Ray. "Why Not in My Backyard? A Radical Interpretation of Public Opposition to the Deep Disposal of Radioactive Waste in the United Kingdom." *Environment and Planning A*. 22(1990), 1239–1258.

Love, Brad, Michael Mackert, and Kami Silk. "Consumer Trust in Information Sources: Testing an Interdisciplinary Model." *SAGE* open, April–June, 2013, 1–13. http://sgo.sagepub.com/content/3/2/2158244013492782. Accessed October 16, 2013.

Macfarlane, Allison and Rodney Ewing (eds). *Uncertainty Underground: Yucca Mountain and the Nations High-Level Nuclear Waste*. Cambridge, MA: MIT Press, 2006.

National Research Council. *One Step at a Time: The Staged Development of Geologic Repositories for High Level Radioactive Waste*. Washington, DC: National Academy Press, 2003.

Nuclear Energy Agency. *Reversibility and Retrievability in Planning for a Geological Disposal of Radioactive Waste*. Paris, NEA 6993. 2012 www.oecd-nea.org/rwm/docs/2012/6993. Accessed October 12, 2013.

O'Hare, Michael, Lawrence Bacow, and Debra Sanderson. *Facility Siting and Public Opposition*. New York: Van Nostrand and Reinhold, 1983.

O'Neill, James. "New Jersey in Good Shape to Meet Obama's New Climate-Change Plan." 2014. NorthJersey.com. www.northjersey/com/news/new-jersey-in-good-shape-to-meet-obama-s-new-climate-change-plan-1.1026726?page=all. Accessed August 12, 2015.

Poortinga, Wouter and Nicholas Pidgeon. "Exploring the Dimensionality of Trust in Risk Regulation." *Risk Analysis*. 23(2003), 961–972.

Portney, Kent. *Siting Hazardous Waste Treatment Facilities: The NIMBY Syndrome*. New York: Auburn House, 1991.

Robb, James, William Riebsame, and Hannah Gosnell, eds. *Atlas of the New West: Portrait of a Changing Region*. New York: WW Norton, 1997.

Rudzitis, Gundars and Rosemary Streatfeild. "The Importance of Amenities and Attitudes: A Washington Example." *Journal of Environmental Systems*. 22(1993): 269–277.

Scheufele, Dietram and Bruce Lewenstein. "The Public and Nanotechnology. How Citizens Made Sense of Emerging Technologies." *Journal of Nanotechnology Research*. 7(2005): 659–667.

U.S. GAO. *Commercial Nuclear Waste: Effects of a Termination of the Yucca Mountain Repository Program and Lessons Learned*. GAO-11–229. Washington, DC: GAO, 2011.

Williams, Christine and Jeff Gulati. "Social Networks in Political Campaigns: Facebook and the 2006 Midterm Elections." Paper presented at the 2007 Annual meeting of the American Political Science Association, www.academia.edu/281042/Social_Networks_IN_Politcial_Campaigns. Accessed November 5, 2013.

Part 3

Relational perspectives and the politics of energy

Introduction

Vanesa Castán Broto

Energy infrastructures, for lighting, heating, powering our economies and communications are an essential part of everyday life. Most often, energy is understood from the supply side, especially in relation to the specific fuels needed for different purposes, as well as the various networks that conduct them. Policy makers often talk of 'the industry' as the main actors concerned with energy, as if energy was only a matter for big oil producers, utility providers, and, in the best case, small and medium enterprises trying to develop renewable alternatives. Consumers are at the receiving end of the energy supply chain. Consumers are often imagined as simplified characters, who can be simply characterized by an average consumption of electricity or fuel. Consumers are also people who use energy in different ways according to their daily patterns of action, in houses that interact with their surrounding environments and atmospheres.

Geographers question the models that engineers, planners, and economists use to make sense of a complex arena of energy production and consumption. Dominant models of thinking have developed alongside the material arrangements that constitute contemporary energy systems. Places where narratives of energy have emerged alongside networked systems of electricity provision, such as centralized energy networks, these modes of thinking have been adjusted enough to provide appropriate explanations of why these systems work. However, for the majority of the world, these are gross simplifications of how we meet our energy needs. For example, almost one billion people live in informal settlements in urban or peri-urban areas, with precarious forms of infrastructure provision where fancy notions of green energy and consumers pale in comparison with the creative ways in which energy is accessed. The geography of energy draws attention to the variegated means in which the actual infrastructure arrangements in different geographical locations – particularly those locations unknown to the experts that produce dominant forms of knowledge – defy technocratic models of service delivery and modernist utopias. Geographical analyses of energy have also challenged the notion of consumer, and with that, challenged that energy itself is a necessarily commoditized product characterized by exchange from producers to consumers. While a production–consumption model of energy use may support established, hegemonic, systems of provision, we live in a moment of transition. This means that the global energy challenge is not simply one of adjusting efficiency settings, and synchronizing production constraints with consumers' demands. Rather, the global energy challenge requires reexamining and fundamentally reconfiguring our relations with energy. These goes to the fundamental role that energy plays in our lives, but also to how we configure its role through social interactions – a line of questioning that reveals the social construction of energy (Nadaï and Van Der Horst 2010).

Energy geographies, in alliance with the social sciences, have shown that energy questions cannot be confined to the realm of engineers, planners, geologists, and activists chains (Shaw and Ozaki 2013). Energy geographers, for example, have turned attention to the different practices that create energy needs, the cultures of energy use, and the extent to which there are actions that would reduce the demand, for example, in the relationship between energy practices and the built environment (Ozaki and Shaw 2013, Shove 1998, Lovell 2005). This has emerged as a strong critique of behaviouralism: the idea that an energy transition simply depends on rational consumers learning to adjust their thermostat. These approaches have generated a fertile ground within geography which has helped to understand how cultures shape our expectations about the use of energy (Shove and Walker 2014). Moreover, geographical analyses of practice question particular understandings of what is normal in specific locations from Hong Kong to the UK (e.g. Walker et al. 2015).

The interest in developing alternative perspectives on energy has fostered the development of specific theoretical approaches that reimagine the relationship between energy, things and society. The development of relational approaches to energy has enabled a greater examination of socio-natural relations. Relational approaches transcend analytical dichotomies between humans and things by focusing instead in what connects different aspects of energy system and the relative positions of those things that are connected (Harrison and Popke 2011). A relational perspective to explain and understand human action – from everyday practices to attempts at governing socio-technical systems – can contribute analytical tools to describe the multiplicity of strategic projects that are laid down in energy landscapes. Assemblage theory, for example, focuses on how people and objects find themselves in specific settings in relation to arrangements of material and semiotic things (Rutherford and Coutard 2014). Both people and objects may intervene in the formation of new assemblages, with different degrees of success in the adaption of such assemblages to their specific visions or strategies. In doing so, relational perspectives also raise the question of the political, particularly in relation to how knowledge and experience shape political possibilities in which different futures – perhaps sustainable, perhaps more just – may become possible.

Often the critique against this work is that it is not possible to draw a connection between its insights and its possible implications for energy policy. Nothing could be farther from the truth. These approaches are already generating important new insights, although they often make for uncomfortable reading in economic-oriented or industrial settings. For example, in development studies, the topic of energy has generated discussions about measuring energy poverty in a way that is meaningful to energy users (SEforAll, 2015; Sagar 2005). The complex entanglement of energy with environmental justice debates reveals that issues of energy access and energy poverty are intertwined with overall notions of society's good life (Walker et al. 2013). Relational perspectives have generated new conceptualisations of energy concepts such as energy vulnerability and how energy is implicated in the production of inequality (Bouzarovski and Petrova 2015). By adopting different perspectives on the geographies of publics, communities, and the social creation of energy demands, the cases in this third section engage with the everyday politics of energy.

Themes of Part 3

The collection of chapters in Part 3 includes contributions that challenge hegemonic understandings of energy and engage with alternative politics that could be deployed to address global energy challenges. While not every contribution focuses on delivering new theoretical under-standings of society-energy relations, they all engage with new ways of looking at energy uses in context, and with how we configure new forms of demand in relation to specific material and cultural regimes. Some of these contributions also engage with the possibility of change towards sustainable energy systems.

These are crucial questions particularly in the context of climate change and the ever-pressing need to reduce our carbon emissions. As most carbon emissions are linked to energy production and consumption, much of the effort to achieve a low carbon society are related to interventions in our energy system, from new renewable energy infrastructures to retrofitting programmes and education campaigns to reduce current energy consumption. The range of geographical locations presented in this section serves to explore energy challenges in relation to different cultural and material situations, thereby forcing the development or even reinvention of theoretical tools and methods of analysis.

Many of these papers engage with how energy is perceived and experienced by those who use it, engaging directly with the redefinition of energy uses as a question of demand. Butler and Parkhill focus on the question of energy demand not as behavior change, but rather as a construct which requires active governance. In doing so they present energy demand as a matter for public policy. This contrasts with Fuller's chapter, which puts communities at the centre of the demand arguments, as active agents of change. Sarah Fuller explores, specifically, the extent to which communities can become active agents of transitions to sustainability using a concrete example, Brixton Energy, which shows how energy transitions embed in the community. Read together, these chapters challenge the notion of agency, and how far people can shape their ability to produce or access energy in their everyday life. In relation to Part 2, there are emerging questions about who are the 'communities' that perceive energy-related land transformations as a threat, and those who are actively looking for alternatives to energy consumption.

Jiusto and McCauley's study of energy and place making in informal settlements in South Africa constitutes a challenge to conventional understanding of how energy should be provided to facilitate energy access. They explain how the provision of energy services coevolves with the configuration of informal settlements. The chapter makes a case for pedagogies for engage-ment with energy geographies that facilitate knowledge co-production with the inhabitants of such informal settlements. For Silver, who explores the provision of energy in Accra, Ghana, the question of energy access fosters new understandings of how infrastructure and service provision are implicated in localized forms of incremental urbanism. Both analyses provide a foundation for understanding how energy services are implicated in creating particular urban configurations. Methodologically, they point to cities as relatively uncharted areas for energy geographies.

While these analyses push energy themes beyond the practice and policy implications of a particular context – trying to understand how to reduce energy consumption or how to facilitate energy access – they challenge contemporary ways of imagining energy. This is manifest in Nadaï and Labèussire's study of the socially constructed nature of wind energy. By looking at recent developments in France, they question the notion of renewability, as the sustainability of putatively renewable technologies often relates to the way they are inserted in particular systems of capital and material circulation.

Nowhere is this more evident than in those chapters in which the specific circumstances of energy provision in a particular place become linked to the wider processes of political transformation at the national level. Energy may be the object of these politics, as it is the case of the Petrocasas in Venezuela (Delgado), where the redirection of capital from fossil fuel exploitation to social housing programmes creates a conflictive energy landscape. Equally, the energy transitions that the recent discovery of gas deposits is fueling in Mozambique shows how multiple actors are intervening in an incomplete energy transition, connecting institutions and businesses, rather than people and households (Kirshner). In Macedonia, as Petrova explains, such political project is a neoliberal one: one that does not understand histories or place identities, and which reconfigures identities. Because of the use of fuelwood for heating, energy becomes a way of establishing a new politics. These cases show how energy projects follow broader ideological projects which are materialized locally with crucial consequences for the lives of those who are supposed to benefit.

In every case, practices around energy appear intrinsically linked to power and politics, whether this is the politics of everyday life, through which different forms of discursive and material authority are configured, or to broader forms of political change and geopolitics which are not only manifest on the geographies of resource extraction and land transformation (see Parts 1 and 2), but also in the availability of certain energy services within specific contexts. The focus on the mundane, everyday practices, and the ways demand is configured speak to the very notion of the politics of particularity at the centre of the actual possibilities to bring about energy transitions.

References

Bouzarovski, S. and Petrova, S. 2015. A global perspective on domestic energy deprivation: overcoming the energy poverty–fuel poverty binary. *Energy Research & Social Science*, 10, 31–40.

Cecelski, E., Policy, A. A. E. and Unit, A. A. E. 2000. Enabling equitable access to rural electrification: current thinking and major activities in energy, poverty and gender. *World Development Report*, 1, 2.3.

Harrison, C. and Popke, J. 2011. "Because you got to have heat": The networked assemblage of energy poverty in eastern North Carolina. *Annals of the Association of American Geographers*, 101(4), 949–961.

Lovell, H. 2005. Supply and demand for low energy housing in the UK: insights from a science and technology studies approach. *Housing studies*, 20(5), 815–829.

Nadaï, A. and Van der Horst, D. 2010. Introduction: landscapes of energies. *Landscape Research*, 35(2), 143–155.

Ozaki, R. and Shaw, I. 2013. Entangled practices: governance, sustainable technologies, and energy consumption. *Sociology*, 48(3): 590–605.

Rutherford, J. and Coutard, O. 2014. Urban energy transitions: places, processes and politics of socio-technical change. *Urban Studies*, 51(7), 1353–1377.

Sagar, A. D. 2005. Alleviating energy poverty for the world's poor. *Energy Policy*, 33(11), 1367–1372.

SEforAll. Sustainable Energy for All 2015 - Progress towards sustainable energy. (World Bank/International Energy Agency, Washington DC, 2015).

Shaw, I. and Ozaki, R. 2013. Energy provision and housing development: re-thinking professional and technological relations. *Energy Policy*, 60: 427–430.

Shove, E. 1998. Gaps, barriers and conceptual chasms: theories of technology transfer and energy in buildings. *Energy Policy*, 26(15), 1105–1112.

Shove, Elizabeth, and Gordon Walker. 2014. What is energy for? Social practice and energy demand. *Theory, Culture & Society*, 31(5), 41–58.

Walker, G., Karvonen, A. and Guy, S. 2015. Zero carbon homes and zero carbon living: sociomaterial interdependencies in carbon governance. *Transactions of the Institute of British Geographers*, 40(4), 494–506.

16
Governing transitions in energy demand

Catherine Butler and Karen Parkhill

Introduction

Research and policy on energy systems transitions has often been focused on energy supply and on the technical dimensions of such supply systems. Energy demand is, however, increasingly recognised as central to reaching key global policy aims, such as climate change targets and fuel poverty goals. In this context, energy use has come to be problematised in terms of there being 'too much demand' or for use being concentrated at 'particular times'. While, in the past, energy supply has been the focus of governance in order to meet demand, increasingly energy use is becoming important for governance interventions designed to shift energy demand toward lower or less 'peaky' usage. These processes raise an important set of questions for energy geography. Within the wide-ranging literature that addresses issues of energy demand governance, it is possible to identify multiple ways in which geographical contributions and concerns are brought into view. In particular, significant contributions are identifiable in analyses stemming from the socio-technical transitions literature (e.g. Smith et al., 2005; Geels and Schot, 2007), works orientated to practice theory (e.g. Shove and Walker, 2010; Schatzki, 2015), and analysts mobilising concepts from the governmentality tradition (e.g. Butler, 2010; Bulkeley et al., 2015). These areas of theory have been developed, applied and critiqued in ways that bring an explicitly geographical set of concerns about energy demand governance into view.

Across these broadly defined and interlinked positions that speak to questions about transitions in energy demand, geographical concepts are applied both implicitly and explicitly. Such concepts include but are not restricted to ideas of scale, space, materiality, and territory. To date little work has been undertaken to draw together the thinking on governance of energy demand across different analyses, or to explicate the relevance of geographical modes of thinking within them. This chapter will discuss selected literature within energy transitions research concerning governance for shaping energy demand, highlighting the relevance of geographical concepts for analysis in this area. Our aim is to open up insights into how energy geographies might engage with questions about the role of government in transitions of energy demand, and highlight key agendas and future directions for research in this area. The chapter concludes with a discussion of the implications of current arguments about the governance of demand for energy geographies.

Theoretical perspectives on the governance of demand: geographical concepts and imaginaries

Though multiple works and perspectives are evident within this field of research, here we address selected contributions from three theoretical traditions that have been particularly prevalent within analysis of governance and energy systems. For present purposes these are divided into: (1) those concerned with socio-technical transitions in energy systems, most prevalently articulated through the multi-level perspective; (2) those oriented toward practice theory as a basis for understanding social action and change processes related to energy; and (3) those informed by governmentality as a orienting conceptual tradition for understanding processes of governing within energy systems. These works address energy demand *in particular* to varying degrees, but we argue that all in some way offer insights relevant to thinking about the governance of transitions in energy demand, and to understanding the role of geographical concepts and contributions.

Socio-technical transitions and the multi-level perspective

A large body of research concerning energy has focused on systemic processes of socio-technological transitions, wherein technological and social aspects of systems and innovations are analysed and understood terms of their 'co-determination over time' (Truffer and Coenen, 2012: 4). Though the focus here has not been principally on energy demand or governance *per se*, these concerns do form a key element of discussions about the formation of socio-technical assemblages and how they might be changed. For example, the concept of transition management emerges out of these analyses and articulates a particular stance on how transitions in energy systems might be governed. More broadly, theories from socio-technological transitions are implicitly concerned with understanding the relations between different scales of governance in terms of what this has meant historically and might mean in future for energy system change. In this regard, a key perspective that has gained particular traction over the last decade is the multi-level perspective (MLP), which is closely connected to ideas of transition management.

The MLP, first developed by Rip and Kemp (1998) and Geels (2002), seeks to conceptualise how transitions occur, or are resisted, through interrelations between social and technical components. This conception of transition articulates three levels through which processes interact to enact a transition: (1) The sociotechnical landscape – the background socio-technical fabric that provides context; (2) The sociotechnical regime – the observable patterning of socio-technical development in dominant forms that stabilize overtime and are reproduced through myriad intersecting activities (e.g. regulations and standards, establishment of routines); and (3) Technological niches – micro-level radical niche innovations developed by small networks of 'fringe' actors, that, if successfully developed, disrupt incumbent regimes to produce new trajectories (Geels and Schot, 2007: 400).

To date, the MLP has often been employed to explore past transitions or potential future transitions largely related to energy supply rather than energy demand (Kern and Smith, 2008; Shove and Walker, 2010). One notable exception where MLP has been applied to energy demand is an energy transition project carried out by the Dutch Ministry of Economic Affairs. This project highlighted how it is often the case that experimentation regarding low energy lifestyles is omitted from transition projects (Kern and Smith, 2008). As a methodological schema, however, the MLP does have the *potential* to contribute to our understandings of energy demand governance through, for example, an analysis of the various 'socio-evolutionary' processes (i.e. sociological and evolutionary economic) that influence regime development (see Geels and Schot, 2007). That is to say, if we think of the energy system in regime terms, we are oriented to consider how energy demand emerges out of the coordination and steering of many actors and resources, whether this is an intended or emergent feature of transformation processes (Smith et al., 2005).

As a theory of energy system transitions the MLP has received significant critical attention from geographical scholars (e.g. Bridge et al. 2013; Calvert, 2015; Murphy, 2015). In particular, the MLP has been critiqued for being overtly aspatial and for not addressing the different 'levels' in terms of the implications for space and scale (Bridge et al. 2013; Murphy, 2015). Indeed, geographers have been key in arguing that MLP and transitions scholars more widely require geographical concepts to 'more effectively capture the complex dynamics through which innovation systems, socio-technical niches, and regimes co-evolve' (Murphy, 2015: 76; see also Bridge et al. 2013). Territory, geographical context, scale, and the linkages between these and other components, have all been argued as being essential concepts for understanding change processes:

> [T]ransitions are shaped both by the ways in which socio-technical systems are embedded in particular territorial contexts, and by the multi-scalar relationships linking their heterogeneous elements to actors, materials, and forces situated or emanating from different locations or scales.
>
> (Murphy, 2015: 75)

The MLP can be developed through geographical concepts to create a methodological schema to help identify many of the different actors, institutions and components across multiple scales of governance that are important to affect, steer or limit change in energy demand. In this way, thinking can be oriented toward many of the vertical interdependencies between different components of socio-technical systems, including for example, how policy makers contribute to the creation of the 'rules of the game' (Geels and Schot, 2007: 404) that coordinate economic, material, and technical action spaces. This brings into view key geographical ideas about 'transnational governance' that focus on the relevance of non-state actors in making decisions and enacting governing processes (e.g. see Bulkeley et al., 2012; Bouzarovski et al., 2015). Here, a stronger argument is made for focusing on non-state actors as agents of change in and of themselves, rather than as only influencers of state action. In terms of energy demand governance, non-state actors, such as businesses, communities, and NGOs are recognized as having increasingly important roles in driving and delivering change, and can be seen as important agents in advancing transitions far beyond lobbying roles.

Geographical advances, then, direct us to consider the spatial, as well as the temporal, dimensions of socio-technical transitions and to situate the actors, institutions, and relationships that are involved in change processes in place. For energy demand governance these contributions are particularly relevant given the cultural, social and material embeddedness of energy use within daily life. The geographical lens on socio-technical transitions, as a theory through which to understand energy governance, brings into view issues connected to the distribution of demand (e.g. away from production industries toward consumption in the home and for transport – Bridge et al., 2013); the importance of different scales of action and their complex interrelations for governing (Shove and Walker, 2010); and the significance of different locations for how energy demand is or can be configured, both in material and cultural terms (e.g. differences between rural and urban transitions, cultural specificities of energy services such as heat).

Current debates about the governance of energy demand have seen dialogue between socio-technical transitions, particularly the MLP, and practice theory. In the next section, we explore elements of this dialogue and discuss the geographical contributions that are embedded in analyses using practice theory to understand energy demand governance.

Practice theory and governance

Scholars taking up practice theory as an orienting basis for interpreting social action have made several major contributions to thinking about the governance of energy demand. While governance has, again, not necessarily been the primary focus for analysis, engagements with the socio-technical transitions literature and critiques of practice theorists for failure to deliver implications for policy has prompted several key contributions. The first contribution we discuss arises from critical engagement with the ideas and concepts of governance embedded in the MLP. Shove and Walker (2010) highlight two key dimensions that are neglected by the MLP in terms of understanding energy demand governance. First, given it is largely future oriented, the MLP does not orient us to think about *why* things are the way that they are, and *why* it is that certain resource-intensive practices (e.g. showering daily) have taken hold. Second (and as a consequence of the first), the MLP often fails to explicate the governance of the 'seemingly uncontrollable processes that characterize the emergence, reproduction and disappearance of more and less sustainable patterns of daily life' (ibid.).

These neglected dimensions are a result, in part, of the emphasis the MLP places on vertical, rather than horizontal, relationships between niches, regimes and landscapes, which implicitly positions publics as either the benefactor or victim of policies. Shove and Walker (ibid.) contend that contrary to this depiction of governance, the responses and reactions to a particular policy 'constitute the scheme itself', and as such *all* actors (material, infrastructural and human) are just as deeply implicated in governance as policymakers. Implicit in Shove and Walker's arguments is a concern with geographical concepts of scale and critiques of the treatment of scales as bounded (Betsill and Bulkeley, 2007), as well as of depictions of power as extending from the 'top' down to the 'bottom' scale. Schatzki (2015) too takes up this issue regarding the depiction of scale in terms of different levels of change within his development of a practice theoretical perspective on energy demand governance.

Schatzki highlights that contained in the thesis of practice theory is the principle that social phenomena lie on a single level, i.e. that phenomena are not micro or macro, or on planes or levels. In this sense, to characterize a relationship between governing and governed, or between policy and planning and the social phenomena they aim to manipulate, as top down or bottom-up is positioned as misleading. Instead, Schatzki argues it is more revealing to characterize governing interactions as nested relationships and arrangements that are 'here-there or here and all around' (ibid.: 17).

The concept of scale remains important, then, in this practice oriented notion of 'governing interactions as nested relationships' but is configured differently to the MLP. Schatzki characterises scale as 'the continuum between small and large' and as 'a matter of lesser or greater spatial-temporal spread' between what he terms 'practice-arrangement bundles' – referring to the relationship between linked sayings and doings (practices) and linked materials including people, artefacts, and organisms (arrangements) (ibid.: 17). Largeness (and smallness) is produced through the degree of density of connections within bundles; it is the complexes of relation that mark large phenomena (e.g. a bureaucracy). For Schatzki the MLP offers too 'tidy' an account of change by characterizing the social in terms of regimes and niches. He argues that instead 'nexuses of practices and arrangements are contingent, embrace a myriad of relations and constantly, if unevenly and only infrequently, rapidly evolve' (ibid.: 18). This means that rather than change occurring as a 'regime shift' it rests on 'developments along the entire uneven front of change, not just on the actions taken in power centers' (ibid.).

Analysts working with a practice theory lens have thus delineated a conception of social action and change that entails a shift away from both individualistic and scalar understandings, toward one that recognises practices as complex webs of social actions that produce scale through the extent of their interrelations. This chimes with other geographical analyses of energy transitions that have highlighted the importance of being attentive to the processes through which scale is made and emerges (e.g. Bridge et al., 2013). With this conception of scale, policy-makers 'are by implication themselves part of the patterns, systems and social arrangements they hope to govern: they do not intervene from the outside, nor do their actions have effect in isolation' (Shove et al., 2012: 145). This suggests a conception of change and governance that is concerned with 'the seemingly uncontrollable processes that characterise the emergence, reproduction and disappearance of more and less sustainable patterns of daily life' (Shove and Walker, 2010: 472).

The second contribution of practice theorists to thinking about the governance of energy demand arises, in part, through responses to critiques of practice oriented analysis. In particular, a number of works have begun to respond to criticisms of the practice approach for its apparent inability to achieve purchase in application, and for seemingly being unable to recommend routes to induce change. Though behavioral change programs and transition management approaches may not have delivered significant change, they have nonetheless been applied by governing bodies with aims to engender sustainability (e.g. see Kern and Smith, 2008; Capstick et al., 2015). By contrast, practice theory has been critiqued for a lack of clarity in terms of its messages for policy and what it contributes to understanding of how to steer change processes toward sustainability.

Such criticisms have, in the first instance, been countered by scholars through a line of analysis that sought to explicate what approaches to governance or 'intervention' informed by practice theory and research could look like. For Shove et al. (2012) this means developing a 'modest approach' to policy where the goal is not to try to manipulate or control behaviors in the pursuit of pre-defined outcomes, but instead developing a process-based approach that is 'anchored in and never detached from the details and specificities of the practices in question' (ibid. 145).

Shove et al. (2012) use the example of the Japanese Government's 'Cool Biz programme' (whereby government buildings were not heated or cooled when temperatures were 20–28°C; instead office workers were encouraged to wear less formal business attire) to illustrate how practice oriented policy interventions might be formulated. They highlight how this policy intervention did not emphasize using less energy (in this case reducing air conditioning), instead comfort, and by association what constitutes smart and appropriate work wear was adapted and transformed, leading to changes in work wear practices consistent with lower energy usage. Shove et al. (2012: 156) show how policy interventions are 'emerging effects of the systems that policy makers seek to influence'. Practices circulate, they cross (territorial and sectorial) boundaries, they are (partly) formed by and inform policies, and energy demand interventions must take account of this. As such, interventions should be recognized as needing to be 'continuous and reflexive, historical and cumulative' (Spurling and McMeekin, 2015: 78).

Spurling and McMeekin build on the arguments of Shove et al. (2012), showing how practice approaches can facilitate a change from policy interventions that are focused on technological fixes and behavior change, to ones that more fully take into account the role of policy in shaping and steering practice. They re-orient thinking about the role of policy in engendering change toward three distinctively practice theoretically informed formulations of intervention. First, there is *recrafting* practices, whereby interventions recraft the resource-intensity of practices through changing the elements of which a practice is comprised (Spurling and McMeekin, 2015). These elements are: the materials (including objects, tools, and infrastructures); competences (i.e. knowledge and embodied skills); and meanings (i.e. cultural conventions, expectations, and socially shared meanings). In recrafting the elements of a practice the interventions would challenge its resource intensity, rather than its scale and extent.

Second, unsustainable practices can be *substituted* by more sustainable practices, for example, replacing driving with cycling. Here, it is *how* needs and wants are met that is targeted, rather than making provision for the ways that needs and wants are currently met e.g. by building more roads. Policy interventions using a substitution framing would act 'to change the balance of competition' between unsustainable and sustainable practices (Spurling and McMeekin, 2015: 83). The third practice-informed intervention approach is *connectivity*, whereby the focus is on changing how practices interlock. Materials, competencies and meanings link to form practices, and practices connect with other practices to form regular patterns. These connections are known as 'complexes' and 'bundles', which are performed and therefore (re)produced in everyday life. In this style of policy intervention, the focus is on recrafting the complexes and bundles; the 'spaces' and 'networks' of connectivity. As such, 'the negotiability of need' (i.e. the emergent and contingent qualities of need) is foregrounded.

It is beyond the scope of this chapter to give detailed examples of how the above 'practice informed policy interventions' may manifest (see Spurling and McMeekin, 2015). But even this brief discussion highlights the insights that practice theory brings to thinking about governance processes and the role of policy in energy demand transitions. Crucially, practice theory offers a basis for more radical critique and signals a distinctive way of thinking about the role of government policy in shaping, steering and shifting energy demand. These arguments about the significance of practice theory for thinking about policy have been expanded and reoriented by recent analysis that suggests practice theory – and theory in general – should not be considered in terms of rivalry, but as resources that facilitate debate and reflection (Shove, 2015). The practice theory contribution in this sense is to help generate a more significant debate about how policy has effects in the world, and thereby how transitions in energy demand can be effectively governed.

Practice theory has increasingly been taken up by geographers to explore how contexts and processes important for shaping demand and transitions are 'socially and spatially constituted' (Murphy, 2015: 76). As a theoretical perspective it embeds and orients one toward particular understandings of several geographical concepts. Running through the works discussed above, for example, are notions of scale, territory, space, networks and materiality – though these are often left implicit, they are central to the way practice theory is applied in thinking through energy demand governance. In simple terms we and others (e.g. Walker et al., 2014; Bulkeley et al., 2015) argue that context, materiality, scale, and space, matter when it comes to understanding practices and governance. Several studies within this tradition highlight the cultural differences and contextual dynamics that shape practices associated with energy demand (e.g. Hitchings, 2011; Maller and Strengers, 2013). Other analyses have emphasized the political use of materials in shaping everyday life and practice, with particular reference to the role of 'things' (e.g. Marres, 2010; Bickerstaff et al., 2015) and infrastructures (Walker et al., 2014). Ultimately, then, geographical concepts and sensibilities are key to exploring how practices form and change and how energy demand is governed through the shaping and steering of practices.

Governmentality and energy transitions

In this final section, we focus on discussion of selected geographical works that have utilized concepts stemming from the governmentality tradition to develop insights into the governance of energy system transitions. Governmentality represents a theoretical tradition derived from Foucault that is explicitly concerned with conceptualizing and analyzing governance processes and power more generally (e.g. Burchell et al., 1991; Miller and Rose, 2008). In brief, key concepts for thinking about governance that are posed by this theoretical tradition are those of 'problematisations', 'rationalities' and 'technologies' (Miller and Rose, 2008; see also Rose, 1996). First, governing is seen to involve processes of *rendering things problematic* (Dean, 2010; see also Miller and Rose, 2008). Problems are not pre-given, simply waiting to be revealed, but are constructed and made visible through complex processes of interaction. In contexts of liberal government, problems are often connected to some aspect of individual or collective conduct which must be made amenable to intervention. As Miller and Rose (ibid.: 15) put it '[conduct] had to be susceptible to some more or less rationalised set of techniques or instruments that allowed it to be acted upon and potentially transformed'. They further propose that the activity of problematising is intrinsically linked to devising ways to remedy it since 'there was little point, from the perspective of government, in identifying a problem unless one simultaneously set out some measures to rectify it' (ibid.). In order to relate the intrinsic links between a way of representing and knowing a phenomenon and a way of acting upon it so as to transform it, there exists an analytic distinction between 'rationalities' or 'mentalities' of government and 'technologies' (Miller and Rose, 2008; Dean, 2010).

Rationalities are ways of rendering reality thinkable in such a way that it becomes amenable to calculation and programming. Technologies are assemblages of persons, techniques, institutions and instruments for steering and shaping conduct. This refers to all of the devices, tools, techniques, personnel, materials and apparatuses that enable authorities to act upon the conduct of persons individually and collectively, and often in distant locales. Miller and Rose (2008: 16) explain 'rationalities and technologies, thought and intervention, [are] two in-dissociable dimensions through which one might characterise and analyse governmentalities and begin to open them up to critical judgement'.

Across different uses of the governmentality tradition questions have arisen about how geographical concepts are formulated. While Foucault and his contemporaries bring focus on 'populations' and argue for a form of governing that acts upon those populations, they do not explicitly set out what this means for ideas concerning territory and space (ibid.: Rose-Redwood, 2006). Ó Tuathail (1996), among others, argues that geographical knowledge production and the ordering of space are absolutely central to the formation of modern government. Spatial practices are argued to constitute 'technologies of power concerned with the governmental production and management of territorial space' (ibid.: 7). Geographical concepts of territory and spatial ordering thus offer important intersecting lenses within analysis influenced by ideas arising from governmentality. As with practice theory and the dialogue with the MLP, scale also forms an underlying concern as analysts concern themselves with the ways that processes of scaling form part of the rationalities through which people are entrained to particular courses of action.

In this context, an increasing number of geographical accounts have engaged with governmentality as a basis for understanding 'the mechanisms of knowledge production that states have used to constitute their subjects and territories as governable' (Rose-Redwood, 2006: 469). This wider trend in geographical analyses is reflected in works examining environmental governance in general (e.g. Bulkeley and Watson, 2007; Rutland and Aylett, 2008; Butler, 2010), and energy in particular (e.g. Bulkeley et al., 2015). Authors have examined alignment between the rationalities of governments and particular strategies or approaches that are evident within different national contexts. For example, analysis focused on the UK has engaged critically with behavioral change programs arguing that they operate through logics of self-responsibility and choice that are, in turn, consistent with particular advanced liberal governmental rationalities and technologies of rule (e.g. Butler, 2010). Such 'advanced liberal' rationalities involve a new specification of the subjects of rule wherein citizens are positioned 'as active individuals seeking to maximise their quality of life through acts of rational choice' (Rose, 1996: 57) and give rise to problematisations that position individuals and private lives as the scale at which change happens. Behavioral change programs have been argued to be a technology of government aligned with these contemporary rationalities of government, and critiqued for obscuring other important dimensions of the problem and its concomitant solutions (Butler, 2010). Other work in the US context drawing on this conceptual tradition has examined how local state authorities attempted to achieve their climate change policy objectives by enlisting the self-governing capacities of its residents, similarly mobilising problematisations that focus on individual behaviour (e.g. Rutland and Aylett, 2008). This line of analysis, then, has offered detailed analysis of how it is that governments at different scales and across varying territories act upon their citizens and shape energy demand in particular ways.

Though there has arguably been less dialogue between works on governance using the governmentality perspective and practice theory than socio-technical transitions and practice, there are persistent conceptual themes or ideas within geographical analyses that resonate. For example, across both practice and governmentality based analyses there is critique of individualizing approaches to governing, but the latter gives greater focus to the distinctive liberal political rationalities that tend to influence and shape the nature and direction of governing approaches, processes or strategies. More recently, Bulkeley et al. (2015) have sought to engage directly with both practice theory and governmentality for what they bring in combination to understanding governance of energy demand. They suggest that by applying ideas from governmentality, social practices can be seen as not only emergent but as 'actively constituted through the workings of (various) governmental programs stemming from the efforts of various agencies to intervene in relation to distinct problematics – of social welfare, the working of the economy, or the protection of the environment' (ibid.: 115). Using an analysis of smart grids they argue that ultimately 'everyday practices can respond to intervention; that governing can indeed create new dispositions of things in which what constitutes appropriate conduct is reconfigured' (ibid.: 124).

This set of conclusions opens up interesting lines of questioning that pertain to the governance of practice, the intentions of government and other bodies, and their relative power to achieve ends congruent with governmental aims. They also bring attention to areas of interest that have formed a focus for other analyses, such as the ways that rationalities shape problematisations and thereby the required technologies of government with implications for how (and which) dispositions are (re)created. These dimensions of interest for the debate on the governance of energy demand add to those already opened up throughout this literature discussion. In the following section we pick up threads from the literature and seek to outline key areas through which the debate on energy demand governance could be advanced through geographical imaginaries.

Governance and energy demand: advances in the debate

The scope of the project for energy geographies might sometimes be thought as one concerned with the geopolitical, social, and justice issues associated with supply. Through this chapter we have sought to show how energy demand forms an equally important part of the project for geographers. From our discussion of selected literature within energy research concerning energy demand governance, we have highlighted the ways that debates in this area draw on, and intersect with, a number of core geographical concepts. Concepts of scale, materiality, and territory, in particular, feature prominently, with work addressing questions about the multi-scalar nature of governance relating to energy demand, the role of materials in acting upon people and steering action, the nature of problem formation and resolution within different territories, and the relevance of territories and state boundaries in the processes that shape practices.

Of the approaches we have discussed for addressing questions about the governance of energy demand, there are notable critical engagements with the multi-level perspective for its tendencies to be aspatial and give limited attention to the questions of scale that are implied by the framework (Bridge et al., 2013; Murphy et al., 2015; Calvert, 2015). Added to this are critical engagements that build from practice theory to argue that the MLP fails to adequately characterise energy demand governance by situating it as one of interaction between bounded and vertically organised scales. This means that little consideration has been given to governance as a 'horizontally' organised set of processes, i.e. with *all* actors (material, infrastructural and human) being as deeply implicated in governance as policy-makers (Shove and Walker, 2010). This highlights the importance of geographical concepts of scale, context and spatial embeddedness (Bridge et al., 2013) for thinking about energy demand governance and indicates how thought about change requires reflection on not only temporal but also spatial dimensions.

In addition to the relevance of scale within practice theory work on energy demand governance, materiality has also been an important concern. The role of materials as political objects that enshrine particular modes of governance and entrain particular courses of action has been highlighted within analyses using practice theory (e.g. on smart meters see Strengers, 2013; on cars see Spurling and McMeekin, 2015). Materials form a core element of social practices (Shove et al., 2012) or of practice-arrangement bundles (Schatzki, 2015), to use different terms, and as such the theory embeds attentiveness to the material world in ways resonant of geographical imaginaries. Material politics are central to geographical research agendas (Braun and Whatmore, 2010; Clark et al., 2008), with multiple analyses from differing fields advancing understanding of our entanglements with objects and their abilities to advance or stifle change (Thrift, 2008). So it is here that another important dimension of what geography offers to the analysis of energy demand governance becomes visible.

Concepts for thinking about the governance of energy demand from governmentality suggest the importance of paying attention to the ways that framings of problems entail processes of scaling and operate within and through the constitution of territories and power. This set of ideas gives a basis for identification of further geographical agendas related to energy demand and processes of governance. The first concerns the demarcation of territories for governance within nations. Here, the governance of energy demand can be highlighted as limited to the realm of particular government departments and particular spaces of social action, such as the home or individual. These processes of demarcation arguably obscure the wider material, cultural, and economic changes that are likely to be associated with changes in demand (e.g. changes to infrastructural development and the social processes steering mobilities that are associated with supporting petroleum fueled personal transport). This recognition suggests the importance of thinking more expansively about the governance of energy demand extending our focus from energy *per se* to an analysis of how everyday lives are shaped by governance across scales with implications for energy use.

This line of thinking brings us to questions about the extent to which energy demand governance is a problem of government or of wider institutions expanding far beyond the state. A key criticism that has been levied at sustainability transitions literature is that it has been 'spatially blind' and has overemphasised 'the national level at the expense of other geographical levels' (Truffer and Coenen, 2012: 4). In this respect, though many geographical concepts have clearly been taken up and advanced within the literature on energy demand, less focus can been found on ideas that take analysis beyond national boundaries such as those of 'transnational governance'. These have found greater purchase in analysis focused on climate change and energy system transitions more broadly (e.g. Bulkeley et al., 2012; Bouzarovski et al., 2015) but suggest further ground for analyses focused on governing transitions in energy demand in particular.

Finally, we come to a core recurring issue in works examining governance of energy systems that concerns *intentionality* and the ability of governing actors and institutions to achieve change. A key contribution here has been to suggest that the governance of energy demand is largely unintentional, with many of the historical analyses of transitions in demand highlighting complexity and indeterminacy in such processes. In applying geographical concepts and imaginaries, however, analysts have addressed questions concerning what can be intentionally achieved in relation to energy demand. For example, Schatzki (2015) uses the concept of scale to explain differences in the largeness or smallness of targets of governance and ascribe differential abilities or power to affect action; he applies such geographical sensibilities to explore how the intersections (or lack thereof) between spheres of influence and the spatial-temporal spread of problematic phenomena are important in the potentials that exist for *inducing* change. Working more explicitly from a geographical perspective, Bulkeley et al. (2015) signal the ways that governing actions can (and do) delimit or expand possibilities for change in particular directions. Focusing on smart meters they argue that governmental interventions in socio-material arrangements create 'new dispositions of things in which what constitutes appropriate conduct is reconfigured' (ibid.: 124). These analyses thus highlight again the relevance of concern with materials, territory, and the interrelations between scales, for understanding the governance of energy demand, and in particular for grappling with issues of *intention*.

We have highlighted that there are multiple theoretical lines and concepts from within geography evident in the thinking about governance and energy demand, even if sometimes left implicit. By bringing together the three key conceptual strands we have reviewed in this chapter, we have begun to map how these different theoretical positions intersect with one another and illuminated key dimensions of current debates about energy demand governance that are likely to be important for research agendas within energy geographies.

Acknowledgement

The research was supported by the UK Research Councils under the Engineering and Physical Sciences Research Council award EP/M008150/1.

References

Betsill, M. and Bulkeley, H. (2007) Looking back and thinking ahead: A decade of cities and climate change research, *Local Environment*, 12: 5, 447–456.

Bickerstaff, K., Devine-Wright, P. and Butler, C. (2015) Living with low carbon technologies: An agenda for sharing and comparing qualitative energy research, *Energy Policy*, 84: 241–249.

Bouzarovski, S., Bradshaw, M. and Wochnik, A. (2015) Making territory through infrastructure: The governance of natural gas transit in Europe, *Geoforum*, 64: 217–228.

Braun, B. and Whatmore, S. (eds.) (2010) *Political Matter: Technoscience, Democracy, and Public Life*, London: University of Minnesota Press.

Bridge, G., Bouzarovski, S., Bradshaw, M. and Eyre, N. (2013) Geographies of energy transition: Space, place and the low-carbon economy, *Energy Policy*, 53: 331–340.

Bulkeley H. and Watson, M. (2007) Modes of governing municipal waste, *Environment and Planning A*, 39: 2733–2753.

Bulkeley, H. et al. (2012) Governing climate change transnationally: Assessing the evidence from a database of sixty initiatives, *Environment and Planning C: Government and Policy*, 30(4): 591–612.

Bulkeley, H., Powells, G. and Bell, S. (2015) Smart grids and the governing of energy use: Reconfiguring practices? In: Y. Strengers and C. Maller (eds), *Social Practices, Intervention and Sustainability: Beyond Behaviour Change*, Abingdon: Routledge, pp. 112–126.

Burchell, G., Gordon, C. and Miller, P. (1991) *The Foucault Effect: Studies in Governmentality*, Chicago, IL: University of Chicago Press.

Butler, C. (2010) Morality and climate change: Is leaving your TV on standby a risky behaviour? *Environmental Values*, 19(2): 169–192.

Butler, C., Parkhill, K. A., Shirani, F., Henwood, K. and Pidgeon, N. (2014) Exploring the dynamics of energy demand through a biographical lens, *Nature and Culture*, 9(2): 164–182.

Calvert, K. (2015) From 'energy geography' to 'energy geographies': Perspectives on a fertile academic borderland, *Progress in Human Geography*, 40(1): 105–125.

Capstick, S., Lorenzoni, I., Corner, A., Whitmarch, L. (2015) Prospects for radical emissions reduction through behavior and lifestyle change, *Carbon Management*, 5(4): 429–445.

Clark, N., Massey, D. and Sarre, P. (2008) *Material Geographies: A World in the Making*, Milton Keynes: Open University Press.

Dean, M. (2010) *Governmentality: Power and Rule in Modern Society*, 2nd Edition, London: Sage.

Elzen, B. and van Mierlo, A. (2012) Anchoring of innovations: Assessing Dutch efforts to harvest energy from glasshouses, *Environmental Innovation and Societal Transitions*, 5: 1–18.

Geels, F. W. (2002) Technological transitions as evolutionary reconfiguration processes: A multi-level perspective and a case-study, *Research Policy* 31(8): 1257–1274.

Geels, F. W. and Schot, J. (2007) Typology of sociotechnical transition pathways, *Research Policy*, 36: 399–417.

Hand, M., Shove, E. and Southerton, D. (2005) Explaining showering: A discussion of the material, conventional and temporal dimensions of practice. *Soc. Research Online*, 10(2), www.socresonline.org. uk/10/2/hand.html.

Hansen, T. and Coenen, L. (2015) The geography of sustainability transitions: Review, synthesis and reflections on an emergent research field, *Environmental Innovation and Societal Transitions*, 17: 92–109.

Hitchings, R. (2011) Coping with the immediate experience of climate: Regional variations and indoor trajectories. *WIRes Climate Change*, 2(2):170–178.

Kern, F. and Smith, A. (2008) Restructuring energy systems for sustainability? Energy transition policy in the Netherlands, *Energy Policy*, 36: 4093–4103.

Maller, C. and Strengers, Y. (2013) The global migration of everyday life: Investigating the practice memories of Australian migrants, *Geoforum*, 44: 243–252.

Marres, N. (2010) Front-staging non-humans: Publicity as a constraint on the political activity of things: In B. Braun and S. Whatmore (eds.) *Political Matter: Technoscience, Democracy, and Public Life*, London: University of Minnesota Press.

Miller, P. and Rose, N. (2008) *Governing the Present: Administering Economic, Social and Personal Life*, Cambridge: Polity.

Murphy, J. T. (2015) Human geography and socio-technical transition studies: Promising intersections, *Environmental Innovation and Societal Transitions*, 17: 73–91.

Ó Tuathail, G. (1996) *Critical Geopolitics: The Politics of Writing Global Space*. Minneapolis, MN: University of Minnesota Press.

Rip, A. and Kemp, R. (1998) Technological change. In: Raynor, S. and Malone, E.L. (eds), *Human Choice and Climate Change*, Columbus, OH: Batelle Press.

Rose, N. (1996) Governing advanced liberal democracies. In: A. Barry, T. Osborne and N. Rose (eds.) *Foucault and Political Reason: Liberalism, Neoliberalism and Rationalities of Government*, London: Routledge.

Rose-Redwood, R. S. (2006) Governmentality, geography, and the geo-coded world, *Progress in Human Geography*, 30(4): 469–486.

Rutland, T. and Aylett, A. (2008) 'The work of policy: Actor networks, governmentality, and local action on climate change in Portland, Oregon', *Environment and Planning D: Society and Space*, 26: 627–646.

Schatzki, T. (2015) Practices, governance and sustainability. In: Y. Strengers and C. Maller (eds), *Social Practices, Intervention and Sustainability: Beyond Behaviour Change*, Abingdon: Routledge, pp. 15–30.

Shove, E. (2015) Linking low carbon policy and social practice. In: Y. Strengers and C. Maller (eds), *Social Practices, Intervention and Sustainability: Beyond Behaviour Change*, Abingdon: Routledge, pp. 31–44.

Shove, E. and Walker, G. (2010) Governing transitions in the sustainability of everyday life, *Research Policy*, 39: 471–476.

Shove, E., Pantzar, M. and Watson, M. (2012) *The Dynamics of Social Practice: Everyday Life and How It Changes*, London: SAGE.

Shove, E., Walker, G. and Brown, S. (2014) Material culture, room temperature and the social organisation of thermal energy, *Journal of Material Culture*, 19(2): 113–124.

Smith, A., Stirling, A. and Berhout, F. (2005) The governance of sustainable socio-technical transitions, *Research Policy*, 34: 1491–1510.

Spurling, N. and McMeekin, A. (2015) Interventions in practices: Sustainable mobility policies in England. In: Y. Strengers and C. Maller (eds), *Social Practices, Intervention and Sustainability: Beyond Behaviour Change*, Oxon: Routledge, pp. 78–94.

Strengers, Y. (2013) *Smart Energy Technologies in Everyday Life: Smart Utopia?* Basingstoke: Palgrave Macmillan.

Thrift, N. (2008) Halos: Making more room in the world for new political orders: In B. Braun and S. Whatmore (eds), *Political Matter: Technoscience, Democracy, and Public Life*, London: University of Minnesota Press.

Truffer, B. and Coenen, L. (2012) Environmental innovation and sustainability transitions in regional studies, *Regional Studies*, 46(1): 1–21.

Walker, G., Shove, E. and Brown, S. (2014) How does air conditioning become 'needed'? A case study of routes, rationales and dynamics, *Energy Research and Social Science*, 4: 1–9.

17

Embedding energy transitions in the community

Sara Fuller

Introduction

The concept of community is increasingly employed within debates about UK energy transitions. Emerging from both grassroots and policy agendas, the community has been situated as a space of transformation particularly in relation to the development of renewable energy (Seyfang et al. 2013; Walker and Cass 2007; Parkhill et al. 2015). Such a positioning is significant not only because of the technical, economic and political connections that such community energy infrastructures might engender (Rutherford and Coutard 2014) but also because it raises both conceptual and practical questions for understanding and engaging with the spatial and scalar nature of energy transitions.

While there is increasing interest in the idea of community as part of a low carbon transition, it nonetheless remains a contested term. Evidence suggests that communities are rarely neatly defined or delineated; instead, as Lovell (2015: 1364) argues they 'jostle and vie against each other' as they operate at different sites and scales. Broadly speaking, community is viewed as existing somewhere between the individual and government (Aiken 2012); however such initiatives may also strategically seek to 'territorialise' a low-carbon transition (Bridge et al. 2013). It is thus important to understand why community has been put forward as an appropriate site and scale for addressing energy transitions.

In the UK policy arena, the focus has predominantly been on area-based communities. The 2009 Low Carbon Transition Plan aimed to 'make it easier for local authorities, businesses or community groups to generate electricity at community scale' (HM Government 2009: 96) while the more recent Community Energy Strategy set out roles for communities in terms of energy production, reducing energy use, managing energy demand and purchasing energy. In so doing, the strategy articulated that community-led action 'can often tackle challenges more effectively than government alone, developing solutions to meet local needs, and involving local people' (Department of Energy and Climate Change 2014: 7).

From a grassroots perspective, the 'relocalisation' agenda has been manifest through initiatives such as Transition Towns. The Transition movement originated in late 2005 with Transition Town Totnes (Devon, UK) as a community-based response to the challenges of peak oil and the need to build local resilience to climate change (Hopkins and Lipman 2009). Since that time the Transition movement has grown across the world, with over 1000 initiatives in 43 countries which seek to reduce dependency on global markets and mobilise community action around a low carbon transition through a series of 'ingredients' or steps (North and Longhurst 2013; Seyfang and Haxeltine 2012; Feola and Nunes 2014). In practice, this comprises locally based collective activities such community-owned renewable energy, locally grown food and encouraging energy conservation.

Such relocalisation is often positioned as 'a positive and fulfilling process that can be achieved through the combined weight of innumerable community-scale actions' (Bailey et al. 2010: 602). In this context, locally embedded community energy projects are seen to deliver benefits such as reduced carbon emissions, local economic development and democratic decision making (Rae and Bradley 2012; Mulugetta et al. 2010). However, such community level action also warrants careful consideration not only in terms of potential risks associated with energy security and affordability, but more broadly in relation to a 'just transition' (Bulkeley and Fuller 2012).

This chapter traces the evolution of one community energy project in London – Brixton Energy – and considers the challenges and opportunities for embedding this project into the community. In so doing, it considers how and why community has been positioned as a site for energy transitions, how such energy schemes become embedded within the community and more broadly, how the embedding of energy transitions into the community raises questions for justice. The chapter develops a threefold understanding of embeddedness – social, financial and technical – and argues that unpacking the distinctions between the community and the local, understanding the dynamics of individual and collective ownership and considering how such community energy projects connect with existing infrastructure systems provides a framework for understanding the moral and material dimensions of community energy in practice.

Positioning community as a site for energy transitions

For the past two decades there has been a growing emphasis on communities as a site for the generation and ownership of renewable energy projects (Walker et al. 2007). Over this time, a diverse range of approaches – in terms of technology, technical configurations and ownership – have emerged. The following section reviews the multiple ways that community is understood, considers how the concept of community is operationalised in the context of energy schemes and considers some opportunities and challenges for embedding local energy projects into the community.

In debates about energy transitions, community has multiple meanings and roles and can be seen as actor, scale, place, network, process and identity (Walker 2011). A key distinction often drawn in the literature is between communities of place, or area-based communities, and those going beyond a relationship with locality, such as communities of interest (Park 2012; Parkhill et al. 2015). Area-based communities have a place-based identity, shared history and values, shared infrastructure, and political and administrative power (Heiskanen et al. 2010; Rae and Bradley 2012) while communities of interest draw together participants with shared interests and practices (Heiskanen et al. 2010). Importantly these categories are not mutually exclusive but may work to complement each other with, for example, initiatives such as Transition Towns, drawing together both communities of place and interest (Aiken 2012). This multiplicity becomes particularly apparent in urban areas which comprise a greater diversity of communities than rural spaces (Smith 2011). While this is often seen to create greater complexity in terms of progressing a community approach to low carbon transitions, it may also create productive opportunities for valuing the 'urban' as part of such initiatives (North and Longhurst 2013).

While there has been ongoing reflection over the meaning of community in the academic literature, policy development in this arena has largely focused on area-based communities. Even here, however, there is no fixed definition. A review of UK low carbon community schemes identified a wide variety of types of community, defined by geographic area; by the number of buildings; by the number of households or residents; or simply by the communities themselves (Bulkeley and Fuller 2012; Fuller and Bulkeley 2013). Thus while community is recognised as a site for intervention, there is little coherence concerning what 'community' should constitute and instead it is 'a socially, culturally, politically constructed concept which is strategically deployed in diverse and complex forms' (Park 2012: 392).

Given this lack of consensus, a precise definition of 'community energy' is problematic (Seyfang et al. 2013; Walker and Devine-Wright 2008). As Walker and Cass (2007: 461) note community energy can be seen to encompass 'processes of project development that are to some degree local and collective in nature, and/or beneficial project outcomes (economic and social) that are also to some degree local and collective, rather than distant, individualised or corporate in destination'. In practice, there are a wide range of energy generation and conservation projects initiated by a diverse range of groups with multiple configurations of 'hardware' and 'software' (Seyfang et al. 2013; Armstrong and Bulkeley 2014; Walker and Cass 2007).

Importantly, such community renewable energy schemes are not simply infrastructural interventions; rather any 'hardware' is intertwined with the 'software' of its social organisation to give meaning and purpose (Walker and Cass 2007; Bomberg and McEwen 2012). This draws attention to the contested nature of the socio-material processes and practices which are entwined with the technological interventions (Armstrong and Bulkeley 2014; Rydin et al. 2013). These include factors such as the degree of community participation; the manner of governance; the structure of ownership and whether or not locally generated energy is consumed locally (Hoffman and High-Pippert 2010; Gormally et al. 2014).

Beyond the initiation of such projects, it is apparent that embedding such projects also encompasses economic, material and cultural aspects (Bridge et al. 2013). In economic terms, for example, projects need to be financially viable not only in their development and implementation but also in their operation and maintenance (Rydin et al. 2015). In social and cultural terms this is particularly significant as evidence suggests that sustaining participation in community initiatives is often challenging (Hoffman and High-Pippert 2010), especially given that such communities are built on the expectation of significant commitments to new forms of technology and/or to changes in behaviour which may be difficult to achieve.

A key element of community energy projects is the expectation on communities to work co-operatively to make decisions about energy generation and use. In this context, community identity is critical for enabling forms of co-operation and collaboration by overcoming constraints on individual behaviour change (Heiskanen et al. 2010). However, such an expectation may place a burden on communities if they lack control, expertise, skills or resources (Seyfang 2010). It also assumes that 'communities are willing and able to work successfully together to achieve a common goal, whereas in reality communities can be complex with differing dynamics and levels of cohesion' (Gormally et al. 2014: 918). As such, an area-based community does not immediately imply the ready existence of a 'community' and how and for whom communities are defined is likely to have implications for how individuals within these communities are able to work collaboratively.

Beyond the expectations of collaborative working, there are also opportunities for community energy projects to directly address questions of justice in terms of their outcomes (Bulkeley and Fuller 2012). For example, local energy generation has potential to improve access to affordable energy for low-income households. However, this is dependent on the model of development employed and whether attempts are made to address fuel poverty (Walker 2008; Bomberg and

McEwen 2012). It is thus important to consider who or what is included in community energy schemes and the potential costs and benefits.

In summary, community is positioned as both a means through which transitions in social practices and behaviours to produce less carbon intensive lifestyles can be achieved and as a site at which appropriate forms of technology may be developed and deployed (Heiskanen et al. 2010; Middlemiss and Parrish 2010; Moloney et al. 2010; Walker 2011). In practice, community energy projects are the outcome of ongoing and contested interactions between their social and material elements (Rydin et al. 2013; Bridge et al. 2013). Critically however, we need to understand how such energy initiatives become embedded within the community and how community represents a critical site for asking questions about justice in the context of energy transitions. The chapter now turns to examine a case study of Brixton Energy to explore these issues in more detail.

Introducing Brixton Energy

Brixton is an inner-city area of South London and part of the London Borough of Lambeth. Established in 2011, Brixton Energy was the first urban energy co-operative in the UK. The idea of a community based energy project in Brixton has been developed over the last decade, through a combination of grassroots and policy activity (see Figure 17.1).

The initial driver for Brixton Energy came from the Buildings and Energy Group within Transition Town Brixton (TTB). As part of the wider Transition Town movement, TTB was launched in 2007 as the first urban Transition Town in the UK. Based on initial activity facilitated through the Lambeth Climate Action Group, TTB has sought to raise local awareness of climate change and peak oil and plan a transition to a better low energy future through 're-localising'

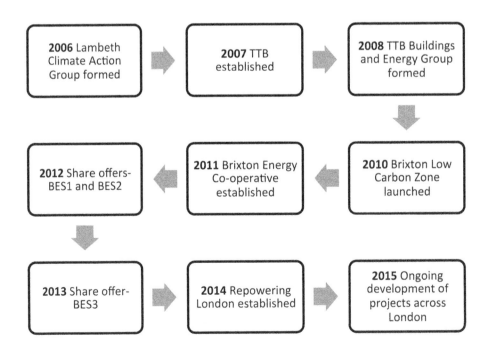

Figure 17.1 The development of Brixton Energy

and building local community resilience. Since the inception of TTB and through the remit of TTB's Building and Energy Group, there have been ongoing discussions about developing a community energy project in Brixton. More recently, the development of Brixton Energy as a social enterprise was supported by the Brixton Low Carbon Zone (LCZ), one of ten such zones in London funded by the Mayor of London from 2010–2012 to identify innovative ways of reducing carbon emissions across the city. The Brixton LCZ contained around 3500 properties including 10 high rise and 36 low rise blocks, street properties and commercial and public sector buildings.

As a result of this parallel grassroots and policy activity, Brixton Energy was established as a co-operative in 2011. The first share offer – Brixton Energy solar 1 (BES1), to install solar panels on the Loughborough Estate – took place in 2012 (Table 17.1) with minimum investment set at £250. While open to all, priority was given to local residents in the event of oversubscription. Since the success of the first share offer, similar share offers (BES2 and BES3) have resulted in the installation of solar panels on two other buildings in Brixton, with a fourth project – Brixton Energy Solar 4 – in the planning phase (Repowering London 2015).

A defining feature of the project is a focus on issues of justice and equity, as manifest in Brixton Energy's overarching aims, where the vision is to create a greener future by: generating energy in Brixton; increasing Brixton's energy resilience and security; raising awareness about energy efficiency and tackling fuel poverty and providing training and employment for local people (Brixton Energy 2013).

Since 2011, Brixton Energy has become more prominent, not only in Brixton itself but also across the city of London. This is demonstrated through Brixton Energy's contributions to locally based schemes but also to a wider network – Repowering London – an organisation which specialises in co-producing energy programmes with community groups and local authorities. Recently, for example, Hackney Energy in north east London has emerged with a similar structure and range of projects to that of Brixton Energy.

The chapter now turns to consider how this energy project has become embedded in the community. In so doing, it draws on fieldwork which comprised interviews with key respondents

Table 17.1 Brixton Energy Projects

Project name	Date	Location	Share offer	Overview
Brixton Energy Solar 1	2012	Elmore House	£58,000 raised with 103 investors	37.24 kW solar array on the roof of Elmore House in the Loughborough Estate.
Brixton Energy Solar 2	2012	Styles Gardens	£60,000 raised with 78 investors	45kW photovoltaic panels on the roofs of Styles Gardens, five of the housing blocks in the Loughborough Estate.
Brixton Energy Solar 3	2013	Roupell Park	£65,650 raised with 71 investors	52kW of photovoltaic panels on four buildings in Roupell Park Estate: Hyperion House, Fairview House, Warnham House and Community Office.

Source: Compiled from Repowering London 2015.

associated with the low carbon agenda in Brixton, including Transition Town Brixton, Brixton Energy, Lambeth Council and local housing providers. Through an analysis of the case study material, it reflects on three interconnected ways that influence how Brixton Energy is embedded within the community:

- Social embeddedness: delimiting 'community'
- Financial embeddedness: developing collective ownership
- Technical embeddedness: connections with existing infrastructure

It draws across these three key dimensions to make the case that the *moral* embedding within the community is intrinsically connected to its *material* embedding; namely that we cannot understand how issues of justice and equity are bought to the forefront without understanding the material ways that such energy projects are developed.

Embedding energy transitions in Brixton

Social embeddedness: delimiting 'community'

In relation to the social embedding of Brixton Energy, a key consideration was the process by which the term 'community' was defined and implemented, particularly in the context of a delineation between 'community' and 'local'. From its inception, there have been multiple models of community in co-existence, both geographic communities such as those in Brixton or the Loughborough housing estate, alongside communities of interest such as TTB, which are not distinct and are in fact 'are all layered on top of each other, to a large extent' (interview, local authority). As such, organic grassroots communities co-exist with specific models and ideals of community that have arisen from policy interventions.

It is apparent therefore that due to particular funding models, the mobilisation of community in relation to Brixton Energy has been both organic and instrumental. While TTB has a wider remit and a more grassroots notion of community, the LCZ model of community was based on a specific geographical area and number of buildings. It was acknowledged however that this may not be the most progressive approach; respondents indicated that 'you can't just geographically define communities' (interview, local authority). Thus while Brixton Energy partially emerged from a novel policy approach, there was also widespread recognition of the limitations of this approach. This intersection between 'local' and the multiplicity of communities thus becomes highly significant in terms of mobilising community energy.

Within these multiple communities, one distinct community is the individuals involved in creating, developing and implementing the project. The creation and mobilisation of this specific community relied on people with vision and energy, with respondents talking of the need to get 'buoyed up in other people's enthusiasm but also picking up when people mean business and they have the knowledge . . . and also very positive and optimistic' (interview, Brixton Energy). This enthusiasm facilitated Brixton Energy to move forward quickly through weekly meetings. A number of factors contributing to the success of the group in terms of making rapid progress were identified, including: a dynamic and diverse group of people, the level of knowledge and expertise within the group, the existence of trust and respect within the group and the ability to recognise progress and achievements.

Outside this group of active and enthusiastic residents, there was an initial challenge in reaching out to the wider, local community. In this sense, Brixton Energy benefited from two distinct opportunities to engage residents. The LCZ had a community engagement officer whose role was to facilitate community projects and connect people while TTB as a grassroots organisation had longstanding community links. Nonetheless, time and resource constraints meant that 'usual suspects' were often targeted and it was acknowledged that where populations are living with fuel poverty, it was harder to 'get people going' (interview, local authority).

Beyond the initial enthusiasm and mobilisation, the continual process of re-engaging the community is important and is one that takes time and energy. In this context, the material presence of the solar panels plays a role; respondents noted that 'there was not always a need to talk' (interview, Brixton Energy) and instead local residents could be engaged by seeing the projects in practice. In this sense, while climate change was sometimes seen to be an abstract agenda, a community energy project was viewed to be more tangible. As one respondent noted: 'If you get involved in all of that, it almost paralyses you. So actually, focusing on much more practical, day-to-day issues actually is a better way to motivate people . . . and it also makes very practical sense' (interview, Brixton Energy). As such, it is clear that embedding a community energy project is an ongoing process.

Finally, as noted in the literature, the case of Brixton energy demonstrates the need to unpack the assumption that a community energy project is necessarily in tandem with the concept of community purely of its geographic scale. In Brixton, this was evident in a number of forms. First was that while the roots of the project had emerged from 'the community', the project itself was very specific in terms of the idea of community that was being mobilised. In addition, there was a lack of widespread engagement from across all sectors of the community. For example, despite the focus on lower income residents, many of these perceived 'hard to reach' were not directly involved.

Financial embeddedness: developing collective ownership

One way of understanding financial embeddedness is the relationship between individual and collective ownership and hence collective sharing of risks and benefits. In the context of Brixton Energy, this is couched in terms of income in terms of making the scheme accessible to those on low incomes and saving money, as well as wider social benefits.

Brixton Energy operates as a mutual society and gathers investment through the sale of shares to community members. The first share offer in 2012 was undertaken on the basis of a 25-year investment. There are three key phases to the investment model, summarised below:

- The community invests in a co-operative (1 member = 1 vote)
- The co-operative installs new renewable energy project on local buildings
- The technology generates an income which is used to pay for a community energy efficiency fund, an annual dividend for shareholders and administration costs (Brixton Energy 2013)

In the context of this model, issues of justice come to the forefront. The Loughborough estate (the site of the first share offers) is located in one of the most deprived wards in England, with high unemployment, high levels of fuel poverty and the majority of properties in the social housing sector. A minimum level of investment was set for each share offer to cover administrative costs but there was careful consideration to ensure that this did not become exclusionary. As one respondent noted: 'you don't want to stick it too high because . . . it would be nice if the people in the estate invested. The problem is that. . . if you set it at £250, are you excluding everybody or just a few people or what' (interview, Brixton Energy). The relationship between the opportunities for individual action and collective benefits thus become appararent when the level of investment in the scheme is considered.

One of the ways in which the social dividend from the project operates is that funding is generated to pay for draughtproofing and energy advice sessions. Facilitating this social dividend involved a careful process of calculating the overall percentage dividend that investors would receive to ensure that there would be enough funding to also support the energy efficiency measures. As one respondent noted, the initial dividend was calculated at an average of 4 per cent return each year over 25 years. However, further calculations suggested this would not generate enough profit to enable training sessions, advice and draughtproofing and as a result, it was decided that a suitable rate of return for investors would instead be 3 per cent.

This financial embeddedness in terms of collective ownership is also underpinned by a less tangible set of calculations; in other words, the emphasis placed on building community capacity and local resilience means that benefits of the scheme are more than simply financial. From the perspective of Brixton Energy, the reason why is it deemed a 'community' project is because there are both social and economic outcomes: 'It's not just putting the panels on the roof, it's the social outcomes, which is why it's a community project and not just a money making project' (interview, Brixton Energy). As such, there was a clear understanding that people investing in the project were not purely interested in financial gains, but were doing so because they supported renewable energy and community development. This approach was reinforced by local housing providers who felt that while issues of efficiency were important 'it wasn't so much financial savings that wasn't the thing driving us, it's really about . . . the enrichment of the community' (interview, housing provider). This involved not only building existing skills but also identifying appropriate opportunities for how such skillsets could be used within the community.

Alongside providing individual and collective benefits, there was also an understanding that the collective nature of the project reduced individual risks. One respondent identified a number of potential risks — some of which were covered by insurance while others were much more uncertain:

> There is insurance to cover the installation, to cover the limited liability, there is no insurance about if we enter the ice age in the next 25 years! So if they don't produce enough. So that has to be put down on the risk register. The government could refuse to give any decent Feed in Tariffs for community groups . . . They can't take it away from you once you've done it but they could make it more difficult to set up others.
>
> (Interview, Brixton Energy)

As such, while there were risks associated with the project this was able to be managed via the dynamics of individual and collective ownership which brings the notion of community to the forefront.

Technical embeddedness: connections with existing infrastructure

The final dimension of embeddedness relates to configurations of technology and specifically the connection within existing systems of infrastructure. In contrast to many rural schemes, a key challenge with an urban energy project of this type is the complexities of building ownership. In the case of Brixton Energy, some buildings are owned directly by the London Borough of Lambeth, some are managed by social housing providers while others are privately owned – either residential or commercial. This meant that Brixton Energy had to engage with a number of systems of property ownership and energy provision.

For social housing providers, there are several direct benefits from the installation of solar panels. One such benefit is that they will be able to access cheaper electricity from which they can power communal lighting and lifts. This will place them in a position to be able to reduce, or at least not raise, the service charges and rent of tenants. However, given the complexity of building ownership, it has been difficult to ensure that residents living in the buildings where the solar panels are physically located can receive the financial benefits of the scheme. This is in part because the building contains a mixture of tenants and leaseholders; as one respondent noted:

> If it was all leaseholders, you could say to them, well now the communal lighting and lifts are going to be paid for or sourced from a cheaper rate of energy so you could reduce everybody's service charges. But because you've got tenants it doesn't work and it's too complicated.
>
> (Interview, Brixton Energy)

Thus while the installation of solar panels means that the housing provider receives cheaper energy, there is no immediate corresponding reduction in energy costs for residents of the building.

This technical embedding within existing infrastructure systems means that some elements remain outside of the control of Brixton Energy which potentially poses a risk. The lack of direct control over housing providers to pass on savings to their tenants relies on their goodwill to act in a responsible way. One way of trying to ensure this happens is to get residents involved to put pressure on the housing provider. As one respondent noted: 'once you get them involved, they can start putting pressure, because there is a Residents Association. They will start saying, how is this being passed on?'(interview, Brixton Energy). This risk was also partly mitigated through the social dividend for the project which supported fuel poverty measures. For example, undertaking draughtbusting activities (such as improving insulation) in individual flats is the only guaranteed way to see a reduction in individual energy bills. However, this also serves to create a potential tension where collective outcomes depend on individual actions and responsibilities.

More recently there have been discussions about enabling a more direct connection between the solar panels and reduced energy bills. There is work ongoing – in collaboration with residents in the blocks – to develop a model which would allow residents to have energy from the solar panels supplied directly to their flats. In practice, this would involve collective switching of energy suppliers whereby 100 per cent of residents would have to change their energy supplier. This not only raises logistical and technical challenges but also potential concerns about trust in terms of persuading people on low incomes to change suppliers.

There are also ongoing questions about the community share offer and whether it is appropriate to go outside the community for investment. This has implications for community as it was acknowledged that residents and investors may operate in different spheres specifically because of the urban nature of the project. One interviewee explained that 'investors are not going to be the same as the people who are going to benefit in one sense . . . if you've got a small village, they are all going to get not only their dividend but their cheaper energy. And they're the same people. But you won't get that here' (interview, Brixton Energy). As such, there is some disconnect between the community of investors and the community of residents.

Conclusions

This chapter has examined how a local energy project is intertwined with notions of community. In so doing, it has drawn out three interconnected dimensions of embeddedness: social embeddedness in terms of delimiting 'community'; financial embeddedness in relation to developing collective ownership; and technical embeddedness associated with infrastructural configurations. Unpacking the distinctions between the community and the local, the dynamics of individual and collective ownership and the connections with existing infrastructure systems allows an understanding of both the scalar and spatial dimensions of the case of Brixton Energy.

In relation to questions of scale, community energy projects are often understood as those that have decision processes or project outcomes that are to some extent local and collective (Walker and Cass 2007). However, in order to understand the process of embedding, we need to consider how the local is itself comprised of a number of different interlocking scales of activity that bring the role of the individual into sharp relief as part of wider collective action. It is also clear that these dimensions of embeddedness are interrelated in numerous ways, which supports existing research suggesting that the materiality of energy resources works to shape the constitution of community (Armstrong and Bulkeley 2014). The case study highlights that the financial and technical embeddedness of the scheme depends on the model of community employed, suggesting that social embeddedness is the core success factor in such community energy projects.

Moreover, the spatiality of the project is clearly significant. While the urban location of Brixton Energy brought a number of challenges, in terms of developing connections with existing energy infrastructures and delineating multiple communities, it also provided an opportunity for issues of justice to come to the forefront in ways that might not otherwise be possible. The Brixton Energy case study demonstrates that embedding a local energy project into the community comprises both material and moral elements. These are not mutually exclusive processes, more that the materiality of a community energy project, through elements such as solar panels, is fundamentally underpinned by a set of moral values that have led to the development of this project. In the case of Brixton, these questions have been addressed in practice via the benefits arising from the share offer and the clear focus on assisting with issues of fuel poverty. In summary, this moral embedding directly connects to the spatiality of the project and its explicit connection with wider structural conditions.

In conclusion, there is a clear need to critically interrogate how notions of community are being mobilised and enacted around energy projects. There is no single fixed definition of community and no default correlation between 'the community' and 'the local'. A careful consideration of the scalar and spatial dimensions of community energy infrastructures is required in order to understand how such interventions may serve to progress a just low carbon transition.

Acknowledgements

This chapter reflects an ongoing collaboration with Harriet Bulkeley, Durham University. It particularly draws on work undertaken for a Thinkpiece commissioned by the Joseph Rowntree Foundation, *Assessing the potential for socially just low carbon communities*, by Harriet Bulkeley and Sara Fuller (available at www.jrf.org.uk/publications/low-carbon-communities-social-justice).

References

Aiken, Gerald. 'Community Transitions to Low Carbon Futures in the Transition Towns Network (TTN)'. *Geography Compass* 6 (2012): 89–99.

Armstrong, Andrea, and Harriet Bulkeley. 'Micro-Hydro Politics: Producing and Contesting Community Energy in the North of England'. *Geoforum* 55 (2014): 66–76.

Bailey, Ian, Rob Hopkins, and Geoff Wilson. 'Some Things Old, Some Things New: The Spatial Representations and Politics of Change of the Peak Oil Relocalisation Movement'. *Geoforum* 41, 4 (2010): 595–605.

Bomberg, Elizabeth, and Nicola McEwen. 'Mobilizing Community Energy'. *Energy Policy* 51 (2012): 435–44.

Bridge, Gavin, Stefan Bouzarovski, Michael Bradshaw, and Nick Eyre. 'Geographies of Energy Transition: Space, Place and the Low-Carbon Economy'. *Energy Policy* 53 (2013): 331–40.

Brixton Energy. 'Community Owned Solar Power in Brixton'. 2013, at https://brixtonenergy.co.uk (accessed 6 July 2016).

Bulkeley, Harriet, and Sara Fuller. *Low Carbon Communities and Social Justice*. York: Joseph Rowntree Foundation Viewpoint. 2012.

Department of Energy and Climate Change, *Community Energy Strategy*. London: Department of Energy and Climate Change. 2014.

Feola, Giuseppe, and Richard Nunes. 'Success and Failure of Grassroots Innovations for Addressing Climate Change: The Case of the Transition Movement'. *Global Environmental Change* 24 (2014): 232–50.

Fuller, Sara, and Harriet Bulkeley. 'Energy Justice and the Low Carbon Transition: Assessing Low Carbon Community Programmes in the UK'. In *Energy Justice in a Changing Climate*, edited by Karen Bickerstaff, Gordon Walker, and Harriet Bulkeley. London: Zed Books, 2013: 61–78.

Gormally, Alexandra, Colin Pooley, Duncan Whyatt, and Roger Timmis. ' "They Made Gunpowder . . . Yes down by the River There, That's Your Energy Source": Attitudes towards Community Renewable Energy in Cumbria'. *Local Environment* 19 (2014): 915–32.

Heiskanen, Eva, Mikael Johnson, Simon Robinson, Edina Vadovics, and Mika Saastamoinen. 'Low-Carbon Communities as a Context for Individual Behavioural Change'. *Energy Policy* 38, 12 (2010): 7586–95.

HM Government, *The UK Low Carbon Transition Plan*. London: The Stationery Office. 2009.

Hoffman, Steven M., and Angela High-Pippert. 'From Private Lives to Collective Action: Recruitment and Participation Incentives for a Community Energy Program'. *Energy Policy* 38, 12 (2010): 7567–74.

Hopkins, Rob, and Peter Lipman, *Who We Are and What We Do*. Totnes: Transition Network. 2009.

Lovell, Heather. 'The Multiple Communities of Low-Carbon Transition: An Assessment of Communities Involved in Forest Carbon Measurement'. *Local Environment* 20 (2015): 1363–82.

Middlemiss, Lucie, and Bradley Parrish. 'Building Capacity for Low-Carbon Communities: The Role of Grassroots Initiatives'. *Energy Policy* 38, 12 (2010): 7559–66.

Moloney, Susie, Ralph E. Horne, and John Fien. 'Transitioning to Low Carbon Communities–from Behaviour Change to Systemic Change: Lessons from Australia'. *Energy Policy* 38, 12 (2010): 7614–23.

Mulugetta, Yacob, Tim Jackson, and Dan van der Horst. 'Carbon Reduction at Community Scale'. *Energy Policy* 38, 12 (2010): 7541–45.

North, Peter, and Noel Longhurst. 'Grassroots Localisation? The Scalar Potential of and Limits of the "Transition" Approach to Climate Change and Resource Constraint.' *Urban Studies* 50 (2013): 1423–38.

Park, Jung Jin. 'Fostering Community Energy and Equal Opportunities between Communities'. *Local Environment* 17, 4 (2012): 387–408.

Parkhill, Karen Anne, Fiona Shirani, Catherine Butler, Karen Henwood, Chris Groves, and Nick F Pidgeon. ' "We Are a Community [but] That Takes a Certain Amount of Energy": Exploring Shared Visions, Social Action, and Resilience in Place-Based Community-Led Energy Initiatives'. *Environmental Science and Policy* 53 (2015): 60–69.

Rae, Callum, and Fiona Bradley. 'Energy Autonomy in Sustainable Communities—A Review of Key Issues'. *Renewable and Sustainable Energy Reviews* 16 (2012): 6497–506.

Repowering London. 'Projects'. 2015, at www.repowering.org.uk/projects (accessed 6 July 2016).

Rutherford, Jonathan, and Olivier Coutard. 'Urban Energy Transitions: Places, Processes and Politics of Socio-Technical Change'. *Urban Studies* 51 (2014): 1353–77.

Rydin, Yvonne, Catalina Turcu, Simon Guy, and Patrick Austin. 'Mapping the Coevolution of Urban Energy Systems: Pathways of Change'. *Environment and Planning A* 45 (2013): 634–49.

Rydin, Yvonne, Simon Guy, Chris Goodier, Ksenia Chmutina, Patrick Devine-Wright, and Bouke Wiersma. 'The Financial Entanglements of Local Energy Projects'. *Geoforum* 59 (2015): 1–11.

Seyfang, Gill. 'Community Action for Sustainable Housing: Building a Low-Carbon Future'. *Energy Policy* 38, no. 12 (2010): 7624–33.

Seyfang, Gill, and Alex Haxeltine. 'Growing Grassroots Innovations: Exploring the Role of Community-Based Initiatives in Governing Sustainable Energy Transitions'. *Environment and Planning C* 30 (2012): 381–400.

Seyfang, Gill, Jung Jin Park, and Adrian Smith. 'A Thousand Flowers Blooming? An Examination of Community Energy in the UK'. *Energy Policy* 61 (2013): 977–89.

Smith, Amanda. 'The Transition Town Network: A Review of Current Evolutions and Renaissance'. *Social Movement Studies* 10 (2011): 99–105.

Walker, Gordon. 'Decentralised Systems and Fuel Poverty: Are There Any Links or Risks?'. *Energy Policy* 36 (2008): 4514–17.

Walker, Gordon. 'The Role for "Community" in Carbon Governance'. *Interdisciplinary Reviews Climate Change* 2 (2011): 777–82.

Walker, Gordon, and Noel Cass. 'Carbon Reduction, "the Public" and Renewable Energy: Engaging with Socio-Technical Configurations'. *Area* 39, 4 (2007): 458–69.

Walker, Gordon, and Patrick Devine-Wright. 'Community Renewable Energy: What Should It Mean?'. *Energy Policy* 36 (2008): 497–500.

Walker, Gordon, Sue Hunter, Patrick Devine-Wright, Bob Evans, and Helen Fay. 'Harnessing Community Energies: Explaining and Evaluating Community-Based Localism in Renewable Energy Policy in the UK'. *Global Environmental Politics* 7 (2007): 64–82.

18

Energy and place-making in informal settlements

A view from Cape Town

Scott Jiusto and Stephen M. McCauley

Introduction

> There was this open space where we decided to squat. I said, Oh, that's my chance to have
> my own shack . . . But the city came to bulldoze our shacks . . . And then we told those
> people, if they doesn't need a trouble, they must get off of those bulldozers, before we burn
> [them]. But we didn't burn the bulldozers, we said, We need the bosses . . . No shacks must
> be bulldozed again here, because the people, they need a place to stay.
>
> (Buyiswa, community leader, on the founding of Monwabisi Park
informal settlement, Khayelitsha, in 1997)

In the burgeoning cities of Africa, Asia, South America and throughout the "global south,"
nearly a billion people are living in and actively transforming the social and physical landscapes
of places referred to as slums, informal settlements, squatter camps, favelas, or other local variants
(UN-HABITAT 2012). In South Africa, rural migrants seeking access to urban jobs, education,
excitement and possibility, along with urban poor of many circumstances, have over decades
created these foothold communities on the geographic and social margins of the city: along
highways, railway reserves, river banks and sewage works; on steep and unstable hillsides; within
floodplains and landfills. In their first incarnation, their homes reflect the necessity to build fast
and cheap – sheet metal, scrap wood, plastic sheeting, cardboard – but over time, shacks often
grow to accommodate newborn children, greater material comforts, and family, friends and
other sojourners on the move from place to place within the city and to and from rural, ancestral
lands. Local economies and institutions emerge within and alongside shack homes, as residents
and others establish salons, churches, shops, roadside barbeques, taverns, gangs, daycare centers,
civic associations and more. And as these marginalized urban dwellers gradually transform their
settlements, they and their allies increasingly pressure government to provide equitable access
to water and sanitation, roads, transport, education, healthcare and other essential public
services, with electrical energy services a very high priority (Huchzermeyer and Karam 2006).
They also demand that, as communities, they be meaningfully involved in "co-producing" with
government how services are provided and how their places will evolve (Mitlin 2008).

Figure 18.1 Monwabisi Park informal settlement

But as this case study from Cape Town, South Africa will show, even where the political will and financial resources exist to provide access to electricity and other basic services, it is rarely possible to do so simply by extending to informal settlements social and technical strategies that work reasonably well elsewhere (e.g., in established urban neighborhoods or rural areas). New strategies are needed because many of the fundamentally geographical challenges to sustainable energy development that this book highlights are especially daunting in informal settlements, which typically are:

- Highly dynamic and contested places in which changes in energy systems can have strong, immediate, highly visible and often contradictory impacts on vulnerable individuals, community life, urban spatial configurations, housing stock, and other aspects of society . . . and vice versa. A deep understanding of how livelihood strategies and everyday energy dynamics and the energy landscape all interact and might co-evolve is therefore required (Mehlwana 1997).
- Places where improving energy systems significantly requires substantial, simultaneous changes in how energy is consumed, produced and supplied (Swilling 2014).
- Places where multiple "energy publics" and diverse other actors operate at overlapping spatial scales, simultaneously constructing those scales and shaping the "realm of the possible" for system evolution, from individual householders operating their own informal electrical distribution utility to vertically integrated, state-owned national electrical power utilities.

All in all, informal settlements are cauldrons of social innovation and unrest central to the transformation of cities and the pursuit of justice and sustainability in the 21st century. Yet despite a deep literature on informality and urban development (e.g., Mitlin and Satterthwaite 2004; Roy 2005; Huchzermeyer and Karam 2006), and increasing interest in related energy issues (e.g., Visagie 2008; Swilling 2014; Silver 2014), there is very little practical guidance available to help communities, government, business, civil society and academics collaborate to implement sustainable energy systems in these difficult environments, especially when compared to the urgent need and the vastly greater resources dedicated to sustainable energy transition in more developed contexts. Our core argument is that more robust, collaborative effort and greater commitment to experimentation and collective learning is needed to better support the co-evolution of healthier, more sustainable, more empowering energy systems and informal settlement communities.

The case study explores questions such as, what key factors affect the co-evolution of informal settlements and energy systems? How do broader sustainable energy goals articulate with policy and planning around informal settlements? From a community perspective – what systems exist, what's hoped for, what's feared, and what moments create opportunities for change and improvement? And fundamentally, how can diverse partnerships together build knowledge networks and energy supply chains to improve the lives of urban slum dwellers?

The chapter begins with an overview of informal settlements and introduces themes drawn from the literature and our observations in Cape Town. The body of the chapter then traces a narrative arc familiar to anyone actively striving to improve conditions in informal settlements: a story that begins with scenes of what people do to secure energy services and the issues that arise when operating without access to improved energy infrastructure, followed by scenes that describe the advent of a major system transition – the introduction of electricity – and then tracing, almost literally, stories extending from the electrical box along wires to nearby neighbors and then out to the periphery of the settlement, exploring along the way different kinds of social relations and energy system considerations, including the sense of entropy that can all too quickly envelope and begin transforming the "transition to modernity" into a more complex, fraught, fragile, sometimes retrograde process yielding mixed blessings and curses and overlapping narratives. We conclude with thoughts about how academics and other "outsiders" might contribute further to the realization of more sustainable energy systems for urban slum dwellers.

Action-oriented research strategy

> As the seat of delivery of basic services (land, water, energy, sanitation and solid waste), the platform for redistribution and the sphere closest to the communities it serves, local government is the key site for enormous opportunity to develop pioneering, "learn by doing" and transformative solutions. However, innovative, responsive and transformative shifts at this level – while crucial – are not sufficient . . . Change also needs to occur from a macro-perspective through an overarching framework that enables the systems and the institutional form to align and support implementation at the local level. Thus 20 years on and in spite of government's firm commitment to poverty alleviation and to advance development, the problems remain deeply rooted, whereby apartheid spatial form and poverty persist and inequality has deepened.
>
> (SEA 2014)

Many of the insights discussed in this chapter derive from "shared action learning" community development projects conducted in Cape Town over the past nine years by undergraduate students from Worcester Polytechnic Institute (WPI) (e.g., Church et al. 2014) in close collaboration with local non-governmental and community-based organizations, City of Cape Town and Stellenbosch Municipality agencies, community "co-researchers," and us and other WPI faculty contributing as advisors and project strategists (Jiusto et al. 2013). The aim has been to advance the kind of "learning by doing" of the sort called for by Sustainable Energy Africa that can make incremental change in communities while informing transformational policy at higher scales.

The research methodology is built on our annual, two month long, intensive collaborations in which students and local partners have sought to advance specific, locally defined initiatives, with more general insights about informal settlement dynamics emerging through the process (Elmes et al. 2012). We've left ourselves in the story to illustrate, in a small way, what such grounded work looks like and to reflect our particular positionality, including often blindnesses to much of what is happening around us, even as we work with these diverse partners to pursue better collective understandings and help "co-produce" improved conditions in local communities (Mitlin 2008). More importantly, we also include the voices of other stakeholders, especially informal settlement dwellers whose perspectives and participation are often missing from formal planning and development processes.

Informal settlements in Cape Town: an introduction

In South Africa, informal settlements reflect the legacy of state-enforced racialized land use segregation under colonialism and apartheid, inflected since the advent of democracy in 1994 by a complex mix of both privatization and de-regulation and pro-poor interventions (Huchzermeyer 2004; Marais and Ntema 2013). What's brewing at the ground level are intensely complex, contested, contradictory places.

Of Cape Town's approximately one million households and 3.7 million population in 2013, an estimated 193,000 households were living in shacks in the city's 200 or so informal settlements and another 75,000 households were "backyard" shack dwellers in more formal areas (Housing Development Agency 2013). These settlements are highly diverse: they range in population from just a few dozen to over 20,000 people; some are comprised largely of "Black" Xhosa-speaking migrants from the Eastern Cape Province and their offspring, some are predominantly "Coloured," and others reflect the ethnic and cultural diversity of not only South Africa but of an increasingly mobile and integrating African continent. Some settlements have existed for decades, others for just weeks or months; some are recognized and serviced to varying degrees by government, while others are slated for eradication; some have reasonably clear, if informal and always contested governance structures, while others are free-for-alls; and some are home to residents who share a sense of identity and community, while others are the antithesis of community, human dumping grounds pervaded by isolation, loneliness and precariousness. This social diversity, expressed within equally complex material landscapes, makes energy system planning and development difficult.

Heterogeneity notwithstanding, informal settlements also share common features that bear on the pursuit of energy sustainability. They are almost by definition transformative spaces, where people are actively seeking a toehold in the city and transforming "unused" wetlands, hillsides, parkland, utility buffers and the like into residential areas. Problems of high unemployment and poor access to water and sanitation, quality education, health care, nutrition, and other basic social amenities are endemic, as is an all-pervasive feeling of vulnerability to crime, flooding, fire, sexual assault, theft, vandalism, drug and alcohol abuse, disrespect, disease and, in bitter irony, of being forcibly removed to somewhere worse, more isolating and without ready recourse

Figure 18.2 Shacks in Monwabisi Park

to the network of family and friends through which the poor sustain themselves. And yet, they are also at the same time places of hope, love, friendship, striving, economic and creative activity, partying, and vibrancy, places where many people live their entire lives and where some who have "made it" into the middle class choose to stay.

Informal settlements in South Africa also share a communal belief that the state should be more actively engaged in addressing their needs and in the project of transforming settlements into places of proper homes, with proper services, including electricity. The legally ambiguous status of informal settlements, with residents and entire communities holding only weak or informal rights of tenure and occupation, greatly complicates the question of whether and how local governments might respond to these entreaties. As others have shown in the context of informal settlements in other parts of the world (e.g., Roy 2005), informality demands a discretionary mode of governance in which regulations are flexibly interpreted or suspended. Pursuit of a uniform, universally serviced city is a distant ideal; in practice the state apparatus enters into a precarious, negotiated governance with informal settlement communities, addressing service needs on an ad hoc basis as crises erupt or opportunities arise.

Yet upon the advent of democracy in 1994, the African National Congress-led government, through the Reconstruction and Development Programme (RDP), aimed for a wholesale upgrade of service provision, making the "housing" problem a central concern of national policy. The promise of free RDP housing for the poor on a massive scale as *the* answer to informality is widely viewed now as unworkable: too slow, dislocating, disruptive to social networks, ineffective in generating healthy communities, and disempowering by encouraging people to wait for government to "solve" their problems. A broad consensus has emerged among national government and diverse groups and social movements such as those we partner with in Cape Town – e.g., the South Africa Shack Dweller International Alliance, local government, academics – that slums are best viewed not as outlaw places to be eradicated and (possibly) replaced, but as emergent communities to be supported through incremental, *in situ* upgrading understood broadly as cooperative efforts among residents and diverse stakeholders to improve basic services and community well-being (Mitlin 2008).

While energy services are essential for economic development and human well-being, providing safe, reliable, and affordable access to electricity and other clean fuels for cooking and heating is daunting. In informal settlements, not only are the three oft-theorized overlapping domains of sustainability – social, environmental, and economic – urgently and immediately in play, so too are difficult challenges of institutional and technological sustainability. The maelstrom of vandalism and theft in informal settlements renders many common assumptions regarding technical choices invalid; lengthy negotiations and political jockeying may be required to determine which communities will be serviced first, or who will provide various services and benefit from job opportunities, and which shacks may need to be moved to allow placement and servicing of power poles. As we'll see below, the kind of embeddedness of electricity infrastructures in power relations that Hughes (1983) and others have illuminated in a Western context, play out robustly and sometimes violently in informal settlements, at scales ranging from the household to the community to the city and state, involving myriad actors vying for power and influence (Harber 2011).

Off-Grid Energy, Township Style

We round the corner into Flamingo Crescent and squeeze our small car into the open space in front of two shacks, navigating around a handful of people sitting on crates and other improvised seats. The center of their attention is an imbaula, a small tin bucket cookstove burning wood and boiling water for coffee and a morning meal. Imbaulas are a foundational energy technology in many informal settlements, and the smells and sounds of cooking on these open flames – mealie pap boiling in Xhosa communities, curries in "Coloured" communities, fried meat "braai" anywhere when there is enough income, or homemade beer brewed in 200-liter barrels over open wood fire – pungently punctuate the energy landscape. Maintaining flows around food practices, including constructing supply chains for fuels, establishes an underlying rhythm of this landscape. This wood, we learn, is acquired by young men in the community, entrepreneurs who have disassembled pallets and collected wood scraps from a nearby salvage yard, an urban twist on fuel procurement practices in many unelectrified rural communities, where women usually bear the burden of daily fuel collection and other household chores. As we settle into conversation, our hosts continue tending the imbaulas, regularly reaching for scraps of wood as smoke billows from buckets, a reminder that energy landscapes are also embodied landscapes.

> I grew up going to the field to fetch wood and all that . . . and when I got married, I thought to myself, That life is finished for me, I'm not going to do that anymore, making a gulley [imbaula] outside and cooking outside . . . black pots and black cans and black hands.
> No! I don't want [that] no more.

> (Mari, Flamingo Crescent Resident)

Figure 18.3 Cooking with an imbaula in Flamingo Crescent

Figure 18.4 Buyiswa cooking

Inside a double-wide shack that serves as home, spaza shop selling household goods, and meeting place with community partners, we see clear plastic bottles of paraffin, fuel for stoves recognizable to us as "kerosene" camping stoves. Easier, safer, more controllable and the least expensive improved alternative to imbaulas for cooking and heating, paraffin fuel and stoves are ubiquitous in informal settlements. Usually purchased from local spaza shops, and sometimes acquired on credit or by calling on neighbors for a small cupful in moments of need, the fuel is a staple for those who can afford stoves, yet it also carries serious, well-known risks – fire, fumes, and fatal ingestion by children (and tellingly, more often by adults). Stacked here, along with candles, matches, lighters, cookies, cigarettes, chips, cool drinks, and fruit, the bottles of paraffin are a store of relative plentitude and wealth, a commodity for sale and income.

It's Friday, we are using paraffin to get warm, people are getting drunk, it can burn down our shacks.

(Anonymous, Imizamo Yethu settlement)

Inside most shacks, lighting is poor even during daylight, which enters mainly from an open door, windows being few due to weather and security risks and their limited effectiveness amid densely packed shacks. Absent electricity, candles are the lighting fixture of choice, even though reading, studying and working by them is difficult. Outside, street lighting is impossible without electricity and people deeply fear being robbed or raped if caught walking along paths and roadways at night. Most people avoid nighttime trips alone, collecting "night soil" until morning when it can be safely disposed.

Figure 18.5 Shack fire

Figure 18.6 After a fire in Langrug

Seasonal weather patterns present other challenges. In the summer sun, corrugated iron roofs heat up poorly ventilated, minimally insulated shacks. In the cool, damp winters, shacks often flood and drop to grindingly cold temperatures. Imbaulas are commonly moved inside overnight during winter for a secondary role of space heating, the easy access and low cost of wood justifying for many the attendant increase in fire hazard, even when other forms of heating are available.

> The huge fire we had in twenty-oh, twenty-five? And then we had another fire here where one of the, where, ah, one kid burned out. And then we had a huge fire there at the back which came all the way till where I'm staying now, and that was huge. Because it was gas exploding, it was paraffin. Ew! It was like . . . don't wanna go through that again. [Afterward], I was moving in with you, and you were moving in with me . . ., you keep my things there in your new place, so it was like that. Everybody was, like, putting up their shacks again, and re-build and re-construct.
>
> (Mari, Flamingo Crescent)

Only weeks after this interview, fire tears through another community where we have been working, destroying 70 shacks, leaving many people possessionless and killing two. In 2008, two fires destroyed two dozen shacks and a community center complex in Khayelitsha that was the base of operations for our student-community projects, including, ironically, one team working on safer energy options. The City of Cape Town reported 3480 shacks damaged or destroyed by fire and 105 killed in just one year, 2012 (City of Cape Town 2013). Local residents are urgently aware of the collective conundrum they face: open flame heating and illumination are energy technologies that work within existing supply chains and social relations, but at the cost of increasing everyone's vulnerability to fire and the general sense of precariousness that punctuates the everyday rhythms of life in informal settlements. Ironically, some fires provide opportunities to reorganize shacks in layouts more conducive to roads, sewers and electrical grids, but the window of opportunity is brief, as many residents will of necessity begin rebuilding literally before the smoke has fully cleared.

When not bringing violent change, open flame cooking and heating bring a slower, more grinding exposure to pollutants and particulates that combine in toxic irony with dampness and mold from leaky, poorly heated homes to make asthma, tuberculosis and other respiratory diseases endemic. People living off-grid also often feel excluded from the promises of modernity (Robinson 2006), with its TVs, refrigerators, computers, cell phone chargers, lights, and appliances. Electricity provision, along with water and sanitation services, are thus front and center in political protests and advocacy for settlement upgrading.

Electricity cometh!

We return to Flamingo Crescent in October 2014, just four months since a "re-blocking" effort started in earnest, and already it is a landscape transformed. Re-blocking is a strategy to improve services, safety and quality of life in informal settlements by upgrading existing shacks, one by one or in small groups, while introducing public services such as water, sewer, roads, public lighting, and electricity in shacks. It has been embraced as a vehicle for co-production of the informal settlement communities, bringing together several offices within the City of Cape Town, anti-poverty and community development NGOs, community leaders, and in some cases national and international donors and NGOs.

The assemblage of 100 shacks that had evolved over 15 years has been replaced by ordered rows of sturdy, fire-resistant tin shacks, each with a concrete outhouse and water supply, but still without foundations and thus finessing regulatory restrictions about the permanence of this settlement. Each shack has been painted with an official ERF number (street address), symbolizing a new level of [in]formality and land tenure, including the right to formal electricity and postal service. Roads, finished with brick blocks, establish a tight grid through the community, allowing access for the trucks that come periodically to complete water hook-ups, waste management systems, construction, and soon, electricity. The pre-school crèche is just completed and, along with public space amenities, provides space for children to gather and play, an alternative to the adjacent busy city road that had been the children's main gathering place (though the road still sees much activity). All visible signs of shack fires are gone, and residents have eased into new rhythms created within the new spatiality of the community. Clothes lines are fashioned over and through the block roads, gardens grow in scant open patches on the community periphery and abutting shacks. The charred tin imbaulas are still focal points for cooking and socializing, but electricity poles now stand promisingly around the community awaiting connection; the energy landscape is now animated by discussions about hopes and dreams, plans and impending changes, security, development, and modernization.

Gridded electricity is the most immediate sign of the slums' integration and visibility within the city's ever-changing energy landscape. Conventional development theories posit a predictable, clear transition from primary fuels (wood or paraffin for cooking and heating, candles for light) to cleaner fuels (gas) and electricity, as urbanization rates and national and/or settlement incomes rise symbiotically. More grounded theoretical approaches challenge such models, pointing to pervasive complexities that disrupt a simple linear urban-energy-income transition pathway for many, if not most, slum dwellers. Mehlwana (1997) points to five complicating factors: urbanization is not always a one-way movement, and often involves oscillating migration that prevents permanent urbanization; the correlation between how long a place has been urbanized and the uptake of "modern" energy systems is often weak; income is not the only determinant of energy use; complex social relations underlie household management decisions that are often understood as rational and financially influenced decisions; and gender and power dynamics within households are invisible in the rational actor transition model.

People hold diverse, often contradictory, views regarding electrical power access and upgrading in informal settlements. Some see electricity first as an opportunity to modernize cooking practices, though in practice cooking relies on other fuels long after electricity access is acquired. Some see great hope in public safety improvements and social change, where others are far less optimistic. Still, gaining access to electricity stands as a practically universal aspiration, a transformative moment in slum dwellers' energy landscape, one emblematic of a community's recognition by the state and the promise of incremental upgrading.

On a Tuesday afternoon, with great import but little fanfare (that would await the mayor's ribbon cutting ceremony), the lights came on in Flamingo Crescent. Street lights mounted high above the shacks were the first connection to the power grid, amply illuminating public spaces and the recently blocked roads, and immediately shifting flows in the community, bringing many activities into the light, and pushing some further into dark corners. Residents now eagerly anticipate the connection of their own shacks. Electricity boxes installed in each home will soon be connected, ready to be activated by prepaid cards with electrical "units" (kilowatt-hours) purchased at local convenience stores. The shacks contain one or two overhead lights and outlets for other appliances, though to use an electric stove residents will have to purchase fuses of a higher wattage. Much of the chatter this day revolves around the prepaid cards – how much power they will provide, how to maximize their value, and what obstacles inevitably await. Training sessions offered by the City have reportedly been helpful, though the depth of prior knowledge about these nuances, in a community that has not had gridded electricity itself, reflects the kind of conversations and personal experiences with electricity that flow through informal settlements.

Across town, in informal settlements where electrical service has been present for some time, the community moves to a different soundtrack – gospel, hip hop, and reggae spill from local shabeens (taverns) and homes, often to the chagrin of neighbors whose thin metal shack cladding performs as poorly as a sound barrier as a thermal barrier. Barber shops and beauty salons announce their presence with lighted displays, spaza shops and shabeens run refrigerators, youth read by electric light, clinging more realistically to dreams of completing high school and college, and everywhere cell phones are pulled out casually and often. Nighttime roads are still dangerous, but less so.

At the household scale, what it means to be a "residential electricity consumer" is far more varied than in wealthier communities, including those in which most planners' imaginations are grounded. Access may come all at once to an area, formally such as through re-blocking, or piecemeal via myriad strategies and gradients of [in]formality. Usage varies widely within and between households with change in employment, household make-up, shack size, financial windfalls and disasters, appliances bought/sold/stolen/damaged, seasonal temperature, and modes of accessing power (SEA 2014, p. 23).

For those with a legal box, prepaid cards allow purchase of units in small increments, as little as 10 rand (~$0.80 USD). "Poor" households are eligible to receive 50 kilowatt-hours free of charge each month under the national Free Basic Electricity (FBE) policy administered through local government, though important elements of the benefit are still being refined (SEA 2014).

Beyond serving one's own home, the box is for many an important asset, allowing households to in effect become micro-distribution utilities to nearby neighbors lacking access.

Stretching the line between formal and informal

Sizwe has lived in the settlement of Nyanga for 20 years, most recently alone in his 15-square-meter shack following the deaths of his mother from tuberculosis and a cousin from HIV/AIDS, and the relocation of his twin sister to Johannesburg. Though holding a regular job with a local NGO, Sizwe chooses to stay in a settlement reputed to have the highest murder rate in South Africa because he is familiar with the area and can live rent-free. He describes how a lively, varied, and often exploitive market in electricity services sprang up as street lights appeared in the early 2000s along with service to established structures with ERF numbers, such as apartheid-era hostels of four families each and newer RDP houses.

Sizwe reports that when he began a construction job in 2005 and needed a cellphone, he implored his mother to get electricity in their shack, but she said, "We will wait, because it is out of the law and if we connect what will stop us from doing the next thing out of the law, and the next?" Only when a trusted neighbor received an electrical box did she relent, closeness with the box-holder – spatially and socially – providing sufficient moral and practical grounding for proceeding.

The first step in connecting illegally, "township-style," to a box or power pole is to hire a local electrician – not just anyone, but the person specifically, if informally, authorized to do the work by the local powers that be (e.g., street committee, ward councilor, civic association, gangsters, cops). Once connected, Sizwe no longer had to leave his phone at a neighbor's home to charge, and his family quickly acquired and enjoyed new appliances. Meanwhile, however, three other households were also tapping the same box, causing overloaded circuit breakers to trip, and it was soon decided that two households would continue tapping the box while the other two tapped the pole directly.

The economics of accessing power through such "micro-utilities" are complex and often much more expensive and uncertain than obtaining power legally. All four households not only paid their electrician 250 rand (~$20 USD) for the initial connection, they also pay for repairs to wires that are cut, disconnected, or stolen, sometimes more than once a month. They also pay the box holder a R50 fee monthly. Some micro-utility operators pay for their entire electricity bill from sales to neighbors, and may even profit. Beyond power supply costs, Sizwe replaces fuses every two months and once repaired his refrigerator three times in a single month and his television three times last year. When local connection points become overly complex, a neighborhood electricity manager emerges through what Sizwe calls 'slum governance', to establish some temporary solution. But excessive loading and power fluctuation are problems not only at the micro-scale, but as we will see, also at regional and national scales.

Backyards and beyond: densification and extensification on the slippery slopes of modernization

"I see you have a washer machine."

"Yes, but she's using that basket [laundry tub] because she's saving the electricity, because the thing is, she cannot use it because it's gonna use a lot of energy and the people that we share that box with, you understand, they're gonna complain because . . . that electricity, it's expensive."

(Anonymous)

Electricity service can also alter local landscapes by making settlements more appealing, inducing in-migration, causing existing structures to be enlarged and new shacks to be added in the "backyard" spaces surrounding them. Tenants and backyarders, whether relatives, friends or strangers, typically provide rental income and/or share in managing the household and associated costs and challenges. But as new arrivals connect to the grid, risks can increase. More lines are spliced and longer extension cords run through increasingly complicated terrain. Settlement densification increases, with more shacks, closer together, sometimes to the point that moving about requires navigating narrow, labyrinthine pathways. When access for firefighters and other emergency services is thus impeded, an unintended effect of gaining electricity service in an area can be to *increase* rather than *decrease* fire risk, as fires may spread faster, farther and less controllably. Electrocution is also a risk.

While backyard effects densify the existing settlement footprint, electrification and other upgrading activities can also attract new settlement and extensification along the settlement's periphery. If new arrivals can quickly set up and resist local government or private landowners "demolishing" their shacks (the term of art), then new areas or entire settlements can rapidly form. One such case is in Langrug, an informal settlement formed by agriculture laborers in South Africa's famed wineland district near Cape Town.

Langrug is built on a steep slope immediately above the long established, generally low income Groendal community. Brick houses on the lower slopes reflect formality and a fragile sense of permanence; informal dwellings surging up the slope above them, along with some public services – one paved and a few graded dirt roads, ablution blocks and water standpipes, refuse collection points, and electricity connections – reflect the upgrading efforts of area residents, Stellenbosch Municipality, South African Shack Dwellers International Alliance and other partners. As the Langrug community grew upslope, the area's landed class that depends on cheap agricultural labor put pressure on the municipality to declare a hard boundary above which further services would not be extended, making arguments about landscape aesthetics, viewshed degradation, and negative impacts on the tourist economy – arguments not unlike those pertaining to energy landscape controversies elsewhere. As a result, the 600 or so residents of Zwelitsha, the uppermost section of Langrug, enjoy spectacular views but feel themselves perched at the edge of modernity, drawing as they can on electrical and other basic services from below.

And from that metaphorical vantage point, the outlook is anything but clear. At the national level, South Africa is still just two decades on from an intense civil struggle that led to a democracy in which the vast majority of the population could finally reside, work, move, vote, and associate freely across racial lines once drawn and policed by the apartheid state. On the promising side, South Africa has received considerable praise for adopting and to some extent implementing some of the most progressive policies for informal settlement upgrading and improving energy access for the urban poor. Nearly three million RDP houses with basic services have been provided free of charge to poor citizens. Social welfare grants for the elderly, disabled, and unemployed women with children have reduced extreme poverty and, along with new progressive energy subsidies, reduced energy poverty. Access to electricity has grown remarkably, from 36 percent of households in 1994 to 87 percent today, and national policy calls for all citizens to have access by 2025 to "modern" energy sources – electricity and renewables – that it is hoped will also reduce use of paraffin, candles, firewood, and other unhealthy, unsafe and environmentally unsustainable resources (SEA 2014). The will and capacity to produce such results is impressive, especially given the enormous demands of revamping all levels of government to serve those who for generations had been systematically discriminated against, and to integrate members of these groups into government, despite educational and experiential deficits.

At the same time, much RDP housing is of very poor quality, located in sprawling, low density, bedroom communities far from job opportunities and lacking most basic social amenities of true community and often reinforcing the spatial disparities of apartheid. Only about half the households eligible for FBE are actually benefitting (SEA 2014). Urbanization and population rates continue to surpass efforts to provide affordable housing, so more households live in informal dwellings today than ever before. While the rate of informal dwellings growth nationally seems to be slowing, this is not the case in Cape Town (ibid.), perhaps in part because of the relative success of local government in providing basic services.

Meanwhile, the national electrical generation and transmission system operated by the public utility ESKOM that for decades produced excess coal-fired electricity for sale throughout southern Africa has become a pariah, as "load-shedding" – rolling blackouts lasting two hours or more – has become a frequent, debilitating and costly occurrence across the country. Despite substantial wind and solar potential, these renewables account for a miniscule 0.1 percent of generation, versus 87 percent fossil fuels, mostly coal (Energy Information Administration 2015). And the ANC's commitment to developing nuclear power would entail an enormous capital outlay, one that would surely delay investment in cheaper, quicker, safer, more job-producing and less corruption-prone energy options, to say nothing about options that can be integrated with informal settlement upgrading strategies.

There is thus a palpable unease, shared by many across class and racial divides, that the promise and social energy borne of the democratic revolution may be dissipating, or lagging behind the hopes and expectations of those in desperate need. Thus the pursuit of energy sustainability in informal settlements is just that – a pursuit, a chasing, a constant effort to grasp the ephemeral, stabilize the transitory, and make tenable the human demands for energy within environmental limits and social realities.

> In the time of apartheid, there was never load-shedding. Although the apartheid was wrong, but they never had load-shedding. Why can't this government cope today? Although there is extra people getting electric today, even so.
>
> (Anonymous middle-class suburb resident)

Pissing into the fiery wind

> Keeping the place [upgraded Flamingo Crescent] safe and clean, I don't think it's gonna happen, because people do other things with their money, they don't think about their safety.
>
> (Mari)

Our contributions in this complex environment have revolved around working on new models for community development, models that allow us to support our local partners' initiatives, such as re-blocking in informal settlements, with dialogue, reflection and collaborative prototyping, while also stressing de-colonizing forms of engagement. In particular, our students work closely with community co-researchers to promote their involvement in imagining, planning and implementing change processes, with an eye toward community-generated and -sustained initiatives supported by local partners. In the energy space, this has involved projects addressing safe home lighting and cooking strategies, low carbon and efficient electricity practices, and cultivation of entrepreneurship around energy products and services. A critical conundrum we have only dimly begun to understand and respond effectively to is that of building a sustainable "supply-and-knowledge" chain that brings together material and social elements of locally sustainable energy [sub-]systems.

Local agents are essential both because they are deeply versed in the realities that are likely to make the difference between success and failure of sustainable energy efforts in informal settlements, and because every investment in informal settlements is also an opportunity to create jobs and build skills. But there are as yet few successful examples in South Africa of supply chains that reach deeply into informal settlements to provide the urban poor access to more suitable energy products and services, whether at the retail level through spaza shops, or through government agencies or other distribution points.

Yet, opportunities seem to abound for such interventions in communities both with and without electricity. Efficiency is highly valued for cost saving, and cooking products like the Wonderbag – an insulated sack that allows for the slow completion of cooking without fuel – hold promise for reducing the use of combustion fuels and creating job opportunities (Church et al. 2014). Another low energy, low cost product, the Litre of Light, provides nearly free home daylighting by repurposing plastic bottles built into shack roofs, while the IShack, a micro-franchise initiative, uses solar powered hubs for charging phones and other appliances (Keller 2012). Experiments continue with low cost insulation and even fire retardant shack exteriors. And the goal of low cost solar energy production, that may power a small shack, support enterprises or even provide a node for a micro-grid, is being pursued by the City and a range of NGO's and entrepreneurs. Still, diffusion of these products is negligible, for reasons that are not fully understood. Some community partners describe an aversion to stop-gap interventions designed for the poor; they reserve their efforts for "real" access to modern energy. It is also hard to overstate the role of vandalism and theft as barriers to material interventions; external structures like solar panels are particularly exposed, but even securing tools and materials during installations requires extreme vigilance. Still, community members and partner organizations strive for innovation around sustainable energy practices, motivated largely by ambitions around safety, health, access and opportunity, and often also attuned to concerns about climate change and resilience. One thing seems clear, however – the range of experimentation remains inadequate to the need.

Conclusion

This chapter has explored energy issues in the context of efforts to transform informal settlements in Cape Town, South Africa into healthier, safer, more satisfying places for people to live. We've drawn on different voices and experiences from around Cape Town to sketch out scenes and tell a loose narrative about challenges of energy provisioning and place-making in these marginalized, peri-urban communities. The story traces two parallel, idealized arcs, one an historical/social arc that begins with informal settlement origin stories and moves on to scenes of daily life and energy provisioning, then to actions taken informally to improve access to electricity, then to the arrival of formal, state-supported electrical power services as an anticipated force for modernization and social transformation, and finally to scenes that illustrate unforeseen consequences and unravelings that can follow in its wake. The other is a spatial/social arc, from individuals and families and their shack homes, to neighbors sharing and sometimes contesting energy services, to neighborhood power brokers controlling illegal grid connection points, to settlement-wide dynamics of densification and expansion driven by forces such as fire, fear, hope, and electrification, to city-wide and national policies, actors and socio-technical systems, all of which are embedded, as are we and the chapters in this book, in global discourses of energy sustainability and development.

A not surprising finding from our work is that there is a great need for more rigorous research and for deep practical commitments to help co-produce improved energy services in informal settlements. We have explored some of the texture and complexity around energy relations in these communities, and the conditionality of these findings suggests that any could be pursued in much greater depth. Even more challenging is to develop policy and implementation strategies responsive to this social and material complexity. The difficulty of translating insight into effective intervention is illustrated by how little the nuanced anthropological work done in this same location of Cape Town some 20 years ago (Ross 1993; Mehlwana 1997) seems to be informing policy and program development. For example, the recent report *Tackling Urban Energy Poverty in South Africa* (SEA 2014) provides a fine, comprehensive overview of national and regional energy policy and consumption trends, and a cogent analysis of urban energy poverty and the challenges of developmental local government in the delivery of energy services. But the report says little about the sub-surface texture and complexities, the anthropologies of energy, in these communities and about what kinds of "bottom-up" strategies might be an effective, indeed necessary, complement if the promise of a relatively progressive state policy environment is to be realized in this complex milieu. This detached quality is surely not for a lack of understanding, as these challenges are evident to any researcher who ventures into the informal sector, but instead suggests limits in conventional modes of both policy development and academic practice. Absent a deep and sustained commitment to policy experimentation and socio-technical innovation geared precisely to informal settlement environments, the haphazard importation of strategies from rural or more developed urban contexts are likely to continue to produce sub-optimal results and lead to the steady erosion of incremental upgrading gains.

Highlighting this entropy is not an indictment of slum dwellers, nor of practitioners and policymakers who struggle against these deeply structural, institutional and historical tendencies. Rather, the agency of these stakeholders in constructing workable energy solutions is central to this narrative. We suggest that making headway in this context requires grappling with fundamental questions around the knowledge/practice interface. What kinds of knowledge could lead most immediately to improvements around these difficult challenges? What kinds of knowledge formation strategies can produce such outcomes? What types of knowledge/practice communities can best flush out the promising strategies of intervention and collaboration? Our modest efforts in Cape Town with students and local partners have sought answers to these questions through projects that also seek to support community members' creative impulses in these areas. Much more collaborative experimentation, "learning (and sometimes failing) by doing" is needed, however, so that more promising, workable models can be nurtured and improved over time (e.g., Keller 2012; Gaunt et al. 2012; Castán Broto et al. 2015).

Any specific pathway to improving energy access and quality in informal settlements is uncertain, but the broad approach of co-produced, incremental upgrading that is being explored in Cape Town and elsewhere seems worth pursuing. It signals a commitment by a diversity of actors – state and non-state, local to global, across race – to engage collaboratively and move together into the space of experimentation and mutual learning. Fundamentally, *in situ* upgrading supports established communities, recognizing the legitimacy of residents and their agency in creating a place and community and mobilizing the flow of basic public services. As the chapter describes, electricity plays a critical role in this process of incremental urbanism, unleashing possibilities for entrepreneurial activity, entertainment and communication and also generating myriad new social relations and geographies.

Beyond questions of praxis and community development strategies in informal settlements, the chapter contributes to expanding theories about the geography of energy in informal settlements. As geographers and others seek to understand the role of networked infrastructures in constructing cities and creating possibilities for sustainable urban futures, the struggle for modern energy in informal settlements reminds us that, for a huge portion of the human population, the city is constructed over smoky imbaulas, over a tangle of spliced wires, through the vagaries of slum governance, and the endless claims-making to municipalities. The simultaneous densification and extensification of settlements in response to electricity access is a spatial reflection of the intense social negotiations and innovations that will continue to shape the pursuit of sustainability in energy systems and informal settlements.

References

Castán Broto, V., Stevens, L. and Salazar, D., 2015. *Energy access and urban poverty: Energy and everyday life in an informal settlement in Maputo, Mozambique (Poor People's Energy Briefing)*. Rugby, UK: Practical Action Publishing.

Church, A., Huet de Bacellar, T., Kennedy, R. and Wu, J., 2014. *An entrepreneurial initiative for distributing energy efficient products in low income communities*. Worcester Polytechnic Institute Cape Town Project Centre.

City of Cape Town, 2013. Challenges – informal settlement incidents. *Fire Safety Symposium – Goodwood*. Cape Town: City of Cape Town. www.westerncape.gov.za/text/2013/March/fires-in-informal-settlements.pdf.

Elmes, M. B., Jiusto, S., Whiteman, G., Hersh, R. and Guthey, G. T., 2012. Teaching social entrepreneurship and innovation from the perspective of place and place-making. *Academy of Management Learning & Education*, 11(4), 533–554.

Energy Information Administration, 2015. *South Africa: International energy data and analysis* [online]. Energy Information Administration, US DOE. Available from: www.eia.gov/beta/international/analysis _includes/countries_long/South_Africa/south_africa.pdf.

Gaunt, T., Salida, M., Macfarlane, R., Maboda, S., Reddy, Y. and Borchers, M., 2012. *Informal electrification in South Africa: Experiences, opportunities, and challenges*. Cape Town: Sustainable Energy Africa.

Harber, A., 2011. *Diepsloot*. Johannesburg: Jonathan Ball.

Housing Development Agency, 2013. *Western Cape: Informal settlements status (2013)*. Johannesburg: Housing Development Agency.

Huchzermeyer, M., 2004. *Unlawful occupation: Informal settlements and urban policy in South Africa and Brazil*. Trenton, NJ: Africa World Press.

Huchzermeyer, M. and Karam, A., 2006. *Informal settlements: A perpetual challenge?* Cape Town: UCT Press.

Hughes, T. P., 1983. *Networks of power: Electrification in western society 1880–1930*. Baltimore, MD: Johns Hopkins University Press.

Jiusto, S., McCauley, S. and Stephens, J. C., 2013. Integrating shared action learning into higher education for sustainability. *Journal of Sustainability Education* [online], 5. Available from: www.jsedimensions.org/ wordpress/content/integrating-shared-action-learning-into-higher-education-for-sustainability_2013_06/ [accessed May 2013].

Keller, A., 2012. *Conceptualising a sustainable energy solution for in situ informal settlement upgrading*. (MPhil thesis). Stellenbosch University.

Marais, L. and Ntema, J., 2013. The upgrading of an informal settlement in South Africa: Two decades onwards. *Habitat International*, 39, 85–95.

Mehlwana, A. M., 1997. The anthropology of fuels: Situational analysis and energy use in urban low-income townships of South Africa. *Energy for Sustainable Development*, 3(5), 5–15.

Mitlin, D. 2008. With and beyond the state — co-production as a route to political influence, power and transformation for grassroots organizations. *Environment and Urbanization*, 20(2), 339–360.

Mitlin, D. and Satterthwaite, D., 2004. *Empowering squatter citizen: Local government, civil society, and urban poverty reduction*. London, Sterling, VA: Earthscan.

Robinson, J., 2006. *Ordinary cities: Between modernity and development*. London: Routledge.

Ross, F., 1993. Transforming transition: Exploring transition theories in the light of fuel use in a squatter settlement. *Journal of Energy in Southern Africa*, 4(2), 44–48.

Roy, A., 2005. Urban informality: Toward an epistemology of planning. *Journal of the American Planning Association*, 71(2), 147–158.

SEA, 2014. *Tackling urban energy poverty in South Africa*. Heinrich Boll Stiftung Southern Africa.

Silver, J., 2014. Incremental infrastructures: Material improvisation and social collaboration across post-colonial Accra. *Urban Geography*, 35(6), 788–804.

Swilling, M., 2014. Contesting inclusive urbanism in a divided city: The limits to the neoliberalisation of Cape Town's energy system. *Urban Studies*, 51(15), 3180–3197.

UN-HABITAT, 2012. *State of the world's cities 2012/2013: Prosperity of cities*. UN-HABITAT.

Visagie, E., 2008. The supply of clean energy services to the urban and peri-urban poor in South Africa. *Energy for Sustainable Development*, 12(4), 14–21.

19

The energy geographies of incremental infrastructures in Ga Mashie, Accra

Jonathan Silver

Introduction

Across the low income neighbourhoods of the rapidly growing coastal West African city of Accra shifting energy geographies of a noticeably incremental nature are present and a part of everyday urban life. Incremental infrastructures can be understood in the context of energy as urban systems (including but also beyond electricity) in the making, undergoing constant adjustment and reconfiguration through testing and experimenting as urban dwellers seek to shift energy flows and circulations to address current conditions of socio-environmental inequality and shape future energy possibilities. This incrementalism is interwoven into the lived experiences and daily practices of people and energy across both formal and informal networks in the city generating important questions for how policymakers, planners and communities come to understand and act upon urban energy imperatives in the African and wider global South context. Paying attention to these incremental geographies and the ways that energy becomes embedded in people's lives opens up insights into how we problematise networked systems and the everyday life of energy. Without acknowledging these widespread incremental urbanisms within the growing field of energy geography scholars will miss the important urban energy dynamics in contexts outside the universal, standardised infrastructures that are operated in global North cities (Graham and Marvin, 2001; Luque-Ayala and Silver, 2016). It is these notions of what an energy system constitutes (integrated, networked and with full access) that continue to shape much of our wider knowledges, policy orientations and developmental visions concerning energy yet ones ill-suited to the multiple terrains and shifting (energy) modernities of 21st century urbanisation across the global South.

This chapter seeks to contribute to the growing literature on energy geographies to better account for the lived experiences of infrastructure and the incremental reshaping of urban worlds. It does this through a focus on the specific geographical context of postcolonial Accra and its electricity system. This is undertaken through an examination of incremental and lived experiences of energy in one particular neighbourhood of the city, Ga Mashie, a low income, centrally located 'popular neighbourhood' with both formal and informal infrastructure conditions. It does this in order to open up discussions across a series of conversations, many of which are captured within this volume that seek to better interrogate the scales, flows and circulations of energy, open up attention to the energy terrains across different contexts and the resulting geographies of these shifting infrastructures. The next section outlines the imperatives associated with urban energy in sub-Saharan Africa and the necessity of researching and contributing toward debates focused on the incremental nature of energy geographies. Section three provides an overview of the neighbourhood of Ga Mashie. Section four examines the lived experiences of infrastructure in the area in further detail before the chapter concludes by arguing that the development of an explicitly geographical perspective on energy must take seriously such incrementalism

Why energy geographies need to engage with the incremental

Despite sub-Saharan Africa having the lowest urbanisation levels of all global regions rapid urban growth on the continent is expected to see over 700 million urban dwellers by 2030 (UN-Habitat, 2008) and 1.2 billion by 2050 (UN-DESA, 2009) in what Parnell and Pieterse (2014) term 'Africa's Urban Revolution'. This rapid urbanisation will account for nearly all population growth in sub-Saharan Africa including in Ghana over the coming decades, creating multiple policy, financial and social challenges for governments as urban infrastructure systems are increasingly placed at the centre of development efforts (Khennas, 2012). One of the most crucial of these service provision demands is the distribution of energy services to power the burgeoning towns and cities of the continent (Madlener and Sunak, 2011).

As numerous policy publications make explicit (Infrastructure Consortium for Africa, 2009; UN-Habitat, 2014) the investment deficit for infrastructure including energy is significant (Simone, 2010) producing widespread disruption (Khennas, 2012) and shortages in accessing technologies and essential services (UN-Habitat, 2014) necessary for economic growth and social reproduction. Attention, resources and policies are increasingly being focused on new investment to develop widespread urban energy provision and are increasingly well explored in the energy geography literature (see Introduction to this volume). At a national level, countries such as Ghana are broadening their ambitions from electrification to engage with wider issues of carbon reduction, climate change mitigation and renewable energy (Bulkeley et al., 2010; Ghana Energy Commission, 2011). What these and the plethora of initiatives, projects, financing and emerging institutional arrangements show is that energy has over the last few decades has become an important concern to (multi-scalar) urban governance actors and the broader development aspirations of countries such as Ghana (Kebede et al., 2010; Sokona et al., 2012). And urban contexts are becoming increasingly important sites in these processes with global goals around energy being reframed at multiple scales and across geographically stretched networks that implicate the everyday operations of infrastructure across and beyond cities. It is within the context of these multiple, shifting energy geographies across the continent that the need to understand the particular contexts, dynamics and trajectories becomes a vital prerequisite to addressing the multiple developmental challenges through which energy plays a mediating role.

There are of course important differences in the ways that the urban energy demands, infrastructures and plans across sub-Saharan Africa are (re)shaped by the regions distinct and diverse energy geographies, and these are very different in turn to the development of modern urban energy services in the global North (Rutherford and Coutard, 2014; Luque-Ayala and Silver, 2016). The growing numbers of urban poor (Parnell and Walawege, 2011) demand access to affordable, clean energy from informal settlements that constitute large parts of many cities and likely to remain so for future decades. As Rutherford and Courtard (2014: 1356) comment, urban energy transition in the South thus clearly means something very different from the North, combining issues around governance, access to finance, trade and supply chains with everyday concerns of, among other things, very low basic household incomes, availability of cooking fuel and indoor air pollution.

It is important then for scholars and of course policymakers and planners to consider further the relations between energy and geography within global South urban contexts such as Accra. One way to undertake such an examination that in the words of the Editors of this volume, 'challenges established assumptions' is to consider further the incrementalism of urban energy systems that characterise many of these urban spaces and the energy circulations that become, 'enmeshed with the grain of everyday life'.

A focus on the incremental nature of energy systems can draw on a growing body of recent work across postcolonial urban studies in sub-Saharan Africa specifically and global South infrastructure space more broadly that is interrogating the circulations, geographies and city building in neighbourhood and household spaces (De Boeck and Plissart, 2004; Diouf, 2003; Edjabe and Pierterse, 2010, 2011; Simone, 2004, 2010). This work provides important ways to link micro-scales of everyday life; household, neighbourhood and so forth to wider circulations of people, technologies and means of inhabiting cities. While there are important differences between these perspectives, particularly concerning the role of the state in instigating and supporting incremental upgrading from these everyday processes this research points to very different urban trajectories and futures than those of the global North. They offer us a landscape of multiple energy infrastructure beyond the formal grid. To engage with this series of propositions is to address the lived and experienced ways in which these urban energy systems are operated, maintained and adjusted alongside the potentially agency of urban dwellers to shape these conditions and the wider energy geographies of urban worlds.

Mbembe and Nuttal (2004: 369), considering Johannesburg, suggest a useful way of thinking through the urban life of cities such as Accra 'like the continent itself, it is an amalgam of often disjointed circulatory processes . . . it has become, in spite of itself, a place of intermingling and improvisation.' This widely held notion of city life in the South suggests that simultaneously and alongside the ways in which energy geographies are produced through formal planning, state and large scale private sector investment we must also pay attention to related forms of the reshaping of infrastructure space that can be considered as in the making, lived, provisional and often temporary in nature. Work on incremental and lived infrastructures has grown out of the socio-technical grounding of research on infrastructure and offers ways to analyse how residents seek to improve household or neighbourhood conditions through multiple and unfolding intersections and experiences with infrastructure (Graham and McFarlane, 2014; Simone, 2008, 2011; McFarlane, 2011).

This relatively recent work on the incremental in postcolonial urban studies has drawn on a longer tradition of scholarly attention in development studies research across regions including Latin America, South East Asia and sub-Saharan Africa. This broad and extensive literature has particularly focused on housing (Turner, 1972; Afshar, 1991; Kallus and Law-Yone, 1997; Mukhija, 2001) that has helped to shape development discourse and practice (see for instance Turner, 1996; Satterthwaite and Mitlin 2013) especially around notions of incremental upgrading of existing informal settlements and associated strategies of poverty reduction. This work and associated debates across the academic, policy and civic society worlds of development have extended notions of incrementalism from the initial focus on housing to wider infrastructure systems and the right to various urban services including of course energy (Abbot, 2002; Choguill et al., 1993; Hasan and Vaidya, 1986).

Together these historic debates in development studies and contemporary work across postcolonial urban studies offer an important resource to the emerging field of energy geographies. Here, we can posit that across infrastructure space (and in contrast to larger scale investment by either market or state mechanisms) residents are engaged in a series of encounters and lived experiences with energy networks and associated socio-environmental conditions. These socio-technical, energy relations can be framed as a constant and unfolding dialectic of adjustment and readjustment (Turner, 1972) that may be unsanctioned and often, in arbitrary ways illegal (Roy, 2009b). Such incremental interventions are predicated on addressing conditions of poverty, making possible flows of energy into households or simply navigate the challenges of everyday issues of energy security, disruption, poverty and viability (Castán Broto et al., 2014). They lay on the periphery of the official, mapped and regulated circuits of energy and electricity plans of city planners, engineers and policymakers (Silver, 2014). It is across these infrastructure spaces and through these incremental activities that the urban poor may themselves generate new energy geographies through shifting material and social relations, either within the home or the wider neighbourhood. Such incremental infrastructures open up ways to generate income, to keep the lights on for homework or to improve tomorrow's prospects. Through these acts, residents in low income neighbourhoods work hard to sustain not only survival in the city but bring forth new infrastructural worlds and energy futures. Thus, while state investment may provide electrification, and subsequent operation and maintenance for growing numbers of the urban population this reality remains far from a universal experience in many global South cities (Furlong, 2014). Here the role of urban dwellers and their lived experiences of shifting and incrementally adjusting energy systems should be given greater attention.

Researching Accra's incremental energy geographies

Accra is a fast-growing city situated on the Gulf of Guinea and one of West Africa's most important economic hubs reflecting its position at the centre of the country's rapidly growing economy in which, 'Ghana has become a rising star and is one of the recent success stories in Africa' (Breisinger et al., 2009: 3). Ghana's urban population, now estimated at 51 per cent (UN-Habitat, 2008) includes many new urban dwellers living in conditions of poverty and socio-environmental marginalisation (Songsore, 2009). In response to these dynamics commentators have characterised such processes as the, 'urbanization of poverty' (Ravallion, 2005). This rapid urbanisation (of poverty) will account for nearly all population growth in Africa and generates a series of competing energy demands. These include the delivery of basic service provision such as electrification in urban spaces visibly lacking networked systems, the upgrading of existing networks and economic support to residents unable to sustain flows of electricity into poor households. Despite recent attention focused on its growth and development Accra remains a city divided by socio-environmental inequality and networked services such as electricity continue to reinforce unequal relations that link into longer colonial and post-colonial histories of exclusion and control (see Silver, 2015 for a history of the electricity network in the city). Despite these networked inequalities or 'splintered urbanisms' (Graham and Marvin, 2001) urban dwellers may find ways to incrementally reshape these conditions in order to support the precarious lives they experience in the city.

To examine these incremental infrastructures fieldwork was conducted over six months in 2010 and 2011 in Accra to develop a detailed ethnography of the intersections between the housing conditions, the energy network and urban dwellers in low income neighbourhoods. The main site of this work was the community of Ga Mashie, together with insights gained through living in Osu, another centrally located neighbourhood. Working with a research assistant we arranged seven discussion groups across different parts of Ga Mashie, spent much time walking and drifting around the neighbourhood and learning about the intersections of infrastructure and urban life for residents. Brought together the research sought to build up a detailed analysis of the incremental energy geographies of the area. This ethnography was supported by a number of complimentary research methods that helped to broaden the perspective of the research and included 15 interviews with key stakeholders both in this neighbourhood and working with assorted urban intermediaries in the wider city. A qualitative, 35 household survey also helped to test emerging ideas and analysis and draw in a wider section of research participants. As Merrifield (2002: 14) argues, 'truth claims about cities must be conceived from the bottom upward, must be located and grounded in the street, in urban public space' and the methodology has sought to reflect such an approach in seeking to understand these incremental energy geographies.

The popular neighbourhood of Ga Mashie

James Town or Ga Mashie as it is known locally (and used throughout this chapter) is a long established 'popular neighbourhood' in central Accra often termed 'Old Accra' due to its place as the nexus of colonial activity and function as the main port in the city and plays an important role in the history of energy in the city. As a local politician (#2) explains, 'James Town is a historical point of interest for energy in Ghana. It was one of the first places to have electricity but not for the community until the Nkrumah era.' Yet during the colonial era the residents of Ga Mashie experienced urban life without functioning services or decent housing and in contrast

Figure 19.1 The incremental energy geographies of Ga Mashie

to the emerging networked services in the residential and adjacent commercial spaces of the British coloniser. From the 1930s the neighbourhood became an important centre for the burgeoning independence movement. Protests by war veterans at the harbour and the election in 1951 of Kwame Nkrumah add to the historical importance of the neighbourhood in the wider life of the city and the nation. After independence President Nkrumah's modernisation vision for the country includes the development of a central business district in Accra and Ga Mashie is repeatedly targeted for slum clearance to make way for these plans. Hess (2000: 54) explains that, 'The administration repeatedly instigated slum clearance in James Town and Ussher Town, but relocation and demolition efforts were consistently resisted by Ga leaders and members of the opposition'. The spectre of demolition continues to haunt residents in Ga Mashie showing an overlapping history of eviction and uncertainty throughout different eras of the city history and the precarious nature of urban life in the neighbourhood. Although the neighbourhood is popular it is also classified by the Ghana Statistical Service, (2012) as a slum with severe overcrowding among its estimated 125,000 residents. These socio-environmental conditions are experienced by up to 50 percent of Accra's residents and complicating standard definitions of formality and informality. Like other popular neighbourhoods in the capital Ga Mashie experiences a broad number of socio-environmental hazards that place it in the 'severe' category of environmental burdens for areas in the Greater Accra Metropolitan Area. These include according to Songster (2009:1), 'inadequate potable water supply, unsanitary conditions, insect infestation, uncollected garbage, poor waste water disposal, smoky kitchens, crowding and shelter poverty'.

Older people in Ga Mashie remember the installation of the electricity network into the neighbourhood as the spearhead for a wider programme of modernisation in Ghana during the early 1960s. As one participant of a workshop (#3) elaborated, 'When Nkrumah was President he made promises to the nation that we would have the electrical power and he constructed the dam. Because of the importance of the harbour for the country then electrical power came here from around 1965.' The area now has an ageing electricity network that is managed by the state owned Electric Company of Ghana (ECG) and provides extensive distribution across Ga Mashie, including into some parts of the higher density, informal sections. The operation of this electricity network is facing considerable strain due to the increasing population of the area and resulting usage, particularly in the context of under-investment in maintenance and vital upgrading. Alongside the ever-increasing cost of electricity and intricately tied to the national crisis over generation (that has come to be known by the long-suffering public as 'Dumsor') energy access or security of supply are far from certain in Ga Mashie.

The household survey showed an average spend of between 6 and 20 Cedis (£1–£4) per month on pre-paid electricity credit and revealed that many households struggle to afford a constant flow of energy to meet daily needs with just under 80 percent of household respondents sometimes or often struggling to afford electricity credit for the home. This energy poverty was a key concern for participants in a workshop, 'In James Town they must bring the bills down' (Participant, resident workshop #4). The cost of electricity constitutes a financial burden for most of Ga Mashie's residents many of whom survive below the poverty line. Although a lifeline tariff of around 6 Cedis (£1) per month is supposed to be provided by the national government as an energy subsidy to the urban poor this payment to the utility companies it is often unpaid. The retrofitting of pre-paid meters (PPM) technology across the neighborhood from 2008 resulted in significant tension between the utility company and residents and illustrated the ongoing struggles over the governing of infrastructure space in Ga Mashie. Together with the ongoing 'Dumsor' crisis and the breakdown and malfunction of ageing infrastructure residents in Ga Mashie face multiple difficulties in accessing and sustaining vital flows of electricity into their homes and businesses. It is these urban energy geographies that provide the basis from which residents in the neighborhood seek to incrementally adjust and reshape conditions of energy poverty and wider marginalisation.

Ga Mashie's incremental electricity network

In Ga Mashie the reconfiguration of the electricity system is noticeably incremental and provisional and the network is in a state of constant movement and flux. During workshops in the neighbourhood participants articulated the difficulty of sustaining flows of electricity into households. The lack of large-scale investment into the energy network to respond to these issues by the state or market means that households have to use a range of ways of intervening across the infrastructure in order to connect and access flows of electricity. Many households in Ga Mashie are involved in a series of ongoing acts to figure out new network spaces and flows beyond the geographies of the official system and the often restrictive rates of the Electric Company of Ghana (ECG). These ways of incrementally reshaping the electricity network shape and mediate the circulations of energy into the neighbourhood in an often experimental and certainly shifting maze of reconfiguration and as a response to the relative cost of electricity.

The research identified a range of these moments of adjustment that incrementally reshape the energy network in Ga Mashie including a series of clandestine connections, which are common across many urban poor areas throughout the city and beyond. As a local politician (#2) explains 'Electricity in James Town is a problem. They say it costs too much and they are not able to afford it, so most of us use the illegal connection.' It is suggested by both residents and stakeholders such as at the local ECG payment office that many people are involved in clandestine connections to the electricity network. It can involve households that have already secured an official connection to the network (but are perhaps unable or unwilling to pay), alongside households in the informal parts of Ga Mashie that to do currently enjoy access to the network. Such reconfiguration is normally undertaken with the support of electricians who are sympathetic to the need of households to access the flows of electricity that remain unaffordable to many. A connection can cost up to 50 Cedis (£40) and the payment to an electrician can seem like a large upfront investment for most families and is usually negotiated in advance with payments perhaps over a number of months.

Another incremental adjustment of the network in Ga Mashie involves households or businesses, working again with electricians, to configure a 'split' electrical supply system. This involves residents registering and paying for a proportion of their electricity through the ECG network and then also using a clandestine connection to access further flows of energy. Here, the appearance of paying for electricity is constructed through this interaction with the network and keeps investigation by the ECG limited. This adjustment is popular among the residents of the neighbourhood, especially to aid economic activity with flows of electricity considered too expensive to sustain small locally based enterprises and the tight margins in which they operate. These actions provide a strategy to counter the increasing unaffordablity of electricity generated to create revenues for the underfunded and often crisis hit Volta River Authority as generators and the ECG as distributors of electricity in Ghana.

Alongside such connections to the network that reconfigure the neighbourhood's electricity network on an ad hoc, unplanned basis, other incremental interventions serve to support households in accessing flows of electricity in response to structural transformations in the flow and circulation of electricity. Perhaps most importantly has been the introduction of the PPM technology which began in Accra as a pilot project for the country and in Ga Mashie from around 2008. Estimates from a community leader suggest that over 90 percent of households now have this technology installed. Opinion is divided about the use of PPM technology.

The household survey showed that 65 percent of respondent households preferred these PPM compared to the old credit meter system, suggesting it brings benefits to the area such as the inability of unscrupulous landlords to now present artificially high electricity bills and providing a level of control over the cost of electricity for the household. For poorer households with limited financial resources the need to pay upfront for electricity in the form of credit means interruptions in supply during periods when the household struggles to raise the required finances for the purchasing of more credit. In the past, the sustaining of electricity flows into the household continued until billing at the end of the month or even longer. As one resident rather accurately comments, 'If you have money you can get a card. If you don't you sit in the dark' (Workshop #2). Furthermore, in the large households of Ga Mashie it is difficult to keep track of who is using the electricity and thus responsible for further purchasing of credit for the PPM. These new energy geographies show both the everyday circulations of energy in Ga Mashie but they also show the infrastructural conditions upon which residents incrementally reshape the network.

The new technologies imposed on energy access via the PPM have, like the earlier 'post-paid' meter system, become a focus of attention by residents that seek to resist the effects of neoliberal management of the electricity system that aims to increase revenue for the utility companies. These moments of incremental adjustment involve two main activities. First, households without PPM are working together by using the same meter as neighbours and seeking to confuse the ECG in relation to who has responsibility for the bill. Yet this strategy is being recognised by the ECG who are trying to stop joint usage of meters mainly through the introduction of PPM which has made such activity less likely to produce incremental gains and forces the residents to find new pathways of reconfiguration. Thus, another way for accessing underpaid or free electricity is to ask an electrician to come and adjust the meter to stop it going up or measuring energy usage, something that can be done with both the PPM and the older meter. Residents of Ga Mashie may be charged from two to ten Cedis for this network reconfiguration from an electrician which like the previously described incremental adjustments can create a transformation in the ability of households to sustain the metabolic flow of electricity.

These incremental ways of countering energy poverty are not always successful or long lasting. Improvements in accessing energy can be reversed, provide the requirement for a bribe to a public official, or even result in criminal proceedings against the household, showing the precariousness that exists in the space of the everyday. The ECG employs technicians to go house to house to find these clandestine connections to the network in a dialectical dance of connection, disconnection, reconfiguration and such like that plays out across the electricity network in a way that produces an unmappable infrastructure of movement and flux, becoming and unbecoming. The ECG undertake such activities as these new configurations of infrastructure flows threaten its revenues, sustainability and future investment plans. If evidence of these clandestine connections is found then the ECG may cut the connection, in some cases informally demand money from the family in order to sustain such configurations or even prosecute those without the finances to pay the bill (or bribe).

Those unable to make such payments, i.e. the poorest in the community, are most vulnerable to processes of criminalisation through such clandestine connections, something which is likely to further compound the poverty of the household. These multiple strategies developed by residents generate an ongoing, low intensity conflict between community members and ECG, a relationship that plays out through everyday energy interactions of these different urban actors and the network spaces they seek to direct and control. Residents in Ga Mashie suggest they are forced to engage in such modifications due to energy poverty, the peripheral status in the urban economy and the introduction of PPM technologies that mediates the flows of energy to the household. Many households in the neighbourhood tamper with the PPM to access free or lower cost flows of electricity. Thus a dialectical relation exists between the historically mediated production of energy infrastructure in the city (Silver, 2015) and the incremental responses by urban dwellers to these socio-environmental conditions that show the city to be a space of ongoing flow and circulation, reconfiguration and adjustment. A range of other ways of intervening in the electricity network by residents in Ga Mashie are generated, experimented with and often upscaled as older methods become obsolete, service providers develop new technologies or the need for energy increases dramatically. These incremental shifts in the circulation of electricity across the neighbourhood may also be read as demanding a future with lower tariffs or even free energy for marginalised communities, an attempt by the neighbourhood to not just envisage but bring about a future and differentiated energy geography of this 'popular neighbourhood'.

It is across the networked, urban poor spaces of Accra in neighbourhoods such as Ga Mashie that it is possible to draw out a growing sense of incremental energy geographies through a seemingly similar set of ways in which urban dwellers in such neighbourhoods seek to interact, experiment, experience and intervene with the energy network. These incremental ways of reconfiguring infrastructure show a remarkable tenacity despite the operations of urban service providers and other urban governance actors in seeking to ensure that flows of electricity are accounted and paid for. New technologies such as PPM may limit the incremental gains made by a particular intervention, such is the nature of this dialectic across everyday life. Simultaneously new strategies emerge and are experimented with to counter such foreclosure and provide a new and incremental configuration. This is process of ceaseless, circulating and experimental micro-scale transformations of the urban energy network that moves between and across the geographies of the city's neighbourhoods.

Conclusion: an incremental energy geographies?

It is worth reflecting upon the incremental nature of the infrastructure space described above. First, that incremental adjustment is undertaken in response to the current geographies of energy and its generation, distribution, operation and maintenance across Accra. The lived experiences of low income communities prompt attempts to reshape network space to sustain everyday life through the inequalities of contemporary Accra. Second, these ways of intervening across energy conditions are generated through everyday social reproduction and the need to open up new possibilities for socio-economic improvement. Without the capacity to amass large scale investment for network operation, upgrading and repair due to limited access to capital, resources or political connections many urban dwellers need to find ways to intersect interests through and with the energy network. This social reproduction of urban life is entwined with the need to sustain circulations of electricity into the household to support health, economic activity, education and so forth. Third, the everyday becomes a site in which these urban dwellers are required to prefigure, imagine and bring about reassembled infrastructure as a way to reproduce urban life in popular neighbourhoods. Fourth, that these energy geographies resonate not just in Accra but across the multiple and shifting forms of global urbanism both in the South and increasingly in the North. Roy's (2009a: 829) critique and subsequent call for action that, 'the world is not flat, and it is time to produce a more contoured knowledge of its cities' needs to be explored further in researching the socio-spatial distribution of energy activities. Crucial to such a task is to take seriously the incremental nature of energy networks in rapidly growing cities such as Accra and to acknowledge the lived experiences of everyday energy usage. To do so involves shifting beyond the linear (Euro-American) ways of explaining energy transition and the ways that infrastructure space is (re)shaped.

Addressing these incremental infrastructures of energy and the ways that these incremental systems are operated, maintained and reshaped through lived experiences helps to situate energy geographies within the everyday space of the network.

Finally, attention needs to be given to the agency of urban dwellers in contexts that leave them without sustained urban service provision of energy into the household. Understanding how urban populations shift the socio-materialities of these networks through everyday interactions is vital to both understanding and supporting such attempts to re-wire the city. This lived experience of infrastructure space from which urban dwellers negotiate and navigate the inequalities reflected and reinforced through energy distribution, operation and maintenance is vital to how energy imperatives are taken forward in cities such as Accra. Beyond the mechanisms of state and large-scale private investment incremental adjustments to infrastructure are part of everyday life for residents in neighbourhoods like Ga Mashie as they seek to steer, direct and to open up (incrementally) new possibilities for fairer urban service provision of energy.

References

Abbott, John. 2002. An analysis of informal settlement upgrading and critique of existing methodological approaches. *Habitat International*, 26(3), 303–315.

Afshar, Farokh. 1991. Affordable housing in Pakistan. *Habitat International*, 15(4), 131–152.

Breisinger, Clemens, Diao, Xinshen, Schweickert, Rainer, & Wiebelt, Manfred. 2009. Managing Future Oil Revenues in Ghana – An Assessment of Alternative Allocation Options. Kiel Working Papers 1518. Retrieved from: www.ifpri.org/sites/default/files/publications/ifpridp00893.pdf.

Bulkeley, Harriet, Castán Broto, Vanesa, Hodson, Mike and Marvin, Simon (Eds.). 2010. *Cities and Low Carbon Transitions*. London: Routledge.

Castán Broto, V., Salazar, D. and Adams, K. 2014. Communities and urban energy landscapes in Maputo, Mozambique. *People, Place, Policy*, 8(3), 192–207.

Choguill, Charles L., Franceys, Richard and Cotton, Andrew. 1993. Building community infrastructure in the 1990s: progressive improvement. *Habitat International*, 17(4), 1–12.

De Boeck, Filip and Plissart, Marie-Françoise. 2004. *Kinshasa: Tales of the Invisible City*. Antwerp: Ludon.

Diouf, Mamadou. 2003. Engaging postcolonial cultures: African youth and public space. *African Studies Review* 46(2), 1–12.

Edjabe, Ntone and Pieterse, Edgar (Eds.). 2010. *African Cities Reader: Pan-African Practices*. Cape Town: Chimurenga Press and African Centre for Cities.

Edjabe, Ntone and Pieterse, Edgar (Eds.). 2011. *African Cities Reader II: Pan-African Practices*. Cape Town: Chimurenga Press & African Centre for Cities.

Furlong, K. 2014. STS beyond the 'modern infrastructure ideal': extending theory by engaging with infrastructure challenges in the South. *Technology in Society*, 38, 139–147.

Ghana Energy Commission (2011) Energy Statistics 2000-2009 (Online) www.energycom.gov.gh/pages/docs/energy_statistics.pdf (Accessed 02.05.11).

Ghana Statistical Service. 2012. *Population and Housing Census: Summary Report of Final Results*. Accra: Sakoa Press Limited.

Graham, Stephen, and Marvin, Simon. 2001. *Splintered Urbanism*. New York: Routledge.

Graham, Stephen, and McFarlane, Colin (Eds.). 2014. *Infrastructural Lives: Urban Infrastructure in Context*. London: Routledge.

Hasan, Arif, and Vaidya, Chetan. 1986. Two approaches to the improvement of low-income urban areas—Madras and Orangi. *Habitat International*, 10(3), 225–234.

Hess, Janet. 2000. The structure of nationalism in Accra, Ghana. *Africa Today*, 47(2), 35–58.

Infrastructure Consortium for Africa. 2009. *Annual Report 2009*. Tunis Belvedere, Tunisia: Infrastructure Consortium for Africa.

Ingold, Tim. 2000. *The Perception of the Environment: Essays in Livelihood, Dwelling and Skill*. London: Routledge.

Kallus, Rachel, and Law-Yone, Hubert. 1997. Neighborhood—the metamorphosis of an idea. *Journal of Architectural and Planning Research*, 14(2), 107–125.

Kebede, Ellene, Kagochi, John and Jolly, Curtis M. 2010. Energy consumption and economic development in Sub-Sahara Africa. *Energy Economics*, 32(3), 532–537.

Khennas, Smail. 2012. Understanding the political economy and key drivers of energy access in addressing national energy access priorities and policies: African perspectives. *Energy Policy*, 47(1), 21–26.

Luque-Ayala, Andres and Silver, Jonathan (Eds.). 2016. *Energy, Power and Protest on the Urban Grid: Geographies of the Electric City*. London: Routledge.

McFarlane, C. 2011. *Learning the City: Knowledge and Translocal Assemblage*. Oxford: Wiley Blackwell.

Madlener, R. and Sunak, Y. 2011. Impacts of urbanization on urban structures and energy demand: what can we learn for urban energy planning and urbanization management? *Sustainable Cities and Society*, 1(1), 45–53.

Mbembe, Achille and Nuttal, Sarah. 2004. Writing the world from an African metropolis. *Public Culture*, 16(3), 347–372.

Merrifield, Andy. 2002. *Dialectical Urbanism*. New York: Monthly Review Press.

Mukhija, Vinit. 2001. Upgrading housing settlements in developing countries: the impact of existing physical conditions. *Cities*, 18(4), 213–222.

Parnell, Susan and Pieterse, Edgar (Eds.). 2014. *Africa's Urban Revolution*. London, Zed Books.

Parnell, S. and Walawege, R. 2011. Sub-Saharan African urbanisation and global environmental change. *Global Environmental Change*, 21, S12–S20.

Ravallion, Martin. 2005. On the urbanization of poverty. *Journal of Development Economics*, 68(2), 435–442.

Roy, Ananya. 2009a. The 21st-century metropolis: new geographies of theory. *Regional Studies*, 43(6), 819–830.

Roy, Ananya. 2009b. Why India cannot plan its cities: informality, insurgence and the idiom of urbanization. *Planning Theory*, 8(1), 76–87.

Rutherford, Jonathan, and Coutard, Olivier. 2014. Urban energy transitions: places, processes and politics of socio-technical change. *Urban Studies*, 51(7), 1353–1377.

Satterthwaite, David, and Mitlin, Diana. 2013. *Empowering Squatter Citizen: Local Government, Civil Society and Urban Poverty Reduction*. London: Routledge.

Silver, Jonathan. 2014. Incremental infrastructures: material improvisation and social collaboration across post-colonial Accra. *Urban Geography*, 35(6), 788–804.

Silver, Jonathan. 2015. Disrupted infrastructures: an urban political ecology of interrupted electricity in Accra. *International Journal of Urban Regional Research*, 39(5), 984–1003.

Simone, A. 2004. People as infrastructure: intersecting fragments in Johannesburg. *Public Culture*, 16(3), 407–429.

Simone, Abdou Malique. 2008. The politics of the possible: making urban life in Phnom Pehn. *Singapore Journal of Tropical Geography*, 29(2), 186–204.

Simone, Abdou Malique. 2010. *City Life from Dakar to Jakarta*. New York: Routledge.

Sokona, Youba, Yacob Mulugetta, and Haruna Gujba. 2012. Widening energy access in Africa: Towards energy transition. *Energy Policy*, 47(1), 3–10.

Songsore, Jacob. 2009. *Environmental Health Watch and Disaster Monitoring in the Greater Accra Metropolitan Area (GAMA)*. Accra: Ghana Universities Press.

Turner, John. 1972. Housing as a verb. In J. F. C. Turner and R. Fichter (Eds.), *Freedom to Build*. New York: Macmillan, pp.148–175.

Turner, John. 1976. Tools for building community: an examination of 13 hypotheses. *Habitat International*, 20(3), 339–347.

UN DESA. 2009. *Urban and Rural Areas 2009*. (Online) www.un.org/en/development/desa/population/publications/pdf/urbanization/urbanization-wallchart2009.pdf (accessed 12.06.14).

UN-Habitat. 2008. *The State of African Cities*. Nairobi: UN-Habitat.

UN-Habitat. 2014. *The State of African Cities*. Nairobi: UN-Habitat.

20

Exhaustible-renewable wind power

Alain Nadaï and Olivier Labussière

Introduction

Policy debates concerning the development of non-fossil energies have mostly been framed in technological terms, letting us believe that if we were to address and solve technological issues, alternative energy patterns would be at hand. Within this technological rhetoric, natural entities such as wind or solar radiation, which are commonly conceived as potential renewable energy resources, have been quantitatively measured and mapped so as to assess the related technological potentials – e.g. wind power or solar PV production potentials (GWh or GW).

As emphasised quite a while ago by Shove (1998), this way of approaching energy issues casts a singular perspective on the development of energy technologies, because the social dimension is then conceived and approached as a social/institutional barrier to the realisation of a (pre)given technological potential: the possibility for social actors to take part in the construction of this technological potential is sidetracked from the beginning. More broadly, by conceiving such an abstract physical potential as a guide to energy change, it is a whole set of actual issues and of messy but decisive socio-material relations involved in the development of energy projects that are not properly accounted for. With them, it is the so-called 'externalities' and the sustainability – i.e. the social or environmental consequences – involved in changing our ways of dealing with energy(ies) that are not fully addressed.

Ready-made dichotomies such as 'renewable'/'non-renewable', 'non-fossil'/'fossil' energy, serve this state of affairs. They suggest that such a qualification mirrors a matter of fact, a natural qualification, instead of pointing to the rich web of socio-material relations and practices which underlay sustainability. The first category of energies (i.e. the 'renewable' and 'non-fossil' energies) is supposed to be sustainable, while the second is not ('non-renewable' and 'fossil' energies): wind or solar energies are sustainable; oil is not. Not to say that oil could be – it has never really been developed to be so. However, now that so-called 'renewable' energy technologies and finance have been industrialised and globalised, the question of whether and under which conditions they are – or not – sustainable is a current and an actual issue. It calls for re-opening and exploring the ways in which we extract, concentrate, circulate and consume these energies: this is a task for a human geography of energy transitions.

This chapter builds on previous contributions pointing at the constructed nature of wind power potential, in order to introduce the reader to a human geography of wind power. Drawing from a set of case studies of wind power development and a (theoretical) discussion of the sociotechnical dimension of wind power technology and of various types of commons involved, assembled and re-composed in its development, we challenge the idea that wind energy is renewable per se. Rather, we defend the idea that concern for the social and situated dimension of wind power development and due processes can endow wind power with the property of being a renewable energy.

The first part presents our approach. The second part presents cases of renewable and non-renewable wind power developments in three parts of France – Aveyron, Seine et Marne, Narbonnaise (see Figure 20.1) – and draws up conditions for wind power to be developed as a renewable energy. The third part discusses these results emphasising the importance of accounting for the material and relational dimension of wind energy.

Figure 20.1 Case study locations

Walking the hidden 'worlds' of renewable energy resources, entering a commodity chain approach

During the past decade, following and analysing the emerging wind power landscapes in France, we walked the field with civil servants and local inhabitants. If innovative solutions sometimes took place, local opposition to wind power also became vivid: it can be regarded as a symptom of this emerging energy world and its difficulties to generate genuinely shared dimensions.

One session of fieldwork, at the origin of our desire to write this chapter, was especially disconcerting: local inhabitants refused to be interviewed, whereas they had agreed to do so during a former field session, two years before. Their refusal was justified by the turn taken by local wind power development, a process which they felt powerless to influence, and one that denied locally shared practices and values. Beyond the usual considerations about the impact of wind power, such a refusal faced us with the problem of its potential unsustainability. Wind power development could exhaust itself by affecting and exhausting for a long time the energies of the local inhabitants. As stated by an interviewee: 'you better not talk about wind power around here for at least a generation'.

If we are serious about such a testimony, this fieldwork experience calls for a re-examination of the fundamental notions of 'renewable'/'non-renewable'. Wind energy is presented as a 'renewable' energy because of the physical characteristics of the wind resource: intermittent, yet flowing and abundant. This suggests that wind is unlimited and may be appropriated under free access. The development of industrial wind farms depends, however, on the use of other resources such as land, electrical network, local actors' involvement, local social synergies . . . These are limited. They may be held and controlled to a great extent by one or some individual(s). From this perspective, the appropriation of the wind is not free and might be better reflected upon by considering the assets engaged in its harnessing as a 'bundle of resources'. This image is, however, only partially satisfactory. It suggests that the resources involved in the development of wind power are brought together in one place and on one single scale, which never is the case. The challenge is much more to investigate the long chain of institutions, groups and devices through which these resources are – or not – taken into account at different levels.

Several works in social sciences help us to face this challenge. Some point out the progressive *fossilisation* of renewable energies as they are developed by historical energy operators and adapted to fossil energy institutions and infrastructures (Evrard, 2013; Raman, 2013). Raman also looks at the material supply chain of renewable energy technologies and underlines their increasing use in rare earths. Other works (Castree, 2003; Bakker & Bridge, 2006; Bridge, 2010) insist on the economic, social and environmental consequences of reducing energy resources to their physicality. Hartwick (1998) proposes a commodity chain approach in order to account for the consequences of transforming entities into commodities, from production to consumption.

These analyses suggest analysing further the full chain through which 'renewable energy' is made available for use as an energy commodity in order to understand how far the notion of 'renewability' may be considered as a part and a product of this commodification. One can, indeed, reasonably assume that the new economy of energy is framing the resource as an abstract flow – renewable per se – for renewability not to be conditioned upon the complexities of the development of the resource. Laying bare the web of relations and entities as well as the transformations that are engaged in the process of commodification of wind energy is thus a way to deconstruct renewability.

The notion of commodity and commodification has been developed in critical geography so as to highlight that the status of commodity is not intrinsic to the entities commodified. It is assigned and this assignment has consequences. In particular, capitalist commodification has been broadly associated with a set of dimensions – not all being required – (Castree, 2003) such as:

- *Privatisation* – i.e. assigning rights to a named individual, group or institution.
- *Alienability* – i.e. the possibility for the commodity to be physically and morally separated from their sellers.
- *Individuation* – i.e. the representational and physical act of separation from a supporting context, such as water from its environment.
- *Abstraction* – i.e. the qualitative specificity of a thing is assimilated to the qualitative homogeneity of a broader type or process (allows for instance for unproblematic equivalence, such as a wetland here is made replaceable by a wetland elsewhere).
- *Valuation* – how things take on specific form of value (for instance, blindly profit-driven in capitalist society).
- *Displacement* – how something appears as other than itself (spatio-temporal separation of production and consumption, so that you cannot see the exploitation of south African workers included in Italian handmade gold jewellery).

These dimensions suggest looking more specifically, in our case, along the chain from wind to ReN-kWh, at the transformations, attachment/detachment, framings, displacements of the resource in order to make it into a consumable kWh. In order to do so, we approach wind power as a sociotechnical system. This means that looking at it as an isolated technical artefact is a fallacy, for it ignores the networks and collectives engaged in its production. Works in the sociology of science and technology have contributed to highlighting this hybrid and ramified dimensions of sociotechnical systems (Akrich, 1988, 1989, 1993). These works are rooted in Gilbert Simondon's (1989) approach to the co-genesis of technical objects and their 'associated milieu', seen as a comprehensive set of relationships that gives to this object a specific existence and effectiveness.

We approach the commodification of wind energy as a chain of transformations from the untamed kinetic energy of the wind to the consumption of a standard electric ('renewable' or not) kilowatt-hour. In a first approach, this chain may be described by six operational links:

- *depositing* (turning the untamed wind into a deposit, i.e. a potential energy resource);
- *harnessing* (capturing the untamed wind);
- *extracting* (the energy from the wind);
- *provisioning* (entering the extracted energy into a provisioning system; i.e., the grid);
- *distributing* (transporting the energy to the places where it is to be used); and
- *commercialising* (exchanging the energy for money with the final user).

Such a chain focuses the attention on the materiality of the energy resources, their changing reality while passing from one state to another, as well as on the collectives that are engaged in these transformations. In the case of wind energy, our proposal is to follow the transformations of energy from a state of heterogeneity and intermittency to one homogeneous and stable commodity that is suitable for commercialisation while – at the same time – describing the strategies and the rivalries through which emerging sociotechnical collectives try to take advantage of the possibilities (concentrations, accumulations, re-allocations . . .) offered by the chain. In the end, this will allow us to bring together in an encompassing view the ways in which the renewability of wind energy is (or is not) held stable in a context of locally contested sustainability of wind power developments (see above).

Our empirical material is based on the analysis of the French wind power policy and of various local case studies (Labussière & Nadaï, 2014), and we focus here on three of them. This does not allow an evenly comprehensive examination of the commodity chain, because emphasis has been put on somewhat upstream links of the chain, at the level of the planning and siting of the projects (depositing ∞ harnessing). However, it allows for a critical examination of the entities included in or put aside by the construction of this chain and after all considered as externalities of the wind power policy. It also allows for a comprehension of the way in which the notion of 'renewability' is hold as a stable attribute of wind kWh in a context of contested, not always sustainable, wind power developments.

X-ble wind energy, a commodification account

Wind power is not exhaustible or renewable per se: it is X-ble, 'X' indicating that its fate and sustainability depends on how it is socially and materially assembled. In order to explore this assemblage along the commodity chain of wind energy, we now turn to examining a few case studies, located in France, in the light of a commodity chain approach. We first consider the framing of wind power development by national policies. This allows us to cast light on how wind energy has been articulated with the current commodification of electricity – i.e. its construction as a standard kWh – and how 'renewability' as an attribute of wind energy is constructed, stabilised and circulated along the commodity chain. We then turn to three different case studies of local wind power development. We follow the bundle of resources that are engaged into it and the way in which they are – or not – accounted for in this development.

French wind power

Since the end of the 1990s, the European Union has endowed energy and climate policies with an unprecedented regulatory basis.

In gradually implementing this regulatory framework, France has deeply modified its energy sector. It unbundled its former monopoly – Electricité de France (EDF) – thus allowing for other electricity providers to sell electricity on the French market. The French government also sustained a diversification of its electricity mix, by adopting feed-in tariffs for renewable electricity (FR, 2000), by reforming its energy policy programing law (POPE law) (FR 2005, 2009a, 2009b) and by undertaking an unprecedented stakeholders' consultation: the Grenelle Environment Forum (COMOP 2007; FR 2008, 2009c, 2009d). This resulted, among other things, in the adoption of medium-term objectives for ReN development and in a development of installed capacity (see Table 20.1).

On the aggregate, this increase in the installed wind power capacity, which is significant for France, has barely followed an economic rationale. It started with the windiest/most profitable zones and progressed towards less windy/profitable areas, in spite of all the heterogeneities involved in the development of projects on the local level.

Table 20.1 Cumulated installed wind power capacity in France (megawatts)

Year	2000	2005	2010	2013
MW	48	873	5,979	7,821

Sources: SER, L'énergie éolienne en France – Panorama 2013, www.enr.fr/docs/2013122234_SERCarteEolien20132.pdf, consulted 8 July 2014.

This development has leaned on the progressive structuration of a commodity chain which allowed for the conversion of untamed wind into a marketable ReN kWh. It was supported by the progressive adoption of a policy framework which first included as key devices a feed-in tariff (2001) and an authorisation for electricity production[1] and grid connection. These two procedures framed French wind power as a privately owned/privately developed activity and a grid connected form of energy production: private developers could benefit from a fixed price – a tariff higher than the price of the standard 'non-renewable' kWh – in exchange for each wind power kWh injected into the grid.

Grid injection imposes a physical transformation of the electrical current which comes out of the turbine apparatus. Horizontal axis turbines, the most developed technology, use large blades to harness the wind. When the wind blows, the blades are forced round, driving a gear system which allows a shaft alternator (1500tr/mn) to generate electricity. A transformer located in the tower of the turbine increases the voltage up to 20kV for grid injection.[2] This physical transformation of wind energy underlay a change in the status of the electron. Once injected into the grid, it becomes part of the electrical flow like any other electron, renewable or not. The 'renewable' origin of this electron is physically blurred. It would be lost, had a system of 'guarantee of origin' not been first set in place in 2006 in France, then updated (FR, 2012) and harmonised at the EU level starting 2009, with the adoption of the Renewable Energy Directive (2009). The 'guarantee of origin' allows for a circulation and a trading of the 'renewable' qualification of wind power electricity. This quality is detached from the physical electrical current, abstracted as an informational asset – a certificate/computer file – traded and re-bundled with conventional electrical current as a commercial product: the 'green electricity' proposed to end users by electricity producers.

Individuation of wind electricity – in Castree's sense (identified above) of a separation of this electricity from its supporting environment and context – is at the core of this chain of operations. Yet, individuation here results in a pooling of the electrons and a loss in singularity for the electrons coming out of the same wind farm: their collective origin as the outcome of a singular wind farm is made imprecise in the process. It is detached from this origin and from the landscape in which the production of electricity is taking place, and redefined as a generic attribute that can be re-bundled with any standard kWh. This displacement is not hidden to the final electricity consumer: commercial contracts blankly assert that by contracting green electricity final users do not actually buy a green electron or a green kWh, they only contribute to the extent of their purchase to the remuneration of renewable electricity producers.

Yet, the reach of this displacement is made undiscernible by the displacement itself, for the definition of 'renewable' is itself brought into a referential in which it loses its situated origin. Indeed, the quality of being renewable is (re)defined through the decree that sets the 'guarantee of origin' as an institutional and trading system. The eligibility to the guarantee as a renewable energy production infrastructure is ruled by the administrative authorisation for electricity production, itself referring to the code of energy for the definition of what is a renewable energy.[3] The corresponding definition clearly is essential: the code merely lists the types of energies considered renewable, without any consideration or provision regarding the ways in which these are actually assembled as energy resources. Therefore, while the displacement operated by the guarantee of origin is transparent to the final consumer, the meaning of this displacement – that is, the extent to which the wind power project that is remunerated through the renewable certificate was actually developed in a sustainable way or not – is not traceable. Among other things, the relation to landscape is no longer traceable.

In the end the commodity chain is roughly divided into two parts. A downstream part is made up of 'provisioning ∞ distributing ∞ commercialising'. It is the part whereby the kinetic energy of the wind is extracted through the turbine as an appropriated uneven electrical energy (alternator) and enters a genuine commodification process. This part of the chain includes both the physical transformation of the energy (transformer, merging with electrical flow) and its abstraction as informational assets (standard kWh; certificate of guarantee) that can be separately traded and re-bundled as marketable product ('green kWh').

The upstream part of the commodity chain, made up of 'depositing ∞ harnessing ∞ extracting', is the one whereby the untamed wind is turned into appropriated uneven electrical energy, to be injected in the grid. On a practical level, it covers the mapping of wind energy deposit on various scales, the development of the wind farm project including its siting, and the extraction/conversion of the kinetic energy of the wind by the turbines. It is thus the part of the chain in which territorial resources such as land, landscape, local collectives are engaged into project development and planning. In France, the structure of this part of the commodity chain has also been supported by public action – e.g. since the end of 1990s the French energy agency has coordinated the mapping of wind deposit on a regional scale – and the progressive setting up of a policy framework for project authorisation, which included: permitting, impact studies, public inquiries, spatial planning instruments (Wind Power Development Zones or WDZ), and a wind power tax in order to redistribute part of the benefits from wind power to local territories.

The politics of the process through which this framing emerged and was evolved has been analysed elsewhere (Nadaï, 2007; Nadaï et al., 2015; Labussière & Nadaï, 2015). It has been described as a 'backward planning process', symptomatic of the difficulty besetting French politics with respect to decentralising (wind) energy policy and managing the politicisation of wind power. The adoption of feed-in tariffs (2001), in the absence of any planning or permitting framework, witnessed French regulators' belief that economic incentives and private business alone could take charge of developing wind as a renewable energy resource. A few years of contested wind power development and rising local opposition to the energy landscapes they produced, were enough to nurture an emerging controversy about the relevance of wind power as a renewable energy technology. In 2005, a stormy parliamentary debate on the energy programming law (Pope Law) resulted in the adoption of WDZ, construction permit (including impact studies and public inquiries) and a redistributive tax. These were aimed at regulating the so-called environmental and landscape 'impact' of wind farms and bringing their development into spatial planning.

The adoption of WDZ bounced on ongoing local experiments with the spatial planning of wind power (Nadaï & Labussière, 2014; Nadaï & Debourdeau, 2013). It was aimed at enticing local authorities to pool together and devise spatial planning. WDZ had to be proposed by towns or groups of towns. Only the projects sited in these zones became eligible to feed-in tariffs. It could thus potentially bridge the upstream and the downstream of the wind power commodity chain and rearticulate landscape planning issues with economic incentives (feed-in tariff). However, it also carried out the political ambivalence under which it was born. Statutorily, the French WDZ was no more than an electricity supply contract, that had to be approved by the departmental prefect (state representative). It was not an urban planning document. Endowing WDZ with such a status – as was the case in Germany or Denmark – would have transferred the decisions about wind power projects and landscapes into the hands of the local democracy (e.g. community councils).

WDZ has been contested from its inception. It was first contested by wind developers and the renewable energy lobby who pointed at it as an additional institutional barrier to renewable energy development. More recently, wind power opponents' networks have attacked several wind power development zones in regional courts in order to block the development of wind power. Eventually, in 2015, the WDZ setting was cancelled on a national level. This withdrawal occurred in a controversial context in which certain groups of actors lobbied for streamlining authorisation procedures, others for making them more demanding. These interacted in a complex way. Wind power ended up being listed as an environmentally risky facility (ICPE, literally 'Classified Installation for Environmental Protection') and subjected as such to dedicated risk assessment/authorisation procedures. This outcome can be regarded as symptomatic of a political will to reduce the rising political complexities surrounding wind power development by addressing them through the formal, procedural and objective appraisal of risk assessment. Public inquiries still remain part of the process. Yet, as standardisation and formal appraisal gain room in the upstream portion of the chain, the quality of being 'renewable' comes to be defined in a more objective, stable way. It becomes detached from the process of its emergence and from the energy landscape in which the renewable energy is produced.

It is thus interesting to examine in more detail this upstream part of the chain, so as to shed light on the multiple fates that the resources, which enter the production of wind energy, can follow on the local level.

Before considering the local case studies, Figure 20.2 (below) sums up the commodity chain, from depositing to commercialising. The upper first line describes the steps in the chain. The second line brings together the successive states, status and transformations of the property of being 'renewable', using the language of commodification. The third line points at the way in which the energy landscape is made present in the chain. The fourth line points at the collectives of actors in each of the steps of the chain.

Aveyron, exhausting a collective

Aveyron (South West France) is one of the French departments with the best wind potential. Wind power development started in the Aveyron in 1999. No wind power planning whatsoever was in place before then. In order to cope with the increasing number of projects submitted for approval, the local administration decided to set an inter-services platform (in 2000) and to start devising a planning scheme. At that time, the Regional Natural Park of the Grands Causses (RNPGC), a non-state actor, had suggested approaching wind power planning on the scale of the 'massifs'. The suggestion was that massif entities – i.e. local high plateaus, part of Aveyron, called the 'Causses' (Causse Méjean, Causse Noir, Causse du Larzac), offered a framing that was more compatible with collective action since local mayors could collaborate on this scale in planning wind power. The Causses scale also made it possible to better take into account issues of landscape (far-reaching co-visibilities) and proximity. In 2000, the idea was discarded by the Aveyron prefecture as being too complicated because the massifs overlapped the administrative divides.

The outcome was a first wind power planning scheme issued in 2005. The approach translated wind power issues into zoning through several operations such as the definition of landscape 'types' based on morphology and heritage values, the mapping of regulatory constraints and the addition of buffer zones so as to compensate for regulatory insufficiencies in the face of the exceptionally far-reaching co-visibilities imposed by industrial wind turbines. This gradual shift from a qualitative landscape issue to a zoning logic (favourable, unfavourable or negative) certainly answered to administrative instructors' need for rationality and objectivity in the face of the pressure from wind power developers (Nadaï & Labussière, 2009). Superimposing the map of the wind speeds and this planning scheme provided the developers with emerging visions of profitability (the best deposit . . .) and renewability (. . . with the less administrative 'constraints') (Figure 20.3).

	Depositing	Harnessing	Extracting	Provisioning	Distributing	Commercialising
Chain of commodification	turning the untamed wind into a deposit	capturing the untamed wind	extracting the energy from the wind	entering the extracted energy into a provisioning system	transporting the energy to the places where it is to be used	exchanging the energy for money with the final user
	from untamed wind mesures / maps of potential	from testing poles to wind turbine blades	from blades to grid connection [upstream local transformer]	from local transformer to electricity market/trading places	from electricity market/trading places to distribution grid (upstream meter)	meter and downstream meter
Renewability	*Abstraction* untamed wind brought into representation through map m/s + some crit.	*Individuation* untamed wind framed as laminar wind, partly separable from supporting context *Abstraction* untamed wind made 'renewable' energy through permit authorisation (incl. impact studies, risk assessment , ICPE)	*Appropriation / alienability* wind on blades, kinetic energy turned into uneven alternative elect. energy (alternator), privately owned	Connexion author. Feed-in tariffs + Garantee of origin	*Individ . / abstraction* uneven elec. energ. turned into grid compatible alternat. elec. current, kWh made physically all the same (and standard kWh) through elec. transformer. *Abstraction* ReN kWh quality detached and circulated as information, (guarantee of origin) *Displacement* ReN quality redefined as generic attribute of kWh, and rebundled into commercial product ('green kWh') *Valuation* through trading betw. utilities on market places	
Energy landcapes	Mapping of physical components of the landscape (electric grid, wind speed ...)	Socio-spatial landscape (project siting, autoris. Proced., publ. inqu., planning...)	Absent	Absent	Absent	Absent

Figure 20.2 The commodity chain of French wind power

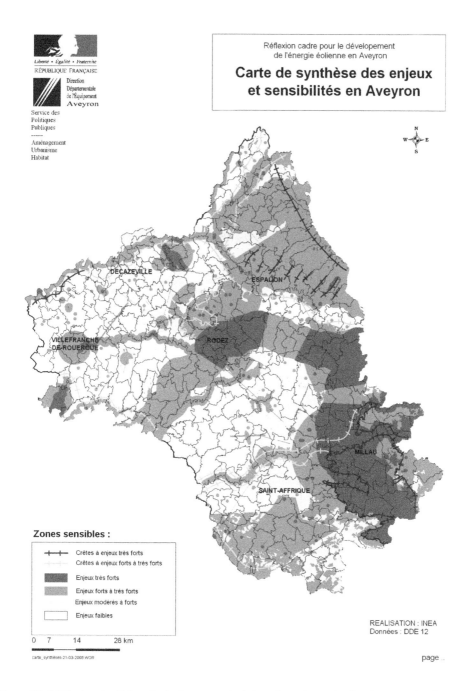

Figure 20.3 Wind potential and wind power administrative scheme in the Aveyron (April, 2005)

Source: Préfecture de l'Aveyron, 2005.

Figure 20.4 The Lévezou landscape (a view over the lake of Salle Curan)
Source: Nadaï & Labussière

However, the development inside the favourable zones was left unplanned and the pressure for project development was not really regulated. As the local administration was unused to communicate figures about projects under consideration (accepted, under acceptance, refused), word-of-mouth made up for the lack of information. Inhabitants of a hamlet in the Massif of Lévezou (Figure 20.4) started to go door-to-door in order to cross-reference information. Doing so, they joined private concerns into a network covering the Lévezou, in which they quantified more than 200 wind turbines under consideration for project development. In other words, wind power development was reaching a tipping point and compromising the entire massif of the Lévezou, threatening to convert it to a contested energy landscape. In order to structure a resistance against wind power, they created a league ('Levezou in peril') so as to tie threads (heritage, proximity, landscape) that were kept separate by the administration. Thus the local opposition endowed massif entities with a political existence. At the same time, landscape protection was facing the limits of the first Aveyron wind power plan. Co-visibilities between the projects sited in these zones and the zones deemed to be protected for wind power visual impact could not be avoided. The rapid technological development of wind energy technology (increase in size, power and economic profitability in wind areas deemed unprofitable a few years ago) soon made landscape choices obsolete. Eventually, this first administrative wind power scheme proved unable to bring wind power projects on track for assembling local resources in a way that proved sustainable. These projects clearly contained seeds of unforeseen environmental and social externalities. 'Renewability', if any, was here defined in a way that was rather procedural – i.e. attached to project/electricity production authorisation and detached from the field.

In 2006, WDZ had just entered into implementation phase at the national level and provided the local administration with a legitimacy to revise the existing wind power plan. The Aveyron prefect was replaced. The new prefect imposed a temporary moratorium on wind power permits until all WDZ could be submitted to the Aveyron administration by local authorities (towns or groups of towns). With the financial and methodological support of the RNPGC, new wind power basins were designed by coordinating the WDZ processes on the scale of the massifs, which were thus endowed with a political and relational existence. The process, which is still under way, has highlighted the unexpected potential of highlands (former commons used for grazing during the nineteenth century) at the outermost of the Causses.

The topography of these highlands limits the co-visibilities between the wind farms and the villages. Their status makes it easier for communities to share the financial benefits from wind power. While this new way of approaching wind power planning and projects embodied a different – more debated and potentially more durable – way of assembling local resources in the development of wind power, it did not live up to its promise. In the years before, private developers had already intensively prospected the countryside. They had made contacts, offers, if not contracted with local mayors and land owners: they had shown promise of significant revenues from wind power developments. Stimulated by a high feed-in tariff, favoured by the neo-liberal start of French wind power policy and aligned with a procedural-definition of 'renewability', most of these seeds of projects had been developed in a way that was lacking transparency, delaying or limiting their becoming public in order to avoid debates and controversies as regards to their territorialisation. The expectations that had been weaved around the seeds of these projects resulted in small coalitions spread out over the territory. They did not agree with conditioning their future revenues upon a more collective approach and a large scale negotiation. The lack of transparency certainly played a role in the rise of local oppositions to wind power, partly because they fell apart and nurtured animosity and conflicts in the process of recasting the planning of wind power, partly because collective attempts of wind power planning ran into the *de facto* development of these projects. As local social synergies in villages fell short, people's capacity to engage in the devising of a territorial project for wind power was in certain cases merely exhausted. As one of our interviewee admitted it, a generation must passed by wind energy to be tamed – i.e. for wind power to break away from passed conflicts.

Seine-et-Marne, trampling on commons

Seine-et-Marne is situated in the south-eastern Ile-de-France region, one hour distant from Paris (see Figure 20.5). It is an area occupied by industrial agriculture. The population is both rural and neo-rural, i.e., former urbanites who have left the city in search of a better quality of life. Wind power development is faced with lively opposition (11 projects under way, 7 projects stopped, 10 anti-wind power NGOs in 2010, and 6 MW approved in 2013).

Opposition to wind power has been described by outsiders (i.e., private developers) as rooted in NIMBY ('not in my back yard') concerns and neo-rural populations. Our analysis adopts an insider's perspective on the construction of local opposition (Nadaï & Labussière, 2014). We follow the development of networks of so-called 'opponents' to wind power in the village of Ventville[4] in the southern part of Seine-et-Marne.

Figure 20.5 The Seine-et-Marne agricultural landscape
Source: Nadaï & Labussière

Before wind power, Ventville was famous for its ambiance. Agricultural families used to work as community organisers. Farmers and their families were personally committed to building a public life that could be shared with the other inhabitants. Farmers' families used to lead local NGO's and had, through generations of mayors, been in charge of local politics. Being the mayor relied on the everyday art of working with (and for) neighbours, who are also the voters. This collective management of the local life can be regarded as a first 'common' shared between the municipality, local institutions, NGO's and inhabitants. Another 'common' relies on a shared agricultural space, called the 'plain'. The 'plain' is a continuous entity of fields. These fields are free of infrastructure (i.e., no high voltage lines, no motorways, only a small road for access). They are easy to plough, crop and, eventually, switch parcels of land to consolidate more continuous and coherent individual parcels. The continued agricultural quality of the 'plain' has been maintained for generations of farmers by individual management of each parcel and by their periodic exchange through land consolidation. In both instances, the common good resulted in a wise entanglement of private and collective concerns so as to maintain over time the agricultural potential of the plain.

In 2003, the arrival of wind power started to rend these social synergies. A project was initiated by two farmers on their own lands, in the 'plain'. As one of them was the mayor of the village and omitted informing the other inhabitants, a conflict of interest ensued. The first inhabitants to hear of the project, Mr and Mrs Why, did not have a pre-defined stand on wind power. They took a one-day trip to 'experience how it felt to be in a wind power landscape' (at Janville,

100 km from Ventville). Later, during a city council meeting, they advised the mayor and two of his assistants to go there and experience for themselves a wind power landscape. The mayor and his assistants did so, but maintained their refusal to open a public debate about the Ventville project. Mr and Mrs Why organised a consultation at their house, where people could come, read documentation about wind power, talk and eventually sign a petition asking for a referendum. Yet the municipality had already submitted a proposal for a wind power development zone (WDZ) for administrative approval and the mayor did not follow up on the petition. He put forward the democratic legitimacy of his mandate as a basis for deciding on the project. Inhabitants asking for a public debate about the project had no choice but to become engaged in the local opposition to wind power.

In the end, growing tensions led to ruptures in long-established social networks and to social violence, such insults and muggings. The perspective of a wind power project also affected the traditional management of the 'plain' because of the uncertainties and constraints produced by the devotion of parcels to fixed infrastructures (i.e., wind turbines, underground cabling). Thus, after several attempts to open the process to public consultation and to allow for a collective appraisal of wind power issues (i.e., sharing the experiencing of a wind power landscape with the city council, organising a home-made consultation, petitioning for a referendum), the dispute moved to electoral ground in 2008. For the first time in the history of Ventville, the mayor faced an opposition list in the local election. He was re-elected and publicly framed his success as the sign of political support to the wind power project and a tacit acceptance of the landscapes changes that would result.

In this context, the departmental-level (Seine-et-Marne) wind power planning exacerbated the strategic dimension of local conflicts. The local administration undertook a usual sieve mapping exercise that directed the potential wind power development towards three sectors on the margins of Seine-et-Marne. These sectors included Ventville, where landscape sensitivity to wind power development was expected to be low. Such a framing increased local tensions, since the first wind power projects to be authorised would turn potential sectors into actual wind power zones for additional developments.

Local opponents networked so as to coordinate areas of vigilance in the south of Seine-et-Marne. Myriad local NGOs pooled their resources on a new scale and engaged in collective action (see Figure 20.6 for an example of an opponent's area of vigilance). Such a meso-territorial level allowed this new community to seize the sectors targeted by the administration, to discuss in a more-than-a-project perspective the groundings of French wind power policy. The challenge was to foster a debate on landscape protection and wind power policy on a scale congruent with landscape issues. As this new web was not limited in space, the politicisation of wind power policy had a chance to proceed and progress gradually over the frontiers of the Ile-de-France region.

In spite of local and more than local opposition as well as legal proceedings, the project was eventually granted administrative authorisation. Hence, it was officially registered as a renewable energy project eligible to feed-in tariffs as any other wind power project.

The case of Ventville underlines how hard it can be for a municipality to endow a wind power project with a collective dimension. At the national level, the French government asserts that wind power is endowed with public interest and considers as renewable any project granted with administrative authorisation. Yet the Ventville case proves that, for many reasons, the public dimension of wind power is not a given.[5] It must be reconstructed on a project-by-project basis. Still, in the Ventville case, French institutions prove weak in sustaining this reconstruction: they leave local authorities with a critical role in bringing the wind power on track for sustainability and failures to do so nurture local opposition. In Ventville, where the agricultural

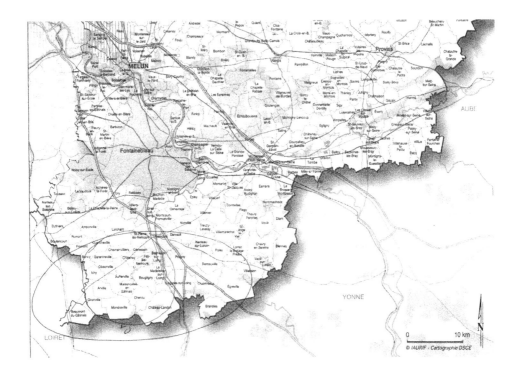

Figure 20.6 Area of vigilance against wind power, as mapped by an opponent (Ventville): 'All this [hand-drawn ellipse] is my area of vigilance, it is the area in which I oppose wind power. So, it is regional. I have walked this area' (member of a network of local opposition to wind power)

Source: Interview by the authors. Basemap produced by the Institut of Urbanism and Landplanning of the Paris Region, available at www.la-seine-et-marne.com/wp-content/uploads/2013/12/iaurif-77.pdf.

and social 'commons' have historically been strong, the mayor has in the end been allowed to harness the wind through a narrow coalition with a private developer.

Narbonnaise, venturing in ontologies

The Regional Natural Park of the Narbonnaise (PNRN) covers the eastern part of the Aude department (Languedoc-Roussillon, south-western France), which stretches along the Mediterranean coast just north of the Pyrenean mountains. Since the adoption of fixed tariffs in France, in 2000, the Aude and especially the windy Narbonnaise have been pioneer sectors in the development of French wind power. By the end of 2007, ten wind farms (110.2 MW, 92 turbines) were installed on the territory of PNRN and various planning documents (at the regional, departmental and PNRN levels[6]) had already attempted to regulate wind power development.

In 2002, the area included some of the allegedly worst examples of wind farm siting in France and the PNRN was faced with an increasing number of wind power projects. It decided to commission a landscape company to devise a planning scheme for wind power; a process which proved quite innovative (Nadaï & Labussière, 2013). The PNRN wind power charter

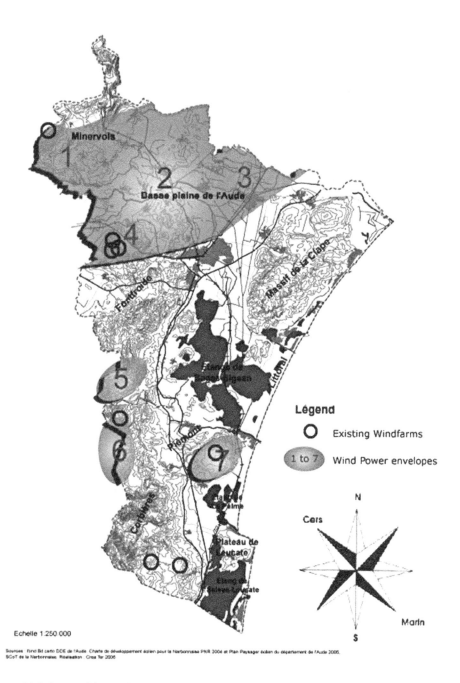

Figure 20.7 Favourable envelopes for wind farm development, according the PNRN wind
power charter

Source: SYCOT, 2006, Schéma de cohérence territoriale de la Narbonnaise – Document d'orientations Générales,
November, Narbonne, p. 46.

was adopted by the PNRN steering committee in 2003. This planning process brought together mayors, wind power developers, NGOs and ministerial field services in a consultation. The aim was to set the boundaries for favourable and non-favourable envelopes for wind power developments, together with specific landscape recommendations ('re-powering',[7] densification, dismantling). The resulting envelopes were thus sectors in which it was felt acceptable to set wind farms (cf. Figure 20.7).

Surprisingly, they mainly targeted small plateaus, such as the Plateau de Garrigue Haute, which overhang villages. The graphic design of these envelopes carefully avoided territorial delimitations (such as administrative borders, communal or private land limits) in order for planning to ward off appropriation strategies and be geared towards a multidimensional and relational appraisal of the wind resource (see Figure 20.6). Concretely, this meant providing wind power developers with areas in which project development was thought to be acceptable as well as with recommendations for approaching the siting of wind farms in these areas. Yet, the plan did not prescribe any specific design for the siting of the project. Rather, it enticed developers to go into the field and further explore the likely relations with – or recomposition of – local resources (such as landscape, land, wildlife . . .) that their project could induce. Unlike much normative planning that produces zonings for the development of a generic technology, the Narbonnaise planning process thus maintained its openness to the multiplicity of situations. The wind power potential was therefore not only a technical one but it also included territorial dimensions. Two processes are illustrative of these dimensions.

The first process pertains to a project sited in an envelope covering the Plateau de Garrigue Haute (Nadaï & Labussière, 2013). In the 1990s, the Plateau de Garrigue Haute welcomed the first industrial wind farm in France. In 2010, this project was entering its repowering phase, again the first such instance in France. Repowering provided an occasion for reconsidering the siting of the project and for articulating it on various scales and dimensions of the landscape. On the large scale, the landscape company in charge of the wind power charter emphasised the need to account for landscape relations and align the turbines with a major historical axis in the landscape, an old Roman road parallel to the seashore (i.e., Via Domitia). The plateau also emerged during the devising of the charter in the form of a piece of common land that was traditionally used by several farmers and villages for sheep grazing ('biens sectionnaires'). The status and location of these common lands allowed the surrounding municipalities to depart from administrative frontiers, join and take part in siting turbines, which could be sited as one coherent wind farm on the plateau. On the small scale, the existing wind turbines provided birdwatchers with an opportunity to depart from categories of protected species in order to follow and observe migrating birds' strategies in relation to the presence of wind turbines (Nadaï and Labussière, 2010). Birdwatchers devised an innovative method called 'micro-siting' that allowed them to translate bird strategies into statistical and spatial representations congruent with planning categories, so as to negotiate the siting of turbines with the wind power developer (see Figure 20.8).

The second process (Nadaï & Labussière, 2010) relates to a neighbouring small plateau (Villesèque-des-Corbières), former grazing land that had been invaded by a garrigue cover. Boar hunters and protectors of raptors had joined together and set a up a European Life project[8] in order to re-introduce sheep grazing and reopen the habitat for small game (such as rabbits and hares, which are prey for boars and raptors). As the local branch of the French Bird protection NGO –the Ligue de Protection des Oiseaux (LPO) – survey showed, the repowering on the Plateau de Garrigue Haute was likely to deprive raptors of part of their hunting territory. It was thus decided to use part of the wind power benefits for environmental compensation (habitat creation for small game in a nearby area). Concretely, this meant additional financial support

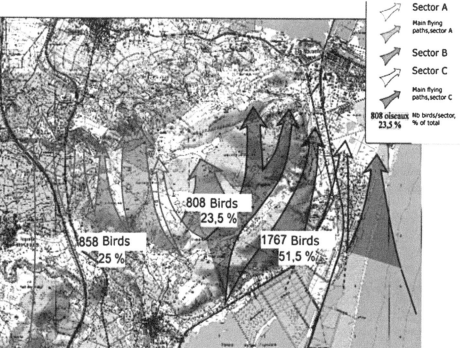

Figure 20.8 Micrositing on the Plateau de Garrigue Haute

Source: Ligue de Protection des Oiseaux, 2001.

for a shepherd in order to bring it to economically viable size. Therefore, wind power development not only contributed to ecological management and the revival of a traditional agricultural activity, but also was interwoven into traditional structures of landscape management.

In both of these processes, we clearly see that capturing an untamed wind calls for a reappraisal of ontologies (e.g. birds), of their attributes (e.g. protected) and of relations to the resource (e.g. the ability of birds to play with the wind). Repopulating the wind can be regarded as a strategy, which avoids reducing the wind to its physicality and appropriating it in an un-renewable way.

In this case, the renewability of emerging wind power projects – and potential – is constructed through specific rules of reciprocity between birds, birdwatchers, game hunters and the private developers. Indeed, LPO made its collaboration on this wind power project conditional upon a follow up and monitoring of birds death on the wind farms over the years.

A few years after this innovative experiment, the Plateau de Garrigue Haute repowering project still was being processed for administrative permitting. The local LPO deciding to stop its collaboration with other wind power developers and projects in the Narbonnaise area because of two strong a pressure for wind power development in this area. Therefore, while this experience had proved very innovative, it could not be followed up on a bigger scale because of the lack of regulation on wind power development in the Narbonnaise.

The contestable sustainability of wind power

What commodification does to wind power

The three cases that we have presented illustrate the difficulties faced in attempting to assemble wind power in a sustainable way at the local level, as wind power policy gears its construction towards private profit and market making. Seen from a local perspective, the difficulties faced in the development of wind power projects are often perceived – and thematised (Geraint et al., 2009) – as resulting from planning shortcomings in the face of wind power developers pressure to get new projects on the ground.

The commodity chain analysis sheds a renewed and more incisive light about these difficulties and the role of French institutions. The wind power commodity chain is divided into two parts by the feed-in tariffs and the guarantee of origin, with an associated rupture in the way renewability is constructed. In the upstream part of the chain, the bundling of resources raises issues. While the term 'renewable' is never explicitly challenged, the sustainability of wind energy is at stake and under construction. In the downstream part of the chain, renewability is detached it from its conditions of production and made tradable. This allows for a large-scale trading and commercial re-bundling of renewability into 'green' electricity, which results in an upwards economic incentive on project development. This does more than just accelerate the pace of wind power development. It brings this development on tracks that favour market-based rather than territorial assemblages, and gives priority to the type of project development that most efficiently maximises private profit.

In this approach, project development acquires a performative dimension, because returns on experience from projects development and exploitation – their material dimension – end up grounding certain relational assemblages as more realistic and performant than others. They better serve policy objective. The actors in charge of these projects, their way of practising are endowed with economic and institutional powers, not the least because of the money transfer sustained by economic instruments such as feed-in tariffs. These actors are thus set in a position to give policy formulation an inflexion, which ends up privileging what counts in their way of developing projects.

As the materiality of wind power shapes and is shaped within this process, it can be said to be relational. Asserting that wind energy is renewable *per se* because of its physicality (flow energy) – as does the French energy code – is not right or wrong. It is performative: it gears project construction towards a minimal way of assembling the wind as an energy resource, a way that does not exhaust the physical flow.

This construction of wind power around a physical definition of the renewable resource only holds to the extent that projects assembled on these bases do not raise too much opposition. In France, it is proving to be a fragile and ill functioning construction. The case studies that we have presented in above illustrate tensions and (failed) attempts at proceeding differently. On a national level, the issue is also recurrent, as witnessed by the progressive judicialisation of both wind power projects and WDZ. Interestingly, with the recent withdrawing of French WDZ, wind power developers have shown increased interest in devising charters of good practices for project development. Differences and tensions prove that the issue of which resources to account for and how to bundle them in order to make wind a sustainable energy is left unsolved.

Sustainability and the relational materiality of wind power

Wind resources, with their associated materialities and relational interdependencies, may either support or resist this project. This results in certain side effects, which may generate tensions (externalities). We point out here some of these external effects in relation to the materiality of the resource.

Non-storable energy and the unique electricity market

The fact that electricity still cannot be stored has contributed, at least in the EU, to a neoliberal account by which grid interconnection and a unique electricity market would sustain the energy transition because it would allow offsetting ReN electricity intermittency through European-wide market pooling. French policy which constructs the ReN kWh as a standard tradable kWh (plus a ReN tradable quality, a guarantee of origin) sits alongside this neoliberal project of unique electricity market. It abstracts the renewable quality (see above) and transforms the ReN kWh into a standard kWh that can be traded like any other kWh on the electricity market. So 'provisioning', in the case of wind energy, is obtained by market-pooling and market-based regulation.

Intermittence as a resistance of the resource

However, untamed wind, the unpredictability of wind, still resists commodification. As we have shown, wind energy is circulated along the commodity chain in various forms: intermittent ReN electricity is one of them. As intermittency had not been accounted for in the first years of ReN development in France and in the EU, it has soon raised issues by interfering with the existing practices of valuation on the electricity grid and market, provoking negative electricity prices. This first happened in Spain, then across the rest of the EU, and it became critical in Denmark as the share of wind electricity has risen. This was dealt with in various ways in order to have wind power developers integrate the cost induced by their electricity production. In this case, the resource resists commodification to the extent that its temporal materiality cannot be fully tamed through storing: the attempt to make it into a stock-like resource, so as to regulate the electrical commodity price, fails in the face of lack of sociotechnical solutions that could take charge of a flow to stock conversion on a relevant economic scale.

WDZ as a bridging device

The extracting of electrical energy operated by the rotor in the wind turbines is the step at which energy is actually appropriated by wind power developers: the administrative production and connection authorisations are the key devices allowing for the appropriation of the energy of the wind, because they register the property of the turbines and allocate the right to feed-in tariff to the corresponding owner. As these authorisations are delivered upon purely technical grounds, the bundling of other (non-electrical) resources is not accounted for and there would be no reason for ReN to be developed in a sustainable way.

Interestingly, the French WDZ was supposed to operate as such a bridge. It was aimed at conditioning the eligibility of wind power projects to feed-in tariff upon the inclusion of these projects into a potentially multi-dimensional spatial planning (landscape, environment . . .). Yet, marred by the ambiguity of the political process that overarched on their adoption (Nadaï, 2007), the devising of WDZ left room for interpretation of the way in which they should be implemented. Early on, WDZ were contested by wind power developers and professional unions, who casted them as a barrier to wind power development. Later on, they were strategically contested in the courts by opponents to wind power in order to stop the development of this energy. WDZ ended being legally cancelled in 2013, under the political lobbying of a coalition gathering members of the French Green party and the wind power developers' union.

In the end, there is currently no longer any articulation between the upstream part of the commodity chain – depositing and harnessing – over which territorial and social resources are engaged in the development of wind power projects, and its downstream part that reconstructs renewability as a stable attribute of standard electricity commodity.

These few examples illustrate how unaccounted resources or effects constantly challenge the assemblage of wind energy. In this process, the materiality of the wind resource can resist or go along with the attempts at commodifying it as energy. In doing so, the materiality of wind and electricity takes part in the development of web of relations that may or may not contribute in accounting for all the other resources that are engaged in the construction of wind as energy resource. The sustainability of wind energy depends on the extent to which this relational potential is made representable and/or debatable, so as to be brought into politics for arbitrage. Entities are recomposed in this process, as with the example of birds in Narbonnaise. Birds are not just qualified as skilled. Qualifying them so, brings them into a process that is performative: it allows for a siting of the windfarm that makes it possible for migrating birds to fly over/under or through the windfarm. Birds and wind power are both recomposed in the process: they become compatible beings. All this means that the process of assemblage of a sociotechnical system has an ontological reach and it is this process which makes an energy sustainable or not.

Conclusion

Renewable energies are not necessarily sustainable. Sustainability depends on the way in which resources are assembled as energy resources.

Taking the case of French wind power, we have looked at the way through which it is being assembled as an energy resource and an energy commodity. The analysis has allowed us to point at the key role of policy instruments – such as production authorisation, grid connection authorisation, feed-in tariff, guarantees of origin – in articulating the construction of renewable energy to a legal, often physical definition of renewability.

The reduction of renewability to such an essential (physical) attribute of flow-energies disregards the way in which other, often shared, resources are engaged in the construction of renewable energies. This also eases market-making as it detaches the qualification of such energies from their territorially embedded and sometimes conflicting construction and origin.

The commodity chain analysis has allowed to explore these interactions and cast a new, more political, perspective on the relation between the upstream (depositing, siting, grid connection) and downstream (distributing, trading, marketing) parts of the chain. It has allowed us to point at various ways in which wind, as a material resource resists or goes along with commodification.

This way of constructing renewability generates intense market pressure (upwards) onto non-wind resources. As shown by different case studies, it makes it extremely difficult for local actors to bring the development of wind farms on sustainable tracks. Side effects ensue, which only part of them are accounted for by planning institutions or market design. Renewable energies, if wrongly assembled, can exhaust their potential for development. Sustainability is a possibility for rather than an attribute of renewable energies.

French wind power is just one instance of renewable energy. Human geography shall contribute further in exploring the words of these energies by reopening the reductive 'renewable' qualifications of other resources such as solar, biomass or hydro energy.

Notes

1 www.developpement-durable.gouv.fr/Le-regime-d-autorisation-d.html, consulted September 9, 2015.
2 This voltage is further increased by a transformer located in the grid system, for the electrical current to reach 200 to 400kW and circulate in high voltage (transportation) lines.
3 Definition of renewable energy sources: Art. L. 211–2, energy code.
4 'Ventville' is a pseudonym, chosen by the authors for the purposes of this paper.
5 Including the fact that France has a low-CO_2 electricity mix (dominance of nuclear energy).
6 Région Languedoc Roussillon (2003) *Schéma régional éolien*, 4 volumes, Narbonne; PNRN, 2003 Charte du Développement Eolien – Projet de Parc Naturel Régional de la Narbonnaise en Méditerranée (2003), available (on 08/25/08), www.parc-naturel-narbonnaise.fr/en_actions/maitrise_de_l_energie_et_energies_renouvelables/charte_eolienne; Préfecture de l'Aude (2005) *Plan de gestion des paysages de l'Aude vis-à-vis des projets éoliens*, Narbonne.
7 'Re-powering' consists in dismantling an existing wind farm and in increasing its capacity by installing new, bigger and more powerful wind turbines. It is currently the way that countries such as Germany and Denmark increase their wind power capacity.
8 http://aude.lpo.fr/life-consavicor/accueil.htm.

References

Akrich, Madeleine. 1988. "La recherche pour l'innovation ou l'innovation pour la recherche? Le développement du photovoltaïque en Polynésie", *Culture Technique*, 18, pp. 318–329.

Akrich, Madeleine. 1989. "La construction d'un système socio-technique. Esquisse pour une anthropologie des techniques", *Anthropologie et Sociétés*, 13(2), pp. 31–54.

Akrich, Madeleine. 1993. "Les formes de la médiation technique", *Réseaux*, 11(60), pp. 87–98.

Bakker, K., and G. Bridge. 2006. "Material worlds? Resource geographies and the 'matter of nature'". *Progress in Human Geography*, 30(1), pp. 5–27.

Bridge, G. 2010. Resource geographies 1: Making carbon economies, old and new", *Progress in Human Geography*, 35(6), pp. 820–834.

Castree. 2003. "Commodifying what nature?" *Progress in Human Geography*, 27, pp. 273–297.

COMOP. 2007. COMOP10, Plan de développement des EnR à haute qualité environnementale, rapport à Borloo J-L. Paris.

Evrard A. 2013. *Contre vents et marées. Politiques des énergies renouvelables en Europe*, Presses de Sciences Po, Académique, Paris.

FR. 2000. Loi n°2000–108 du 10 février 2000 relative à la modernisation et au développement du service public de l'électricité. Paris.

FR. 2005. Loi n° 2005–781 du 13 juillet 2005 de programme fixant les orientations de la politique énergétique, JO n° 163 du 14 juillet 2005:11570.

FR. 2008. Plan national de développement des énergies renouvelables de la France, ministère de l'Écologie, de l'Énergie, du Développement durable et de l'Aménagement du territoire, 17 novembre 2008. Paris.

FR. 2009a. Circulaire du 23 mars 2009 relative à la territorialisation de la mise en œuvre du Grenelle de l'environnement. Paris.

FR. 2009b. Circulaire du 19 mai 2009 relative à la planification du développement de l'énergie éolienne terrestre adressée par la Direction de l'énergie et du climat. Paris.

FR. 2009c. Loi "Grenelle I" n°2009–967 du 3 août 2009 de programmation relative à la mise en oeuvre du Grenelle de l'environnement. Paris.

FR. 2009d. Projet de loi "Grenelle II" portant engagement national pour l'environnement, n°155, déposé le 12 janvier. Paris.

FR. 2012. Decree n° 2012–62 du 20 janvier 2012 modifying the decree n° 2006–1118 du 5 septembre 2006 that sets the guarantee of origin, in order to update it with the EU directive 2009/28/CE.

Geraint, E., Cowell, R., Warren, C., Strachan, P., Szarka, J., Hadwin, R., Miner, P., Wolsink, M. and Nadaï, A. 2009. "Wind power: Is there a 'planning problem'? Expanding wind power: A problem of planning, or of perception? The problems of planning—A developer's perspectivewind farms: More respectful and open debate needed, not less planning: Problem 'carrier' or problem 'source'? 'Innovative' wind power planning", *Planning Theory & Practice*, 10(4), pp. 521–547, http://dx.doi.org/10.1080/14649350903441555.

Hartwick, E. 1998. Geographies of consumption: A commodity chain approach. *Environment and Planning D*, 16, pp. 423–437.

Labussière O. & Nadaï A. 2014. "Unexpected wind power 'potentials': The art of planning with inherited socio-geographical configurations (France)", *Scottish Geographical Journal*, http://dx.doi.org/10.1080/14702541.2014.922210.

Labussière O. and Nadaï A. 2015. "Wind power landscapes in France: Landscape and energy decentralisation" in *Renewable Energies and European Landscapes: Lessons from Southern European Cases*, Frolova, M., Prados, M.J. and Nadaï A., eds, pp. 81–93, Springer, Dordrecht, www.springer.com/us/book/9789401798426.

Nadaï, A. 2007. "Planning, siting and the local acceptance of wind power: Some lessons from the French case", *Energy Policy*, 35, pp. 2715–2726, www.sciencedirect.com/science/article/pii/S030142150600485X.

Nadaï, A. and Debourdeau A. 2013. "Actions, séquences, épreuves de transition' dans les Crêtes Pré-Ardennaises (CCPA)", In *La sociologie de l'énergie* – Tome 1 *Gouvernance et concepts*, Beslay, C., Zélem, M.C. (dirs), Paris, Ed CNRS, collection alpha, pp. 63–72.

Nadaï, A. and Labussière, O. 2009. "Wind power planning in France (Aveyron): From State regulation to local experimentation", *Land Use Policy*, 26(3), pp. 744–754, www.sciencedirect.com/science/article/pii/S0264837708001208.

Nadaï, A. and Labussière, O. 2010. "Birds, turbines and the making of wind power landscape in South France (Aude)", *Landscape Research*, 35(2), pp. 209–233, www.tandfonline.com/doi/abs/10.1080/01426390903557964#.UqJQNydstLs.

Nadaï, A. and Labussière, O. 2013. "Playing with the line, channelling multiplicity – wind power planning in the Narbonnaise (Aude, France)", *Environment and Planning D*, 31(1), pp. 116–139, www.envplan.com/abstract.cgi?id=d22610.

Nadaï, A. and Labussière, O. 2014. "Communs paysagers et devenirs éoliens opposés: Le cas de la Seine-et-Marne (France)", *Projet de Paysage*. Dossier thématique n°10, "Le paysage à l'épreuve de la transition énergétique" (S. Briffaud coord.).

Nadaï, A., Krauss, W., Afonso Ana, I., Dracklé, D., Hinkelbein, O., Labussière, O. and Mendes, C. 2015. "A comparison of the emergence of wind energy landscapes in France, Germany and Portugal", in *Landscape and Sustainable Development. The French Perspective*, Luginbühl, Y., Howard, P. and Terrasson, D. eds, pp. 133–144, Ashgate, Farnham.

Préfecture de l'Aveyron, 2005. Réflexion cadre pour un développement de l'énergie éolienne en Aveyron, August, Rodez, France.

Raman, S. 2013. "Fossilizing renewable energies". *Science as Culture*, 22(2), 172–180.

Shove, E. 1998. Gaps, barriers and conceptual chasms: Theories of technology transfer and energy in buildings. *Energy Policy*, 26(15), pp. 1105–1112.

Simondon, G. 1989, *Du mode d'existence des objets techniques*, Paris, Aubier.

UE. 2009 Commission européenne, Directive 2009/28/CE du Parlement européen et du Conseil du 23 avril 2009 relative à la promotion de l'utilisation de l'énergie produite à partir de sources renouvelables [modifiant et abrogeant les directives 2001/77/CE et 2003/30/CE], Brussels.

21

Conflictive energy landscapes

Petrocasas and the petrochemical revolution in Venezuela

Elvin Delgado

Introduction

On Sunday September 23, 2007 late Venezuelan President Hugo Chávez Frías dedicated his television program *Aló Presidente* number 295 to the petrochemical corporation PEQUIVEN (Petroquímica de Venezuela, S.A.). This show took place in CPET (*Complejo Petroquímico El Tablazo*, El Tablazo Petrochemical Complex) now CPAMC (*Complejo Petroquímico Ana María Campos*, Ana Maria Campos Petrochemical Complex), on the northeastern shore of Lake Maracaibo. During this transmission, President Chávez explained how almost everything we use in our everyday life is based on products derived from hydrocarbons processed in the petrochemical industry. "Fertilizers, ink, mattresses, security helmets, textiles, diapers, among many other things come from petrochemical products," he said. His argument, "*somos petroquímicos*" ("we are petrochemical beings"), emphasized the importance of the petrochemical industry in the life of Venezuelans. President Chávez called it: *La Revolución Petroquímica Socialista* (the Socialist Petrochemical Revolution), to transform hydrocarbons into social development. This petrochemical revolution is one of the components of *Plan Siembra Petrolera* (Plan to Sow the Oil), directed towards the re-distribution of oil's wealth among Venezuelans. The President highlighted that the main objectives of the petrochemical revolution are to develop the nation's reserves of natural gas in order to promote the petrochemical sector; promote the economic and social development of the country; and transform the national production model in order to satisfy the needs of Venezuelans. In short, focusing on the production of natural gas, the petrochemical revolution attempted to make Venezuela one of the world leaders in the production of fertilizers, plastic derivates, and polyethylene.[1] These petrochemical products will be used in the textile, agricultural, health, construction, and foods industries around the country.

State-led development of the petrochemical sector in the mid-20th century was, and still is, an important component of Venezuela's modernization strategy to "sow the nation's oil". Access to and control over natural gas deposits – which are characterized by a high degree of temporal and spatial variation as well as the efficient transportation and transformation of this fossilized nature – was pivotal to this process. Yet, national and international interests throughout the

20th century focused on the exploration, extraction and production of Venezuelan oil and relegated natural gas to a second place. Indeed, the main goal of the oil industry (before and after its nationalization) was to generate the largest amount of revenue in the shortest amount of time. To achieve this, the oil industry invested more capital in upstream activities (i.e. exploration and extraction) while neglecting downstream parts of production and marketing. This pattern of investment was adopted by PDVSA (*Petróleos de Venezuela, S.A.*). For example, PDVSA – following production practices from previous companies (national and foreign) – used associated gas for oil extraction processes. This meant the re-injection of gas to increase the pressure of the wells allowing for the extraction of more oil per active well. Therefore, the use of natural gas, which was primarily intended for petrochemical processes, revolved around oil production activities instead. This put PEQUIVEN in a vulnerable position because it lacked the appropriate influx of natural gas needed for its industrial processes. As a result, the petrochemical sector – in the words of President Chávez – was known as "*la industria olvidada*" (the forgotten industry), which in turn inspired the creation of the petrochemical revolution.

The petrochemical revolution can be conceptualized as a government's attempt to modernize the petrochemical sector – an industry historically neglected by the state – to bring it back to its feet. This modernization strategy will allow the government to develop the downstream sector of the country and transform hydrocarbons into manufactured goods. As José Khan, Minister of Mining and Basic Industry, stated during the launch of Venezuela's petrochemical revolution, "We have to break the policy that Venezuela had of being simply exporters of raw materials. Now this will mean that the country can enter the international market generating added wealth to the nation" (Venezuelanalysis.com 2007). The Venezuelan strategy to produce social goods derived from fossil fuels is contingent upon their capacity to *integrate* the upstream and downstream sectors of the hydrocarbon commodity chain and their ability to manipulate the physical and chemical characteristics of oil and natural gas in order to *transform* them into different commodities. These two processes could only be achieved through the investment of massive amounts of capital to modernize the infrastructure and technology needed in this economic sector.

The petrochemical revolution's strategy to rework space and nature in order to turn fossil fuels into social development demands a critical exploration. Which actors are involved in the development of the petrochemical revolution and how do their relationships define this process? What specific raw materials are needed in the process and how are they commodified in the petrochemical sector? Which outcomes are produced by the petrochemical revolution? In other words, what are the social, environmental, and geographical implications of this strategy at different scales? Finally, what particular type of energy landscape is produced with the petrochemical revolution?

The primary objective of this chapter is to provide a critical exploration of Hugo Chávez's petrochemical revolution.[2] In particular, I want to use a project known as *Petrocasas* (Petrohouses – houses made out of plastic) as a case study to analyze the particular type of conflictive energy landscape that is produced in the transformation of fossil fuels into social development in Venezuela. To pursue this objective, I use notions of materiality of natural resources, commodification,[3] and metabolism of commodity production,[4] to examine the different kinds of socio-ecological impacts produced at the downstream stage of the hydrocarbon commodity chain. Focusing on this particular moment of production has two important advantages. First, it provides an especially suitable case study to observe "the metabolism of commodity production" as an articulation of capital investment, technology, nature, and society (Bridge 2000: 239). Second, it presents a unique opportunity to explore the tensions that emerge from "an incompatibility between the different imperatives governing the technological forces, social relations, and environmental conditions of production" embedded in the petrochemical industry (Bridge 2000: 238).

The central claim of this chapter is that not only does the petrochemical sector play an important role in the commodification of fossil fuels, but that it is also one of the most important actors in the modernization strategies of most petro-states today. I further suggest that petrochemical industries have become both vehicles of capital conversion and accumulation. The chapter illustrates how capital investments directed towards the development of complex gas pipeline systems and petrochemical complexes in Venezuela facilitate the transportation and transformation of fossil fuels into different forms of social development. However, it will also demonstrate how these petrochemical transformations at their molecular level re-produce a contested energy landscape as a set of social and ecological tensions embedded within this stage of production.

I divide this chapter into five sections. First, I review concepts of materiality, commodification, and metabolism of commodity production, and consider their utility in understanding the extensive spatial reach of fossil fuels. Second, I examine the complex infrastructure developed by the petrochemical revolution to secure the flow of natural gas into CPAMC for the production of plastics and highlight the way in which this infrastructure is embedded in territories and local communities. Third, I provide a detailed description of *Petrocasas* and emphasize the social and political rationale of this project as a government's attempt to solve the housing deficit in Venezuela. In doing so, I argue that the social and ecological tensions associated with the creation of *Petrocasas* are obscured by the commodification process itself. Fourth, I focus on the chlor-alkali plant in CPAMC and describe the metabolic process involved in the production of PVC to create *Petrocasas*. In particular, I pay attention to the social and ecological impacts of this process upon fishing communities in the area. Fifth, I conclude with a summary of the main ideas and processes described along this chapter and suggest that conceptualizing all the actors involved in the petrochemical revolution as individual nodes in an energy network will allow us to critically analyze the extensive spatial reach of fossil fuels and the way in which their commodification produce a conflictive energy landscape in Venezuela.

Materiality, commodification, and spatiality of fossil fuels

At the center of the petrochemical revolution lie three important factors of production, namely: natural resources, infrastructure, and technology. First, the state needs to secure access to and control over natural resources in order to maintain a consistent flow of raw materials necessary in petrochemical activities. Second, the state needs to build the necessary infrastructure to transport not only the extracted natural resources, but also the different commodities derived from their petrochemical transformations. Finally, the state has to develop and adopt new types of technologies and the relevant scientific knowledge to transform these natural resources into different types of social goods. In doing so, the Venezuelan government has to take into consideration how the specific physical and chemical characteristics of natural resources define the way in which they can be extracted, transported, and transformed into different commodities.

Recognizing the importance of the physical and chemical properties of natural resources and the physical and biochemical processes that transform them has been a central concern of resource geographers in the last decades (Kloppenburg 1988; Boyd et al. 2001; Bakker 2004; Bridge 2004). Along these lines, for example, Castree (2003: 275) rightly asserts that "the physical properties of those things [commodities] are evidently important." Castree's statement echoes Georgescu-Roegen's (1979: 1039) famous statement that "matter matters, too." In this context, commodities are objects that comprise and combine use and exchange value and are produced, distributed, purchased and consumed along a commodity chain (Leslie and Reimer 1999;

Prudham 2009). It is important to understand, however, that "not all commodities are equal" (Thrift 2000: 96, quoted in Castree 2003: 275), and that the process of commodification may be different depending on what kind of natural resources are being commodified (Castree 2003).

The commodification of oil and natural gas into final products is at the heart of the petrochemical revolution. In this context, it is important to highlight that the commodification of fossil fuels, widely distributed across networks of production, distribution, exchange, and consumption leave behind traces of different kinds with "important and intertwined social and environmental implications" (Prudham 2009: 123). This is especially true for oil and natural gas – both of which are natural resources with an *extensive spatial reach*.

Here, I want to make two observations about the spatially extensive character of oil and natural gas in the context of the social, economic, and environmental impacts associated with their extraction, transportation, transformation and consumption. First, the biophysical characteristics of oil (a subterranean fluid) and natural gas (a volatile and expansive fossil fuel) allows them to move through the subsoil across the surface of the earth as commodities with specific physical and chemical properties. Their characteristics also enable them to circulate in many different forms (e.g., *Petrocasas*, gasoline, fertilizers, medicines) across social, political, and natural spheres. The capacity that fossil fuels have to be transformed into different kinds of social goods affects each one of these spheres once they enter in contact with them. For instance, the transformation of oil and natural gas into *Petrocasas* or medicines might be beneficial to the person receiving the house or using the medicine, thus having a positive impact to society. Yet, the toxic byproducts associated with these transformations can also be detrimental to the environment and communities living in close proximity to the petrochemical industry. It is important to remember that these transformations do not occur spontaneously. Oil and natural gas need to be transported through pipelines and tankers. Moreover, their transformation can only happen in relation to each other and to other minerals such as salt, under specific controlled conditions, and with the use of advanced technology characteristic of refineries and petrochemical industries. In short, we need to understand that oil and natural gas are not only physical, but also social, political, and economic in nature (Delgado 2016).

Second, the material transformations of fossil fuels into other products – a process called the "metabolism of commodity production" (Bridge 2000: 239) are subject to a particular set of social relations and environmental consequences. In the context of oil and natural gas, these transformations are articulated within a division of labor and as such are "the bearer of particular relations of production" (Watts 2004: 280). This connection is important because, even though commodities have physical properties, "[W]e cannot forget the socio-ecological conditions under which commodities are produced" (Bridge and McManus 2000: 16). In addition, it is important to highlight that the flow of fossil fuels from extraction sites to petrochemical industries as well as their transformations into other commodities produces particular places of social and ecological distress at each stage of production along the hydrocarbon commodity chain. In the case of Venezuela, for example, this process re-produces conflictive energy landscapes.

While the market for oil and natural gas is global, the locality of the deposits, their extraction, and physical flow of these commodities has a strong regional dimension (Bridge 2008). In the context of the petrochemical revolution, the social and environmental implications associated with the commodification of oil and natural gas can be conceptualized by looking at the material flows of these resources in a constant interconnection of commodity circuits[5] that are embedded in an energy network. Central to these interconnections are the dynamic transformations of these commodities when their trajectories intersect, on the one hand, and the way in which these transformations reconfigure socio-environmental spaces along this energy network, on the other hand.

The petrochemical revolution requires the state's careful attention to the metabolism[6] of economy and nature by mediating the relationship between the space-time dimensions of petro-capitalism and the space–time dimensions of ecology along the hydrocarbon commodity chain. This aspect of the petrochemical revolution is particularly important in light of the accelerating process by which oil, and especially natural gas, becomes commodified through the deepening metabolic interactions between social, ecological, political and economic spheres. This process requires the development of the specific infrastructure necessary to transport oil and natural gas from extraction sites to petrochemical facilities, a process that I will examine below.

Re-scaling the geographies of natural gas

The launch of the petrochemical revolution is a government's attempt to connect the upstream and downstream sectors of the hydrocarbon commodity chain and secure the uninterrupted flow of natural gas to the petrochemical industry. This process requires the construction of a complex gas pipeline network systems and the modernization of petrochemical facilities around the country. The former, when it is completed will not only increase the transferability of natural gas from extraction sites to petrochemical facilities, but it will also interconnect areas of gas surplus with those experiencing scarcity by compressing space between these two regions. The latter will maximize the petrochemical transformations of fossil fuels, especially natural gas, into different social goods.

The Venezuelan petro-state used a multi-scalar geographical approach to accomplish the socioeconomic agenda associated with the petrochemical revolution. In doing so, it developed complex oil and gas pipeline transportation systems and highly specialized petrochemical facilities to support the transformation of fossil fuels into different forms of social development (see Table 21.1).

Table 21.1 Projects developed by the petrochemical revolution

Project	Objective	Type	Investment ($Million)
Plataforma Deltana	Produce 1.470 mcf/d[1]	Gas development	3,810
Mariscal Sucre	Produce 1.200 mcf/d and the construction of LNG[2] plant	Gas development	2,700
Rafael Urdaneta	Produce 1.000 mcf/d	Gas development	2,900
Gas Anaco	Produce 2.400 mcf/d	Gas development	2,433
Criogénico de Oriente	Extraction of 62 mb/d[3] of ethanol	Processing	926
José 250	Increase production of NGL[4]	Processing	664
Project ICO	Supply natural gas to the western region	Transportation	530
Sistema Nor-Oriental	Management of offshore gas	Transportation	1,066
Gasificación Nacional	Supply gas to 2.6 million families	Distribution	2,334
GNV	Supply gas to 450,000 vehicles	Distribution	921
TOTAL	18,284		

[1] Million cubic feet per day
[2] Liquified Natural Gas
[3] Million barrels per day
[4] Natural Gas Liquids
Source: Adapted from Caro et al. (2009: 112).

The location of these facilities around the country follows a "top-down" scalar approach to make sure that the products derived from these petrochemical processes will benefit the country at different geographic levels (i.e. regions, States, districts, municipalities, capitals, and *parroquias*). This initiative, which is divided in two stages (2007–2013 and 2014–2021), includes the construction of 87 large projects from which 35 will be devoted to the production of raw materials to supply another 52 factories for the production of agricultural fertilizers (including nitrogen and phosphate-based fertilizers), urea, ammonia, detergents, plastics, and other finished products (Aló Presidente 2007). The principal petrochemical complexes are geographically spread out across five of Venezuela's northern coastal States: Zulia, Falcón, Carabobo, Anzoátegui, and Barinas (see Figure 21.1).

The development of the petrochemical sector in Venezuela (through the auspices of the petrochemical revolution) has been quite impressive. However, for the purpose of this chapter, I would like to focus on Project ICO (*Proyecto Interconección Centro Occidente*, Interconnection Center Occident Project), as it relates to the production of *Petrocasas* in CPAMC. The main objective of Project ICO is to connect the natural gas transmission systems from Anaco city in the State of Anzoátegui (eastern region) and Barquisimeto city in the State of Lara (central region) with transmission systems located at the western side, specifically Ulé and CPAMC on Lake

Figure 21.1 Main petrochemical companies in Venezuela

Source: Cartography by Elvin Delgado.

Maracaibo. The estimated investment for this project, which is provided by the state, is $884 million (PDVSA 2009). The project consists of a 300 km gas pipeline system (30"–36" diameter) and three compression stations, namely: (1) Los Morros (72,000 Hp) in the State of Aragua; (2) Altagracia (54,000 Hp) in the State of Guárico; and (3) Morón (54,000 Hp) in the State of Carabobo, to connect the Anaco-Barquisimeto System, which consists of 858 km of gas pipelines with the CRP (*Centro de Refinación Paraguaná*, Refining Center Paraguana) (pdvsa.com).[7] From CRP the gas will be transported to the Ulé-Amuay System and CPAMC in Lake Maracaibo. This project has the capacity to transport approximately 520 million cubic feet per day (mcf/d) of natural gas for domestic and industrial consumption.[8] This spatially extensive project moves natural gas across two deposits located at opposites ends of the country.

Although, the infrastructure developed by the petrochemical revolution is extensive and complex, we should not forget that this infrastructure, the natural gas that flows through it, and the technology used to transform it is embedded in the territory and affect local communities. It is important to highlight that gas, as a natural resource, "is both naturally endowed and socially produced, as nature (re)produces itself and as humans both directly and indirectly intervene with nature" (Kaup 2008: 1735). Swyngedouw (2004: 131), helps us conceptualize this process by suggesting that "[S]ocio-spatial relations operate over a certain distance and produce scalar configurations." Indeed, if oil and gas extraction landscapes are inherently invasive and affect ecosystems and human health at their upstream stage, then the transportation of these fossil fuels may present a similar set of social and ecological consequences at their midstream and downstream stages of production. For instance, some of the risks involved in the transportation of fossil fuels are leaks along all segments of the transport system; the possibility of explosions at each point of transfer; destruction of ecosystems; the displacement and under-representation of communities; and pollution of drinking water, among many others. As such, we should not overlook the ways in which their materiality not only shapes "the political-economic relations within which a commodity becomes embedded" but also reconfigure socio-ecological spaces along their trajectories (Bakker and Bridge 2006: 14). Now that we have described the magnitude of the infrastructure developed by the petrochemical revolution, let us turn to the creation of *Petrocasas*.

Petrocasas: metabolism of fossil fuels' production

One of the most important aspects of the petrochemical revolution is the transformation of fossil fuels into different social goods. Through this process, the Venezuelan government had developed a system to produce houses made out of PVC known as *Petrocasas* as a way to address the country's housing deficit. Currently, more than 60 per cent of the Venezuelan population lives in and around the capital city of Caracas with a large majority living in what Venezuelans call "*ranchos*" (slums) (Venezuelanalysis.com 2011a) (see Figure 21.2).

During the inauguration of a law-decree to address the issues of overpopulation and homelessness in Venezuela on 20 January 2011, President Chávez pointed out that "[T]his is an old story, Caracas is super populated and the poor are living in their shacks on the hillsides ... The national government must take on, with greater depth, the cancellation of this social debt" (Venezuelanalysis.com 2011b). Through this project, the state expects to build two million houses in the following seven years (Venezuelanalysis.com 2011a) and replace every "*rancho*" with a proper house provided by a *Petrocasa*. It would be a massive, if indirect, energy landscape.

This project is only one example of the many initiatives developed by the petrochemical revolution to sow fossil fuels in Venezuelan. In doing so, the government has created a mixed company subsidiary of PEQUIVEN under the name of *Petrocasas* to increase the production of

Figure 21.2 Traditional rustic house known as "rancho" in the city of Cabimas on the eastern shore of Lake Maracaibo

Source: Author's personal archive. Fieldwork, 26 June 2010.

PVC, propylene and other plastics. This social project developed an alternate building system for public housing using panels made out of PVC as a key material.

The technology used for the development of *Petrocasas* was adopted from Austria, Brasil and Germany, but the patent is owned by PEQUIVEN. All the basic building materials needed for the construction of each *Petrocasa*, together with the bathroom furniture as well as other components made of synthetic wood (e.g., doors, beds, chairs, kitchen cabinets, book shelves) are made out of petrochemical products produced at CPAMC and other petrochemical facilities around the country (Boletín Pequiven 2007).

Each unit consists of an 18-piece kit with an instruction manual to assemble a 70-square meter house with three bedrooms, two bathrooms, kitchen, living room and dining room as well as all the basic installations for electricity, water, gas, and telephone services (Comunicaciones Pequiven 2009). Each one of the 18 panels has a code to facilitate the assemblage of the house. Once assembled, the panels are filled with concrete. It takes approximately 12 days and a team of 10 workers to finish the construction of each unit.[9] Several advantages are associated with this construction style. For example, each *Petrocasa* is approximately 50 per cent cheaper than traditional construction styles; they are biologically, chemically and fire resistant, durable and easy to assemble; they require low maintenance costs; and provide sound and thermal insulation (Comunicaciones Pequiven 2008). The social component of the project is of equal importance. The state facilitates a multidisciplinary team composed of engineers, construction managers, social workers, and other organizers, who provide training and logistical support to the community where each project takes place. This social organization creates construction brigades that engage all the members of the community in the construction of each home.

This rationale was explained by President Chávez during the inauguration of Venezuela's *Gran Misión de Vivienda* (Great Housing Mission) on 30 April 2011 in the city of Coro as "[N]obody knows how to build houses better than the people. . . if we want to put an end to poverty, then we must give power to the poor. This is the main principle of the socialist revolution" (Venezuelanalysis.com 2011a). The state also guarantees security and provides the transportation of materials through the National Guard. So far, more than one million families have registered to receive a *Petrocasa* (Aló Presidente N° 375).

The material and symbolic benefits associated with this project may be obvious to the recipient of each *Petrocasa*. Not only do they have a new and comfortable home provided by the state, but also this infrastructure provides the security they need to raise their families. Along these lines, Kaika and Swyngedouw (2000: 123) point out that "[C]ommodities do not only carry their materiality, but also the promise and the dream of a better society" (quoted in Prudham 2009: 134). What is not obvious, however, is the social and environmental tensions at the production sites or the kind of conflictive activities that are being supported and reproduced via the purchase and circulation of these commodities (Prudham 2009). Harvey (1990: 423) provides a useful way to conceptualize the social relations that are revealed or concealed by the production, circulation, and exchange of commodities: "[t]he grapes that sit upon the supermarket shelves are mute; we cannot see the fingerprints of exploitation upon them or tell immediately what part of the world they are from." Do the new homeowners of *Petrocasas* know the carcinogenic effects of producing PVC in CPAMC? Are they aware of the toxic pollution generated by CPAMC in the use of mercury and chlorine in the chlor-alkali plant for the production of PVC? Finally, do they know the potential impacts that these processes exert upon fishing communities located near CPAMC? My intention is not to blame the recipients of these houses, but rather to unveil the socio-ecological impacts associated with the production of *Petrocasas* in Lake Maracaibo.

This is one of the reasons why Watts (1999: 306) argues that "there is more to commodities than their physical properties." I am not trying to expand on the idea of *the fetishism of commodities* elaborated by Marx (1967) and explicitly advocated by Harvey (1990) and Hartwick (2000), but rather my intent is to underscore the importance of including both the social and material aspect of commodities in order to better conceptualize the socio-ecological consequences together with all the social tensions produced in the commodification of fossil fuels. Considering this point is important because it not only recognizes that fossil fuels, and the commodities produced through their transformations, "have their origins in social relations that are largely obscured by the commodity form," but it also illustrates how commodities, through their metabolisms, "perform myriad social functions as 'things in motion'" (Bakker and Bridge 2006: 12). As such, "we need to reconnect back to the material reality of the producer" and the communities located around the production sites (Hartwick 2000: 1190). We should not only look at the social and environmental risks associated with petrochemical processes within the confines of the industry, but it is also important to trace the trajectories of these commodities and their socio-ecological impacts beyond the walls of the petrochemical complex. These effects are particularly visible in the context of PVC production in CPAMC, examined below.

Socio-environmental implications of PVC production in CPAMC

Like alchemists searching for gold, those in the petrochemical industries seek to manipulate the biophysical characteristics of natural resources in order to transform them into more valuable commodities. The metabolic process of producing PVC in the chlor-alkali plant for the creation of *Petrocasas* involves: (i) the inflow of intermediate products derived from oil, natural gas, and salt; (ii) the combination and conversion of each one of these raw materials using specialized petrochemical technologies; and (iii) the outflow of different commodities, heat, and waste. Let us take a closer look at the production of PVC in CPAMC. Understanding this process is important because it allow us to identify the different resources needed to produce PVC, the process involved in its production, and the socio-ecological impacts associated with this process at multiple scales.

Once oil and natural gas are extracted, they are transported to refineries and/or petrochemical facilities. At this stage, refineries produce pure chemicals, called feedstocks, from crude oil (e.g., naptha) or natural gas (e.g., methane, ethane, butane, and propane), which are sold to petrochemical industries. Another important raw material used in petrochemical processes is salt because it plays a major role as an active transforming agent in the production of other commodities. Once salt is transported into the chlor-alkali plant in CPAMC, it is separated into its basic components through a process known as electrolytic chlorine production. In short, by moving a current of electricity through saturated salt brine, salt is broken into gaseous chlorine (Cl_2) at the anode and sodium (Na) at the cathode. The co-products of this process are caustic soda (NaOH) and hydrogen (H_2). There are three different cell technologies to produce chlorine, all of which require electrolysis, namely: (i) diaphragm; (ii) membrane; and (iii) mercury. The process used in CPAMC is the mercury cell technology. Therefore, massive amounts of mercury are used in the production of PVC in the CPAMC. Once chlorine is separated from sodium, it is combined with ethylene (C_2H_4) – a product from oil refineries and petrochemical industries – to produce an intermediate chemical known as ethylene dichloride ($C_2H_4Cl_2$) or EDC (Laszlo 2001). Finally, EDC is transformed into vinyl chloride monomer (VCM) – the basic building block to form chains of PVC through a process known as "polymerization".[10]

The substantial socio-environmental consequences produced by petrochemical industries around the world are well documented.[11] Some of these consequences are: (i) impacts to the health of workers and communities located near petrochemical complexes; (ii) chemical pollution; and (iii) physical pollution. For instance, chemical pollution can occur through emissions to the air and release of toxic chemicals into the water or land. Subsequently, physical pollution is created by the production of solid waste as the output of different petrochemical processes. By looking at petrochemical activities we can analyze the ways in which the metabolic transformation of fossil fuels generates waste and pollution and how these outputs interact with ecological processes and social landscapes produced, ultimately, form the energy wealth of the country. Toxic outputs such as pollution and waste, then, provides the primary lens through which petrochemical industries can be conceptualized not only as the center stage of fossil fuels' transformations, but also as the precursors of social, economic, and environmental distress.

Two main problems can be identified in Lake Maracaibo as a result of PVC production activities performed in the CPAMC. First, there are significant pollutant discharges of chlorine and its products during material conversions. Second, some mercury is lost in production to the environment (becoming toxic pollution) while the rest is accumulated in semi-solid or solidified sludge from wastewater treatments at the chlor-alkali plants. A study conducted by ICLAM (*Instituto para el Control y la Conservación de la Cuenca del Lago de Maracaibo*, Institute for the Control and Conservation of Lake Maracaibo's Watershed) in 1985, found that for each ton of sodium hydroxide (caustic soda) produced in the chlor-alkali plants in CPAMC, 250 grams of mercury are disposed of into the environment (ICLAM 1985). These production and disposal practices have been carried out since the inauguration of this petrochemical complex in El Tablazo Bay in 1969. Not only is mercury lost in production, but also some of it finds its way into the lake. The same study found that since 1979, an average of 55.11 kilograms of mercury lost in production was dumped into the lake by CPAMC (ICLAM 1985: 3). Moreover, another study found that the concentrations of mercury in the water near CPAMC exceeded the maximum allowed by law (0.405 ppb at the surface and 0.991 ppb at the bottom of the lake) (ICLAM 1991a: 13).

Needless to say, the socio-environmental impacts of mercury accumulation in anaerobic bacteria and marine food chains have been a major problem in Lake Maracaibo. For example, dangerous concentrations of the substance have been detected in the tissue of fish, aquatic birds, sediment, mollusks, crustaceans, and larvae by local scientists from the University of Zulia and the ICLAM (ICLAM 1991b, 1997; Colina and Romero 1992). Not only have the flora and fauna been affected by these emissions, but fishing communities located near CPAMC have also felt (directly or indirectly) the negative consequences associated with petrochemical waste and pollution. One example can be seen in the fishing community of Lagunillas de Agua located in the northeast shore of Lake Maracaibo. In this case, high concentrations of toxic heavy metals such as mercury (Hg), lead (Pb) and vanadium (V) were found in the kidneys and liver of anencephalic fetuses in the Pedro María Clara Hospital in the city of Ojeda from mothers living in this fishing community (Barrios et al. 1995; Moreno-Fuenmayor et al. 1996). Anencephaly is a congenital absence of most of the brain and spinal cord of an unborn baby. This congenital disease is attributed to maternal exposure to high levels of toxic heavy metals, which are byproducts characteristic of oil and petrochemical activities (Tahan et al. 1993). The expected international rate of anencephaly cases is one case for every 1,000 births (WHO 1997). In the state of Zulia, where these fishing communities are located, the rate of anencephaly cases has increased from 0.9 cases for every 1,000 births in 1969 to 5.1 for every 1,000 births in 1994 (Tahan et al. 1993).

As explained earlier, the production of PVC requires the use of great amounts of pure quality industrial salt. Therefore, if we trace the material flow of the salt used in the chlor-alkali plant in CPAMC back to its production site, we will be able to uncover how the industrialization of this natural resource is affecting the livelihoods and health of fishers in the community of Ancón de Iturre. The intensification of PVC production led to the modernization of the salt industry in Venezuela. As a result, PEQUIVEN created a joint venture agreement with Cargill in 1995 to create PRODUSAL (*Productora de Sal, C.A.*) to industrialize solar salt production and secure a constant flow of pure quality salt for petrochemical processes. PRODUSAL is located in the southern section of Los Olivitos Wildlife Refuge and Fishing Reserve (Los Olivitos) in the municipality of Miranda (northeastern shore of the lake). This company took advantage of the geographic characteristics of Los Olivitos[12] and transformed approximately 25 per cent of the southern section of this Ramsar site[13] (around 5,000 hectares of mangroves) into large-scale industrial salt evaporation ponds, salt concentrators, and crystallizers. This required the installation of dikes, which cover a perimeter of approximately 30 kilometers upstream, to block the passage of fresh water from the Cocuiza and Palmar rivers into the mangrove (see Figure 21.3).

This process increased the salinity of this ecosystem, which in turn reduced evaporation time and increased production. The production of industrial solar salt involves pumping large quantities of water (approximately 120,000 litres per minute) from Los Olivitos into the evaporator ponds. Unfortunately, the amount of water pumped exceeds the capacity that the wetland has to recover from that loss. This process lowers the normal water levels in the wetland, which in turn has a detrimental impact on the ecosystem and the livelihoods of fishing communities in the area, especially Ancón de Iturre. An even greater danger than the extensions of the dikes to block the passage of fresh water is the way in which PRODUSAL discharge the bitters[14] (*amargos* in Spanish) accumulated in their industrial production directly into the

Figure 21.3 Palmar River

Source: Author's personal archive. Fieldwork, 2 February 2007.

wetland (Kneen 2002). To put this in perspective, every ton of salt produced by PRODUSAL, creates one ton of bitterns. The company has the capacity to produce 800 thousand tons of salt per year.[15]

Historically, the production of salt before the creation of PRODUSAL was in the hands of traditional artisanal producers in the fishing community of Ancón de Iturre, which used to produce salt during the dry season and fish during the rainy season. Fishers from Ancón de Iturre were displaced from the area and dispossessed from their livelihoods with the creation of PRODUSAL. In short, the impacts this company has had on the community include: (i) the unemployment of more than two thousand *salineros* (salt workers), with the process of space appropriation; (ii) the decline of fish populations resulting from the negative environmental effects on the ecosystem, which in turn affected the livelihoods of the community; and (iii) a surge of health problems directly related to the chemicals and pollutants produced by the industrial salt process (Algarra et al. 1989, 1994; Delgado 2012).

The metabolism of fossil fuel production in CPAMC to produce plastics presents two problems with scalar characteristics. First, petrochemical transformations are rooted in an inherent contradiction within the confinement of the industry perpetuated in the form of pollution and toxic waste. Second, the tensions and forces that flow with the production of PVC move beyond the industry through the water, the fish, the air, the body, and the region, and shape a conflictive energy landscape, "organized in a perpetually shifting and contested scalar configuration" of spaces of social and ecological distress (Swyngedouw 2004: 147).

Conclusion

The primary objective of this chapter has been to explore the particular conflictive energy landscape produced by the petrochemical revolution in Venezuela. To pursue this objective, this chapter analyzed how the metabolic transformations of oil, natural gas, and salt into plastics re-produces spaces of social and ecological distress at multiple scales, which materialize as a set of tensions embedded within this particular mode of production. This chapter has sought to interpret these processes as an articulation of particular political, economic, and technological processes by which former modes of economic activity were organized for the modernization of the petrochemical industry. In summary, using the example of *Petrocasas*, this chapter elucidated: (i) the social, political, and economic rationale behind the petrochemical revolution; (ii) the influx of capital, natural resources, technology, scientific knowledge, and infrastructure necessary for the implementation of this modernization strategy; (iii) the way in which the materiality of fossil fuels not only shapes the development of this project, but also how their transformation into *Petrocasas* creates a specific set of socio-ecological tensions that may be obscured by the commodification process itself; and (iv) the metabolic processes involved in the petrochemical production of PVC.

No matter where oil or natural gas are extracted in Venezuela, sooner or later they will be transported to petrochemical complexes before they continue their trajectories in the production chain. The same can be said about salt. Once it is produced, either accumulated through solar evaporation or extracted from mines, it will end up in petrochemical industries as well. Thus, we can conceptualize petrochemical industries as "mega-magnets" fixed in place that attract different kinds of natural resources from distant places for their production processes. More importantly, as I have argued, not only does the petrochemical sector plays an important role in the commodification of fossil fuels, but it is also one of the most important actors in the modernization strategies of most petro-states today. Furthermore, through the transformation of natural resources into different social goods, petrochemical industries become vehicles of capital conversion. For instance, natural gas may be expansive and volatile, stubborn and "uncooperative" (Bakker 2004),[16] but it is also a "highly strategic resource for capital accumulation" (Bakker 2002: 771). Just as the commodification of the seed (Kloppenburg 1988), the hybrid corn (Fitzgerald 1990), and the industrial chicken (Boyd et al. 2001), the transformation of oil, natural gas, and salt into *Petrocasas* portrays the different ways in which nature – and natural landscapes – can be refashioned and incorporated into faster circuits of capital. Castree and Braun (1998: 4) echoed this statement by suggesting that "[W]here the juggernaut of capitalist development has not yet made this complete, technological innovations – increasingly in the form of "technoscience" – stand to finish the task." The petrochemical metabolism of fossil fuel production, then, helps us "see commodities as complex, mutable, and mobile sites of social relations, cultural identity, and economic power" (Castree 2001: 1520). Moreover, the contested making of intermediate materials such as naphtha, methane, caustic soda, and chlorine, as well as their further transformations in petrochemical facilities "similarly fuse physical-environmental metabolisms with sociocultural, and political-economic relations" (Swyngedouw 2004: 129). Therefore, whether we consider the production of affordable houses made out of plastic such as *Petrocasas*, the making of an advanced and highly technological petrochemical complex, the construction of massive pipelines systems such as Project ICO to move stranded gas from well-head to petrochemical industries, they all highlight the particular social, environmental, and spatial relations embedded in energy landscapes.

Questions of environmental degradation and social conflict are at the heart of the petrochemical revolution. Of particular importance, is the way in which the economic, social, and environmental consequences associated with this modernization strategy were distributed unevenly. For example, while the material benefits derived from the production of *Petrocasas* have diffused at a national scale, the socio-environmental impacts associated with this mode of production have concentrated heavily at a local scale in the Lake Maracaibo region.

Indeed, the development of the petrochemical industry in Venezuela has generated a series of social and environmental conflicts in which fishing communities affected by the extraction, transportation, and transformation of fossil fuels and salt have occupied center stage. The asymmetrical distribution of these conflicts in Lake Maracaibo have produced a rather contentious energy landscape at each stage of production along the hydrocarbon commodity chain. For instance, the high incidence of anencephalic babies from mothers exposed to toxic chemicals in Lagunillas de Agua or moments in which environment-related social marginalization and conflicts over the ecological conditions of livelihoods experienced by fishers in Ancón de Iturre as a result of the industrialization of salt in Los Olivitos would support this interpretation.

These cases suggest how oil, natural gas, and salt, on the one hand, and the different institutional arrangements coordinated by the state to extract, transport, and transform them into different social goods, on the other hand, "are constituted as networks of interwoven processes that are human and natural, real and fictional, mechanical and organic" (Swyngedouw 2004: 129). Conceptualizing each one of these actors as individual nodes embedded in an energy network, allows us a useful way to critically analyze the extensive spatial reach of fossil fuels and the way in which their commodification produce a conflictive energy landscape in the form of different spaces of socio-ecological distress at multiple scales in Venezuela.

Notes

1 Oil is also an important input of production in petrochemical processes. However, the petrochemical revolution focuses on securing natural gas to meet the goals intended by President Chávez.

2 The development of the petrochimal revolution still continues under current President Nicolás Maduro Moros.

3 In the context of this chapter I treat oil and natural gas as underground resources in their crude form that have not been altered by any external force. In this case, they are just materials from the earth and not yet commodities. They become commodities only when they enter in contact with technology, science, economics, labor, and law – all of which happen above ground. Therefore, oil and natural gas become commodities through a process known as commodification. That is, the transformation of a resource from something that has use value to something that is transformed and therefore has exchange value (e.g., gasoline, fertilizers, or refined oil).

4 The metabolism of commodity production is a concept used by social scientists, especially energy geographers, to explain the process involved in transforming natural resources into different commodities. In the case of fossil fuels, for example, this process involves the inflows of raw materials (e.g., oil, natural gas, water, salt, electricity), the conversions of energy, and the outcomes of this process in the form of different commodities (e.g., plastics, fertilizers, medicines) together with pollution, heat, and waste (see also Benton 1989, 1992; Leff 1994; O'Connor 1998).

5 The trajectories and valences of the 'lives' of commodities from a geographical perspective have been conceptualized in terms of chains, networks or circuits. I do not discuss these conceptualizations here, but instead I recommend a review of Winson (1993), Hartwick (1998), and Robbins (1999).

6 In political economic theory, metabolism is a concept developed by Karl Marx to explain the relationship that exist between humans and nature as they use their labor to transform natural resources into something they can use and also exchange in the market. In this case, the Venezuelan state has to mediate this interaction between natural resources (i.e., oil and natural gas) which are part of nature, the institutions that govern them, oil workers, and the economic system under which these resources are refined and transformed into commodities to be sold in national and international markets.

7 Online at www.pdvsa.com/index.php?tpl=interface.sp/design/readmenu.tpl.html&newsid_obj_id= 7609&newsid_temas=84 [Accessed 15 March 2015].

8 To put it in perspective, the Tennessee Gas Pipeline's Broad Run Flexibility Project moves 590 mcf/d of natural gas from West Virginia to the Gulf Coast states (EIA 2016).

9 *Instructivo de armado vivienda Petrocasa* (Instruction manual to assemble a Petrocasa) online at www.petrocasa.com.ve/construccion/#1 [Accessed 10 June 2015].

10 How is PVC made? Available online at www.pvc.org/What-is-PVC/How-is-PVC-made [Accessed 1 July 2015].

11 For example, in 1989, an explosion at the Phillips Petroleum plastic plant in Pasadena, Texas, killed 23 workers (Schneider 1991). Another example of a petrochemical disaster goes back to 1956 in Minamata, Japan where the release of mercury in the industrial wastewater from the Chisso Corporation's chemical factory produced one of the worst cases of mercury poisoning in the world. Approximately 900 people died and 2265 people were certified as having directly suffered from mercury poisoning, known as Minamata disease (Ui 1992).

12 The wetland of Los Olivitos has ideal geographical conditions for the production of solar salt. It receives brackish water from El Tablazo Bay and salt water via a direct connection to the Gulf of Venezuela. The evaporation rate in Lake Maracaibo's watershed is high due to its close proximity to the equator (10° 47' N Latitude), which in turn increases the salinity of this marine ecosystem.

13 Los Olivitos was listed as a wetland of international importance according to the Ramsar Treaty. The Convention on Wetlands of International Importance, called the Ramsar Convention, is an inter-governmental treaty that provides the framework for national action and international co-operation for the conservation and wise use of wetlands and their resources. Available online at https://rsis.ramsar.org/ris/859 [Accessed 19 May 2015].

14 The bitter water solution of bromides, magnesium and calcium salts remaining after sodium chloride is crystallized out of seawater in concentrators and evaporators.

15 Online at www.productoradesal.com/produsal/quienes-somos/index.htm [Accessed 25 July 2015].

16 Even though Bakker (2004) was referring to the privatization of water in England and Wales, this characteristic is equally true of natural gas.

References

Algarra, M., Sulbaran, C., Brito, M. and Casler, C. (1994). *Territorio, ambiente y calidad de vida: esquema de uso social para Los Olivitos*. Universidad del Zulia (LUZ), Consejo de Investigaciones Científicas y Humanisticas (CONDES).

Algarra, M., Sulbaran, C., Casler, C. and Brito, M. (1989). *Impacto económico, social y ambiental de la expansión industrial de las Salinas de Los Olivitos*. Universidad del Zulia.

Aló Presidente. (2007). "Presidente Chávez: Venezuela sera una potencia petroquímica". By Erika Guerrero, 23 September, Online at www.alopresidente.gob.ve/informacion/2/880/presidente_chuevezvenezuela_seruuna.html [Accessed, 2 August 2015].

Aló Presidente N° 375. (5 June 2011). Aló Presidente 375. Online at www.alopresidente.gob.ve/resumen_al%C3%B3/4/2123/alupresidente_375.html [Accessed 2 August 2015].

Bakker, K. (2002). "From state to market? Water mercantilización in Spain". *Environment and Planning A*, 34(5): 767–790.

Bakker, K. (2004). *An Uncooperative Commodity: Privatizing Water in England and Wales*. Oxford: Oxford University Press.

Bakker, K. and Bridge, G. (2006). "Material worlds? Resource geographies and the 'matter of nature'". *Progress in Human Geography*, 30(1): 5–27.

Barrios, L., Tahán, J., Marcano, L., Granadillo, V., Cubillán, H., Sánchez, J., Rodríguez, M. Gil de Salazar, F., Salgado, O. and Romero, R. (1995). "Factores socio-sanitarios de la anencefalia en la costa oriental del lago de Maracaibo (Venezuela) y contaminación metálica". *CIENCIA*, 3(1): 49–58.

Benton, T. (1989). "Marxism and natural limits: An ecological critique and reconstruction". *New Left Review*, 178: 51–86.

Benton, T. (1992). "Ecology, socialism and the mastery of nature: A reply to Reiner Grundman". *New Left Review*, 194: 55–74.

Benton, T. (1996). *The Greening of Marxism*. New York: Guilford Press.

Boletín Pequiven. (2007). "Las viviendas de Petrocasas se levantan en Nuestra Señora de Coromoto". Julio, Año 9, N° 51, p. 7. Online at www.pequiven.com/pqv_new/pdf/Boletin51.pdf [Accessed 14 March 2015].

Boyd, W., Prudham, W.S. and Schurman, R.A. (2001). "Industrial dynamics and the problem of nature". *Society and Natural Resources*, 14(7): 555–570.

Bridge, G. (2000). "The social regulation of resource access and environmental impact: Production, nature and contradiction in the US copper industry". *Geoforum*, 31(2): 237–256.

Bridge, G. (2004). "Contested terrain: Mining and the environment". *The Annual Review of Environment and Resources*, 29(1): 205–259.

Bridge, G. (2008). "Global production networks and the extractive sector: Governing resource-based development". *Journal of Economic Geography*, 8(3): 389–419.

Bridge, G. and P. McManus. (2000). "Stick and stones: Environmental narratives and discursive regulation in the forestry and mining sectors". *Antipode*, 32(1): 10–47.

Castree, N. (2001). "Commodity fetishism, geographical imagination and imaginative geographies". *Environment and Planning A*, 33(9): 1519–1525.

Castree, N. (2003). "The production of nature," pp. 275–289. In Shepard, E. and Barnes, T. (eds), *A Companion to Economic Geography*. Oxford: Blackwell.

Castree, N. and Braun, B. (1998). "The construction of nature and the nature of construction: Analytical and political tools for building survivable futures", pp. 3–42. In Braun, B. and Castree, N. (eds), *Remaking Reality: Nature at the Millenium*. London: Routledge.

Colina, M. and Romero, R.A. (1992). "Mercury determination by cold vapour atomic absorption spectrometry in several biological indicators from Lake Maracaibo, Venezuela". *Analyst*, 117: 645–647.

Comunicaciones Pequiven. (2008). "Más de 2 mil familias de Carabobo habitarán sus Petrocasas". Sede Corporativa, Enero-Marzo, Año 10, N° 53, pp. 8–9. Online at www.pequiven.com/pqv_new/pdf/Boletin%2053.pdf [Accessed 19 March 2015].

Comunicaciones Pequiven. (2009). "Petrocasa diversifica los tipos de vivienda en Venezuela". Sede Corporativa, Junio, Año 12, N° 58, pp. 26–27. Online at www.pequiven.com/pqv_new/pdf/Comunicaciones%2058.pdf [Accessed 25 March 2015].

Delgado, E. (2012). "Spaces of socio-ecological distress: Fossil fuels, solar salt, and fishing communities in Lake Maracaibo, Venezuela" (PhD Diss, Syracuse University).

Delgado, E. (2016). "Energy geographies: Thinking critically about energy issues in the classroom". *Journal of Geography in Higher Education*, 40(1): 39–54.

EIA. (2016). "New Northeast pipelines help boost gas production 18%". Online at www.eia.gov/naturalgas/weekly/archive/2016/02_11/index.cfm#tabs-rigs-2 [Accessed 12 February 2016].

Fitzgerald, D. (1990). *The Business of Breeding: Hybrid Corn in Illinois, 1890–1940*. Ithaca, NY: Cornell University Press.

Georgescu-Roegen, N. (1979). "Energy analysis and economic valuation". *Southern Economic Journal*, 45(4): 1023–1058.

Hartwick, E. (1998). "Geographies of consumption: A commodity-chain approach". *Environment and Planning D – Society and Space*, 16(4): 423–437.

Hartwick, E. (2000). "Towards a geographical politics of consumption". *Environment and Planning A*, 32(7): 1177–1192.

Harvey, D. (1990). "Between space and time: Reflections on the geographical imagination". *Annals of the Association of American Geographers*, 80(3): 418–434.

ICLAM. (1985). *Detección de mercurio en el estrecho del Lago de Maracaibo y la Bahía El Tablazo*. Instituto para el Control y la Conservación del Lago de Maracaibo.

ICLAM. (1991a). *Evaluación de la calidad de agua en areas petroleras y petrquímicas del Lago de Maracaibo 1990–1991*. Convenio COIC – ICLAM. Instituto para el Control y la Conservación del Lago de Maracaibo.

ICLAM. (1991b). *Monitoreo en el estrecho del Lago de Maracaibo para la determinación de mercurio*. Instituto para el Control y la Conservación del Lago de Maracaibo.

ICLAM. (1993). *Detección de mercurio (Hg) en peces en la zona del estrecho del Lago de Maracaibo*. Instituto para el Control y la Conservación del Lago de Maracaibo.

ICLAM. (1997). *Evalucaición de la concentración en peces, moluscos, crstáceos y aves de sustancias tóxicas presentes en el lago: Mercurio, cobre y vanadio*. Instituto para el Control y la Conservación del Lago de Maracaibo.

Kaika, M. and Swyngedouw, E. (2000). "Fetishizing the modern city: The phantasmagoria of urban technological networks". *International Journal of Urban and Regional Research*, 24(1): 120–138.

Kaup, B. (2008). "Negotiating through nature: The resistant materiality and materiliaty of resistance in Bolivia's natural gas sector". *Geoforum*, 39(5): 1734–1742.

345

Kloppenburg, J. R. (1988). *First the Seed: The Political Economy of Plant Biotechnology, 1492–2000*. Cambridge: Cambridge University Press.

Kneen, B. (2002). *Invisible Giant: Cargill and its Transnational Strategies*. 2nd edn. London: Pluto Press.

Laszlo, P. (2001). *Salt Grain of Life*. New York: Columbia University Press.

Leff, E. (1994). *Green Production: Towards an Environmental Rationality*. New York: Guilford Press.

Leslie, D. and Reimer, S. (1999). "Spatializing commodity chains". *Progress in Human Geography*, 23(3): 401–420.

Marx, K. (1967). *Capital*, Vol. I. New York: International.

Moreno-Fuenmayor, H., Valera, V., Socorro-Candanoza, L., Bracho, A., Herrera, M., Rodríguez, Z. and Concho, E. (1996). "Programa preventivo de defectos del nacimiento: Incidencia de anencefalia en Maracaibo, Venezuela. Periodo 1993–96". *Investigación Clínica*, 37(4): 271–278.

O'Connor, J. (1998). *Natural Causes: Essays in Ecological Marxism*. New York: Guilford Press.

PDVSA. (2009). Balace de la Gestión Social y Ambiental – Año 2009. Online at www.pdvsa.com/interface. sp/database/fichero/free/5889/1049.PDF [Accessed 3 April 2015].

Prudham, S. (2009). "Commodification," pp. 123–142. In Noel Castree, David Demeritt, Diana Liverman and Bruce Rhoads (eds), *Companion to Environmental Geography*. Hoboken, NJ: Wiley-Blackwell.

Robbins, P. (1999). "Meat matters: Cultural politics along the commodity chain in India". *Cultural Geographies*, 6(4): 399–423.

Schneider, K. (1991). "Petrochemical disaster raise alarm in industry". *New York Times*, 19 June 1991, Online at www.nytimes.com/1991/06/19/us/petrochemical-disasters-raise-alarm-in-industry.html [Accessed 13 March 2015].

Swyngedouw, E. (2004). "Scaled geographies: Nature, place, and the politics of scale", pp. 129–153. In E. Sheppard and R. McMaster (eds), *Scale and Geographic Inquiry: Nature, Society, and Method*. Oxford: Blackwell.

Tahan, J.E., Granadillo, V.A., Sánchez, J.M., Cubillan, H.S. and Romero, R.A. (1993). "Mineralization of biological materials prior to determination of total mercury by cold vapour atomic absorption spectrometry". *Journal of Analytical Atomic Spectrometry*, 8(7): 1005–1010.

Thrift, N. (2000). "Commodity", pp. 95–96. In Johnston, R.J., Gregory, D. and Watts, M. (eds), *The Dictionary of Human Geography* (4th edn). Oxford: Blackwell.

Ui, J. (1992). *Industrial Pollution in Japan*. Tokyo, Japan: United Nations University Press.

Venezuelanalysis.com. (2007). "Venezuela officially launches 'Petrochemical Revolution'". By Chris Carlson, 24 September. Online at http://venezuelanalysis.com/news/2641 [Accessed 18 May 2015].

Venezuelanalysis.com. (2011a). "Venezuelan government launches massive new housing mission". By Rachael Boothroyd, 2 May. Online at http://venezuelanalysis.com/news/6167 [Accessed 18 May 2015].

Venezuelanalysis.com. (2011b). "President Chávez uses legislative authority to create 'Law for Dignified Refuge' in Venezuela". By Juan Reardon, 20 January 2011. Online at http://venezuelanalysis.com/news/5948 [Accessed 18 May 2015].

Watts, M. (1999). "Commodities", pp. 287–304. In M. Watts and D. Goodman (eds), *Introducing Human Geographies*, Routledge: New York.

Watts, M. (2004). "Violent environments: Petroleum conflict and the political ecology of rule in the Niger Delta, Nigeria". In Peet, Richard and Michael Watts, *Liberation Ecologies: Environment, Development, Social Movements* (2nd edn). London: Routledge.

WHO (1997). Anencephaly, live and still births (L+S). www.who.int/genomics/about/en/anencephaly.pdf [Accessed 1 August 2015].

Winson, A. (1993). *The Intimate Commodity: Food and the Development of the Agro-Industrial Complex in Canada*. Toronto: Garamond Press.

22

A Luta Continua

Contending high and low carbon energy transitions in Mozambique

Joshua Kirshner

Introduction

With the discovery of significant coal and gas resources in the past decade, Mozambique has gained attention as an important energy frontier globally. The coal and gas bonanza, however, presents complex political and planning challenges for the country. Just over two decades ago a protracted civil war ended, and today Mozambique is widely viewed as an African post-conflict success story (Vines, 2013). Its economy grew at an average rate of 7.5 per cent annually for much of the past two decades, joining South Africa and Nigeria as one of the top three recipients of foreign direct investment in sub-Saharan Africa in 2013 (Castel-Branco, 2015). Yet the growth has been highly unequal, with limited reduction in relative or absolute levels of poverty, and the country is still very poor by nearly any standard.

Recent offshore gas discoveries in the Rovuma Basin, on the northern border with Tanzania, could make Mozambique one of the world's largest exporters of liquefied natural gas. In parallel, landlocked Tete province has become Africa's second most important coal-producing region (IEA, 2014). The coal and gas rush is expected to spur export growth but also investments in domestic power generation, potentially reworking the country's energy system in the coming years.

Alongside the interest in Mozambique's hydrocarbons resources, there are mounting concerns over climate change. With one of Africa's longest coastlines, Mozambique is vulnerable to climate risks, including flooding, cyclones, sea level rise and rapid urban migration (Castán Broto et al., 2015). These pressures have raised the profile of renewable energy technologies (RETs) with the government seeking donors and investors in this area (Cuamba et al., 2013). Yet as in other African countries, the state's interest in RETs is tempered by the push to reduce energy poverty more broadly. Mozambique has among sub-Saharan Africa's lowest electrification rates, with 21 per cent of its population of 23 million estimated to have regular access to on-grid electricity by 2011 (EdM, 2011). A further 11 per cent is believed to have access through off-grid sources (FUNAE, 2012) while 80 per cent rely on biomass as their sole source of energy (Cuvilas et al., 2010).

Frelimo (*Frente de Libertação de Moçambique*), the party of government since Mozambique gained independence in 1975, aims to balance maximising the benefits of resource extraction for export-led growth with reducing domestic energy poverty. Frelimo claims it seeks to keep energy affordable for ordinary citizens and energy-intensive users alike. The latter consists of several megaprojects concentrated near the capital, Maputo, and in emerging extractive zones in the north. The state's focus is on grid-based electrification, but increasingly it sees potential for distributed and small-scale renewables, particularly for dispersed rural populations.

Among alternative energies, solar PV is the most important and is playing a growing role in the country's electrification. Between 2000 and 2011, the Energy Fund (FUNAE), a state agency, with the support of European and Asian donor organisations, installed over 1.2 MW of solar PV capacity. FUNAE calculates that some 1.5 million residents benefited from the systems by 2012, representing about 0.8 Wp per person (FUNAE, 2012). While the quantity in wattage is small, its ability to improve rural health and education is deemed significant (Cuamba et al., 2013). Solar PV installations have focused on these rural services, with former Energy Minister Salvador Namburete claiming that in 2014 micro-scale solar PV projects were used to electrify 700 schools, 600 health centres and 800 other public buildings in rural areas, at a total cost of US$51 million (MacauHub, 2014). Additionally, FUNAE has estimated over 60 potential micro-hydropower projects with up to 1000 MW capacity. Despite growing interest, only a handful of micro-hydro systems were implemented in recent years (Mika, 2014).

This chapter offers a sketch of Mozambique's contending pathways to expanding energy access. It addresses the following question: in the context of significant fossil fuel resources discoveries, what is the role of RETs and why are they being promoted? The chapter begins by reviewing the concept of energy transitions and outlining some of the recent high-carbon developments in the country. It then examines changes in Mozambique's energy provision system and identifies the actors engaged in efforts to reach the energy poor. This sets a context for an empirical exploration of four small-scale renewable energy projects, representing a range of technologies and sources of finance, in the third section. A final section presents a discussion of these projects and reflects on Mozambique's incomplete energy transition.

Methodologically, the chapter draws on a series of interviews with government officials, policymakers, representatives of utilities, donors, businesses and NGOs conducted in Maputo and Beira in 2013–2014, along with site visits to energy service projects, part of a larger study funded by the UK's ESRC.[1] The analysis builds on insights developed in studies of energy access in low-income countries that suggest energy transitions cannot be achieved in isolation from other aspects of socio-economic development (Ulsrud et al., 2015; Tawney et al., 2015; Murphy, 2001). It also engages with research that argues historical shifts, or transitions, in energy systems can be understood in spatial and place-specific terms as well as temporal ones (Bridge et al., 2013; Hansen and Coenen, 2014). This chapter argues that energy interventions from the state and donors in Mozambique—including on-grid and off-grid initiatives—are largely led by technology. As such, they do not adequately consider the social, cultural and economic conditions of energy users, especially in rural areas. With some exceptions, these interventions have adopted a centralised management approach focused on connecting institutions rather than providing direct access for people.

Is there a high carbon transition in Mozambique?

Energy transition is often defined within policy circles as moving towards a low carbon economy, for instance the UK government's aim to reduce carbon emissions by 80 per cent by 2050. Within energy studies more broadly, the notion of transition is used analytically to address major historical shifts in energy systems at national and global scales (Smil, 2005). This work shows that historical energy transitions have been associated with broad social change, such as industrialisation and urbanisation (Geels, 2002; Bridge et al., 2013).

Yet in many parts of the global south, energy transition implies a significant increase in the availability and affordability of modern energy services. In some contexts, this may entail an increase in carbon intensity, such as through the switch from household fuelwood to grid electricity (Bridge et al., 2013). As elsewhere in sub-Saharan Africa, historically limited energy provision has meant fossil fuels are not embedded in the social and economic landscape of Mozambique to the same extent as wealthier countries (Bridge et al., 2013). Many of the built environments and accoutrements of fossil fuel capitalism, however, such as gas pipelines and thermal power plants, are gaining a foothold in Mozambique amid the extractive resources boom.

The pathway to development the Mozambican state has pursued since the early 1990s is centred on energy and mineral extractive megaprojects, which have prioritized the needs of heavy industrial users and exports to regional energy markets, particularly South Africa's. This has proceeded alongside the large-scale privatisation of state assets, as more than one thousand state-owned enterprises and state shares in many companies were sold off to private investors in the early 1990s (Castel-Branco, 2015).

Unlike neighbouring South Africa and Zambia, Mozambique is not a historically important minerals producer, existing below the radar for global mining companies. This changed in 2008, when visiting geologists in Tete province confirmed the coal seam beneath Moatize basin forms part of the world's largest untapped coal deposit, with over 23 billion metric tonnes of coal, or enough to fire all the coal plants in the USA for 25 years (Besharati, 2012). The deposits are rich in coking coal, used for steel production, and thermal coal, which lacks export value but can be used for power generation (see Figure 22.1). Benefitting from buoyant world market conditions in the 2000s, Tete attracted a surge of investment with coal becoming Mozambique's second largest export earner by 2012 (ibid.).

But the coal operations have faced ongoing logistical and infrastructural gaps and declining global coal prices since 2013. The two largest investors, Brazil's Vale and UK-based Rio Tinto, sold all or part of their concessions in 2014 (Castel-Branco, 2015). The coal complex in Moatize risks becoming an extractive resource-based enclave with weak linkages to local enterprises, foreign ownership of capital, and export of goods with limited or no value added (Bloch and Owusu, 2012; Magrin and Perrier-Bruslé, 2011). These factors combine to create a downward cycle for local development, raising costs of living and creating few jobs, despite high levels of capital investment and elevated local expectations, or imaginaries, of a resources boom (Arias et al., 2014; Büscher, 2015; Kirshner and Power, 2015).

Nevertheless, several investors plan to build coal-fired power plants at mining sites, an effort supported by the state-owned electricity utility *Electricidade de Moçambique* (EdM). These energy generation projects underpin growing trade and investment ties, particularly with the emerging powers, such as China, Brazil and India (Dadwal, 2011). According to EdM, these thermal power plants[2] will fill a supply gap until new hydropower and gas facilities become available in the 2020s, enabling electrification in underserved regions without raising energy prices. Mozambique's electricity pricing is currently among the highest in the SADC region, despite being the second largest energy producer in SADC (Nhamire and Mosca, 2014).

The proposed coal-fired power stations, if completed, would commit Mozambique to a high-carbon development pathway that may be difficult to reverse. Arguing for the pragmatic necessity of coal to stabilise the supply of energy in Mozambique, a Vale official explained:

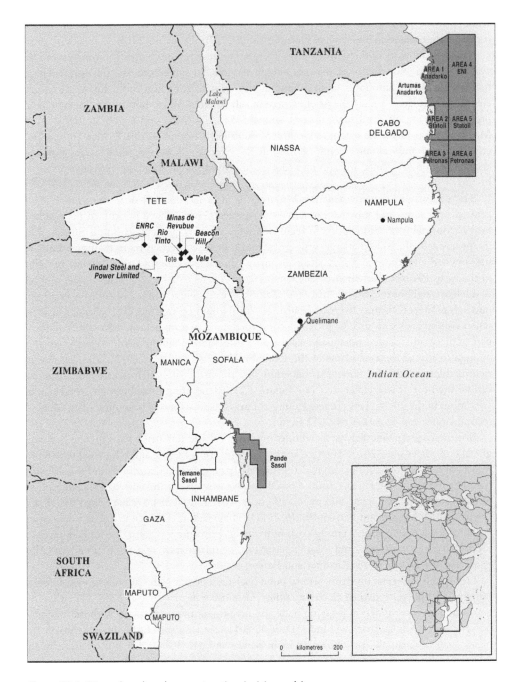

Figure 22.1 Sites of coal and gas extraction in Mozambique

Source: Cartographic Unit, Durham University.

Coal will allow Mozambique to have competitive power. You have to respect international standards for environmental emissions. But coal is a good in-between. You can build the power plant and then generate gas and hydro facilities. There is a big gap. Only coal can fill it. The gas and new hydro will only come on-stream in 2018 or 2020, and meanwhile Mozambique is growing a lot. So coal is unavoidable.

(Lamarie, 2013)

In this view, the persistence of coal in Mozambique's political economy of energy does not preclude the development of cleaner alternatives. The boom in the country's extractives sector could potentially reshape state budgets and spending priorities, in turn supporting alternative energy technologies. Using hydrocarbon revenue windfalls to fund RETs could increase energy access rates in rural areas, adding employment and reversing jobless development. Thus far however the state's campaign to reduce energy poverty has focused on grid extension, with small-scale RETs pursued in remote locales with low levels of consumption and where the grid is not expected to reach. Significant socio-economic and regional inequalities are thus reflected in access to the grid.

Mozambique's changing energy landscape

Following a history of Portuguese colonial rule, anti-colonial struggle and civil war, Mozambique's energy sector remains underdeveloped. Yet it has expanded in recent years amid rapid—albeit unequal—economic growth. Set up by the state in 1977, EdM is responsible for transmission, distribution and some generation. Before independence in 1975, thermal power plants were the major source of electricity, with some input from hydropower supplying localised grids (Cuamba et al., 2013).

This changed with the opening of Cahora Bassa dam in 1974, completed in the last year of Portuguese rule, which reduced thermal in favour of hydropower. Cahora Bassa is located on the Zambezi River in Tete province, until then viewed by colonial planners as a remote outpost (Newitt, 1995). One of the last major infrastructure projects built in Africa during the turbulent period of decolonization, Cahora Bassa remains the largest hydroelectric scheme in southern Africa with 2075 MW capacity (Cuamba et al., 2013). The independent power producer *Hidroeléctrica de Cahora Bassa* (HCB) began operations in 1977, supplying power to the South African and Zimbabwean grids while producing some 90 per cent of the electricity consumed in Mozambique. In 2007—over 30 years after independence—HCB passed from Portuguese to majority (85 per cent) Mozambican ownership, when Portugal (needing funds to reduce its budget deficit to meet EU regulations) agreed to sell most of its equity.

Mozambique produces a huge surplus of hydroelectricity at Cahora Bassa, but it must observe pre-independence agreements between Portugal and South Africa that committed over 85 per cent of the dam's output to supply South Africa's Eskom as a means to finance the project (Isaacman and Isaacman, 2013). Some 65 per cent of HCB's production, or 1500 MW of electricity, is currently exported to South Africa via Apollo substation in Mpumalanga, with a portion re-imported into southern Mozambique on lines owned by Eskom (KPMG, 2013).[3] What resources there are for domestic consumption are highly unevenly distributed, with the BHP Billiton Mozal aluminium smelter, near the South African border, consuming fully two-thirds of all electricity generated in Mozambique. Moreover the existing system for electricity transmission and distribution is obsolete, after decades of limited financial investment (see Figure 22.2).

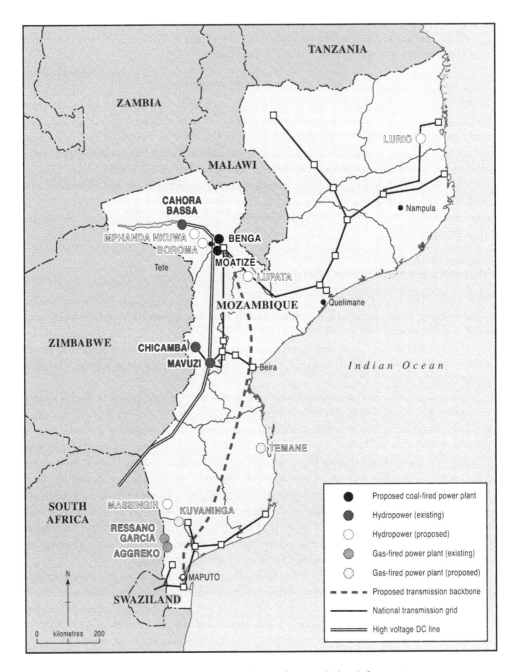

Figure 22.2 Mozambique's electricity generation and transmission infrastructure
Source: Cartographic Unit, Durham University.

The government has responded to these constraints by promoting the construction of several large hydropower facilities. The largest is Mphanda Nkuwa, sited 60 km downstream from Cahora Bassa on the Zambezi River, at an estimated cost of US$2.2 billion (KPMG, 2013). The project has World Bank support and a mix of South African, Brazilian and Chinese investors. The state-owned China State Grid Corporation gained a majority stake in late 2013 through a process that lacked competitive bidding (Ribeiro, 2015), while the dam's construction depends on Eskom's commitment to buy most of its electricity. The project is expected to displace 1400 households and indirectly affect 200,000 Mozambicans' livelihoods once completed (Morrissey, 2013; Isaacman and Sneddon, 2003).

Facing a growing supply shortfall, EdM has begun purchasing more expensive power from Gigawatt Mozambique, a private firm that is 40 per cent owned by the South African firm Gigajoule. Gigawatt invested $320 million in a 110 MW gas-fired power plant in Ressano Garcia, on the South African border (KPMG, 2013; Nhamire and Mosca, 2014). A further 300 MW of gas-fired plants are planned for the southern region, reflecting Mozambique's newfound gas wealth and the state's aim to increase domestic use (Gqada, 2013). In parallel, EdM has invested in comprehensive grid extension since 2009, with Swedish and Norwegian technical support, aiming to reach all district capitals (Cipriano et al., 2015). Along with upgrading Maputo's electricity infrastructure, the grid connected to 125 of 128 district capitals by 2014 (Ouchim, 2013).

Despite this expansion of the centralised network, Mozambique's electrification has been geographically uneven, with grid extension focused on urban spaces and exports to regional markets. Access to grid electricity in Mozambique thus reflects the colonial-era development of natural resources, regional transport corridors, ports and urban centres. Given the large, scattered rural population and high unit costs of grid extension, low electrification rates prevail. The prioritisation of some places and forms of electricity deemed worthy of supply also shapes what constitutes 'off grid' (cf. Ferguson, 2005).

Increasingly, the Mozambican government supports decentralised energy for off-grid spaces, such as solar PV mini-grids, which can reach further into rural areas and create new possibilities for sustainable energy access. But as in other low-income countries, the resources and attention dedicated to improving rural household energy use seem incommensurate with the problem's magnitude (Sagar, 2005). Further, as noted elsewhere in this book, evidence that small-scale renewables can actually facilitate energy access remains patchy and limited.

Mobilizing renewable energies

At the heart of how RETs have been mobilised in Mozambique is the state agency responsible for expanding energy access for rural people, *Fundo de Energia* (National Energy Fund, FUNAE). FUNAE was established in 1997 with Danish assistance. It initially supplied diesel generators and kerosene as a means of addressing energy poverty in rural communities, but its focus has shifted to promoting, supplying and financing renewable sources of energy. While FUNAE has developed some micro-hydro and pilot wind projects, its main focus is on solar PV systems. The agency is funded through the state budget, with revenues from taxes and levies from petroleum and electricity concessions, along with donor support.[4] By agreement with EdM, FUNAE works in areas projected to be >10 km from the grid within five years (Ouchim, 2013). Inter-agency planning, coordination and capacity building, however, have been minimal.

Acting independently of FUNAE, *Deutsche Gesellschaft für Internationale Zusammenarbeit* (German International Cooperation, GIZ) is also supporting RETs in Mozambique under the Energising Development (EnDev) programme. EnDev is a partnership between the Netherlands,

Germany, Norway, Australia, UK and Switzerland that cooperates in several African, Latin American and Asian countries. GIZ serves as lead agency in implementing EnDev. In Mozambique, EnDev is called *Programa de Acesso ao Serviço de Energia Moderna em Moçambique* (Access to Modern Energy Services – Mozambique, AMES-M). It focuses on rehabilitating colonial-era micro hydro mills, improving their capacity and efficiency. GIZ's work with RETs in Mozambique began in the early 1990s (Mika, 2014).

Since the 2000s, FUNAE has worked with a number of donors and development agencies in deploying RETs in projects for supplying electricity for social uses in off-grid environments, including school lighting, water pumps and vaccine refrigeration. These projects use localised mini-grids or stand-alone systems, usually supplied by solar PV, which provide electricity to schools, clinics, administrative posts[5] and other rural institutions. FUNAE has also supplied micro-hydro systems for decentralised generation, but only in upland Manica and Zambézia provinces (Mika, 2014). FUNAE has replicated its model for solar PV across the country, extending its geographic reach but also limiting local involvement in project design. Scholars have cautioned against uncritically accepting local participation as a universal good (e.g. Mohan and Stokke, 2000). Yet evidence suggests devolution and local input in energy projects offers consumers a stake in deciding how electricity can be used and made relevant to their lives (Ulsrud et al., 2015).

The growing public use of solar PV (and to a lesser extent micro-hydro) has become a testing ground for its further development as a means to provide domestic energy services. Since 2010, FUNAE has initiated projects focused on households, small businesses and villages, which include mini-grids and stand-alone solar home systems (SHS) in homes and businesses. Here, FUNAE is actively involved in the production of solar PV within the rural economy, procuring systems of less than 100w that can be bought by households and commercial enterprises through long-term loans, which are underwritten by FUNAE. Beneficiaries view solar energy as appropriate for lighting, phone charging and refrigeration. Solar electricity has not replaced the need for biomass energy, however, as capacity is normally insufficient for an electric cooker coil's current (cf. Mavhunga, 2013). Given difficulties in maintenance and repair, many of FUNAE's systems have been abandoned. Theft of solar panels is a frequent problem, reflected in a second-hand trade in panels found in some municipal markets.

Until recently, Mozambique lacked capacity in solar manufacturing, with systems and components all imported. This mostly consisted of Chinese equipment, often sourced from South Africa by cross-border traders and sold in urban market stalls or electronics shops. Chinese components are cheapest on the market but widely viewed as unreliable. The need to import PV equipment has decreased with FUNAE's development of Mozambique's first solar module assembly plant, supported by a US$13 million concessional loan from India. The FUNAE plant, which opened in Beluluane, outside Maputo in late 2013, aims to produce 5MW of capacity annually, reducing imports and lowering costs in the solar PV systems, which are primarily intended for use by FUNAE in its projects (Namburete, 2014).

FUNAE has also sought to establish favourable contracts in order to diversify the range of solar PV available. In 2013, it awarded a contract to the German firm Fosera to install pico solar systems (5 volts max) in schools and homes in Manica province (Pedro, 2013). Fosera established a subsidiary in Maputo assembling pico-solar units from parts manufactured in Thailand, including solar lanterns and solar phone chargers (Cuamba, 2014). More recently, FUNAE managed the development of PV mini-grids in northern Niassa province, a project financed by the South Korean government. As solar PV has increased in both scope and visibility, it has come to be regarded as a status symbol in many rural settings, such that there is a growing market demand for installation extending beyond FUNAE's programmes.

The following section explores the deployment and uses of RETs in Mozambique through a focus on four small-scale energy projects in the central provinces of Manica and Zambézia (see Figure 22.1). Four key points emerge from the discussion. First, the emphasis in these projects, especially those supported by FUNAE, has been on connecting rural institutions rather than households. Second, despite technical challenges with solar energy, solar panels are taking on new uses in many rural communities. Third, many of the projects are centrally managed, with limited local input in design and operations. The projects are thus led by technology rather than by the people they aim to serve. This has especially been the case with FUNAE while GIZ has taken a more bottom-up approach, albeit on a smaller scale. Fourth, solar energy has not replaced 'non-modern' fuels such as charcoal for cooking and heating. Micro hydro systems fulfil a range of basic energy needs, given their higher capacities, but are limited to sites with feasible terrain. Overall, these RET projects are reshaping Mozambique's energy landscape, proceeding alongside but also constrained by conventional grid expansion.

Observing four off-grid energy projects

Chinhambuzi, Manica province

In March 2013, FUNAE inaugurated a solar PV mini-grid, with financial support from Belgium, in Chinhambuzi, Manica province. Belgian Technical Cooperation (BTC), having cooperated with FUNAE for several years, installed the mini-grid system. The solar mini-grid supplies power to a primary school, clinic, police station, several teachers' houses, the local authority's residence, and 10 *bancas fixas*.[6] But ordinary residents in the administrative post (Chinhambuzi sede) and several nearby villages remain unconnected.

Many residents wish to be connected but capacity constraints have prevented it. The system has 3.6kWp of capacity, or up to 5kW, which is insufficient for linking households to the 3-km network. The majority of residents use wood and charcoal for cooking and heating water. The teachers' houses are furnished with lamps and a refrigerator, with electricity provided until 8:00 PM. The shopkeepers pay 70 MZN per month ($2.50) for electricity, a tariff used to support a security guard. The tariff appears unrelated to operations costs while payments go to the village chief who chairs an oversight committee. The teachers and local authority's staff do not pay any tariff for their electricity, while the shopkeepers' usage is unmetered. Given their lack of access, some residents have considered connecting spontaneously, although this might stress the system and cause outages for shopkeepers, who use high current-drawing freezers. The load will likely increase, as shopkeepers acquire TVs, electric fans and other small appliances. The system is managed from the provincial capital Chimoio, without means to locally monitor capacity or demand.

Even if they could connect, most PV mini-grids and household PV systems produce low-power direct current (DC) electricity. As noted above, these systems work for lighting and some small appliances but not electric cookers.[7] The demand for solar electricity in Chinhambuzi has outpaced the system's capacity and the mini-grid scarcely accommodates future growth. It will be important to consider the system's configuration and pricing so that energy services can provide useful outcomes to a wider group of consumers.

Mavonde, Manica province

Mavonde village is a 1.5-hour drive along an ungraded road from the district capital of Manica, near the Zimbabwean border. Many residents cross the border daily and have kinship and business ties on both sides. The area is affected by gold panning, involving domestic and Chinese operations, and according to residents we consulted, harming river ecosystems and causing soil erosion.

FUNAE manages a solar project in Mavonde, financed by the World Bank's Energy Reform and Access Programme (ERAP). Starting in 2009, FUNAE installed solar stand-alone systems at the primary school, clinic and administrative post headquarters, and in eight shops and households in Mavonde sede and Nhandiro, a smaller village six km up the road. During our visit, however, most of the systems were not working. The beneficiaries had agreed to repay their loans to FUNAE, but many of the systems were down for over a year, and users had stopped payment. FUNAE prepared contracts for the SHS users, but eight months later the contracts were not being honoured by both parties.

FUNAE provides SHS to households in Mavonde through soft loans that it underwrites. If the equipment fails there is some measure of security, as users can request repairs or delay payments. Most residents trust the technology, but many are frustrated by FUNAE's lagging maintenance. Only two of 10 solar-powered street lights were working, both situated near the local authority's residence. Repairs often face lengthy delays, given insufficient staffing in FUNAE's provincial office. We saw public notices for residents interested in SHS to register at the administrative headquarters. The local authority told us they received 65 such requests in one week. A female shopkeeper said she prefers a loan from FUNAE rather than buying panels in Manica, where she must pay cash up-front. At first she was sceptical of solar energy but now recognises its multiple benefits, especially for freezing fish and meat, which otherwise must be dried for storage.

The shopkeepers in Mavonde tend to be more affluent than other villagers, and many have worked seasonally in Chimoio, Beira and as far away as Johannesburg. FUNAE's systems have has the perhaps unintended effect of raising shopkeepers' profits, increasing inequalities. The majority of local residents are dispersed in smaller settlements, preferring to live near their farm plots (*machambas*), which are difficult for FUNAE to reach.

Along with FUNAE's installations, in Nhandiro one sees solar panels purchased independently by residents, suggesting increasing demand for solar electricity. During our visit, a group of men sat around a table drinking beer beside a *banca fixa*, enjoying the shopkeeper's hi-fi system powered by a DIY solar panel. The local chief seemed proud of FUNAE's solar systems, even if the majority of them did not work.[8]

Chua, Manica province

Chua is located in Maridza administrative post, in Manica district. The village has experience with micro hydro power for decades, with several hydraulic grinding wheels dating from the early 1960s. GIZ's AMES-M programme, in collaboration with its local partner organisation Kwaedza Simukai Manica (KSM), aims to rehabilitate eight micro-hydro-powered maize mills in Chua and nearby villages, improving milling and generating electricity to support local development. Following a consultative survey, the project began in 2008 with a €64,000 budget (Zana, 2007).

One of the more successful schemes is in Chua, owned and operated by a middle-aged man, Mr. Lino Ndacada. The system has 22 kW of capacity and connects to a mill and a mini-grid that extends for three km. By day it powers the mill and by night it supplies electricity. The mini-grid connects to 25 houses, with a further 50 houses benefitting from battery charging, thus affecting some 450 people. It also connects to a village business hub with several bancas fixas. Households pay a fixed 200–250 MZN per month ($7.00–8.50) and shops pay 300 MZN

($10) for electricity consumption. Many systems use locally made turbines, while the generators are imported from Germany. GIZ aims to improve the reliability of local turbines. Many of the water canals and tanks, however, are not lined with concrete, with some water lost or silted (Mika, 2014).

GIZ's experience in training local people to develop and operate micro-hydro systems contrasts with FUNAE's more top-down form of management. There were reported productivity gains in milling, along with capacity-building in operations, repairs and management for systems owners and operators (Zana, 2007). The micro-hydro schemes are owned individually, while the mini-grid networks are owned by communities. GIZ and KSM facilitated a hybrid financing scheme with soft loans for hardware such as turbines and grants to support mini-grid installation.

In Chua, the micro-hydro scheme runs through the initiative of local people of very limited resources, with GIZ providing support. Consumers pay for electricity that is managed as a local enterprise. Ndacada has repaid his five-year loan and wants to build an additional system. He has identified a site on the Chua River and wants to employ local residents to develop it. He told us he bought a motorcycle and paid his seven children's school fees with earnings from the micro-hydro system. Overall, the project builds on existing knowledge, uses locally-suited technology and is based on proven demand for electricity. According to several villagers, the system's configuration and prices reflect consumers' needs and expectations.

Majaua-Maia, Zambézia province

In Majaua-Maia, a village in Zambezia province, FUNAE has set up a project to rehabilitate a defunct mini-hydro system. Inaugurated in 2013, the EU provided financing of €2.5 million. The Portuguese engineering firm Canas installed the system.

The project aims to rejuvenate a mini-hydro plant that powered a *fazenda* (plantation) owned by Mr. Maia, a Portuguese landowner, since the 1960s (the nearby village is partly named for him). Mr. Maia allegedly abandoned the fazenda in 1980 during the height of Mozambique's civil war. The main house, a blue-tiled country villa, fell into decline, and the hydro system was neglected. FUNAE's project aims to benefit 20,000 residents (5000 households) in Majaua-Maia and six surrounding villages, three primary schools, a clinic and six grinding mills.

Majaua-Maia is a three-hour drive from the district capital, Milange, on a dirt track that skirts the Malawian border. The facility has 767 kW of capacity and is situated on the Ruo River, a tributary of the Shire. Canas installed a German Ossberger cross-flow turbine while the generator is made by Efacec of Portugal. The turbine sat in a customs terminal in the Beira port for over a year before Canas gained permission to bring it into the country (Canas, 2013). The project's initial phase involved installing a 4-km high-voltage network, to be extended 40 km in a second phase, generating 3500 MWh/year. All homes, schools, a clinic and shops along the existing line are connected. Each household has one light bulb and a power point. During our visit, however, households and shopkeepers were not paying a tariff for electricity, as the project was in test phase (Quelhas, 2014).

Canas sourced most of the system's construction inputs from Portugal. It features state-of-the-art equipment while the powerhouse is an imposing, brightly painted concrete edifice. Apart from supplying energy access as a technical 'fix', however, a community needs assessment and locally suited energy services appeared lacking. Notably, the local grinding mills were not connected to the mini-grid during our visit. Powering these mills is a key priority for residents, particularly women; ground cornmeal (*xima*) is a staple food in the region. Women must wade across the Ruo to use a diesel-generator-powered mill across the border in Malawi, often with children in tow. FUNAE intends to connect the local mills, but had not yet done so.

The project's *fiscal* (supervisor), Mr. Sala, is from the city of Beira, and his two young assistants are from Nampula and Quelimane, provincial centres. This seems a missed opportunity to train local youth for such positions. Operators will likely be recruited from Maputo, Beira or Quelimane, and it might prove difficult to attract skilled personnel to this isolated community with poor roads, or retain them for long. Regarding uses of electricity, we observed three video clubs showing movies, mainly Jackie Chan films dubbed in the local language, *chiChewa*. School-age children frequented these clubs, with no admission charged. We heard sound systems competing from a cluster of bancas fixas. Only one had a refrigerator powered by the mini-grid, along with a desktop computer to burn music CDs and videos.

Limited capacity building has occurred around uses of electricity thus far, despite potential economic opportunities requiring relatively little capital investment, such as milling, welding or food processing. One household was distilling liquor from corn and sugarcane, but without electricity. Our interviews suggest most households cook with charcoal and wood, and some struggle to afford charcoal. This raises questions as to what kind of changes to local energy systems are most important to rural people and how energy access ranks in comparison with other improvements, such as in agriculture, health care or education (cf. Murphy, 2001).

Electricity consumption in Majaua-Maia is unmetered, with residents benefiting from free electricity, at least in the short term. Introducing a tariff may prove difficult, given the possible costs. FUNAE had not tested a payment system or undertaken consultative research; it was unclear if non-payers will eventually be disconnected. When fully operational, Majaua-Maia's network will be one of the largest mini-grids in southern Africa (Quelhas, 2014). The project is ambitious, but gaps in local engagement and capacity building around the technology remain. Greater involvement in early design stages might better inform energy planners about local priorities, constraints and expectations.

Conclusion

The Mozambican government aims to expand access to affordable, reliable and sustainable energy services for the population while also increasing the use of renewables, two goals that can be in tension with one another. Led by FUNAE, RETs have emerged in relation to the changing landscape of the electricity grid, on the one hand, and demands for multiple forms of energy service, on the other.

While solar PV as a decentralised system can seemingly be installed in any rural area sufficiently distant from the existing grid, in practice this process of installation has been highly uneven. The goals of claiming access to electricity in different administrative districts may obscure the uneven way it is installed. Such projects, for example, are rarely sited in the most remote areas of the district, serving to make rural electrification appear socially and spatially more inclusive than it actually is in practice.[9] This dynamic is found in the case studies of Chinhambuzi and Mavonde. The use of micro hydro, in contrast, is limited to places with suitable terrain. Yet it can achieve higher capacities to support a range of energy needs.

Despite its emphasis on connecting rural institutions, FUNAE has begun to include household energy in its projects through mini-grids and SHS. The approach is broadly pro-poor, with price supports and soft loans for end-users, but relies on state procurement and donor financing. FUNAE's operations are coordinated within a single body, suggesting a centralised form of management for delivering *decentralised* energy services, with resulting gaps in local participation in project design and implementation. In Majaua-Maia, for example, women's priorities were side-lined while free electricity provision may create unrealistic expectations, as witnessed in the three video clubs operating simultaneously. Training of local youth in the project's development in Majaua-Maia was minimal while economic opportunities created by the new technology were largely overlooked, giving the project an 'enclave' quality, with little connection to the surrounding economy and everyday practices (cf. Ferguson, 2005). A similar pattern can be observed with solar PV in Chinhambuzi and Mavonde, which has mostly benefitted government workers and shopkeepers rather than ordinary residents.

GIZ has taken a different approach that is designed to scale-up based on local demand. While also reliant on donor financing and technical expertise, it emphasises local entrepreneurship, training and stakeholder engagement around micro hydro generation, as seen in Chua. GIZ has an innovative business model, underscoring the value of local engagement through demand-based energy planning. Overall, fostering locally suited technologies and business models is an ongoing challenge.

In tandem, expansion of the national grid is further shaping the available space for off-grid renewables. Extending the grid remains an important aspect of state-led modernisation efforts, particularly following newly exploited coal and gas resources in the past decade. Rapid grid extension could disrupt the scaling up of distributed energy as more localities are connected, slowing the transition to renewables.

Despite these challenges, solar PV and micro hydro are increasingly viewed as sources of power that can deliver basic needs, from lighting for education to refrigeration for vaccines, the operations of rural government offices and increasingly the mobile phone economy. Yet such forms of energy services leave much of what constitutes energy demand in Mozambique unchanged. In the case of solar PV, currently the primary source of alternative energy, the systems are ill-equipped to power electric cooking, appliances of various kinds or even significant levels of computing or entertainment technologies. Often developed as one-off installations, recurring problems with maintenance, operating costs and theft further limit the ways these technologies are embedded in the rural economy. RETs may still be far from dislodging fossil fuels in Mozambique, yet they have clearly joined growing public debates about the country's energy future.

This chapter has presented a key challenge for researchers of energy transitions: what is the relationship between high and low carbon energy development in a context of limited energy access? More contextually sensitive and comparative research in rural and urban settings is needed on mobilising renewables as development technologies in order to address this question. This might include examining supply chains, increasing local content provision, and efforts to reduce costs and increase the security of supply for RETs. Further research should also engage with the uses of renewables in Mozambique's expanding peri-urban areas, thus far overlooked by state- and donor-supported energy service projects, and examine what energy transitions look like in these emerging sites. Finally, following Huber (2015), there is a need to scale up the analysis to consider the ways that uneven energy access reflects and intersects with larger patterns of social inequality.

Notes

1 The Rising Powers, Clean Development and the Low Carbon Transition in Sub-Saharan Africa (Ref: ES/J01270X/1).
2 Mining firms planning coal-fired power generation in Tete include Vale of Brazil, India's International Coal Ventures and UK's Ncondezi Energy, which entered a joint venture in 2015 with China's Shanghai Electric Power to develop a 300MW plant (Zitamar News, 2015).
3 Isaacman and Isaacman (2015) have observed Cahora Bassa was the largest dam in the world constructed for the specific purpose of exporting energy out of the country.
4 Donors include the World Bank, the European Union and the Governments of Belgium, the Netherlands, Portugal, India and South Korea (Menezes, 2013).

5 The administrative post (*posto administrativo*) is Mozambique's smallest territorial unit, below the district and province. Here it refers to the local authority's residence and government offices.

6 A *banca fixa*, or 'fixed stall,' is sanctioned by the local authorities and differs from ambulatory vendors or informal *bancas*, which can be harassed by police as being illegal (see Lindell, 2008).

7 Even if power levels were sufficient for cooking, PV users would need to purchase an inverter to convert the electricity from direct current (DC) to alternating current (AC) (Murphy, 2001).

8 Daniel Jones, University of Edinburgh, personal communication.

9 Thanks to Marcus Power for this insight.

References

Arias, Martín, Atienza, Miguel, and Cademartori, Jan. 'Large mining enterprises and regional development in Chile: between the enclave and the cluster.' *Journal of Economic Geography* 14 (2014): 73–95.

Besharati, Neissan. Raising Mozambique: Development through coal. South African Institute for International Affairs, Governance of Africa's Resources Programme, Policy Briefing No. 56, Johannesburg (2012).

Bloch, Robin and Owusu, George. 'Linkages in Ghana's gold mining industry: challenging the enclave thesis.' *Resources Policy* 37 (2012): 434–442.

Bridge, Gavin, Bouzarovski, Stefan, Bradshaw, Michael and Eyre, Nick. 'Geographies of energy transition: space, place and the low-carbon economy.' *Energy Policy* 53 (2013): 331–340.

Büscher, Bram. 'Investing in irony? Development, improvement and dispossession in southern African coal spaces.' *European Journal of Development Research* 27, no. 5 (2015): 727–744.

Canas, Victor. Canas, interview with author, 5 November, 2013.

Castán Broto, Vanesa, Ensor, Jonathan, Boyd, Emily and Allen, Charlotte. *Participatory Planning for Climate Compatible Development in Maputo, Mozambique*. London: UCL Press, 2015.

Castel-Branco, Carlos Nuno. 'Growth, capital accumulation and economic porosity in Mozambique: social losses, private gains.' *Review of African Political Economy* 41, no. 1 (2015): 26–48.

Cipriano, Amílcar, Waugh, Colin and Matos, Mathikizana. *The Electricity Sector in Mozambique: An Analysis of the Power Sector Crisis and its Impact on the Business Environment*. Maputo, USAID Mozambique, 2015.

Cuamba, Boaventura. University of Eduardo Mondlane, interview with author, 3 September, 2014.

Cuamba, Boaventura, Cipriano, Amílcar, Jaime Turatsinze and Ruth Henrique. *Investment Incentives for Renewable Energy in Southern Africa: The Case of Mozambique*. Winnipeg: International Institute for Sustainable Development, Trade Knowledge Network, 2013.

Cuvilas, Carlos Alberto, Jirjis, Raida and Lucas, Carlos. 'Energy situation in Mozambique: a review.' *Renewable and Sustainable Energy Reviews* 14 (2010): 2139–2146.

Dadwal, Shebonti. *India and Africa: Towards a Sustainable Energy Partnership*. Johannesburg, South African Institute for International Affairs, Occasional Paper No. 75, 2011.

EdM. *Relatório Anual de Estadística*. Maputo: EdM, 2011.

Ferguson, James. 'Seeing like an oil company: Space, security, and global capital in neoliberal Africa.' *American Anthropologist* 107, no. 3 (2005): 377–382.

FUNAE. *Bodas de Cristal*. Maputo: FUNAE, 2012.

Geels, Frank. 'Technological transitions as evolutionary reconfiguration processes: a multi-level perspective and case study.' *Research Policy* 31 (2002): 1257–1274.

Gqada, Ichumile. *A boom for whom? Mozambique's natural gas and the new development opportunity*. Johannesburg: South African Institute for International Affairs, Occasional Paper No. 151, 2013.

Hansen, Teis and Coenen, Lars. 'The geography of sustainability transitions: review, synthesis and reflections on an emerging research field.' *Environmental Innovation and Societal Transitions* (2014): 1–18.

Huber, Matthew. 'Theorising energy geographies.' *Geography Compass* 9, no. 6 (2015): 327–338.

IEA. *Africa Energy Outlook – A Focus on Energy Prospects in sub-Saharan Africa*. Paris: IEA, 2014.

Isaacman, Allen and Isaacman, Barbara. *Dams, Displacement and the Delusion of Development: Cahora Bassa and its legacies in Mozambique, 1967–2007*. Athens: Ohio University Press, 2013.

Isaacman, Allen and Sneddon, Chris. 'Portuguese colonial intervention, regional conflict and post-colonial amnesia: Cahora Bassa dam, Mozambique 1965–2002.' *Portuguese Studies Review* 11, no. 1 (2003): 207–236.

Kirshner, Joshua and Power, Marcus. 'Mining and extractive urbanism: postdevelopment in a Mozambican boomtown.' *Geoforum* 61 (2015): 67–78.

KPMG. *Mozambique – Country Mining Guide*. Cape Town: KPMG International, 2013.

Lamarie, Cedric. Vale Mozambique, interview with author, 6 October, 2013.

Lindell, Ilda. 'The multiple sites of urban governance: insights from an African city.' *Urban Studies* 45, no. 9 (2008): 1879–1901.

MacauHub (2014) Mozambique to spend US$530 in rural electrification over five years, www.macauhub.com.mo/en/2014/04/25/mozambique-to-spend-us530-million-in-rural-electrification-over-five-years/ (Accessed August 6th 2015).

Magrin, Géraud and Perrier-Bruslé, Laetitia. 'New geographies of resource extraction.' *Echo Géo* 17 (2011): 1–11.

Mavhunga, Clapperton Chakanetsa. '*Cidades Esfumaçadas*: Energy and the Rural-Urban Connection in Mozambique.' *Public Culture* 25, no. 2 (2013): 261–271.

Menezes, Miquelina. Director, FUNAE, interview with author, 23 September, 2013.

Mika, Lasten. Practical Action, interview with author, 25 August, 2014.

Mohan, Giles and Stokke, Kristian. 'Participatory development and empowerment: the dangers of localism.' *Third World Quarterly* 21, no. 2 (2000): 247–268.

Morrissey, James. 'Zuma, Guebuza ignore effect of new hydro dam on residents.' *Mail & Guardian*, November 22, 2013. At: http://mg.co.za/article/2013–11–22–00-zuma-guebuza-ignore-effect-of-new-hydro-dam-on-residents (Accessed June 11, 2014).

Murphy, James. 'Making the energy transition in rural East Africa: is leapfrogging an alternative?' *Technological Forecasting & Social Change* 68 (2001): 173–193.

Namburete, Fernando. FUNAE, interview with author, 7 August, 2014.

Newitt, Malyn. *A History of Mozambique*. Bloomington, IN: University of Indiana Press, 1995.

Nhamire, Borges and Mosca, João. *Electricidade de Moçambique: Mau Serviço, Não Transparente e Politizada*. Maputo: Centro de Integridade Pública, 2014.

Ouchim, Joaquim. EdM, interview with author, 8 November, 2013.

Pedro, Valeriano. Fosera Southern Africa, interview with author, 7 November, 2013.

Quelhas, Jaime. FUNAE delegation in Manica, interview with author, 10 August, 2014.

Ribeiro, Daniel. 'Mozambique: Energy for whom and for what?' Presentation at Energy for Development in Mozambique, workshop hosted by Durham University and University of Eduardo Mondlane, Maputo, 17 March, 2015.

Sagar, Ambuj. 'Alleviating energy poverty for the world's poor.' *Energy Policy* 33, no. 11 (2005): 1367–1372.

Smil, Vaclav. *Energy at the Crossroads: Global Perspectives and Uncertainties*. Cambridge: MIT Press, 2005.

Tawney, Letha, Miller, Mackay and Bazilian, Morgan. 'Innovation for sustainable energy from a pro-poor perspective.' *Climate Policy* 15, no. 1 (2015): 146–162.

Ulsrud, Kirsten, Winther, Tanja, Palit, Debajit and Rohracher, Harald. 'Village-level solar power in Africa: Accelerating access to electricity services through a socio-technical design in Kenya.' *Energy Research and Social Science* 5 (2015): 34–44.

Vines, Alex. *Mozambique's 20 Years of Peace at Risk?* London: Chatham House, 11 April, 2013.

Zana, Crispen. *Technical Appraisal for Micro Hydro Project Proposal in Chua Village, Manica District, Manica Province: a project between GTZ-AMES-M and Kwaedza Simukai Manica*. Maputo: GTZ, 2007.

Zitamar News. 'Shanghai Electric closing in on 300MW Tete power plant deal,' December 11th, 2015, at: http://zitamar.com/shanghai-electric-closing-in-on-300mw-tete-power-plant-deal/ (Accessed December 16th 2015).

23

The politics of forests

Community dimensions of energy resource use

Saska Petrova

Introduction

This chapter unpacks the multi-faceted relationships among energy resource use, community practices, and the legal regulation of forests. It highlights the complex meanings of 'legality' in contexts where forests have been subject to processes of neoliberal economic restructuring, while being enmeshed in the everyday life of local people. I also argue that resources from forests are key elements of the constitution and experience of intra- and inter-community relations in instances where fuelwood accounts for an important part of the energy mix. In advancing these claims, I wish to address the marginalization of the pathways through which neoliberal regulation has been implicated in the micro-level articulation of forest policies (Castree 2011; Pokorny et al. 2012). I also speak to debates focused on the everyday dimensions of forest use within the context of 'community energy' (Link et al. 2012; Seyfang et al. 2014; Castán Broto et al. 2014).

The chapter explores the implications of these processes in the Republic of Macedonia – a country that has been undergoing rapid dynamics of socio-economic and technological change during the last 25 years, as a result of the movement away from communist central planning and a one-party political system. Macedonia lies within Southeastern Europe (or SEE, also known as 'the Balkans'), whose constituent countries have seen the extensive entry of neoliberal economic practices in the management of the forestry sector. This has involved the restructuring of formerly centralized, hierarchical, vertically and horizontally integrated forest monopolies into a myriad of diverse commercial enterprises outside the control of the state (Stahl 2010). But the reform process has exerted a major impact on the livelihoods of local populations whose day-to-day existence is highly linked with forest governance. In part, this is due to the fact that forests account for more than 40 per cent of the total landmass of some SEE states, while representing a valuable economic resource in economically marginal regions. Forests also account for a significant part of the energy mix in SEE, with up to 50 per cent of households in some countries relying on fuelwood as the primary source of heating (International Energy Agency 2008).

One of the consequences of the neoliberalization of the forestry sector – and the subsequent curtailment of everyday informal economic practices in this domain – has been the widespread rise of 'illegal' logging. For the purpose of this chapter, I understand 'illegal logging' to mean the removal of wood in contravention with the relevant regulatory acts applying to the territory where such a practice is undertaken, while noting that the academic literature has frequently challenged the notion that a single legality around natural resource exploitation exists in the articulation of everyday life (Brown 2006). Understood in the narrow legalistic sense at least, illegal logging is particularly common in SEE, to the extent that in some instances it has even been accompanied by local armed conflicts that have been reported in the local media (Nova Makedonija 2011). The expansion of clandestine forest exploitation – in its multiple guises and forms – has been intensified by the increase of energy prices in the region, as a result of which growing numbers of households have been forced to rely on illegally supplied fuelwood in order to maintain thermal comfort in the home (Buzar 2007). Understanding the political economy of illegal logging, as well as the discourses and institutions that frame it, can therefore provide important insights into the micro-scale performance of neoliberal policies in the management of natural resources (Stahl 2010).

I employ the opinions and aspirations of individuals affected by, and implicated in, 'illegal' energy resource use – forestry officials, experts, local people, NGO activists[1] – as a basis for interrogating the different ways in which the privatization and marketization of forest resources have come to be entangled with the 'articulation of multiple resource extractive processes, which the state has been incapable of entirely capturing within its regulatory vision' (Staddon 2009: 173). Inspired by Stark (1996) and Watts (1998), I rely on the notion of 'recombinant capitalism' to explore the institutional context of forestry governance at the community scale, while investigating how neoliberalism has been implicated in dynamics of privatization, corruption and contract outsourcing.

The chapter commences with a brief exploration of some of the main academic debates on the neoliberalization of forestry, and its relation to community-based processes. I then move on to the ramification of these trends for post-communist forest policies in the Balkans. Having overviewed the recent restructuring of the forestry sector in the Republic of Macedonia, the chapter provides an account of the institutional processes of forest neoliberalization by examining the multiple ways in which 'market' relations have entered the domain. The discussion then moves onto perceptions of the constitutive dynamics of corruption – whose intensification, it can be argued, is a specific by-product of recent reforms in the sector – as well as the components and implications of illegal logging, whose emergence in post-communism is seen as a community-level practice of contestation and resistance towards the path chosen by the state in restructuring the forestry sector. The conclusion of the chapter highlights the diverse and recombinant economies – as per the understanding highlighted above (Stenning et al. 2010) – associated with neoliberal forest management policies in the Balkans.

Neoliberalism, forestry and community

The international mechanisms and policies for the large-scale economic development of forestry practices – especially regarding biodiversity protection and climate change mitigation and adaptation – are largely the product of neoliberal thought (Humphreys 2009). For example, the forest precepts of the United Nations Conference on Environment and Development (UNCED) encourage the national and international integration of environmental costs and benefits into market policies and practices, in order to achieve sustainable development and forest conservation. The proposal of the Framework Convention on Climate Change parties for 'avoided deforestation' – viewed as an opportunity for developing countries to receive credits that can be sold to developed states in a global carbon trading scheme by avoiding deforestation over and above an agreed background baseline – indicates the further expansion of market-based governance of the forestry sector (United Nations 2007). This process stimulates the voluntary involvement of the private sector on the trading market, rather than instituting legally binding commitments.

Based on the existence of the principles of marketization, an enhanced role for the private sector, and deregulation and voluntarism, Humphreys (2009) highlights the embeddedness of neoliberal policies within the actiivties of a number of international bodies and forestry programmes. Supplementing such globally oriented work is the wide body of more geographically focused research about the local implications of forest neoliberalization (Carper and Staddon 2009, White and Martin 2002, Prudham 2007, Stahl 2010). While Heynen and Robbins (2005) investigate the reasons for, and consequences of, the privatization of public forests in 'post-Fordist' Milwaukee, McCarthy (2005, 2006) has been studying 'community forestry' – an approach that involves giving local communities the power to manage nearby forest areas, as opposed to state or private ownership and management – in the US and Canada. Although this practice predates the expansion of neoliberalism, its principle of self-management can be understood as operating within the spirit of this paradigm. More recent work has drawn attention to the numerous social, technical and economic difficulties associated with top-down efforts to involve low-income forest managers from the Bolivian, Brazilian, Ecuadorian and Peruvian Amazon in the functioning of global value chains and markets (Pokorny et al. 2012).

However, the everyday articulations of neoliberalism in the domain of social reproduction and alternative economic practice – persuasively researched by authors such as Jarvis et al. (2001) and Stenning et al. (2010) in the context of the household-level reproduction of everyday life – remain insufficiently explored with respect to the utilization and management of nature more generally, and forest resources more specifically. This points to the need for investigating how market-based environmental policies become simultaneously integrated in, and challenged by local community practices and livelihood strategies in the forestry sector. Connecting such micro-level performances of neoliberalism with the operation of state and private institutions at different scales of forestry governance also affords the opportunity for exploring dimensions of distributional and procedural justice, as well as the mechanisms through which constructions of legality are used towards the promotion of particular patterns of social exclusion.

One of the entry points into a deeper understanding of how energy resources from forests become mobilized in the development of community relations is provided by the notion of 'Buen Vivir', which 'includes the classical ideas of quality of life, but with the specific idea that well-being is only possible within a community' (Gudynas 2011: 441). This idea also emphasizes the need for a pluralized understanding of the nature–community nexus, contingent upon cultural, historical and ecological settings. An example of 'Buen Vivir' can be found in Ecuadorian concept of *sumak kawsay*: the wording for a 'life in a community, together with other persons and Nature'. Another integrated socio-ecological understanding of the community-forest relationship exists in the Andean concept of the 'ayllu', where 'well-being encompasses not only persons, but also crops and cattle, and the rest of Nature' (ibid: 444).

At the same time, it should be emphasized that role of 'communities' and 'publics' in the tackling of complex environmental and societal problems has been under a critical spotlight for some time, in both political theory but in the broader social science literature. Žižek (1999) believes that public participation cannot effect change in the dominant system of social relations; rather, he argues, it helps keep such relations in place. In his understanding, it becomes political when a particular demand – e.g. over the exploitation of natural resources – starts to function as a metaphoric condensation of the global opposition against 'those in power, so that the protest is no longer actually just about that demand, but about the universal dimension that resonates in that particular demand' (ibid.: 204). This line of reasoning suggests that politics 'proper' is thus not about reaching consensus (as is the objective of many public participation exercises), but rather the reconfiguration of social space and the transformation of the power relations that underpin it.

The Republic of Macedonia's forestry sector

In the context of my study – the Balkans – woodlands are considered a key natural resource and forestry is a significant economic sector in this part of the world. Under communism, Balkan states treated forest areas as important national heritage and an economic asset, with the state acting as the sole and ultimate owner and manager of the entire forest fund. This was supported by the widespread nationalization of forest land after World War Two. The abandonment of communist central planning and one-party rule resulted in wide and complex socio-economic changes during the 1990s. The promise of EU membership became a *leitmotif* and driving force of the extensive transformation of these countries. The process was accompanied by the establishment of an entirely new set of political, economic and environmental polices and management practices (Buzarovski 2001). It was expected that the restructuring of the forestry sector would start with the restitution of forest areas to their previous owners. State economic enterprises were often privatized, opening new opportunities for the entry of private capital into the forestry market. Consequently, new relationships were generated in the form of reshaped networks of forestry actors, leading to the redistribution of commitments, risks and liabilities.

Avdibegović et al. (2010) have argued that newly established forestry legislation in most former Yugoslav countries is not adequately suited to the emergent ownership and management conditions, being predominantly oriented towards the administration needs of state owned forests. This is despite the fact that the share of private forests is significant at both the national and regional level (for example, in Serbia 47 per cent of the forests are private, while the same share reaches 23 in Croatia and 20 in Bosnia and Herzegovina). At the same time, the restructuring of the forestry sector in these countries has created new uncertainties, especially for those rural communities who were traditionally highly dependent upon locally obtained timber and non-timber forest resources (Tickle 2000; Petrova et al. 2009; Kluvánková-Oravská et al. 2009; Petrova et al. 2011; Petrova 2014a). This is because the new laws and the change of 'traditional' property rights have limited such communities' access to nearby woodland areas.

These issues have been particularly felt in the Republic of Macedonia, a country situated in the central part of the Balkan Peninsula – sharing borders with Greece to the south, Bulgaria to the east, Albania to the west and Kosovo and Serbia to the north. Approximately 37 per cent of Macedonia's territory is covered by forest land, including 910,000 ha of production forests (Ministry of Agriculture, Forestry, and Water Economy 2006), although a large proportion of this is made up of low-quality broadleaf forests or maquis occupying formerly degraded or agricultural land. Timber-grade broadleaf and coniferous forests can be found only in upland mountainous areas (1000 m above sea level) and represent approximately a third of all production forests.

It is little known that the management of natural resources – including forests – in the current territory of the Republic of Macedonia was legally regulated already from the nineteenth century, when this region was part of the Ottoman Empire. The first Ottoman Law on Forestry was adopted in 1870; it classified forests in accordance with land tenure as either public forests owned by the state, endowment forests (e.g. forests given to trusts, churches and monasteries); coppices owned by towns and villages, and private forests. All forest owners were issued '*tapi*' – formal documents confirming their title to the land. While local people were able to use public forests either for free or for a very low price before 1870, the new regulation enforced tax collection, commitment and concession in state owned forests. The alienation of the rights and access of local people towards state owned forests led to the accelerated devastation of many forest areas, especially as many state forests were leased or auctioned for logging to the highest bidder. In order to achieve greater revenues the contractors often cleared the surrounding forests as well. At the end of the lease, many forest areas would be completely degraded.

In 1922, Macedonia became incorporated within the newly established Kingdom of Serbs, Croats and Slovenians (SCS), as part of a larger regional unit – the *Vardarska banovina*. Forest governance in the SCS Kingdom was regulated by a special Law on Forests, which transferred most of the responsibilities for, and the benefits from, the forests to the local and regional levels. This period also saw the creation of the first forest cadastre, in which forest areas were registered as either privately or state owned.

With the creation of Yugoslavia after World War Two, Macedonia became a separate federal republic with an autonomous legal framework. This period also heralded the establishment of a centrally planned economy; the governance of forests under this regime commenced as early as 1940, being marked by the wholesale nationalization of forest land. However, it was only in 1975 that the country saw the establishment of elaborate forest management plans. In a reversal of previous trends, this was the time when 30 local state owned forestry enterprises were created, allowing for forestry management to become possibly the only decentralized activity in communist Yugoslavia. The enterprises ran local forests in accordance with ten year plans, guided by experts and practitioners. It has been argued that this approach not only professionalized the forestry sector, but also helped introduce environmental sustainability in the governance of forests (Avdibegović et al. 2010).

Macedonia managed to increase its forest cover significantly after World War Two. In part, this was due to the establishment of the National Afforestation Fund, which was responsible for the planting of more than 200,000 ha of woodland between 1950 and 1980. Also contributing to the expansion of forests was the abandonment of rural areas, as a result of declining agricultural production and urbanization. As a result, the country's forest cover increased by 9.4 per cent between 1991 and 2005 (Ministry of Agriculture, Forestry, and Water Economy 2006). Several of my interviewees pointed out that forests have continued to expand in recent years, despite several extensive fires (particularly in 2007), as well as the slowing down of urbanization processes and the expansion of illegal logging.

The post-1991 legislative framework for forest management in the Republic of Macedonia has centred on the Law on Forests and the National Strategy on Forests. Other important regulatory acts include the Law on the Environment and the Law on Nature Protection, as well as the Law on Fire Fighting. This legislation has led to the disbandment of local forestry enterprises, and the renewed centralization of forest management in the country. The government now manages state-owned forests through the Ministry of Agriculture, Forestry, and Water Economy (MAFWE), as well as the Ministry of Environment and Physical Planning, the 'Makedonski Shumi' (or 'Macedonian Forests') Public Enterprise, the National Parks and Hunting Service, and the 'Jasen' Public Enterprise. MAFWE has 33 regional units that are poorly staffed in the forestry sector (as most of their employees are within the agriculture domain) and enjoy very little autonomy in relation to the central government. The Ministry also contains the Forestry and Hunting Inspectorate – whose five regional offices and 18 employees mainly control technical compliance with relevant legislation – and the Forestry Police Department, which employs 109 forestry police officers at 19 local stations. The lack of human resources in the forestry sector – especially in relation to the absence of active local administrative representation by the policy-making structures of MAFWE – was identified by several interviewees as one of the main inhibitors of effective communication between local people and the state.

The Forestry and Hunting Inspectorate is also in charge of regulating the Makedonski Shumi state-owned enterprise, which is the only organization in charge of the day-to-day management of forests. This company owns a range of fuelwood reselling depots, while also employing private subcontractors for cutting and transporting biomass, based on the judgements made by its engineers. This means that most local and regional planning activities in the forestry sector are organized by Makedonski Shumi. The price of fuelwood set by this company in its reselling depots is usually significantly lower than that in the private sector depots, and the company also offers fuelwood to low-income households at an even more reduced rate.

Approximately 90 per cent of the forests in Macedonia are formally owned by the state; the remainder are private (State Statistical Office 2010). At the same time, Makedonski Shumi manages both private and public forests. An active process of restituting nationalized property to its former owners has been under way since 1990; as this is not yet complete, it is expected that the percentage of privately owned forests will reach approximately 14 per cent. The country also has a significant share of forests that were owned by religious communities before the communist area – they are also part of the restitution process. Yet there has been no further privatization of forest land, mainly due to the lack of an up-to-date Forest Cadastre, and issues around the current tenure of land owned by religious communities.

Forest activities and the forest industry are a pivotal component of the national economy, providing a key source of employment in rural areas. It is difficult to estimate the number of people employed in forestry activities, because all official statistics jointly refer to the broader 'sector of agriculture, forestry and fishing'. Having said this, it should be mentioned that the total workforce in Macedonia in 2012 was 943,055, out of which 69 per cent were officially employed. Out of this figure, 112,623 people (or 11.9 per cent) were engaged in the 'agriculture, forestry and fishing sector' (State Statistical Office 2012). Estimates provided by Savic et al. (2011) suggest that the number of people employed in the forestry sector only represented approximately 0.8 per cent of the total labour force in 2008. Yet it is thought that forestry-related activities indirectly provide job opportunities to an additional 35 to 40 thousand people through a range of economic multiplication effects. Considering that 40 per cent of the population is estimated to live in rural areas and more than 30 per cent of the demographically active population is formally unemployed – with agricultural households being particularly vulnerable to income poverty – it is evident that forestry can play an important role in determining the quality of living standards in the country (Buzar 2007; State Statistical Office 2010).

The development of neoliberal policies

As noted above, the forestry sector in the Republic of Macedonia was highly localized during communist central planning; it consisted of 30 municipally managed public enterprises, which – while relying on similar natural resource management approaches – achieved widely differing financial results. The restructuring process that was completed in 1997 placed all these firms under the umbrella of a single company (Makedonski Shumi) responsible for operating the entire forestry market in the country. The 1997 reforms, therefore, led to a process of centralization and homogenization, while significantly reshaping the distribution of resources and profit. This also meant that wood and timber prices were unified across the country, although the operating costs of the enterprises were the same. According to one of the experts who I interviewed, the merger of previous local enterprises with Makedonski shumi increased numbers of administrative staff while reducing technical staff (Petrova 2014b).

One of the most significant aspects of the 1997 reorganization was the privatization of collective capital (including sawmills, machines, vehicles and tools) and outsourcing of all woodcutting activities. Basically, woodcutters formerly employed by the 30 local enterprises were offered voluntary severance, accompanied by the provision of publicly owned tools and equipment. Such individuals then were able to establish small private companies, continuing to work for Makedonski Shumi on a contractual basis. This situation transpired despite the fact that Makedonski Shumi still continued to monopolize the sale, and dictate the price, of fuelwood and timber. As such, the restructuring process bears many of the typical hallmarks of 'roll back' neoliberalism (Lockie and Higgins 2007). But the outsourcing of wood exploitation has meant that Makedonski Shumi sometimes struggles with finding contractors for woodcutting activities. The interviewees from the administration of the enterprise thought that the lack of private firms with good quality equipment and well-trained staff was one of the main reasons for this situation. They also emphasized that the employment of small private companies for wood resource supply purposes is not new to the Macedonian forestry sector, since wood cutters from Bosnia – widely regarded as 'the best' in the former Yugoslavia – would have been especially brought for contract-based logging in Macedonia long before Makedonski Shumi was established. The lack of indigenous capacity and skills, it was pointed out, is partly a result of this legacy.

A number of interviewees pointed out that the current management model – whereby logging activities are not meant to be outsourced locally – has led to a situation in which several companies close to the top management of Makedonski Shumi or MAFWE obtain most of the contracts throughout the country. Although 1997 reforms also introduced market relations to the distribution of fuelwood and timber, Makedonski Shumi also controls practically all post-cut sales. This company monopolizes the entire wood market, as it is the exclusive supplier of timber to the households, the wood industry, and a large number of small privately owned companies.

At the same time, non-industrial demand for wood dominates wood consumption in the country – two thirds of the total wood consumption is for fuelwood, and around 25 per cent for sawn wood. While private depots are able to resell the resources provided by Makedonski Shumi, they are also obliged to procure fuelwood at a centrally regulated price. This was deemed too high by the owner of a private furniture making company, who pointed out that it is 'sometimes cheaper to buy timber from Bulgaria than from Makedonski Shumi' (Petrova 2014b). Thus, the silent unbundling and privatization of parts of the sector, combined with the existence of state capital and management means that Makedonski Shumi is neither a true public state owned enterprise nor a private company. The company basically controls the entire wood supply chain in the country; a situation further reinforced by the fact that most logged wood is sold at the national market (as Macedonia is still organizing the regulation on certification, wood export has been very limited).

Interviewees stressed that the current organizational structure of the enterprise, accompanied by the flow of resources and finance within different parts of the country, has turned it into a 'buffer' for solving regional social problems. As explained by an expert who was formerly a representative of an international organization, and currently manages a forestry think thank, Makedonski Shumi is just 'a compensation mechanism' for, inter alia, the 'lack of strategic planning and human resources in MAFWE' (Petrova 2014b). It is also worth noting that although many private forests were nationalized during socialism, some areas stayed in private ownership. With the restitution of the 1990s the percentage of private forests was increased to around 10 per cent of the total forest area. However, unsolved ownership issues and the uneven geographical distribution of private forests prevented private owners from becoming equal actors in the protection and management of forests in the country. This is perhaps one of the reasons why some of the interviewees from the state sector, especially from the Forestry and Hunting Inspectorate, thought that government control over forest areas should remain unchanged.

According to a former representative of the World Food Organization, the main problem is not the governance of forests, but the lack of investment in the sector – including the modernization of equipment – and the inadequate education and training of management and technical staff. However, members of the Association of the Owners of Private Forests and international organizations present in the country had diametrically different standpoints. They argued that the state – embodied in MAFWE and Makedonski Shumi – treats the owners of private forests as an encumbrance and obstacle, rather than a partner. The management of the forestry sector, according to them, favours the interests of the state as represented, in particular, by Makedonski Shumi, which sometimes acts more like an inadequately managed private company than a social agent. Members of an NGO that campaigns for sustainable forestry development also argued that 'Makedonski Shumi sometimes behaves like a private company' by 'logging in private forests later than in the state ones', thus 'lessening competition and controls the price of the fuelwood' (Petrova 2014b).

According to some of my interviewees, one of the most problematic issues is the erasure of the concept of 'communal forests' by the Law on Forests in 1997. The historical background of these areas, as explained above, dates back to the time of the Ottoman Empire, when municipalities and villages were given rights to use nearby coppices. Although such entitlements were formally taken away in 1920, the informal practice of collecting resources from public forest areas near human settlements persisted more or less until the post-communist restructuring of the forestry sector. Changes instituted in the late 1990s alienated local people from forests in their immediate environment, and decreased the amount of care for, and attachment towards, such areas. An academic expert who I interviewed confirmed the general impression that the utilization of local energy resources from forests strengthened communities' sense of ownership and stewardship. This allowed residents to take 'better care of forests' while 'reducing antagonism between rural dwellers and forestry enterprises' (Petrova 2014b).

Forest resource use and corruption

Thus, the entry of neoliberal regulation in 1997, while failing to lead to full fledged privatization, created a specific regulation regime in which a centralized national company acted as a sole mediator, embodiment and regulator of relations between the state and the market. This led to the emergence of a new set of institutional conditions conducive to the rise of alternative economic practices, corruption and illegal logging.

Without going into an in-depth investigation of the complex corruption practices in the Macedonian forestry sector, it is important to consider the role that some of these activities have played in the broader post-communist restructuring of this area of activity, especially in terms of shaping local people's perception of forest management. My public advocacy, expert and NGO interviewees pointed out that corruption has been both the catalyst and outcome of recent structural changes in the forestry sector. Indeed, corruption scandals involving employees from Makedonski Shumi are regularly reported in the national media, and have been on the rise in the recent decade (Nova Makedonija 2009, 2011). Two of the experts who I interviewed thought that such trends were a good indicator of the deeper issues faced by the enterprise. The most commonly identified reasons for this situation were the legal status and management structure of Makedonski Shumi, which it was felt, are mainly there to serve the political and economic interests of governing elites. It was pointed out that most of the personnel and management of both MAFWE and Makedonski Shumi are drawn from the ranks of the political parties in power.

Non-governmental and private sector interviewees went as far as alleging that even the sub-contracting companies used for woodcutting activities are also connected to the circles that exercise economic and political control in the country. They also pointed out that the National Forest Cadastre has not been updated since 1979, which makes it difficult for the state to determine the precise ownership of particular area – thus acting as a source for corruption. This is supplemented by the lack of an updated national forestry plan, as the previous document of this kind expired in 2009. Interviewees kept pointing to the centralized and rigid legal and management system as the main source for corruption in the sector, highlighting the 'the firm, inflexible attitude of the state and Makedonski Shumi, and the high administrative costs for getting formal permission' (Petrova 2014b). Further reasons for the rise of corruption identified by the interviewees included the restrictive manner in which the 1997 and 2009 forestry laws (Ministry of Agriculture, Forestry, and Water Economy 2006) – which have declared all forests either state or private – regulate the traditional rights for the management of forests.

The majority of my interviewees believed that corruption would proceed unhindered if substantial changes to the system were not made. However, there were different opinions about what the scope and nature of the transformations required. While some of the interviewees emphasized the need to improve the state's capacity to exercise its control, administrative and technical functions, others talked about a movement towards greater transparency, decentralization, openness and a further entry of private capital. The representative of an international organization working in the country underlined the need for 'a revitalized system of responsibilities', which would involve 'the improvement of management mechanisms and an integrated approach towards forestry, rural development and tourism' (Petrova 2014b). Interviewees from the MAFWE forestry police had a similar attitude and pointed out that it is very important that the state maintain its regulatory and managerial role. They believed that the modernization of technical equipment (including remote sensing) can help prevent and trace illegal activities in the forests. Academic experts, however, felt that local people should be encouraged to open small and medium sized enterprises. This would help increase competition on the forestry market and reduce the monopolistic position of Makedonski Shumi, especially since 'there is a wide spectrum of forestry activities that can be profitable, such as the trade of non-forest products' (Petrova 2014b).

My interviewees from the NGO sector shared such opinions, while emphasizing the role of private forests in helping reduce corruption. They frequently stressed that private forests are managed in a more sustainable way than state-owned ones, where clear-cutting is more common. The private entrepreneurs who I interviewed shared the feeling that the state is concerned about losing control over the forestry stock, as a result of an entrenched belief that the owners of private forests will revert to aggressive management practices in the absence of state control. With legal changes expected in the near future, however, Makedonski Shumi will only be responsible for nursing forests, and all other activities will be undertaken by privately subcontracted companies. This shift was welcomed by many private sector interviewees, who thought that it will open new opportunities for the entrance of private capital and the improvement of the forestry market in the country. However, representatives of state organizations – and surprisingly, many local people as well – feared that this might create the space for further corruption.

The rise of multiple corruption practices in the forestry sector is thus indicative of broader trends. It points to some of the reasons for, and consequences of, the state's neoliberalization of the forestry sector by 'stealth': A situation that has created a recombinant public-private company via the privatization of collective capital, and the outsourcing of most technical activities. Many corruption activities can be attributed to the exclusive position of this company in the forestry sector, both in terms of its ability to regulate and plan most forestry activities, and its position as the dominant economic actor in this domain.

Challenging neoliberalism via 'illegal' logging

The 1997 Law on Forests does not provide a clear definition of illegal logging, as it only lists that forest-related activities are considered criminal when they 'negatively affect land fertility and endanger forest production or silviculture, or the survival of the forests and their multiple benefits . . . including extensive woodcutting, intensive selective logging and any action causing weed growth or erosion of land by water or wind' (Ministry of Agriculture, Forestry, and Water Economy 2006). There is a dearth of official statistics on the phenomenon, other than the figure that possibly up to 30 per cent of all woodcutting is illegal; it should be pointed out that this only includes activities related to illegal logging (ibid.). Nevertheless, all of my interviewees insisted that multiple forms of clandestine forest exploitation have dramatically expanded during the past 20 years; on average, they estimated that between 30 and 50 per cent more forest resources are being exploited in this manner compared to the situation at the onset of the post-communist transformation.

The majority of interviewees emphasized that there are major differences in the types of illegal logging that occurs, in relation to the scale of wood exploitation, its environmental consequences, as well as the nature of the socio-economic forces and agents behind it. It was often argued that the least ecologically damaging illegal logging is when individuals cut dry branches or trees with diameters up to 8 cm. The main reason for this practice, according to the interviewees, was the need for obtaining fuelwood in order to heat homes. As such, it plays a key role in the articulation of alternative economic practices, as the individuals who exercise such woodcutting have limited financial incomes. The state generally tolerates this behaviour, as its perpetrators are rarely pursued by the MAFWE and Makedonski Shumi police. Most interviewees thought that it is a common part of everyday life in almost all rural areas within the country. In broader terms, this situation indicates the importance of fuelwood as a means for alleviating energy poverty – understood as the inability of households to access socially and materially necessitated levels of energy services in the home (Bouzarovski and Petrova 2015).

Also present is the logging of wood in formerly 'communal' forests that lie in the immediate proximity of villages. As was discussed above, the use of villagers' 'traditional rights' to cut trees from such areas was not considered illegal until 1997. However, the regulatory shift that prohibited this practice did not necessarily translate into a change of behaviours. A number of interviewees emphasized that most villagers still believe that they have rights to use 'their' forests for 'their' needs, deciding to cut the wood for personal use as they see fit, despite frequent conflicts with the forest guards working for Makedonski Shumi (relations with MAFWE's forest police seem to be more cordial) and the threat of legal action. These conflicts have also led to a range of more serious problems, as local people may decide to start fires in nearby forests as a means of protest towards the actions of the authorities, as well as a way of depreciating the value of fuelwood subsequently sold on the market.

According to several interviewees, the most environmentally damaging type of illegal logging takes place in remote mountainous areas and is undertaken by well-trained, highly organized gangs, which sometimes operate across national borders. These groups cut as many trees as possible with the aim of illegally exporting the timber, or reselling it down the commodity chain. According to the individuals I interviewed, however, most of the wood harvested in this manner is eventually used by households for heating. This is because the groups in question lack the technical knowledge to generate construction- or furniture-grade timber. Employees in the forest police emphasized that individuals involved in such activities are often dangerous, armed and difficult to control, in addition to working together with local crime syndicates.

A specific case of illegal logging exists in the case of 'administrative' woodcutting. This is a situation when private owners engage in curing their own forests without formal permission. Many of my interviewees argued that this practice is a result of the existence of a bureaucratic and centralized system of forest management. As Makedonski Shumi is the manager of both publicly and privately owned forests, private owners cannot cut any wood without valid permission from them. In order to obtain such clearance, they have to prepare a documentation file with information confirming the ownership of the forest (if there are more then one owner a verified statement from them is required as well) together with a plan outlining which trees will be felled. A payment of stamp duty is also required in order to submit the request. Given how lengthy and cumbersome this procedure can be, interviewees strongly defended the technically illegal woodcutting that takes place when it is not followed.

There was a lack of consensus among my interviewees regarding the path of forthcoming reforms in the forestry sector; while some of them believed that the state should not loosen its control and regulation of forest areas, others insisted that private owners should be given the freedom to be completely responsible for their forests. Although it was felt that Makedonski Shumi can potentially be much more economically efficient, this will continue to be hampered by the tendency of politicians in government to perceive it as a shelter for their political party clientele, resulting in an excess of administrative personnel and a lack of appropriately educated and trained staff. The situation is further exacerbated by the inadequate level of communication between the main actors in the forestry sector, and the absence of an integrated rural development framework. The latter is, in part, a result of the lack of co-ordination between the Ministry of Environment – which is institutionally responsible for rural development matters – on the one hand, and MAFWE, on the other.

Conclusions

In its entirety, this empirical investigation of the relationship between illegal logging and the 'roll back' neoliberalization of the Macedonian forestry sector has revealed the existence of multiple, parallel and sometimes mutually conflicting practices of forest use in the country. The formal and informal institutional frameworks that allow for the utilization of energy resources from forests are also deeply embedded in the making of communities in cultural, economic and socio-political terms. Practices surrounding the use of energy resources from 'traditional' forest areas in particular expose the multiple meanings of forests as sites of community belonging and self-identification. At the same time, energy resources become enrolled in a variety of political processes, involving a range of *de jure* and *de facto* illicit practices operating at the boundaries of community relations, the public sector and business interests. Of crucial importance in this context is the emergence of a centralized forest management enterprise in charge of not only the state-owned sector, but many aspects of privately owned forest management as well. Thus, the forest–community nexus involves a wide range of multi-sited, diverse and recombinant economies (Stark 1996; Gibson-Graham 2008; Stenning et al. 2010) of energy resource management.

It is in this set of developments, alongside the state's attempt to curtail communal forest use rights, that I locate the rise of practices of illegal logging in the given context. More specifically, their appearance can be traced back to a series of decisions made within the restructuring process of the early 1990s. This involved the restitution of formerly nationalized forest areas in the early 1990s, and attempts to create quasi-market relations by outsourcing the woodcutting business to former employees of Makedonski Shumi. Despite losing their job security, such individuals were forced to act as small-scale entrepreneurs in conditions where informal economic practices were pervasive, and there was little experience with running and managing small-scale companies in a market-based environment. The dispossession of public resources and the disruption of traditional community behaviours stimulated the expansion of corruptive behaviours, while introducing a form of privatization by 'stealth'.

The restructuring process also created a state-run monopoly, which acted as the main contractor and regulator of the forestry market. The clientelistic character of this enterprise – alongside its role as a social policy mechanism thanks to its employment practices and pricing policies – points to the 'strategically selective' (Jessop 2001) role played by the state in the post-communist transformation process: the government enables certain types of market relations while pursuing distinct distributional goals. Illegal logging is a clear result of this 'forest of contradictions' (Robbins and Frasier 2003), having emerged at the nexus of alternative market behaviours, shadow economies, and mixed state–business relations. As such, it plays a significant role in sustaining practices of economic and social reproduction at a variety of scales. In some respects, clandestine forest exploitation can be seen as a bottom-up attempt to challenge the distributional justice implications of the forestry reform process. It brings forth the multiple and contested geographies of legality in relation to natural resource exploitation, where state-led statutory delimitations are problematized by place-based contingencies and practices, creating a regulatory space outside the boundaries of government policy. This 'grey' regulatory realm, however, also encompasses more extensive forms of institutionalised corruption, which involve the wholesale exploitation and extraction of rents from the material assets owned and managed by the state. As such, my findings closely match Robbins's (2000) identification of the extra-legal network of nature-society relations that gives rise to corruption in the Indian context, although the ecosystem implications of such practices in our case remain unclear.

As was pointed out above, the rise of illicit logging practices expresses the presence of a specific case of 'roll back' neoliberalism, where the state – despite retaining its formal ownership and overall control of the sector – has withdrawn from the *de facto* management and ownership of forest resources in favour of private interests and quasi-market relations. The entrance of corruptive and clientelistic relationships has disenfranchised many local communities, who have used illegal logging as a livelihood strategy. Despite the extensive nature of clandestine forest exploitation, the current situation is qualitatively not very different from what existed under communism, when the reliance on traditional rights meant that forest resources played a key role in the articulation of alternative economic practices. This suggests that the extent of recombinant capitalism in the SEE forestry sector may be limited. Still, increasing resource demands – accompanied by conflicts related to illegal logging – may exacerbate pressures on the environment in future years, while further alienating local communities from the governance of energy resources from forests.

Acknowledgements

This chapter is a modified version of Petrova (2014b). Reproduced with permission.

Notes

1   A total of 25 interviews with experts, decision-makers, employees of forestry companies and state agencies, as well as households involved in illegal logging in the Eastern part of Macedonia, were undertaken during April 2011. Not all interviews have been cited here, and many institutional affiliations and all names have been left out in order to ensure confidentiality.

References

Avdibegović, M., Nonić, D., Posavec, S., Petrović, N., Marić, B., Milijić, V., Krajter, S., Ioras, F. and Abrudan, I. V. 2010. Policy options for private forest owners in Western Balkans: A qualitative study. *Notulae Botanicae Horti Agrobotanici Cluj-Napoca* 38: 257–261.

Bouzarovski, S. and Petrova, S. 2015. A global perspective on domestic energy deprivation: Overcoming the energy poverty–fuel poverty binary. *Energy Research & Social Science* 10: 31–40. www.science direct.com/science/article/pii/S221462961500078X (Accessed April 11, 2016).

Brown, D. 2006. Rights and natural resources: Contradictions in claiming rights. In *Human Rights and Poverty Reduction: Realities, Controversies and Strategies*, edited by T. O'Neil. London: Overseas Development Institute. 77–90.

Buzar, S. 2007. *Energy Poverty in Eastern Europe: Hidden Geographies of Deprivation*. Aldershot, UK: Ashgate.

Buzarovski, S. 2001. Local environmental action plans and the 'glocalisation' of post-socialist governance. *GeoJournal* 55: 557–568.

Carper, M. and Staddon, C. 2009. Alternating currents: EU expansion, Bulgarian capitulation and disruptions in the electricity sector of South-east Europe. *Journal of Balkan and Near Eastern Studies* 11: 179–195.

Castán Broto, V., Salazar, D. and Adams, K. 2014. Communities and urban energy landscapes in Maputo, Mozambique. *People, Place and Policy* 8: 192–207. http://extra.shu.ac.uk/ppp-online/wp-content/uploads/2014/12/communities-urban-energy-mozambique.pdf (Accessed February 5, 2015).

Castree, N. 2011. Neoliberalism and the Biophysical Environment 3: Putting theory into practice. *Geography Compass* 5: 35–49.

Gibson-Graham, J. K. 2008. Diverse economies: performative practices for 'other worlds'. *Progress in Human Geography* 32: 613–632. http://phg.sagepub.com/content/32/5/613 (Accessed February 5, 2015).

Gudynas, E. 2011. Buen Vivir: Today's tomorrow. *Development* 54: 441–447. www.palgrave-journals.com/doifinder/10.1057/dev.2011.86 (Accessed December 28, 2015).

Heynen, N. and Robbins, P. 2005. The neoliberalization of nature: Governance, privatization, enclosure and valuation. *Capitalism, Nature, Socialism* 16.

Humphreys, D. 2009. Discourse as ideology: neoliberalism and the limits of international forest policy. *Forest Policy and Economics* 11: 319–325.

International Energy Agency. 2008. *Energy in the Western Balkans: The Path to Reform and Reconstruction*. Paris: IEA.

Jarvis, H., Pratt, A. C. and Cheng-Chong, W. 2001. *The Secret Life of Cities: The Social Reproduction of Everyday Life*. Harlow: Prentice Hall.

Jessop, B. 2001. Institutional re(turns) and the strategic–relational approach. *Environment and Planning A* 33: 1213–1235.

Kluvánková-Oravská, T., Chobotova, V., Banaszak, I., Slavikova, L. and Trifunovova, S. 2009. From government to governance for biodiversity: The perspective of Central and Eastern European transition countries. *Environmental Policy and Governance* 19: 186–196.

Link, C. F., Axinn, W. G. and Ghimire, D. J. 2012. Household energy consumption: Community context and the fuelwood transition. *Social Science Research* 41: 598–611. www.sciencedirect.com/science/article/pii/S0049089X11002353 (Accessed December 2, 2014).

Lockie, S. and Higgins, V. 2007. Roll-out neoliberalism and hybrid practices of regulation in Australian agri-environmental governance. *Journal of Rural Studies* 23: 1–11.

McCarthy, J. 2005. Devolution in the woods: Community forestry as hybrid neoliberalism. *Environment and Planning A* 37: 995–1014.

McCarthy, J. 2006. Neoliberalism and the politics of alternatives: Community forestry in British Columbia and the United States. *Annals of the Association of American Geographers* 96: 84–104.

Ministry of Agriculture, Forestry, and Water Economy. 2006. *Strategy for Sustainable Development of Forestry in the Republic of Macedonia*. Skopje: MAFWE.

Nova Makedonija. 2009. Дрвокрадците сечат, несовесни службеници помагаат. www.novamakedonija.com.mk/NewsDetal.asp?vest=121791034264&id=12&setIzdanie=21864 (Accessed July 20, 2011).

Nova Makedonija. 2011. Во 'Македонски шуми' потрошиле 6,5 илјади евра за виски. www.nova makedonija.com.mk/NewsDetal.asp?vest=12811834475&id=9&setIzdanie=22192 (Accessed 20th July 2011).

Petrova, S. 2014a. *Communities in Transition: Protected Nature and Local People in Eastern and Central Europe*. Aldershot, UK: Ashgate.

Petrova, S. 2014b. Contesting forest neoliberalization: Recombinant geographies of "illegal" logging in the Balkans. *Geoforum* 55: 13–21. www.sciencedirect.com/science/article/pii/S0016718514000876 (Accessed February 5, 2015).

Petrova, S., Bouzarovski, S. and Čihař, M. 2009. From inflexible national legislation to flexible local governance: Management practices in the Pelister National Park, Republic of Macedonia. *GeoJournal* 74: 589–598.

Petrova, S., Čihař, M. and Bouzarovski, S. 2011. Local nuances in the perception of nature protection and place attachment: A tale of two parks. *Area* 43: 327–335.

Pokorny, B., Johnson, J., Medina, G. and Hoch, L. 2012. Market-based conservation of the Amazonian forests: Revisiting win–win expectations. *Geoforum* 43: 387–401.

Prudham, S. 2007. Sustaining sustained yield: class, politics, and post-war forest regulation in British Columbia. *Environment and Planning D: Society and Space* 25: 258–283.

Robbins, P. 2000. The rotten institution: Corruption in natural resource management. *Political Geography* 19: 423–443.

Robbins, P. and Frasier, A. 2003. A forest of contradictions: Producing the landscapes of the Scottish Highlands. *Antipode* 35: 95–118.

Savić, N., Stojanovska, M. and Stojanovski, V. 2011. Analyses of the competitiveness of forest industry in the Republic of Macedonia. *South-east European Forestry* 2: 13–21. http://hrcak.srce.hr/76591?lang=en (Accessed December 28, 2015).

Seyfang, G., Hielscher, S., Hargreaves, T., Martiskainen, M. and Smith, A. 2014. A grassroots sustainable energy niche? Reflections on community energy in the UK. *Environmental Innovation and Societal Transitions* 13: 21–44. www.sciencedirect.com/science/article/pii/S2210422414000227 (Accessed February 5, 2015).

Staddon, C. 2009. Towards a critical political ecology of human–forest interactions: Collecting herbs and mushrooms in a Bulgarian locality. *Transactions of the Institute of British Geographers* 34: 161–176.

Stahl, J. 2010. The rents of illegal logging: The mechanisms behind the rush on forest resources in Southeast Albania. *Conservation and Society* 8: 140–150.

Stark, D. 1996. Recombinant property in East European capitalism. *American Journal of Sociology* 101: 993–1027.

State Statistical Office. 2010. *Statistical Yearbook of the Republic of Macedonia*. Skopje: SSO.

State Statistical Office. 2012. *Labour Force Survey*. Skopje: SSO.

Stenning, A., Smith, A., Rochovskaa, A. and Świątek, D. 2010. *Domesticating Neo-Liberalism: Spaces of Economic Practice and Social Reproduction in Post-Socialist Cities*. Chichester: Wiley-Blackwell.

Tickle, A. 2000. Regulating environmental space in socialist and post-socialist systems: Nature and landscape conservation in the Czech Republic. *Journal of Contemporary European Studies* 8: 57–78.

United Nations. 2007. Decision 2/CP.13. Reducing emissions from deforestation in developing countries: Approaches to stimulate action. FCCC/CP/2007/6/Add.1. Report of the Conference of the Parties on its Thirteenth Session, held in Bali from 3 to 15 December 2007.

Watts, M. 1998. Recombinant capitalism: State, de-collectivisation and the agrarian question in Vietnam. In *Theorising Transition. The Political Economy of Post-Communist Transformation*, edited by A. Pickles and A. Smith. London: Routledge. 450–505.

White, A. and Martin, A. 2002. *Who owns the world's forests? Forest tenure and public forests in transition*. Washington DC: Forest Trends and Centre for International Environmental Law.

Žižek, S. 1999. *The Ticklish Subject: The Absent Centre of Political Ontology*. London: Verso.

Index

Locators in *italics* refer to figures and those in **bold** to tables.

Printed and bound by CPI Group (UK) Ltd, Croydon, CR0 4YY

24/10/2024

01778293-0008